NEUROSCIENCES RESEARCH
SYMPOSIUM SUMMARIES

Volume Six

Work Session Reports by

Edward V. Evarts, Emilio Bizzi,
Robert E. Burke, Mahlon DeLong,
and W. Thomas Thach, Jr.

Richard J. Wurtman

Manfred Eigen and Leo DeMaeyer

Lewis C. Mokrasch, Richard S. Bear,
and Francis O. Schmitt

Detlev Ploog and Theodore Melnechuk

# NEUROSCIENCES RESEARCH SYMPOSIUM SUMMARIES

**Volume Six**

*An Anthology from the Neurosciences Research Program Bulletin*

Edited by
Francis O. Schmitt
George Adelman
Theodore Melnechuk
Frederic G. Worden

Neurosciences Research Program
Massachusetts Institute of Technology

**The M.I.T. Press**
*Massachusetts Institute of Technology*
*Cambridge, Massachusetts, and London, England*

ACKNOWLEDGMENTS OF SPONSORSHIP AND SUPPORT

The Neurosciences Research Program (NRP) is sponsored by the Massachusetts Institute of Technology. The NRP is supported in part by National Institutes of Health, U.S. Public Health Service, Grant No. GM 10211; National Aeronautics and Space Administration, Grant No. NsG 462; Office of Naval Research, Grant Nonr(G) 00011-71; The Rogosin Foundation, and the Neurosciences Research Foundation. Grateful acknowledgment is also made to the following donors: Louis and Eugenie Marron Foundation, The Grant Foundation, Inc., Charles A. Harrington Foundation, and Standard Oil Company of New Jersey.

Library of Congress Card Number: 66-22645
Printed in the United States of America
ISBN 0 262 19107 5

# CONTENTS

# PREFACE

Like its predecessors in the series of *Neurosciences Research Symposium Summaries,* this volume spans a range of scientific disciplines and levels of organization of the nervous system that exceeds the competence of any individual neuroscientist. Such breadth of coverage may well seem unattractive to those with specialistic interests who might prefer that each anthology be more narrowly focused upon a particular topic or area of research. With full appreciation of this risk, the Neurosciences Research Program (NRP) continues the policy of including within one volume topics as diverse as "Carriers and Specificity in Membranes" (Eigen and De Maeyer) and "Are Apes Capable of Language?" (Ploog and Melnechuk) as part of the program's dedication to minimizing the traditional disciplinary barriers and parochialism of method and concept that hinder progress toward achieving a profound scientific understanding of the nervous system, especially the relationship of brain functioning to human behavior. Within the pages of these anthologies, the motivated reader can conveniently educate himself in areas quite disparate to his own training and research, and the unwary browser may hopefully stumble serendipitously upon intellectual surprises of unexpected pertinence to his own scientific needs and potential interest.

One goal of the Neurosciences Research Program (NRP) is to facilitate and promote rapid communication within the field of neuroscience. For the NRP staff a particularly tangible and satisfying expression of this goal was the policy of sending the *NRP Bulletin* free to individual scientists, laboratories, and libraries all over the world. This no-charge distribution policy was maintained for the seven years prior to January 1, 1971. In this way, the proceedings of NRP Work Sessions were issued as public information, rather than being restricted to the twenty or so scientists who participated in each of the two-and-one-half-day meetings. Over these years, the response to the *Bulletin* indicated that it was indeed making an important contribution to working scientists, teachers, and students. It was therefore with some regret that a subscription charge was introduced, largely because, in the context of the reduction of federal financial support for research, it seemed appropriate and necessary to shift some of the cost of the *Bulletins* to their readers.

vii

This sixth annual volume of *Neurosciences Research Symposium Summaries* is the first to republish a series of *NRP Bulletins* for which a subscription has been charged. As such, the volume differs from the preceding five anthologies, which have republished issues of the *Bulletin* that were sent at no cost to readers. For the future, we plan to continue the two forms of publication, anthology and annual subscription.

We wish again to thank the Work Session Chairmen and all of the more than 100 participants for their major contributions to our program and to neuroscience.

## PARTICIPANTS IN REPORTED WORK SESSIONS

Ames, Adelbert, III
Anton-Tay, Fernando
Axelrod, Julius

Bear, Richard S.
Bigelow, Julian H.
Bizzi, Emilio
Blaurock, Alan E.
Bloom, Floyd E.
Branton, Daniel
Brookhart, John M.
Brooks, Vernon B.
Brown, Roger W.
Bruner, Jerome S.
Bullock, Theodore H.
Bunge, Richard P.
Burke, Robert E.

Caspar, Donald L.D.
Chandross, Ronald

Chow, Kao Liang
Coppola, John A.

Dahlström, Annica
Davison, Alan N.
DeLong, Mahlon
De Maeyer, Leo
DeVore, Irven
Donoso, Alfredo O.

Eigen, Manfred
Eisenman, George
Evarts, Edward V.
Eylar, Edwin H.

Fernández-Morán, Humberto
Folch-Pi, Jordi
Frank, Karl
Fuxe, Kjell

Galambos, Robert
Ganong, William F.
Garrett, Merrill F.
Gorski, Roger A.
Grell, Ernst

Henneman, Elwood
Hirano, Asao
Hirsh, Richard
Huang, Ching-hsien
Hubbell, Wayne L.

Ilgenfritz, Georg
Ito, Masao

Karten, Harvey J.
Katchalsky, Aharon
Kennedy, Donald
Kety, Seymour S.

Lenneberg, Eric H.
Levine, Seymour
Lundberg, Anders
Luzzati, Vittorio

McCann, Samuel M.
McEwen, Bruce S.
MacKay, Donald M.
Maickel, Roger P.
Mandel, Paul
Marks, Bernard H.
Marler, Peter
Martini, Luciano
Melnechuk, Theodore
Meyerson, Bengt
Mokrasch, Lewis C.
Moore, Robert Y.
Mountcastle, Vernon B.
Müldner, Heinrich
Müller, Eugenio E.

Nauta, Walle J. H.
Norton, William T.

O'Brien, John S.
Onsager, Lars
Oscarsson, Olov

Phillips, Charles G.
Ploog, Detlev
Pohorecky, Larissa
Premack, David
Pressman, Berton C.

Quarton, Gardner C.

Robertson, J. David

Sawyer, Charles H.
Scharrer, Berta V.
Schmitt, Francis O.
Simon, Wilhelm
Smith, Marion E.
Sweet, William H.

Teuber, Hans-Lukas
Thach, W. Thomas, Jr.

Tosteson, Daniel C.
Towe, Arnold L.
Träuble, Hermann

Urquhart, John
Uzman, Betty G.

Wang, Howard H.
Webster, Henry de F.
Weiss, Paul A.
Whittaker, Victor P.
Wilson, Donald M.
Winkler, Ruthild
Worden, Frederic G.
Wurtman, Richard J.

Zigmond, Michael J.

W. Ross Adey
Professor of Anatomy and Physiology
Director, Space Biology Laboratory
Brain Research Institute
University of California at Los Angeles

Floyd E. Bloom
Acting Chief, Laboratory of
  Neuropharmacology
National Institute of Mental Health

David Bodian
Professor and Director
Department of Anatomy
The Johns Hopkins University
  School of Medicine

Theodore H. Bullock
Professor of Neurosciences
University of California at San Diego
  School of Medicine

Melvin Calvin
Professor of Chemistry
University of California at Berkeley

Leo De Maeyer
Member, Max Planck Institute
  for Biophysical Chemistry
Göttingen-Nikolausberg, West Germany

Mac V. Edds, Jr.
Dean of Natural Sciences
  and Mathematics
University of Massachusetts, Amherst

Gerald M. Edelman
Professor
The Rockefeller University

Manfred Eigen
Director, Max Planck Institute
  for Biophysical Chemistry
Göttingen-Nikolausberg, West Germany

Humberto Fernández-Morán
A.N. Pritzker Professor of Biophysics
Department of Biophysics
University of Chicago

Robert Galambos
Professor of Neurosciences
University of California at San Diego
  School of Medicine

John B. Goodenough
Research Physicist and Group Leader
Lincoln Laboratory
Massachusetts Institute of Technology

Holger V. Hydén
Director, Institute of Neurobiology
University of Göteborg
Faculty of Medicine
Göteborg, Sweden

Aharon Katchalsky
Professor and Head
Polymer Department
Weizmann Institute of Science
Rehovot, Israel

Seymour S. Kety
Chief, Psychiatric Research Laboratories
Massachusetts General Hospital
Professor of Psychiatry
Harvard Medical School

Heinrich Klüver
Sewell L. Avery Distinguished Service
  Professor Emeritus of Biological
  Psychology
The University of Chicago

Albert L. Lehninger
Professor and Director
Department of Physiological Chemistry
The Johns Hopkins University
  School of Medicine

Robert B. Livingston
Professor of Neurosciences
University of California at San Diego
    School of Medicine

H. Christopher Longuet-Higgins
Royal Society Research Professor
Department of Machine Intelligence
    and Perception
University of Edinburgh
Edinburgh, Scotland

Harden M. McConnell
Professor of Chemistry
Stanford University

Donald M. MacKay
Professor of Communication
The University of Keele
Keele, Staffordshire, England

Neal E. Miller
Professor
The Rockefeller University

Frank Morrell
Research Professor of Neurology
    and Psychiatry
New York Medical College

Vernon B. Mountcastle
Professor and Director
Department of Physiology
The Johns Hopkins University
    School of Medicine

Walle J. H. Nauta
Professor of Neuroanatomy
Department of Psychology
Massachusetts Institute of Technology

Marshall Nirenberg
Chief, Laboratory of Biochemical
    Genetics
National Heart and Lung Institute
National Institutes of Health

Lars Onsager
J. Willard Gibbs Professor of
    Theoretical Chemistry
Sterling Chemistry Laboratory
Yale University

Detlev Ploog
Director, Clinical Institute
Max Planck Institute for Psychiatry
Munich, West Germany

Gardner C. Quarton
Director, Mental Health Research
    Institute
Professor of Psychiatry
University of Michigan

Werner E. Reichardt
Director, Max Planck Institute for
    Biological Cybernetics
Tübingen, West Germany

Richard B. Roberts
Chairman, Biophysics Section
Department of Terrestrial Magnetism
Carnegie Institution of Washington

Francis O. Schmitt
Chairman, Neurosciences Research
    Program
Massachusetts Institute of Technology

Richard L. Sidman
Professor of Neuropathology
Harvard Medical School

William H. Sweet
Chief, Neurosurgical Service
Massachusetts General Hospital
Professor of Surgery
Harvard Medical School

Hans-Lukas Teuber
Professor and Head
Department of Psychology
Massachusetts Institute of Technology

Paul A. Weiss
Professor Emeritus
The Rockefeller University

Frederic G. Worden
Executive Director, Neurosciences
    Research Program
Professor of Psychiatry
Massachusetts Institute of Technology

# Central Control of Movement

A report based on an NRP Work Session
held January 11-13, 1970

by

**Edward V. Evarts**
(Work Session Chairman)
National Institute of Mental Health
Bethesda, Maryland

and

**Emilio Bizzi**
Massachusetts Institute of Technology
Cambridge, Massachusetts

**Robert E. Burke**
National Institute of Neurological Diseases and Stroke
Bethesda, Maryland

**Mahlon DeLong**
National Institute of Mental Health
Bethesda, Maryland

**W. Thomas Thach, Jr.**
Massachusetts General Hospital
Boston, Massachusetts

Dorothy W. Bishop
Catherine M. LeBlanc
NRP Writer-Editors

CONTENTS

LIST OF PARTICIPANTS

Mr. Julian H. Bigelow
Institute for Advanced Study
Princeton University
Princeton, New Jersey 08540

Dr. Emilio Bizzi
Department of Psychology
Massachusetts Institute of Technology
Cambridge, Massachusetts 02139

Dr. John M. Brookhart
Department of Physiology
University of Oregon
Medical School
3181 S. W. Sam Jackson Park Road
Portland, Oregon 97201

Dr. Vernon B. Brooks
Department of Physiology
New York Medical College
1 East 106th Street
New York, New York 10029

Dr. Theodore H. Bullock
Department of Neurosciences
University of California, San Diego
    School of Medicine
La Jolla, California 92038

Dr. Robert E. Burke
Laboratory of Neural Control
National Institute of Neurological
    Diseases and Stroke
National Institutes of Health
Bethesda, Maryland 20014

Dr. Kao Liang Chow
Division of Neurology
Stanford University
    School of Medicine
Palo Alto, California 94304

Dr. Mahlon DeLong
Section on Physiology
Laboratory of Clinical Science
National Institute of Mental Health
National Institutes of Health
Bethesda, Maryland 20014

Dr. Edward V. Evarts
Section on Physiology
Laboratory of Clinical Science
National Institute of Mental Health
National Institutes of Health
Bethesda, Maryland 20014

Dr. Karl Frank
Laboratory of Neural Control
National Institute of Neurological
    Diseases and Stroke
National Institutes of Health
Bethesda, Maryland 20014

Dr. Robert Galambos
Department of Neurosciences
University of California, San Diego
    School of Medicine
La Jolla, California 92038

Dr. Elwood Henneman
Department of Physiology
Harvard Medical School
25 Shattuck Street
Boston, Massachusetts 02115

Dr. Masao Ito
Department of Physiology
Faculty of Medicine
Tokyo University
Bunkyo-ku, Tokyo, Japan

Dr. Harvey J. Karten
Department of Psychology
Massachusetts Institute of Technology
Cambridge, Massachusetts 02139

Dr. Donald Kennedy
Department of Biological Sciences
Stanford University
Stanford, California 94305

Dr. Anders Lundberg
Department of Physiology
University of Göteborg
Medicinaregatan 11
Göteborg SV, Sweden

Dr. Peter Marler
The Rockefeller University
New York, New York 10021

Dr. Vernon B. Mountcastle
Department of Physiology
The Johns Hopkins University
    School of Medicine
725 North Wolfe Street
Baltimore, Maryland 21205

Dr. Walle J. H. Nauta
Department of Psychology
Massachusetts Institute of Technology
Cambridge, Massachusetts 02139

Dr. Olov Oscarsson
Institute of Physiology
University of Lund
S - 223 62
Lund, Sweden

Dr. Charles G. Phillips
University Laboratory of Physiology
Oxford, England

Dr. Gardner C. Quarton
Mental Health Research Institute
University of Michigan
Ann Arbor, Michigan 48104

Dr. Francis O. Schmitt
Neurosciences Research Program
280 Newton Street
Brookline, Massachusetts 02146

Dr. William H. Sweet
Massachusetts General Hospital
Boston, Massachusetts 02114

Dr. Hans-Lukas Teuber
Department of Psychology
Massachusetts Institute of Technology
Cambridge, Massachusetts 02139

Dr. W. Thomas Thach, Jr.
Neurology Service
Massachusetts General Hospital
Boston, Massachusetts 02114

Dr. Arnold L. Towe
Department of Physiology and Biophysics
University of Washington
    School of Medicine
Seattle, Washington 98105

Dr. Paul A. Weiss
The Rockefeller University
New York, New York 10021

Dr. Donald M. Wilson
Department of Biology
Stanford University
Stanford, California 94305

Dr. Frederic G. Worden
Neurosciences Research Program
280 Newton Street
Brookline, Massachusetts 02146

Note: NRP Work Session summaries are reviewed and revised by participants prior to publication.

## I. INTRODUCTION

Sixty years ago Sherrington stated that "By combining methods of comparative psychology ... with the methods of experimental physiology, investigation may be expected ere long to furnish new data of importance toward the knowledge of movement as an outcome of the working of the brain" (Sherrington, 1906). Progress in the intervening years has not been so rapid as one might have hoped; but recently a number of neurophysiologists who had previously worked on immobilized preparations have begun to obtain data from moving animals, and the mechanisms whereby the brain controls movement have become topics of increasing interest.

The goal of the Work Session was to identify the critical problems on which neurophysiological research efforts might be concentrated in years ahead, to formulate these problems as clearly as possible, and to outline possible experimental strategies for solution of these problems. In line with these goals, the Work Session format was aimed at eliciting participants' views as to where their research might be heading in the years ahead, rather than providing a review of problems that had already been solved. In order to elicit these views of what might lie ahead, each participant was asked to answer some questions stemming from his own work.

The questions put to Work Session participants fell into several general categories. The largest number of questions asked "How, and to what extent, is motor output controlled by sensory input?" Of course, this is *the* classical theme of almost any general consideration of motor control. As Weiss wrote (1941a), "Nobody in his senses would think of questioning the importance of sensory control of movement. But just what is the precise scope of that control? Is the sensory influx a constructive agent, instrumental in building up the motor patterns, or is it a regulative agent, merely controlling the expression of autonomous patterns without contributing to their differentiation?" Questions falling in this category were put to six of the participants (Brooks, Kennedy, Mountcastle, Oscarsson, Towe, and Wilson*).

A second general category of questions concerned experimental approaches to analyzing the relative functional roles of each of a set of parallel controls to the motoneuron (e.g., the parallel inputs to ocular motoneurons directly from the vestibular system and indirectly from

*Dr. D. M. Wilson died in a boating accident on June 23, 1970.

the vestibular system via the cerebellum). Questions in this general category (parallel controls of the motoneuron) were directed to four participants (Henneman, Ito, Lundberg, and Phillips).

A third category of questions was aimed at generating discussion as to how experimental approaches derived from ethology, psychology, and comparative anatomy might be of value in generating hypotheses on motor control that could then be tested by the neurophysiologist. Questions in this category were directed to Karten, Marler, and Teuber.

The fourth category of questions sought to get at developments that may be expected to result from introduction of certain promising new methods for studying motor control mechanisms. Questions in this category were put to Brookhart and Frank.

So much for the sorts of questions put to the participants. What, then, of the actual questions and answers themselves? Only about half of these will be listed as such in the *Bulletin*. It was decided that rather than listing the individual questions and answers, it would be preferable to organize the material according to the focal points of the Work Session discussion. Thus, instead of following the Work Session format in preparing the *Bulletin*, certain general problem areas were singled out, and chapters were prepared on each of these. Naturally, there are similarities between chapter topics and question categories— but there are also differences.

The first question category was "Translational mechanisms between input and output," and the discussion generated by questions in this category led to two chapters: a first on central patterning of movement and a second dealing more directly with mechanisms whereby sensory input may control motor output. Many participants felt that it might be misleading to speak of sensory input as being "translated" into motor output. Instead, it seemed more fruitful to look upon the sensory input as serving to select one or another stored pattern (or program) of motor behavior. The prominence of Work Session discussion as to how the sensory input might select and modulate the central pattern (or program) is indicated by the fact that two of the five chapters have been devoted to this general topic. Chapter II is concerned primarily with movements that may be centrally autonomous (although they can be triggered by sensory input). Chapter III extends this discussion to movements which, while depending upon a central pattern, are modulated and controlled by sensory inputs during their execution (e.g., a smooth-pursuit eye movement).

Chapter IV deals with the problem of control of reflex mechanisms in the mammalian spinal cord. The focus of attention is here narrowed to consideration of how the flow of afferent information may be controlled within the spinal segment by signals descending from higher centers. Present evidence suggests that the same segmental mechanisms that subserve classically defined reflex functions may also be utilized in some measure in movements specified by central patterning.

The fifth chapter is not based on any one category of questions submitted to Work Session participants, but grew out of trends in Work Session discussion. During the Work Session it became apparent that the narrow definition of corollary discharge as "motor to sensory" and the narrow definition of feedback as "peripheral sensor back to central controller" might no longer be useful. As an alternative, it seemed that one might profitably broaden the definition of these two concepts (feedback and corollary discharge) and consider them as being highly related. It will be the aim of the fifth chapter to consider this broadened definition.

The final chapter will not deal with any one specific conceptual issue, but will discuss several promising research strategies and tactics for studies of central motor control. This chapter stems from questions directed to Brookhart, Frank, Henneman, Ito, and Phillips.

## II. CENTRAL PATTERNING OF MOVEMENT

### by Mahlon DeLong

### Introduction

How the central nervous system produces coordinated or patterned motor output has long been one of the major concerns of neurophysiology. Not surprisingly, many of the presentations and discussions at the Work Session dealt directly with this topic. Because it is now well known that patterned motor output underlying numerous forms of animal behavior is determined by endogenous neural networks that are independent of peripheral feedback (Bullock, 1961; Wilson, 1961, 1964a), there were no debates about the existence of central patterning. However, the applicability of central versus peripheral control, and of interactions between the two, to specific motor systems was considered.

Currently, interest in the nature of neural control systems responsible for animal behavior is widespread among ethologists as well as neurophysiologists, and a considerable body of research has been carried out by workers in both disciplines. The literature pertinent to this topic is extensive, and no attempt has been made to review it completely. Earlier reviews by Bullock (1961) and Wilson (1964a) cover much of the earlier work in greater detail. This section will place emphasis on recent and current work.

### Theoretical Considerations

In order that movements be coordinated, the following basic requirements must be fulfilled by the central nervous system (CNS):

1. The appropriate muscles must be selected.

2. Each participating muscle must be activated or inactivated in proper temporal relationship to the others.

3. The appropriate amount of excitation or inhibition must be exerted on each muscle.

These three basic aspects of coordinated movement may be termed (1) spatial, (2) temporal, and (3) quantitative.

These basic requirements may be met by neural mechanisms in various ways. Until only recently the "peripheral control theory" was the most prevalent view. Derived from Sherringtonian reflexology, it held that coordinated motor output is built up from smaller, discrete phases of movement, linked together by "chain reflexes," with sensory feedback from each phase of the movement reflexly eliciting each subsequent phase. At the other extreme, according to the "central control theory," feedback from the movement is unnecessary for the elaboration of the motor output, because the CNS contains all the information necessary for the patterning, and need not be informed that a particular phase of the movement has occurred in order that the subsequent phase be initiated.

In both theories the spatial requirement, i.e., the selection of appropriate muscles, is met by the specific neural pathways that convey impulses to appropriate motoneurons, either by reflex loops or by interneuronal connections not involving reflex loops. The difference between the two opposing theories lies in the manner in which the timing and quantitative requirements are fulfilled. According to the peripheral control theory, the timing and quantitative requirements are accounted for by varied afferent activity that is routed in feedback loops to provide reinforcement or inhibition of discrete phases of the movement after appropriate time delays. On the other hand, according to the central control theory, the central mechanism contains all the information necessary to specify the fundamental spatial, temporal, and quantitative aspects of the movement. A role for peripheral input in affecting the details of the output pattern and in maintaining the level of central excitability above some threshold level in order that the central mechanism may function is not inconsistent with the concept of central control. That peripheral feedback may provide essential nonspecific excitation to the central mechanism without influencing the output pattern was shown by Wilson (1961) for the locust flight-control mechanism, where removal of certain reflex inputs results in a reduction in overall frequency and amplitude of the movements, without altering the patterning of the movement.

Although examples are known where all aspects of the movement are determined centrally (e.g., the lobster cardiac ganglion; Hagiwara, 1961), an inadequacy is now recognized for either the peripheral control or the central control theory alone to account for many of the motor control systems that have been studied. Interaction between central control mechanisms and peripheral factors is more often the case than either "pure" central or "pure" peripheral control. Bullock (see 1961) discussed the major theoretical ways in which patterned motor output may arise and the variable role of peripheral feedback. These are shown in Figure 1. Basically, motor output may be initiated either by sensory input of a specific type (as in simple relfexes and "triggered movements") or by central commands. In both A and B the motor response is triggered by sensory input, but the patterning of the response is determined by the central network. In A the response occurs only once, while in B the response recurs due to peripheral feedback acting upon the trigger neuron. In C, D, and E the patterned motor output arises from the activity of a central pacemaker. Feedback from the movement may act upon the pacemaker itself (C) and thus influence the frequency of the rhythm, or it may act upon a follower

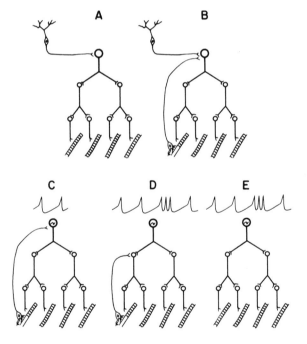

Figure 1. Diagrams of types of pattern formulation. The three levels of neurons are understood to represent branching chains in whose junctions integrative properties may alter the actual impulses and deliver them spatially, as well as temporally, distributed to the effectors (bottom). A and B are shown with receptors C, D, and E, with spontaneous pacemakers giving simple or grouped discharges. B and C have proprioceptive feedback acting on the trigger neuron, D on the shaping of the pattern only. Further explanation in the text. [Bullock, 1961]

neuron (D) and thereby alter the form but not the frequency of the rhythm. A final possibility is that no feedback modulation may be present (E). These hypothetical possibilities are by no means exhaustive, and in reality one may have several mechanisms operative in any particular case.

Because the major difference between peripheral control and central patterning lies in the role of sensory feedback, the most direct means of distinguishing between them is to interrupt feedback loops by sectioning sensory nerves or roots. Several consequences can be imagined:

1. The motor output persists unaltered in spatiotemporal pattern and amplitude.

2. The output persists but is reduced in frequency and/or overall amplitude.

3. The output persists but is altered in form or pattern.

4. No motor output is observed.

In the first two cases the persistence of patterned motor output independent of sensory feedback would be clear evidence for central patterning. The reduction in overall frequency or amplitude in the second case would suggest that a nonspecific tonic effect has been removed. In the third case it appears that there is an interaction between the central mechanisms and peripheral feedback that determines the normal output pattern. In the fourth case one cannot draw any definite conclusion from the observation; although such a result is consistent with peripheral control, it might equally well result from the loss of an essential tonic effect of sensory input upon a central mechanism. As discussed earlier, essential input may be either phasic, resulting from feedback, or maintained spontaneous discharge from receptors, or both. Thus, the technique of interrupting sensory roots may prove to be misleading, or at best inconclusive. Additional techniques are frequently used to overcome these limitations, such as increasing the level of central excitation by random electrical stimulation of the spinal cord or of severed sensory roots. (By this technique one can also study the effects of abnormal or distorted inputs on the motor output and infer certain properties of the mechanism.) Another method of distinguishing between central and peripheral control is to abolish selectively the phasic feedback from movement by sectioning ventral roots and recording from the severed stumps. By leaving sensory roots intact, the contribution of spontaneous activity in receptors to central excitation is retained. A variation of this approach is to paralyze the animal with curare and record the activity in motor nerves. Curarization, by abolishing movement, removes phasic proprioceptive input, yet preserves spontaneous sensory input.

In considering the topic of patterning of motor output, the following three questions seem important:

1. What is the spectrum of motor control between the extremes of "pure" peripheral and "pure" central control?

2. What, in each case, is known about the neural mechanisms responsible for patterned output?

3. What, in each case, is the contribution of peripheral feedback? Is its role to provide timing cues, to modify a basic centrally determined pattern, or to maintain central excitability?

This discussion will deal primarily with those instances where central patterning is of major importance in the production of motor output. Movements for which peripheral feedback plays a greater role will be discussed in Chapter III, which considers, first, examples of central control where sensory feedback plays little or no role at all in the patterning of the output, and then gives instances where the output is modified in various ways by peripheral feedback.

## Central Patterning with Automaticity

Numerous instances of centrally patterned motor output that are not dependent on sensory input or feedback are well known (Bullock, 1961; Wilson, 1964a). In cases where the central mechanism is known, pacemaker neurons usually provide the element of central automaticity. The earliest example was found by Adrian (1931), who demonstrated a respiratory rhythm in the isolated nerve cord of the diving beetle *Dytiscus marginalis.* In the lobster the cardiac ganglion continues to burst even after removal of the ganglia from the animal (Hagiwara, 1961). The patterned bursting can probably originate in a single pacemaker cell that does not receive feedback from the follower neurons it drives. Pacemaker discharge initiates a burst of activity from the ganglion by synaptic and electrotonic excitation of the other cells. The termination of the burst results from fatigue of the individual neurons. Respiratory neurons in the brainstem of the cat discharge rhythmically even when the brainstem is isolated from inputs by sectioning (Salmoiraghi and Baumgarten, 1961). In the crayfish the rhythmic activity of motoneurons innervating the swimmerets originates in the autogenic rhythmicity of neurons in the abdominal ganglia, which persists after isolation of the abdominal cord from the periphery (Hughes and Wiersma, 1960). Central automaticity and patterning at a more complex level is evidenced by the observation that the salamander exhibits coordinated body and limb movements before the sensory pathways are laid down in the embryo (Coghill, 1924).

Examples of a more complex patterned output, which in some species is probably independent of sensory feedback, come from studies on bird vocalizations (Marler, Work Session). Birds of most species produce highly stereotyped and often complex songs. In adults, removal of auditory feedback by deafening does not interfere with performance. Although some bird species, such as the chaffinch (Nottebohm, 1967) and white-crowned sparrow (Konishi, 1965b), fail to develop normal song if deafened shortly after birth or if reared in isolation, other species, such as the domestic chicken (Konishi, 1963) and the ring dove (Nottebohm and Nottebohm, 1971), develop normal calls if deafened after hatching. Between these two extremes is the song sparrow, which can acquire normal song when reared in isolation, yet requires auditory feedback to do so (Mulligan, 1966). The white-crowned sparrow (Marler, 1970; Marler and Tamura, 1964), on the other hand, must hear adult song during a critical period from 10 to 50 days after hatching in order to develop normal adult song later on. Actual singing begins several months after this critical period. Deafening after the critical period, but before singing has begun, results in abnormal song; yet once adult song has been acquired, deafening is without significant effect (Konishi, 1965b). Marler postulates that during the critical period an acoustic memory or "template" of the adult song pattern is formed, which is later used in the actual learning. The motor output is then compared with the stored reference template and gradually molded to it. Once normal song is acquired, the auditory template is no longer required.

Even an untrained, young, white-crowned sparrow seems to possess a crude auditory blueprint or template, for it will develop a song with some normal characteristics if it has been deafened. Marler hypothesizes that this crude template becomes refined by training with normal song of the species. The same crude template may explain the apparent selectivity of learning; the male white-crowned sparrow learns only conspecific songs and rejects those of other species.

The question of whether the patterning of adult song in the absence of auditory feedback is determined entirely by central mechanisms or whether other sources of feedback, such as proprioceptive feedback from the syrinx, might be responsible, has not been answered. Unilateral section of the right hypoglossal nerve in the adult chaffinch has little or no effect on performance, whereas section of the left nerve produces either a substantial dropping out of song elements or considerable distortion of most elements (Nottebohm, 1970). This lateralization of song control appears to be acquired, because if the left nerve is sectioned before full song has emerged, normal song development occurs.

However, unilateral section of the hypoglossal nerve cannot answer the question of whether or not proprioceptive feedback from the syrinx is responsible for the patterning, because feedback via the remaining nerve is possible. One approach to this problem would be to record the activity in the left hypoglossal nerve of an adult, deafened bird and correlate this activity with the recorded song pattern.

The physiology of bird song has been little studied, yet this would seem to be an exceedingly fertile area for neurophysiological investigation. Karten's (1969; Nauta and Karten, 1970) detailed studies of avian neuroanatomy provide an essential foundation for such investigations. His work emphasizes the general similarities between birds and mammals both in specific sensory systems and descending telencephalic efferent systems, making investigation of these motor mechanisms all the more important and relevant to general vertebrate neurophysiology.

## Central Patterning with Triggered Movements

We have considered a number of examples of centrally patterned motor output that is produced autonomously through central commands or pacemakers. Next we shall consider a second class of movements differing from the first in that the triggering comes from the environment. Examples of such "triggered movements" are numerous. Bullock (1961) emphasized the general significance of such movements:

> Now most actions triggered by input signals contain elements of
> pattern not in the input. Consider eye movements, swallowing, cough-
> ing, a cricket's chirp, a grasshopper's hop, a squid's color change, taxes

and instincts, not to speak of more complex behavior. From the feeding movements of a sea anemone to formation of a word in human speech, the predetermined central contribution to the pattern is enormous even if the initiation follows a peripheral cue. Our examples have been confined to motor acts, but is it not just as likely that the output of most central masses of organized neurons constituting an integrating level is also patterned in a way not contained in its input? A preexisting pattern, awaiting first permissive input and then triggering input to release it, may reasonably be inferred to be, not universal, but frequent.

Excellent examples of patterned output that require triggering from sensory cues are found in arthropods. The simplest example is the sound reflex of cicada (Hagiwara and Watanabe, 1956), which is mediated by two sound muscles, each innervated by a single motoneuron. A single stimulus applied to the sensory nerve from the hair sensillae causes an alternating activation of the two muscles. Evidence points to a central driving mechanism acting upon the pair of motoneurons, which may be connected by a simple reciprocal-inhibition network.

Numerous arthropods possess a giant fiber system in the ventral nerve cord that mediates an escape response. These giant fibers are triggered by various forms of sensory input. Wiersma (1947) has shown that a single impulse in any one of the central giant fibers of the crayfish is sufficient to evoke a single flexor escape response. Even when the abdominal cord is isolated from the rest of the nervous system and the sensory nerves cut, a single impulse is able to produce a coordinated motor output (Roberts, 1968). Little is known, however, about the neural mechanism underlying this response. Lack of detailed anatomical information has hindered further understanding of this problem.

Kennedy showed that activity in individual command interneurons in the crayfish cord is able to produce complex, coordinated patterns of behavior involving hundreds of motoneurons (Kennedy et al., 1966b; Kennedy, 1968). Command fibers controlling the postural musculature of the tail, when stimulated electrically at rates of several times per second, produce a coordinated output in the abdominal segments appropriate for either flexion or extension. Although different command fibers may produce qualitatively similar effects, such as flexion or extension, they differ in that some produce generalized flexion or extension, others, primarily rostral or caudal movements. Some compensate for extensor loads via the reflex loop involving the

muscle receptor organs (Fields et al., 1967). Certain command interneurons are capable of releasing more complex behavioral patterns. Larimer
and Kennedy (1969a,b), for example, have identified a specific interneuron in the crayfish abdominal cord that produces a complex cyclic
series of movements in the tail appendages even when stimulated at
constant frequencies; the rhythm of the movements is unrelated to the
rhythm of the stimulation. The output to the individual muscles is
patterned in a rhythmical manner, with dozens of motoneurons
discharging, each at a particular phase of the cycle. The output follows
the same "motor score" from one animal to another, and is undisturbed
by total deafferentation. Again, central connections and automaticity
determine the output.

Each command fiber is viewed as a unique "labeled line" that
specifies a particular movement or geometry. Command fibers do not
act directly upon motoneurons but rather release or unlock activity in
neural networks between the command fiber and the motoneurons.
They are "permissive but not instructive." A number of command
fibers that produce effects on the postural muscles of the abdomen
have been activated by natural stimuli. In each case the motor effect is
identical to that seen with electrical stimulation. Interestingly, however,
the command fibers most effective in evoking complex behavior often
cannot be activated by natural stimulation. These neurons are likely to
be several synapses removed from the receptors and to have more
sophisticated input requirements for activation or triggering. Possibly
they not only have complex receptive fields but also are under the
control of higher nervous centers.

Examples of triggered central patterning in mammals are less
numerous than in arthropods. However, swallowing in mammals provides an excellent example. Doty and his colleagues have shown that
the act of swallowing involves the coordination of nearly 20 different
muscles whose motoneurons are distributed from mesencephalic to posterior medullary levels (Doty, 1967; Doty and Bosma, 1956; Doty et al.,
1967). The patterning of muscular contractions is independent of the
stimulus used to evoke the response, i.e., touching the pharynx, rapid
injection of water into the mouth, or electrical stimulation of the
superior laryngeal nerve. Attempts to alter the pattern by disturbing
feedback loops, i.e., by excision of the participating muscles, fixation
of the hyoid mass, or traction on the tongue, fail to bring about any
significant change in the patterning. The neurons responsible for the
coordination have been shown to be situated bilaterally just dorsal and

rostral to the rostral pole of the inferior olive. These neurons form a "swallowing center," which is viewed as a "functional neuronal grouping interconnected in such a manner as to produce automatically, when it is effectively excited, the inhibitory and excitatory sequences in appropriate motoneurons" (Doty, 1967).

The triggering or initiation of most behavioral acts of any complexity requires a "decision" by the CNS that certain criteria have been met. The "decision" in the case of triggered movement is of an "either-or" nature. One possibility, as exemplified by the Mauthner cell or the giant fiber system of arthropods, is that a single neuron may act as a decision-maker. Convergence of appropriate input on a single cell may initiate a single impulse that releases a complex behavioral response. A second possibility for decision-making within the CNS is a network of mutually interacting neurons, with a threshold for the network that is different from that for any one cell. The excitability of the network as a whole must reach a definite level before it becomes active (Bullock and Horridge, 1965).

Recently, Willows and Hoyle (1969) have identified a neural network in the nudibranch mollusk *Tritonia gilberti* that appears to be of the latter type. Two bilaterally symmetrical groups of approximately 30 cells in the pleural ganglia appear to trigger the swimming escape response, which is normally triggered by contact with the tube feet of certain starfish. When the swimming escape response is triggered by contact with a starfish, it is always preceded by at least one burst of impulses in these neurons. Brief electrical stimulation of one cell of the group only occasionally leads to a response, but weak stimulation of many cells with a surface electrode invariably produces the response. Simultaneous intracellular recordings in several cells have demonstrated that they are interconnected by low resistance pathways, both within the group and across the ganglion. Excitation of any single neuron therefore results in graded stimulation of many others. The excitability of the network as a whole, however, is determined by summed activity in all the neurons, since large depolarizations in single cells or in small groups of cells are usually ineffective in triggering a burst. When contact is made with a few tube feet of the starfish, intracellular recordings have shown nearly synchronous bursts of EPSP's in all cells recorded from. A considerable degree of divergence of sensory neurons or interneurons must occur prior to termination on the pleural ganglia cells. Thus, the function of initiating the response appears to reside with the pleural network, and Willows and Hoyle conclude that it does this by "positive selection of inputs that affect many cells in the network simultaneously and rejection of those that excite only a few components." It would be of interest to know if the swimming response could occur after removal or destruction of the ganglia, or if the response can be initiated by a central command independent of the trigger network.

An example of how a simple behavioral response can be controlled in several ways has been demonstrated by Kupfermann and Kandel (1969) in *Aplysia*. A withdrawal response in external organs of the mantle cavity of this organism occurs in response to tactile stimulation as well as spontaneously. In both instances the response is controlled by five identified motor cells in the abdominal ganglia. These cells receive input from two sources: (1) direct sensory input from receptors on the body, and (2) input from a group of adjacent interneurons. Intracellular study of these neurons has shown that the reflex withdrawal response to tactile stimulation results from direct excitatory input to the motor cells from the sensory receptors, whereas the spontaneous withdrawal response results from a combination of excitatory and inhibitory input to the motor cells from the nearby interneurons. This activity in the interneurons is likened to a central command. Interestingly, the central command may at times be initiated by a tactile stimulus that does not elicit the reflex response.

## Central Control with Peripheral Feedback

### Invertebrates

#### Locusts

Only recently has it become accepted by neurophysiologists that simple forms of movement, such as walking, swimming, or flying, may be centrally patterned. The elegant analysis of the flight-control system in locusts by Wilson (1961, 1964a) discussed by him at the Work Session provides an excellent example of the current approach to this problem. Locusts fly by beating two pairs of wings at the same frequency. Each wing is controlled by fewer than 20 motoneurons, with cell bodies in the thoracic ganglia. Output from these motoneurons consists of nearly synchronous impulses in two small populations, with activity alternating between antagonistic sets of wing muscles. Variation in the number of impulses serves to control both the direction and power of flight. On the afferent side, two sources of phasic proprioceptive feedback are known: (1) a stretch receptor on the hinge of each wing, which indicates wing position, direction of movement relative to the body, and wing velocity; (2) lift receptors located in the wing veins, the cumpaniform sensilla, which provide information about upward force or lift on the wing.

Wilson found that removal of both sources of proprioceptive input caused no qualitative alteration in the pattern of the output, but only a decrease in wing-beat frequency. Stimulation of wind-sensitive receptors on the head caused an increase in frequency. Even when all sensory input was eliminated by isolating the thoracic ganglia, the system could continue to generate patterned output, although at reduced frequency, if other nonphasic stimulation was used. In an isolated preparation it was found that stimulation of the cord with random pulses could drive the normal motor output pattern. Nonphasic stimulation of the stretch receptor fibers could restore the wing-beat frequency to normal. As such an isolated preparation decayed, motor units often dropped out individually, sometimes leaving only a single regularly discharging unit. If a second unit reappeared, it would adopt a normal phase relationship with the other units. From these observations it was clear that the basic coordination of flight was centrally patterned. Phasic feedback information from the stretch receptors was shown to be lost in an averaging process and to serve only a tonic function in maintaining the excitability of the central oscillators.

Borrowing from Hoyle's (1964) models of muscle control systems, Wilson likened the locust flight-control system to a "motor tape," i.e., a centrally stored network for pattern generation that, when activated, plays out in a predetermined manner. However, it was observed that after removal of a single wing, locusts could usually compensate for the asymmetry and fly in a straight path. How could the inherent motor tape be modified when damage to one wing made it no longer appropriate? Observations by Goodman (1965) on locusts suspended in such a way as to allow rolling about the long axis demonstrated that locusts usually showed lack of stability in the dark, tending to roll either clockwise or counterclockwise. When the lights were turned on, however, they flew normally. Stability was found to depend upon two optomotor reactions, a dorsal light reaction and a reaction to the horizon. Wilson found that the instability and direction of rolling is associated with a long-term inherently asymmetrical motor output that is corrected by visual input (Wilson, 1968). Recordings from motor nerves in the dark demonstrate asymmetrical motor outputs correlated with the rolling motions, with rapid compensatory adjustment when the lights are turned on. Such observations indicate that the flight system is not only a motor tape, but also combines the features of a "sensory tape," in Hoyle's terminology, in that it is able to compare exteroceptive feedback with a stored sensory pattern and adjust the output appropriately. Whether the corrective influence is exerted on portions of the motor tape or directly on motoneurons is unclear. Integration of input from the wind-sensitive hairs and the visual system occurs in interneurons in the supraesophageal ganglion that send their axons to the thoracic ganglia. Stimulation of these fibers in the cord of deteriorated preparations has produced high rates of following in motoneurons with latencies consistent with monosynaptic connections (Wilson, 1964b). However, evidence in intact preparations for direct access to motoneurons from higher centers is lacking.

Although more is known about the flight system of the locust than almost any other moderately complex neural system, little is known about the flight-pattern generator in the thoracic ganglia. This is due to the difficulties of recording intracellular synaptic events in arthropod neuropil and the lack of clear anatomical data. One possibility is that the patterned output could be the result of motoneuron interactions alone. This seems unlikely, however, since (1) the same muscles used in flight are also used in walking, jumping, and stridulation, which require markedly different motor scores (Wilson, 1964a), and (2) only weak motoneuron interactions have been demonstrated by antidromic stimulation (Wilson, 1964a) and intracellular recordings (Kendig, 1968).

Evidence for central pacemakers in the locust is lacking at present. Recent attempts by Page and Wilson (1970) to identify pacemakers in the metathoracic ganglion of flying locusts were unsuccessful, although multimodal sensory units responding to visual and tactile stimuli were identified whose discharge was coupled to the wing-beat rhythm. Wilson concludes that the flight-pattern generator is most likely in the form of a diffuse network of neural elements within the thoracic ganglia rather than in the form of a few pacemaker cells. Wilson has constructed model networks that closely simulate the output patterns of the locust flight system (Wilson and Waldron, 1968). He finds that the network which most closely matches the output consists of two small groups of neurons, the members of which are mutually excitatory and produce bursts of impulses separated by periods of silence. The two groups alternate in activity by means of reciprocal inhibition. Frequency of output is proportional only to the quantity of afferent input.

### Crayfish and Lobster

Another example of a simple endogenous motor pattern that is modified by sensory feedback is the swimmeret system of crayfish and lobsters, first studied by Hughes and Wiersma (1960). Four pairs of swimmerets perform rhythmic cyclic beating movements. The motor output to each swimmeret consists of alternating bursts of activity in two antagonistic sets of motoneurons innervating retractors (power-stroke) and protractors (returnstroke). Beating occurs both spontaneously and in response to activity in certain command interneurons (Hughes and Wiersma, 1960). Such patterned activity occurs after isolation of the ganglia from all sensory input, indicating that it is not dependent on timing cues from sensory feedback, but rather is endogenous in origin. As in the locust flight system, each swimmeret nevertheless has receptors that provide phasic sensory feedback. Recent investigations by Davis (1968, 1969a,b,c) have been concerned with the role of this feedback. In contrast to the locust flight system, Davis finds that specific peripheral feedback is not averaged out to provide nonspecific excitation to central oscillators, but acts phasically during each movement. During the powerstroke, for example, peripheral feedback acts positively to reinforce the powerstroke neurons. Excitatory input is simultaneously fed to the returnstroke neurons, but this is totally inhibited by phasic input from other receptors until near the end of the powerstroke.

Intrasegmental reflexes capable of initiating or terminating the powerstroke were found lacking. Thus, the timing of the powerstroke is determined centrally. Moreover, Davis has shown that although intrasegmental reflexes strengthen the linkage between the powerstroke and returnstroke within each movement cycle and may also reinforce the reciprocity between excitor and inhibitor activity, these features are all independently programmed into the CNS (Davis, 1969b). The intrasegmental reflexes act as "amplifying devices" for the preprogrammed motor patterns, influencing the quantitative rather than the temporal aspect of the output.

**Vertebrates**

*Amphibian Locomotion*

The role of peripheral feedback in the locomotor movements of vertebrates has long been a topic of considerable debate. Locomotion in amphibians was investigated by numerous earlier workers (Hering, 1893; Bickel, 1897; Weiss, 1936; Gray and Lissmann, 1940). All observed that deafferentation of one or two limbs of a frog or toad produced no disturbance in the normal diagonal pattern of ambulatory movements, jumping, or swimming. If all four limbs were deafferented, locomotion was still possible, although activity was reduced. Weiss (1936) found that even after total deafferentation a toad still exhibited the basic patterns of motor coordination, although the animal became extremely lethargic. Gray (1939) and Gray and Lissmann (1940, 1946a,b), however, were unable to confirm this. In a careful series of studies they found that walking was dependent upon the integrity of the sensory and motor supply to at least one spinal segment. Gray concluded that locomotor patterning was dependent on rhythmic feedback from the periphery. Evidence that the rhythmicity or timing information was the essential factor of the input from the remaining dorsal root rather than a tonic effect was demonstrated in preparations where all the motor roots were cut except those to a single limb. Leaving at first all the sensory roots intact, weak stimuli to any limb elicited a clear ambulatory rhythm in the intact limb. However, after the sensory roots of the intact limb were cut, the rhythm could no longer be elicited, even with strong stimuli. Thus, with the majority of sensory roots intact, but without phasic input from the muscles, the ambulatory rhythm could not be elicited. Gray decided that "It is difficult to draw any conclusion other than that when proprioceptor impulses from other limbs are effectively excluded, the impulses arising in the proprioceptor endings of the intact limb are essential for the maintenance of the ambulatory rhythm."

Recently this unresolved controversy between Weiss and Gray as to whether or not a deafferented toad can generate a normal loco-motor pattern has been reexamined by Harcombe and Wyman.* They performed total spinal deafferentiations on 28 toads (*Bufo marinus*) and found that following surgery all were capable of producing the normal diagonal locomotor pattern. Their findings are directly contradictory to those of Gray and Lissmann, and support Weiss's contention that the motor pattern does not depend on intact dorsal

*Unpublished data.

roots. They suggest that their findings may be due to the fact that their surgical technique was less traumatic and minimized damage to both the vertebral column and the spinal cord. However, as Harcombe and Wyman point out, the more general question posed by Gray of whether or not any patterned peripheral input is required is not settled by these experiments, because other sources of rhythmic feedback, such as the vestibular, might play a timing role.

Using a variety of other experimental approaches to the study of motor control in amphibians, Weiss (1950) found convincing evidence for the central patterning of locomotor movements. For example, when the forelimbs of adult salamanders were exchanged so that the limbs pointed in the reverse direction, after reinnervation the grafted limbs were observed to move in a coordinated manner, but always in reverse. The markedly distorted peripheral feedback did not disrupt the orderly sequence of muscle activation. If the limb buds were reversed in the embryo, the animals walked in reverse from the beginning and never changed their pattern. Weiss (1941b) also showed that deafferentation of the legs in frog tadpoles produced no impairment of coordinated leg function, demonstrating that sensory pathways were not necessary for the development of normal patterning of movements. From these experiments of Weiss in amphibians, it is clear that the spinal coordinating mechanisms are genetically predetermined, developing without guidance from sensory feedback and functioning according to a set mechanism, even in the face of the distorted feedback and extreme functional inappropriateness produced by the surgical procedures.

Further evidence for central patterning comes from studies by Székely et al. (1969) on the patterning of EMG activity in freely moving normal and deafferented newts, which show that the normal pattern of muscular contraction is maintained following deafferentation of one or both forelimbs. Extensive co-contraction of antagonistic muscle groups was observed, especially in the arm, which was not suspected from direct observation of moving animals. Székely concludes that because the normal pattern of muscular activity is maintained following deafferentation, the coordination is controlled by a central mechanism without afferent information from the moving limb.

*Fish Locomotion*

In higher vertebrates peripheral feedback appears to play a greater role in the patterning of locomotor rhythms. The swimming of the dogfish has been studied in detail by Gray (1933a,b,c) and Lissmann (1946a,b). To determine whether or not the swimming movements are centrally patterned or dependent on peripheral feedback, Lissmann investigated the effects of deafferentation in spinally sectioned fish. He showed that while many dorsal roots could be sectioned without disrupting the rhythm, total deafferentation resulted in immobile preparations. Lissmann concluded that sensory feedback was necessary for the rhythm. However, as discussed earlier, lack of motility following deafferentation may not be taken as evidence for peripheral control, because essential tonic sensory input may have been removed from a central mechanism. Recently Roberts (1969) has attempted to circumvent this by studying the activity in ventral roots in curarized spinal fish. In such preparations he found that spinal motoneurons continued to fire with periodic bursts for several hours. In each spinal segment, motoneurons bursting in one half of the cord alternated with those on the other side. However, each spinal segment discharged without relation to adjacent segments. Thus, although clearly some central patterning as well as central automaticity is seen at the segmental level, intrasegmental coordination is lacking in the absence of phasic proprioceptive input. Moreover, Roberts has shown that changes in the timing of phasic proprioceptive input result in immediate changes in the timing of motor output. By applying oscillations to the body of the fish at varying frequencies, muscle activity in the body was found to follow the frequency of oscillation. Thus, changes in the timing of phasic proprioceptive input influence the timing of motor output. The rhythm of fin movements was also shown to depend on the timing of proprioceptive input from the body during swimming movements. Thus, while the spinal segments are capable of intrinsic activity, this activity is apparently normally overridden by peripheral feedback. These findings conflict with the classical studies of von Holst (1935a,b; 1936a,b), who concluded from careful analysis of individual fin rhythms in various species of fish that fin interrelationships were entirely explainable by central mechanisms.

*Mammalian Locomotion*

Analysis of the role of peripheral feedback in mammalian locomotion has proceeded along several different lines. The techniques of deafferentation were first used by Bickel (1897), who found that bilateral deafferentation of the hindlimbs in the dog did not abolish stepping movements. Later work by Brown (1911a,b; 1914) and Sherrington (1913) provided evidence for central control of stepping in the cat. Brown demonstrated that stepping movements occurred in the hindlimbs of a totally deafferented cat following section of the spinal cord. Moreover, the patterns seen in deafferented preparations differed little from those in which the sensory roots were left intact. Brown (1914) later showed that stepping in the spinal animal would occur under a depth of anesthesia that precluded conduction of impulses in sensory nerves, the so-called "narcosis progression." Sherrington (1913) found evidence for central control of stepping in the decerebrate, totally deafferented cat. Sherrington interrupted all the nerves to the hindlimbs, except those to the main extensors of both knees, and then stimulated the common peroneal nerves bilaterally. Under certain conditions he observed alternating contractions in the two muscles, which he called "reflex stepping." He concluded that the isolated spinal cord was capable of producing a rhythmic pattern as seen in walking.

Recently, Sherrington's experiment has been reevaluated by Egger and Wyman (1969). They used a similar preparation but recorded from ventral roots instead of the extensor muscles in curarized cats. They criticize Sherrington's experiments in that the inductoria used to stimulate the peroneal nerves must have delivered impulses slightly out of phase with each other and thus produced "stepping" at the beat frequencies of the two inductoria. They found that, indeed, when the peroneal nerves were stimulated at unequal frequencies, the legs "stepped" at the beat frequency, and when the peroneals were stimulated at exactly the same frequency, no "stepping" occurred. Thus, considerable doubt is cast upon Sherrington's evidence for "reflex stepping" in the cat. However, the question as to whether or not walking in the intact cat is under central control is not answered by these findings. Graham Brown's experimental findings, although less widely accepted, are difficult to interpret in any other way.

The recent studies on deafferentation in monkeys by Taub and Berman (1968) are relevant here. Contrary to the findings of Mott and Sherrington (1895), Lassek (1953), and Twitchell (1954), they found that following both unilateral and bilateral forelimb deafferentation, although motor deficits are clearly present, monkeys are able to use their limbs adaptively in conditioning experiments and in locomotion. Recovery following bilateral forelimb deafferentation usually took place gradually over several months. In a small number of monkeys after recovery had occurred, total deafferentation was carried out. First the remaining thoracic sensory roots were cut and the animals allowed to recover. No changes in motor performance were observed following this procedure. In the final phase, the remaining lumbosacral roots were sectioned, leaving the animal with its spinal cord totally deafferented. Postoperatively, the animals were unable to use their legs effectively; however, no further disturbance in the use of forelimbs occurred. Unfortunately, none of the animals survived long enough to determine whether or not coordinated leg and arm movements would return. The experiments of Taub and Berman have been confirmed in part by Bossom and Ommaya (1968). These authors have, however, criticized the studies by Taub and Berman on the grounds that (1) their experimental technique may have compromised the vascular supply to the spinal cord, and (2) the thoroughness of the rhizotomies has not been verified by histologic or other techniques except in a few cases. By use of the operating microscope, Bossom and Ommaya have been able to section dorsal roots with less damage to the blood supply of the cord, and to assure that every rootlet is interrupted. They report that their animals usually attempted to use their forelimbs immediately after coming out of anesthesia. Movements were, however, extremely ataxic. Animals that did not attempt to use their arms upon recovery from anesthesia were found to have damage to the spinal cord. Unfortunately, these authors have not yet carried out total deafferentation. It would be most important to determine whether a totally deafferented monkey can move all four limbs in the normal locomotor pattern.

Recent studies by Engberg and Lundberg (1969) on the electromyographic activity in the hindlimb musculature of the unrestrained cat have provided additional evidence for central programming of locomotion. They correlated the electromyographic activity of flexors and extensors with the movements of the limb during normal walking and running. Careful study of the onset of extensor activity during placing of the foot showed that extensor activity preceded the contact of the foot with the ground and could not, therefore, have been produced by proprioceptive feedback from the muscle spindles. They conclude that the basic activity of stepping is centrally programmed activation of extensors and flexors. Proprioceptive feedback may modify this activity in important ways, but does not initiate or maintain it.

## Conclusions

In this chapter consideration has been largely given to the interaction between central patterning and peripheral feedback in the production of coordinated motor output underlying certain basic movement patterns such as respiration, swallowing, and different forms of locomotion. These movements have a common feature in that they lack specific spatial orientation to the environment. In the case of respiration, swallowing, vocalizations, and so on, this is evident. It can be argued as well for the basic locomotor rhythms (e.g., locust wing beating, lobster swimmeret beating, vertebrate walking, and so on). Although locomotion serves to propel the organism through space, the basic pattern generator need not function in relation to the environment. Corrective modification of a basic rhythmic motor-output pattern by proprioceptive and exteroceptive (especially visual) feedback appears to be a general feature of both arthropods and vertebrates.

While the contribution of central patterning is enormous in almost every instance, it is striking how varied is the role of proprioceptive feedback—in some instances exerting only a nonspecific tonic effect (wingbeat-frequency control in the locust flight system), in other cases providing phasic reinforcement of discrete phases of the movement (lobster swimmeret), and elsewhere providing timing cues for the overall patterning (dogfish swimming). It would be satisfying to be able to relate in each case the details of the neural control mechanism to the specific behavioral requirements. Yet such understanding in most cases awaits more detailed observations of the normal behavioral patterns and further studies in other species.

Although the relative contributions of central and peripheral factors is now well understood in many instances, especially in arthropods, the nature of the neural pattern generators is known definitely in but a few cases (e.g., lobster cardiac ganglion). Analogies with "magnetic tapes" and "templates," while providing intellectual stimulation and suggesting further experiments, tell us little about the actual neural mechanisms beyond what experimental observations provide. It is difficult to imagine how a "motor tape" that is "read out" by a "tape head" might look in actual neural circuitry. We have stressed that the basic differences between central and peripheral control lie in the role of peripheral feedback in providing timing cues or phasic reinforcement. The concept of *internal* feedback loops arising throughout the elements in the neural system responsible for motor output was discussed at considerable length at the Work Session and will be described in Chapter V of this report. Is it not conceivable that such internal feedback loops might function in a manner analogous to peripheral feedback loops in providing timing information and phasic reinforcement? Information that a given phase of a rhythmic pattern is occurring could equally well be fed from one portion of the pattern generator to another, which would then initiate a subsequent phase. Evidence for such interactions among groups of neurons has been shown in the majority of examples discussed. The actual neural control mechanisms underlying peripheral control and central patterning need not be as dissimilar as might appear, once one allows for internal feedback.

## III. TRANSLATIONAL MECHANISMS BETWEEN INPUT AND OUTPUT

by Emilio Bizzi and Edward V. Evarts

### Introduction

The previous chapter dealt with movements for which the output was largely (if not entirely) determined by a stored program or "motor tape"—movements that were either "centrally autonomous" or "triggered." The role of sensory input in such movements has been summarized by Bullock (1961) as follows:

> In sum, sensory input is of decisive importance—in creating the permissive steady state centrally (making the frog want to jump), in directing action adaptively (aiming his jump), and in perfecting details during the action in some cases (probably more in mantid fly catching than in frog jumping). But central patterning is the necessary and often the sufficient condition for determining the main characteristic features of almost all actions, whether stimulus triggered or spontaneous.

In concluding, Bullock added,

> Nervous systems are not like present day computers, even complex ones, but have oscillators and built-in stored patterns; they do not give outputs predictable by their inputs or externally controlled "instructions."

Computers also have programs, however; and to say that a central pattern exists in the nervous system does not preclude construction of a transfer function relating sensory input to motor output. Granted that central patterning in the sense that Bullock has used this term is always of critical importance, there may nevertheless be instances in which one can construct equations by virtue of which output becomes predictable from input. Take, for example, the pupillary movement in response to light, as analyzed by Stark (1959) and Stark et al. (1962c). One might argue that the pupil is unique—but actually there are a number of other outputs that, although not as completely under sensory dominance as the pupil, can nevertheless be rather well predicted from the input. Such outputs have classically been referred to as "reflexes." Some of these forms of motor behavior will be considered in this chapter and in the following chapter by Burke, so it would seem worth while to review here the way in which ideas on reflex action developed, and then to consider distinctions between "reflex," "triggered," and "centrally patterned" movements.

Concepts of "Reflex" Movements

A review of the current use of the term "reflex" reveals considerable uncertainty as to what sorts of movement should be included under this term. Should the term "reflex" be limited to responses of the isolated spinal cord, or be extended to all stimulus-response sequences? In the latter case, should we apply the term reflex only to so-called automatic behavior or also to those actions triggered by input stimuli that are arbitrarily related to the output?

Traditionally, reflex action has been viewed as being unlearned (based on inherited neural circuits), predictable from the inputs, uniform, and adjustive or protective in purpose. To these characteristics two other attributes, heavily charged with philosophical implications, have been added: first, that the reflex is "involuntary," and second, that it is not dependent upon consciousness. These views as to the nature of reflex movement are the results of physiological observations of the last 400 years, mixed with the various philosophical beliefs held by the naturalists who have investigated and reflected upon the theme of animal actions.

A brief historical survey of ideas concerning reflex action might therefore be helpful in understanding why certain terminology has become ingrained in current thinking and might, perhaps, suggest a way out of the present state of confusion. We owe to Descartes the first formulation of the reflex as an involuntary and machinelike action. There is no question about the revolutionary value of this assumption, which opened the way to the study of overt animal behavior in terms of the laws of physics and mechanics. However, the sharp distinction between machinelike actions of animals and the volition of man (assumed to be dependent upon the soul) led to a dichotomy between voluntary and involuntary movements—a dichotomy that has lingered through the centuries and has led to an innumerable series of experiments, debates, and speculations. An example of the far-reaching impact of Cartesian thinking can be seen in Sherrington's assertion that the future task of the physiologist was to understand how volition can control spinal reflexes.

The Cartesian concept of reflex action as a neuromuscular response endowed with certain objective characteristics was subsequently (in the 17th and 18th centuries) investigated by many (Willis, Whytt, Swammerdam, to name a few). Their efforts resulted in an indication of the specific anatomical structure involved in reflexes, and in an outline of some of the salient features of reflexes (e.g., their protective character and predictability). Furthermore, certain automatic movements, such as walking and certain postural reactions, were seen to have a resemblance to reflexes. Central to the writings of these authors is the distinction between voluntary and involuntary movement, as well as a tendency to include in the principle of the reflex (conceived as involuntary action) the more common classes of movement, such as motor habits, etc.

During the 19th century the development of anatomical techniques and methods of physiological experimentation led to an understanding of the specific structures involved in the reflex arc, namely, its sensory, central, and motor parts (Bell and Magendie). At the end of this century a great number of reflex arcs had been discovered, e.g., the postural, the vestibular, the tendon, etc. Although there is no question as to the fruitfulness of the intense study of spinal reflexes, it should be pointed out that the approach of "reflexology" led to a tendency to explain the behavior of the intact animal in terms of the characteristics of spinal reflexes. This trend led to the school of thought in which the reflex became the elemental unit of behavior, and according to which mental life, automatic and volitional acts, and even consciousness itself were conceived as a function of the interplay of conditioned reflexes (Pavlov). The Pavlovian point of view involved an abolition of the arbitrary division of animal responses into classes of actions such as automatic, reflex, volitional, and learned. Although this approach is superficially reminiscent of the Cartesian outlook insofar as the reflex is conceived as a machinelike, inevitable reaction, it did in fact represent the end of the Cartesian dichotomy between voluntary movement on one side and reflex (involuntary) movement on the other.

**Concepts of "Reflex," "Triggered," and "Centrally Patterned" Movements**

Following Pavlov, a number of other 20th century physiologists and psychologists shared the view that every action is a result of a sensorimotor process involving a more or less complicated chain of reflex arcs, in which voluntary behavior is not qualitatively different from other forms of behavior. In contrast with these positivistic views, others have questioned the attempt to explain behavior in terms of reflex action. Herrick and the psychologists of the "Gestalt School" and Sperry belong to this group, holding the view that any attempt to derive complex sensorimotor integration from reflexes is futile.

Granted these differences of opinion as to what reflexes are and as to whether complex behavior is built up of combinations of reflexes, it is still necessary to use the term, and before "reflexes" of limb and eye are considered in this chapter, it will be useful to consider Sherrington's ideas and definitions of reflex movement.

Sherrington, in the foreword to the 1947 edition of *The Integrative Action of the Nervous System,* stated that a fundamental property of the reflex was that it could be studied "... free from complication by that type of 'nerve' activity which is called autochthonous (or 'spontaneous') and generates intrinsically arising rhythmic movements, e.g., breathing, etc." One property of a reflex, then, is that it is set off by an external stimulus rather than an internal clock and that its time of initiation depends on the time at which the external stimulus is delivered. This property is possessed by reflexes and triggered movements in common, however, and does not serve to distinguish between the two.

However, another property of the reflex, as Sherrington used the term, was a relationship between *magnitude* of input and *magnitude* of output. Here we have what would seem to be an important distinction between a triggered movement and a Sherringtonian reflex. For triggered movements, the intensity of the initiating stimulus must reach "threshold," but the movement itself will not vary as a function of suprathreshold variations in stimulus intensity. In contrast, the properties of a reflex movement will vary as a function of stimulus intensity above threshold. Another difference between triggered and reflex movements seems to be in the specificity of the stimulus. This distinction is not very sharp, but it is usually held that a reflex has a narrower range of adequate stimuli than a triggered movement. This distinction may or may not be useful, since there are a good many exceptions.

One difficulty in any attempt to define these different sorts of movements is that under normal conditions even simple motor behavior seems to involve elements of all three of the categories of movement. Thus, reflexes contain elements that are triggered, and these triggered elements in turn may be based on an elaborate central pattern. Take, for example, the scratch "reflex." This is called a reflex because the magnitude of the response is related to the magnitude of the stimulus and because there is considerable stimulus specificity. However, the fact that the term "reflex" is used in referring to this form of movement does not imply that the movement is not centrally patterned. Sherrington found that the *locus* and *intensity* of the nociceptive stimulus determined the locus and intensity of scratching—but the scratching movements themselves were clearly under the control of a central pattern, for they survived deafferentation. Thus, Sherrington stated

> The scratch-reflex I find executed without obvious impairment
> of direction or rhythm when all the afferent roots of the scratching
> hind-limb have been cut through. In the execution of these spinal
> reflexes, therefore, the most important afferent factor as regards "local
> sign" is the afferent channel from the place of *initiation* of the reflex.

It would seem that the same movement might be fruitfully investigated either from the standpoint of the role of the motor tape, in relation to mechanisms of triggering, or from the standpoint of the role of reflex factors. Take, for example, the eye movements that are associated with head movements during visual fixation. These eye movements are "reflex" in the sense that the timing, direction, and magnitude of the movements are determined by the input. However, no one could deny the important role of a central program that operates on the input in such a way as to generate the output. In carrying out an experiment, an investigator may choose to vary the sensory input, and observe the corresponding variations of motor output, or he may eliminate sensory input so as to observe the extent to which the movement is centrally determined. Both approaches are fruitful, and it is nonsense to think that success of one approach implies failure of the other.

**Eye and Limb Movements**

In the present chapter we shall consider studies of movement in which emphasis is placed upon the role of sensory input rather than the role of central patterning. The role of sensory input will be considered for two sorts of movements: of the eyes and of the limbs. Work on these two sorts of movement has developed along rather different lines, and will be dealt with in two separate sections of this chapter. Studies of eye movement are highly advanced at the level of systems analysis, but in these analyses the nervous system has usually been treated as a black box; and knowledge of central mechanisms has lagged behind knowledge of input-output relations for eye movements occurring under a wide range of physiological conditions in man. In contrast, studies of cerebral mechanisms of control of limb movements in animals are quite numerous; but these studies have, for the most part, been carried out under conditions in which normal movements were precluded by anesthesia, immobilization, or lesions designed to isolate one or another component of the system.

The studies on the eye provide insights into how the sensory input is utilized, whereas the studies on control of limb movement are concerned in large part with the central neuronal circuits that intervene between stimulus and response. Recently, however, studies have begun relating cortical unit activity both to limb and to eye movements, and they will be discussed in a later section of this chapter.

## Eye Movements

In any consideration of the role of sensory inputs in the control of motor outputs a major question is the way in which the input is sampled. Studies of intermittent versus continuous sampling of the input have shown that for certain sorts of eye movements—smooth-pursuit movements—the eyes are under continuous sensory control, whereas for others—saccadic eye movements—the sensory input acts only intermittently.

### Saccadic Eye Movements

Young and Stark (1963a) proposed that the saccadic response can be modelled by a sampled data control system. They observed that whenever there is a step displacement in target position to one side, with return of the same target to its original position within 100 msec, the subject's eyes respond with a saccade to the first step after a latency of 200 msec, and that the return saccade to the initial position occurs also after 200 msec. Thus, the oculomotor system subserving saccadic eye movements was found to sample for a brief interval and then to be refractory to changes in target position during an intersampling time whose duration was 200 msec.

Although this model has been useful in predicting the response to a variety of target patterns and has stimulated a number of experiments, some aspects of the model have been questioned. First, the duration of the subject's intersampling time has been observed to be less rigidly fixed than was originally thought to be the case. Robinson (1968) has described instances of saccades spaced only 80 msec apart, when subjects were tracking a target at a speed of around 40° per sec. In addition, short intersaccadic times of 130 msec have been found during the so-called corrective saccades that occur when a subject is asked to move his eyes from some initial position to a small target placed 30 to 40° away. Under these circumstances the subject achieves his aim by two saccades: a large one, followed by a small "corrective" jump only 130 msec later. These findings indicate that, at least under certain conditions, the oculomotor system does not behave as a sampled data system having an intersampling period of the order of one 200 msec reaction time (Becker and Fuchs, 1969).

Second, the postulated refractoriness of the oculomotor system to changes in target position occurring within the intersampling time has also been challenged. In fact, Wheeles et al. (1960) have shown that information about target position is accessible to the saccadic motor system from the time of the introduction of the target to up to about 100 msec before saccade initiation. Wheeles and others studied subjects' responses to a light that was stepped first 6° horizontally to one side from center, then 12° in the opposite direction. The results showed that the oculomotor system was sometimes able to cancel a saccade response to the first step and subsequently respond instead to the 12° step. Wheeles also found that the shorter the interval between the first and second (12°) light, the greater the probability for a saccade to be directed to the second light. The response of the oculomotor system can be influenced by the second stimulus during the first 100 msec following the presentation of the first stimulus, while the decision as to what kind of saccade should be made becomes irreversible in the last 100 msec prior to the saccade initiation. According to the Young and Stark model, there should have been a saccade to the first light, followed, after a 200 msec latency, by a second saccade to the 12° light.

It is interesting to consider that in parallel to this irrevocability in saccade decision a distinct decrease in visual perception develops, as Latour (1962) and many others have pointed out. This phenomenon has been investigated electrophysiologically, and the presence of pre-synaptic inhibition of optic tract terminals has been found to occur in concomitance with rapid saccadic eye movements (Bizzi, 1966). However, it is premature to conclude that this decrease in visual perception is due to presynaptic inhibition for two reasons: (1) at the present time this phenomenon has been found only in cats, and (2) depolarization of optic tract terminals has been found to occur only *after* saccadic initiation of eye movements and not prior to it.

A relative suppression (no matter how achieved) of afferent visual information is by no means the only inhibitory event that could take place during a saccade. For instance, it is likely that some attenuation of the vestibular input should also occur during the coordinated eye and head turning, because the induced vestibular impulses act to oppose the movement of the eyes.

Although it is quite clear that afferent visual information is not utilized by the saccadic control system immediately before and during a saccade, we do not know whether proprioceptive information from the eye muscles is also phasically inhibited. Indeed, the possibility that saccades could be influenced by short latency proprioceptive feedback, as proposed by Vossius (1960), cannot be disregarded a priori. As shown by Fuchs and Kornhuber (1969), the vermian portion of the cerebellum (lobuli V, VI, VII) receives short latency (4 msec) proprioceptive impulses. Were the eye muscle proprioceptors to behave in the same way as those of the somatic muscles, which display a maximal afferent activity during contraction (Severin et al., 1967), then an intense barrage of impulses from the eye-muscle proprioceptors would reach the cerebellum and could reflexly influence the terminal portion of a saccade. However, Fillenz's (1955) observation that the afferent impulses from eye muscles do not activate the motoneurons does not fit this hypothesis.

Given this complex picture, it is no wonder that the physiological properties of the oculomotor system and its cortical and subcortical organizations have remained so poorly understood. Although there is little doubt that the main circuits responsible for the development of a saccade are located in the part of the brainstem that is limited rostrally by the nucleus of Darkschewitsch and caudally by the vestibular nuclei, it is not known how cortical and cerebellar impulses affect these brainstem circuits or where the information coming from different sensory modalities is integrated with vestibular and proprioceptive impulses.

**Smooth-Pursuit Eye Movements**

In contrast to the saccadic eye-movement system, smooth-pursuit eye movement can be considered an example of a system in which input signals exert continuous modulation and control over the output. We owe to Robinson (1968) the conclusion that the stimulus-response relation for the smooth-pursuit eye movement system is continuous in nature. He found that subjects could respond to two successive ramp stimuli spaced 150, 100, and 75 msec apart with distinct smooth-pursuit eye movements that were also spaced from one another by 150, 100, and 75 msec. Thus,

> ...one may state with certainty that if the smooth-pursuit system is sampled, its refractory or intersampling time must be less than 75 msec. As pointed out by Rashbass,* it must be noted that the sampled system whose intersampling interval is less than the response time of the mechanical apparatus being controlled is undistinguishable from a continuous nondiscrete system.

Robinson also performed a variable feedback experiment based on the pursuit system alone, by eliminating saccadic eye movement from the feedback signal. He observed only smooth continuous changes in the pursuit eye-movement response. On the basis of these two sets of experiments, Robinson reached the conclusion that the pursuit eye-movement control system was continuous and that a sampled data model like the one proposed by Young and Stark for saccadic movement is not tenable.

The complex neuromuscular control system that performs tracking is modulated by various control loops, among them the visual, the vestibular, and the proprioceptive. The visual channel, however, is normally considered the main feedback loop, particularly when the subject's head is fixed in one position; then the adequate stimulus for initiating a smooth-pursuit eye movement is target velocity. It follows that in the absence of a moving visual stimulus the subjects should be unable to make slow eye movements; paradoxically, exceptions to this rule have been reported (Westheimer and Conover, 1954).

Although it is generally assumed that target velocity is the adequate stimulus for initiating the smooth tracking movement, there are certain questions that should be answered before accepting this concept in toto. For instance, how can the velocity signal be available to the smooth-pursuit control system once the fovea is on the moving target? Even assuming that small error signals would persist during tracking, on what central structure is the discharge of the activated retinal cells impinging? And what kind of interaction would arise when the target being fixed is moved across a rich visual background, which should also represent an adequate retinal stimulus?

---

*See Bibliography: Rashbass (1961).

Besides the visual control loop, the smooth-pursuit eye-movement system also receives afferent inputs from the vestibular system and various proprioceptive impulses arising during head turning. Because vestibular stimulation produces smooth eye movements that have the same characteristics as those generated by the pursuit system, the question arises as to whether the visual and vestibular inputs utilize the same neuronal circuitry.

In the course of the foregoing discussion on eye movements, the saccadic and the smooth-pursuit systems have been conceived as two separate neuronal entities. Behavioral, clinical, pharmacological, and recent neurophysiological data support this concept; and in a later section of this chapter, we will present the results of investigations on single cortical cells showing that their activity is related differently to the two forms of eye movement.

## Limb Movements

The preceding section on eye-movement control was primarily concerned with a study of movement rather than with central neural elements controlling movement. In contrast, studies of limb-movement control have entered rather deeply into the CNS. Five participants at the Work Session have, in fact, studied the limb areas of sensorimotor cortex, and the present section of this chapter will deal with some of the points raised in the Work Session by three partici-pants—Mountcastle, Towe, and Brooks—who were asked questions concerning their ideas as to the sorts of input-output transformations that might take place in sensorimotor cortex. Oscarsson and Phillips, although they have worked extensively on this problem, were asked to give their ideas on different questions, and their views are presented elsewhere in the *Bulletin.*

Of the three participants whose views will be considered here, Mountcastle has worked on the organization of input to the sensory cortex in the monkey, Brooks has investigated the organization of input-output relations in motor cortex of the cat, and Towe has studied both input and output functions in the cat sensorimotor area.

### Sensory-Motor Organization and Movement: A. L. Towe

In speaking of Towe's work, the term "sensorimotor" is used advisedly. As a matter of fact, Towe questioned the value of terms such as "motor" and "sensory" cortex, and objected to posing the question of intercommunication between the two alleged areas, because the mere posing of the question presupposes that there are, in fact, separate "sensory" and "motor" areas. Towe points out that a number of investigators have found a sensory-evoked response in what is called "motor" cortex following a discrete peripheral input; the response is of rather short latency and does not depend for its production on the integrity of primary "sensory" cortex. "Motor" cortex thus has access to "sensory" information via routes other than cerebral cortex, although information also comes from cerebral tissues as well.

Towe also pointed out that, just as the "motor" cortex is not simply motor, so too the pyramidal tract is certainly not simply a motor structure. Actually, the majority of pyramidal tract axons leaving the precentral cortex in the monkey terminate on brainstem and spinal structures that must be called "sensory," if one uses the "sensory/motor" dichotomy.

Towe stated the need for new sorts of experimental approaches to motor organization, approaches that do not start out tied to inaccurate assumptions and nonviable models, and after reading over the first draft of the Work Session program, he was able to point out a number of examples of both (inaccurate assumptions and nonviable models). Thus, the question addressed to Towe at the Work Session sought to get at his ideas on alternative models of motor organization, for it seemed clear that criticism alone would not make the nonviable models die—they must be actively displaced by better models.

Relevant to the problem of alternative models is Towe's work on the cerebral cortex, involving pyramidal tract neurons. His studies on sensory-evoked activity in neurons of the cortex of the cat showed that cortical output via the pyramidal tract could cause enhancement of cortical input by means of facilitatory actions of pyramidal tract neurons on transmission through the dorsal column nuclei. Part of the pyramidal tract arises in the postcentral (or for cat, postcruciate) cortex, and experiments by Towe have delineated the stimulus modalities and response latencies for somatosensory activity evoked in these pyramidal tract neurons. On the basis of several sets of studies, Towe was able to conclude that "... the pyramidal tract—a uniquely mammalian possession that connects the cerebral cortex directly with so many brainstem and spinal neurons, both sensory and motor—constitutes one route by which the cerebral cortex can modify its own afferent input" (Adkins et al., 1966).

Granted that one role of cortical output is the control of its own input, there remains the problem of how this modified input is ultimately able to influence discharge patterns of motoneurons. We asked Towe to give us his ideas on how the afferent signals reaching the cerebral cortex influence the discharge patterns of motoneurons. This was not a very easy question for anyone to answer, but Towe nevertheless prepared the following response:

I would like to say a few words about the evolution of movement, and to suggest what somatic sensorimotor cortex may be about. From the preamble that Ed Evarts provided to my first question, it should be clear that I take exception to a number of popular neurological concepts, largely because the experimental evidence speaks so eloquently against them. To recount the evidence, even in outline form, would require the better part of a day; let a few reminders suffice. All vertebrates move, no matter how little cerebral tissue they may possess, and they continue to move after cerebral insult (Bromiley and Brooks, 1940; Kennard, 1940, 1944; Pitres, 1884; Semmes and Chow, 1955; Travis and Woolsey, 1956). Only mammals possess a pyramidal tract,* but all continue to move after it has been transected (Barron, 1934; Bromiley and Brooks, 1940; Bucy et al., 1966; Buxton and Goodman, 1967; Laursen, 1966; Laursen and Wiesendanger, 1966; Lawrence and Kuypers, 1968; Schafer, 1910; Tower, 1936, 1940; Walker and Richter, 1966; Wiesendanger, 1969; Wiesendanger and Tarnecki, 1966). Further, the motor consequences of cerebral stimulation are altered little (Lewis and Brindley, 1965) or not at all (Towe and Zimmerman, unpublished observations in the domestic cat) following transection of the medullary pyramids. Thinking of primates in particular, eye-hand coordination is not disturbed by disconnection of corticocortical fibers that might link occipital with pericentral tissue (Myers et al., 1962). Nor does removal of somatic motor cortex interfere with a learned latch-box performance (Jacobsen, 1932; Lashley, 1924). Nor does dicing all somatic sensorimotor cortex disturb coordinated movements; habitual move-

*Karten (personal communication) has reported recent findings suggesting that the owl and pigeon possess both a bundle of Bagley and a pyramidal tract, casting new light on the assertions of Obersteiner (1896) about the parrot and of Johnston (1913) about the turtle.

ments are still performed efficiently (Sperry, 1947). Pyramidotomy in the human has a negligible effect on "volitional" movement (Bucy, 1957; Bucy et al., 1964; Guiot and Pecker, 1949; Walker, 1949, 1952).

In the face of these observations alone, one can hardly assign a primary role to the cerebral cortex, and particularly the pyramidal tract, in the initiation and control of movement. Nonetheless, the cerebral cortex clearly *is* involved in the regulation of behavior, and perhaps even of movement; the question is, How? It seems likely that the primary role of mammalian cerebral cortex is to continuously monitor the external environment and to forecast appropriate behaviors, immediate and more distant; past experience would influence that forecast.

All the chordates beyond the sessile tunicates manage to get about, to eat, to respire, and to reproduce; and most are unaided by motor cortex. They are all, by virtue of their life styles, adapted to their environments. Consider the lowly ostracoderms—primitive mud-grubbers inhabiting Silurian and Devonian waters (Romer, 1966; Stensiö, 1958). What little somatic musculature their nervous systems had to regulate was mainly in the postanal tail.* The task was to propel the armored front half, with its fixed mouth and attendant digestive apparatus, hither and yon along the bottom of some favored estuary. A simple prewired network, driven by a single command neuron, would have sufficed for the task. In the case of the crayfish, Don Kennedy (Kennedy et al., 1966a) has demonstrated that activation of a single central neuron yields a reciprocally organized behavior that involves more than 100 motor elements. Such prewired networks must indeed be rife throughout the animal kingdom. And, as observations on the development of singing in birds have so clearly demonstrated (Konishi, 1965a,b; Konishi and Nottebohm, 1969; Lanyon, 1960; Marler and Tamura, 1964; Poulsen, 1951; Thorpe, 1958, 1961), some networks are genetically determined; others are assembled through experience, and still others are derived from a blend of the

*It is to this fact that we might ascribe the bilateral symmetry of the vertebrate brain. An impending asymmetry has appeared in the cerebral cortex of man, with the development of speech and relational ideation.

two factors (Marler, 1970; Marler and Mundinger, 1971; Mulligan, 1966). In the case of our mud-grubbing ostraco- derm, we should like to know what activated the mechanism to propel it along. Perhaps it swam incessantly or in bouts, or perhaps it rested quietly until a hormone or other substance signalled "hunger" to the command neuron. If Alfred Romer is correct (Romer, 1966), then the visage* or some euryp- terid—a predacious arthropod—may have triggered a spurt of swimming, an escape behavior made out of a basic food- gathering behavior. This hypothetical link from the ostraco- derm eye to the command neuron that released swimming was thus one of the earliest sensorimotor "systems" in the vertebrate line.

When finned and bouyant forms appeared on the scene, a new mobility had been added, along with new regulatory requirements. Spatial stabilization could be achieved only through a system that continuously monitored and corrected after a brief delay. Such a system, in continu- ous and intimate contact with its sensory input, contrasts sharply with the network that, when "released" by its command neuron, runs a fixed course. This latter, triggered network cannot be reactivated until it has run its course, and it preempts the stage from other, closely related networks.† The continuously modulated network, based on closed feed- back of error signals, operates even during the discharge of a triggered network. Because it surely shares elements with the triggered network, it is capable of altering the precise play- out of that triggered behavior. In a bony fish, for example, the evasive "flip reaction" reflects a triggered network, whereas the vestibular mechanism constitutes a modulated network. It is easy to imagine how the two might interact; no triggered behavior is likely to be played-out in precisely the same manner each time.

Zoologists such as Paul Weiss and James Gray have long recognized that vertebrate nervous mechanisms are organized into hierarchical systems; and even the earliest ethologists, Charles Otis Whitman (1899) and Oskar Heinroth (1910), recognized the existence of innate, genetically deter-

*A simple "off-unit" in the eye would have sufficed, though it could trigger some false alarms.
†The ethologist will recognize this as a familiar concept, laid bare by the physiological turn of phrase.

mined behaviors. Some behaviors are so specific as to seem reliable taxonomic characters. Meanwhile, we mammalian physiologists focus on the domestic cat and the macaque monkey, with the implicit assumption (in the latter case) that we are studying "little men." We think in terms of either the extreme flexibility of human behavior or the dead-beat character of spinal reflexes, and sometimes attempt to handle the former in terms of the latter. This will forever remain an unsatisfactory approach. When we think of motor behavior, we must include the arena behavior of the Uganda kob, which Helmut Buechner has reported from the Semliki Flats (Buechner, 1963). We must include the sky-pointing display of the blue-footed booby and other sulids that Bryon Nelson reports from the Galapagos (Nelson, 1968). We must include the strutting dance and flight of the male woodcock, which Aldo Leopold described as commencing on warm spring evenings as soon as the luminance level drops to 0.05 foot-candles (Leopold, 1949). We might even try studying the simple act of walking; much could be learned about motor organization from a careful study of locomotor behavior.

Gray and Lissmann have shown that the fully spinal-deafferented toad continues to swim actively, provided that its labyrinthine system remains intact. Addition of bilateral labyrinthectomy abolishes swimming movements (Gray and Lissmann, 1947). On land, the ambulatory pattern of the toad remains after complete dorsal rhizotomy, save one—and that one remaining root may be at any spinal level (Gray and Lissmann, 1946b). So long as some cutaneous input channel remains, the prewired mechanism yielding walking behavior can be activated—the network can be triggered. However, an element of coordination is lost; the pattern of movements remains, but the movements themselves are ungainly. Could this be ascribed to disruption of modulated networks? Gray and Lissmann did not tell us that the toad was rendered hypotonic by the rhizotomy. Had the toad sported a cerebral mantle that had been ablated, Gray would have been compelled to remark on it, to palpate, to manipulate, and to search diligently for the character of the motor deficit. But even so, movement—patterned movement—would have remained.

The results of cerebral ablation in a mammal look suspiciously like the disruption of a modulated network— perhaps a whole collection of them. Let's pursue the idea. Consider the toad, sitting quietly under some shelter, when along trundles a fat and juicy larva. The toad rises up and forward, turning toward the larva until it is aligned. Being then within striking distance, a long and sticky tongue leaps forth to wrap around its target and return with its burden. Consummatory movements follow, unless the taste of the larva triggers rejection movements. The toad then resumes sitting, awaiting the next releasing stimulus, whatever it might be.

The sequence of events is illustrated in Figure 2: E represents the environmental sensory input. At some time the larva operates as a stimulus, S1, to trigger the toad orienting network, N1. The toad orienting behavior, B1, changes the environmental input to E', and in particular brings the larva close enough that it now operates as a stimulus, S2, to trigger the tongue extrusion network, N2. The tongue extrusion and return behavior, B2, creates a new stimulus, S3 (larva in mouth), to trigger swallowing network, N3. Presumably, the act of swallowing does not generate a new releasing stimulus, so that the toad resumes sitting.

An important question arises in connection with this illustration. The total act could be fractionated into several subcomponents that occurred in sequence—a sequence determined by the stimulus configuration produced as a

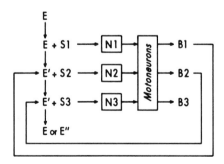

Figure 2. Sequence of behaviors in the capture and ingestion of a larva by a toad. Details in text. [Towe]

consequence of each separate triggered behavior. Did sensory information that entered during the performance of the triggered behavior play any role in the speed of release or precise play-out of the next triggered behavior? Could the information be stored for later use? Chase a skink along an arc, and it will trace out a set of straight line segments, with a pause at each intersection; it will behave as a "sampled data" system. Chase a dog along an arc, and it will trace out a smooth arc; it will behave like a "continuous control" system, with delay. Ah, you say, the differences are merely mechanical! Which suggests that you assume the skink's nervous system is as versatile as the dog's. I doubt it. I suspect the skink does not have the refinements in effector equipment, largely because it has not evolved the nervous mechanism required to make effective use of it. That "nervous mechanism" in all probability includes cerebral cortex as a major component.

Figure 3 outlines the situation visualized for "acerebrate" and "cerebrate" animals. In the "acerebrate" case, shown in Figure 3A, the animal's nervous mechanism consists of a set of prewired, triggered networks, TN, and modulated networks, MN, both acting through the same motor output apparatus, M. The external stimulus configuration, E + S,

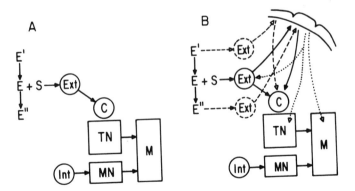

Figure 3. System diagram for vertebrates with little or no cerebral cortex (A) and for those possessing cerebral cortex, whether laminar or striatal (B). Environmental inputs, E', E, E'', occur sequentially in time. Dotted lines in (B) show corticofugal influences in the more elaborate vertebrates; other details in text. [Towe]

contains releasors, S, which are the only "active" external stimuli in the animal's world. These releasors act through the receptive mechanism, Ext, to activate the appropriate command neurons, C, calling a triggered network into activity. The precise play-out of this activity is compromised by the vestibular and proprioceptive receptive mechanism, Int, which operates continuously. Prior, $E'$, and subsequent, $E''$, external inputs can play no role in modifying the triggered behavior. On the other hand, the cerebral tissue of the "cerebrate" animal, shown in Figure 3B, continuously monitors the environmental inputs, $E'$, E, and $E''$, and sets the excitability pattern on command neurons such that the threshold intensity for releasor, S, is lowered (raised for other releasors). A corticofugal modulating system comprising few large, some medium, and many small fibers implies a temporal dispersion of the modulation–a short-term memory. In more elaborate "cerebrate" animals, corticofugal fibers end among elements of the triggered networks, allowing direct modification of their output. Ultimately, corticofugal output gains direct access to the motor output apparatus, giving it potentially independent control. In the ultimate organism, this cerebrum could suppress the operation of any triggered network such that it would have complete autonomy. I doubt that most primates have attained this level of development, though man is close.

    This is what cerebral cortex is probably about. It receives sensory information and sets the excitability of brainstem and spinal elements, both sensory and motor, on the basis of that information and of past events. It keeps constant watch on the external and internal environment, predicts needed output, and readies the subcortical nervous system for quick and efficient rendering of that output as the appropriate stimuli occur. It is a hypermodulating network, built through long selective pressures from a primitive and largely olfactory tissue.

Towe's statement is concerned with the general functions of "sensorimotor" cortex. However, in the primate at least, there are marked functional differences between the portions of the sensorimotor cortex lying anterior to the central fissure and those lying posterior to it. Towe and his colleagues have noted differences in response properties of neurons on the two sides of the "great fissure," having looked at neurons on both sides in the macaque and the domestic cat in chronic waking and sleeping and in fully anesthetized preparations. Furthermore, lesions of these two areas have strikingly different effects (Semmes, personal communication). Also, Mountcastle (although his published work has dealt with postcentral neurons) has recorded from both precentral and postcentral neurons, and has found that pre- and postcentral neurons respond to somatosensory inputs with totally different properties. Finally, Evarts has recorded from precentral (1968, 1969) and postcentral (unpublished) neurons during learned hand movements in the monkey, and has found that the activity of precentral neurons precedes movement and is related to the pattern of muscular contraction, whereas the activity of postcentral neurons does not precede movement but appears to be consequent upon the sensory consequences of movement. Thus, the classical divisions of postcentral sensory and precentral motor cortex may be criticized on grounds raised by Towe, but a distinction between precentral and postcentral gyri is supported by experiments on (1) effects of lesions on sensation and movement, (2) neuronal response properties to somatosensory stimuli, (3) activity of neurons during learned hand movements, (4) different projections of axons leaving the two areas (see Kuypers, 1960, 1962, 1963, 1964; Lawrence and Kuypers, 1965), and (5) different thresholds for movements evoked by electrical stimulation from the two areas. A variety of additional reasons for maintaining a clear distinction between precentral and postcentral cortex might be listed, but the point would seem to have been made adequately.

These five lines of evidence (and some others not cited here) have led virtually every investigator of sensorimotor cortex in man or in nonhuman primates to distinguish between the precentral and the postcentral regions. Certainly, the importance of this distinction has been critical in Mountcastle's work, and the questions directed to Mountcastle at the Work Session were aimed at eliciting hypotheses as to how input signals reaching the postcentral gyrus might influence the output of the precentral gyrus and/or discharge patterns of motoneurons. This question does not presuppose that the precentral gyrus has no input of "its own," independent of the postcentral gyrus (this input will, in fact, be considered later by Vernon Brooks). It is abundantly clear, however, that the most detailed information on tactile and joint inputs reaches the postcentral (rather than precentral) gyrus of the primate, and it seems clear that such information is used to control movement.

The work of Mountcastle and his colleagues has concerned somatic sensation and the steps linking peripheral receptors with sensory cortex. The broad aim of their studies extends beyond the sensory cortex, however; for as Mountcastle, Poggio, and Werner (1963) stated, "We wish, moreover, to determine the way in which those patterns of neural activity are further transformed at the successive neural relays within the cerebral cortex itself, relays which must intervene between its afferent input and its several outputs leading to behavioral responses, and perhaps to appropriately patterned motor activity as well." With this as a background, we asked Mountcastle to respond to the following question:

In what cerebral structures would he search for the next step in elaboration of sensory signals into motor cortex output? To clarify this question, we considered the example of a monkey required to maintain his wrist in a certain position, and therefore to utilize joint-position inputs in order to control motor outputs. In such a situation, let us suppose that we suddenly shift the position of the monkey's wrist and require that he promptly return it to the original position. We asked Mountcastle to give us his ideas on where we should look in our search for the sequential loci involved in the passage of information from sensory cortex to motor cortex.

In response to the first of these questions, Mountcastle proposed that in seeking out the successive transformations, one should begin by looking in the sensory cortex itself. Thus, the work of Mountcastle and colleagues has demonstrated the existence of modality-specific columns of cells in the postcentral gyrus, but the relation between input to the column and output from the column remains to be worked out. This is the first problem to be solved. Mountcastle did not speculate further as to where one would look for further transformations—but the axons of postcentral neurons have many well-known targets (cerebellum, thalamus, and sensory relay nuclei); and when the "intracolumnar" transformations have been worked out, presumably one would trace along the known anatomical pathways in the search for successive transformations.

The second question to Mountcastle was less anatomical and more general, having to do with the possible logical operations according to which (rather than loci at which) sensory input is translated into motor output; this question asked "how" rather than "where." To be sure, there are already a number of transformations between joint receptors and sensory cortex cells receiving joint information—but still, both sorts of cells seem to speak the same general language. Somewhere between the postcentral gyrus and the motoneuron, however, the language changes from the language of stimulus to the language of response. How can we conceive of this translation taking place? What should be the initial experimental approach of a neurophysiologist going into this area?

This second question to Mountcastle raised the question of whether activity in sensory cortex controls motor behavior by "reflex" mechanisms, or whether the motor outputs are best thought of as being "triggered" movements, in which the sensory input provides the information on the basis of which a central pattern of motor output can be selected. In response to this question, Mountcastle leaned in the direction of seeing the inputs to sensory cortex as being more analogous to triggers than to reflex controllers. In this view, he seems to be in general agreement with the ideas presented by Sperry (1969) in a recent paper on "A Modified Concept of Consciousness." Although the input-output processes of the brain were not the primary topic of this paper, Sperry nevertheless considered the problem raised in the second question directed to Mountcastle, and reached the following conclusion: "Any scheme, regardless of its complexity, in which sensory impulses are conceived to be routed through a central network system into a motor response becomes misleading."

**Tight Input-Output Coupling: V. B. Brooks**

Granted that most motor responses should not be conceived as the result of sensory impulses being routed through a central network, there are certain movements that have been viewed as "cortical reflexes," and the questions directed to Vernon Brooks were concerned with the problem of cortical reflexes. Brooks and his colleagues have found that the input to motor cortex from skin and deep structures is funneled into somatotopically organized radial columns (Welt et al., 1967). Asanuma and his colleagues (1968) have found that output to muscle from efferent zones in these columns is related to local skin input. For example, they have observed that "... skin stimulation of the dorsal surface of the paw excites corticofugal neurons that dorsiflex the paw and digits, and, conversely, skin stimulation of the ventral surface excites cells that cause ventroflexion. This input-output relationship is remarkably detailed in stressing digits or parts of the paw where these muscles function and excluding those where they do not." These experimental results have led them to conclude that movements such as the tactile placing reaction are under cortical coordinative dominance to a greater extent than seemed to be the case from the work of Lundberg, who viewed the tactile placing reaction as primarily a spinal reaction under pyramidal control.

On the other hand, Brooks (Welt et al., 1967) has stated, "There must be control devices that can override the stereotyped cortical input-output relations described so far, in order to permit voluntary actions to take place and also to adjust muscle action to limb position. ... For information on the next controlling level we need to turn to experiments where the integrative action of the animal is allowed to modulate the basic flow of information through the minimal building blocks of the motorsensory cortex." (For example, see Brooks et al., 1969.)

Given these findings and conclusions, we asked Brooks to respond to two questions. The first question was as follows: In what general categories of movement might tight input-output coupling (i.e., cortical reflexes) operate in a dominant manner, as contrasted to those movements in which these inputs may be overridden? In response to this question, Brooks prepared the following summary statement:

We do not assume that tight input-output coupling of cortical cells operates dominantly in all reflexes running through area 4 and in those that depend on its intact presence. Let me quote the statement made by my colleagues Asanuma, Stoney, and Abzug (1968). "The present findings suggest that afferent inputs originating from restricted peripheral loci may 'reflexly' induce contraction of a particular muscle or muscles through a loop including the motorsensory cortex. The 'tactile placing reaction' is a reflex-type reaction which is known to be mediated by the pericruciate cortex. It is possible to explain a part of the placing reaction by input-output relationships that we have described. For example, contact with dorsum of the paw leads to paw dorsiflexion, which is one of the initial movements in the placing reaction. However, our findings do not, by themselves, lead to a *complete* understanding of the cortical mechanisms subserving this reaction. For example, a placing reaction can be elicited by contact with any aspect of the forearm, and each of these reactions surely involves different combinations and sequences of muscle contraction."

We are thus beset by the difficulty that too many kinds of contact can lead to too many kinds of movement, to allow a simple input-output relation to explain them all. The particular placing reaction involving the paw dorsum where tight coupling may dominate, was described by Bard (1933) as follows: "If a cat is held in the air with the legs free and dependent and with the head held up (so that it cannot see its forefeet or any object below and in front), the slightest contact of the backs of either pair of feet with the edge of a table results in an immediate and accurate placing of the feet, soles down, on the table close to its edge." Amassian et al. have shown recently (1969, 1970) that this type of contact placing "... is initiated by a lifting-withdrawal phase ... followed by a directed landing phase," which implies that contactual positive feedback may operate during landing, but that it is inhibited during initial lifting.

Let us consider another interpretation of the results reported by Asanuma's group in 1968: "The fact that skin regions that project most heavily to the efferent colonies lie in the pathway of muscle action (and, therefore, in the course of a manipulatory sequence they are likely to be excited following muscle contraction) suggests that the cutaneous input may also subserve a positive feedback function. If this is so, then the input-output configuration which we have described resembles a built-in tracking system which tends to cause a portion of the limb to move toward an impinging tactile stimulus (i.e., to follow the source of stimulation)."

Phillips (Clough et al., 1968) and I (Brooks, 1969) have drawn attention recently to a relevant cortical pursuit reflex which was described by Denny-Brown as "the *instinctive (tactile) grasp response.*" "It is a stereotactic contactual response ..." (Denny-Brown, 1960), "... the adequate stimulus for [which] is stationary light touch ... It is essentially an exploratory palpation directed vertically into space from the point of contact.... [It] results in a succession of extension-flexion movements of the fingers ... accompanied by projection and turning of the whole arm" (Denny-Brown, 1950).

"The motor reaction [is] both delicate and highly integrated. Each movement ... [is] such as to bring a further contact with the stimulus closer to the ... palm of the hand.... All these movements [are] weakly performed and intermittent, as if the further contact with the stimulus resulting from one closing movement [leads] inevitably to a further adjusting movement, and this to repetition of the cycle, thus giving the performance its progressive character. The instinctive grasp reaction therefore necessitates the existence of a mechanism which would appear to involve a highly discriminative sensory component" (Seyffarth and Denny-Brown, 1948).

This clearly fits the cutaneous input-output relation in cat cortex. Phillips and his colleagues (Clough et al., 1968) noted that this tactile-conditioned grasp reflex also fits the cortical support and control of spinal proprioceptive patterns, as seen in the baboon, whose motor cortex seems to receive relatively greater proprioceptive input, compared to cutaneous, than that of the cat. They had shown that cortical activation was strongest for those spinal motoneurons that received the greatest input from muscle spindles. In the present context they stressed that action of intrinsic hand muscles could lead to finger flexion, because of the large heteronymous contribution from the spindles of intrinsic hand muscles to the motoneurons of the long flexors. In the reflex, intrinsic muscle action would be triggered by touch.

The two cortical reflexes, *placing* and *grasping,* were classed together by Denny-Brown, who noted their parallel susceptibility to lesions. "When the parietal lobe is removed, the limb no longer shows placing (or grasping) reactions...." (Denny-Brown, 1956). In contrast, however, "Instinctive tactile *grasping* and *placing* reactions are released by mesial *frontal* lesions ..." (Denny-Brown, 1960).

In what natural functions do the contactual instinctive reflexes occur? Denny-Brown found that "Placing can be demonstrated in man, most easily in children, and is another inborn instinctive exploratory reaction ..." (Denny-Brown, 1956) "[just as is] ... the *instinctive grasping* reaction in the newborn infant.... Reappearance [of grasping] in the adult as a result of [frontal] cerebral damage indicates that the responses have been suppressed in the course of development, and not lost.... They must in some manner enter into normal adult behavior, but we have no evidence as to the part they play. In the patient who presents some ability to control the responses, distraction of attention overcomes that ability. But attention as such is not the sole mechanism of control, for distraction of attention will not alone release the responses in many normal individuals. Nevertheless there are many 'normals' in whom an unmistakable grasp reflex can be obtained when their attention is distracted and there are some whose hand will inevitably palpate and grasp a single object within reaching distance under the same circumstances" (Seyffarth and Denny-Brown, 1948). Twitchell (1965), however, has shown more recently that the grasp response is *acquired* in infants, not inborn.

These descriptions provide *some* answers to the question posed. We can only hazard the guess that tight input-output coupling operates dominantly during certain exploratory and pursuit reactions, and thus perhaps in deliberate manual exploration, or deliberate searching steps in careful walking. The unmodulated in-out relationship may be a sort of built-in program, and this could be analogous to what Professor Weiss said: the actual outcome is a deviation from the basic, more stereotyped patterns.

The second question to Brooks was as follows: What experiments should be carried out to demonstrate the operation of "tight preferential input-output coupling for cortical cells in movement"? For example, how should one go about designing an experiment that would prove that tight input-output coupling occurs in the tactile placing reaction? In response, Brooks prepared this statement:

Tactile placing reactions in general would be poor tests, except possibly for that evoked by touch of the paw dorsum. Experiments could be made with awake animals that perform the contactual cortical pursuit reflex reactions, while records are being taken from muscles and from cortical cells. Cooling the frontal cortex to release the grasp reaction may be a possible experiment. The input-output relations of cortical cells could be checked by using natural peripheral stimulation and intracortical microstimulation. In order to link pyramidal tract cell function to cortical reflex action, three critical parameters need to be established: the threshold, the latency, and the intensity of activation of the cortical cell supposedly participating in the reflex action.

Muscle latency has been shown to be longer than that of pyramidal tract cells, i.e., about 60 msec after the stimulus used to elicit contact placing (Amassian et al., 1969, 1970), or in grasping, even longer. The thresholds are probably also of appropriate levels, as judged by unit recording in awake animals. The intensities of activation can be assessed only by comparing the range of evoked activity by peripheral stimulation with that called forth in execution of a contact reflex.

We are therefore reduced to assessing the intensity of peripheral activation in the cortex and matching that against the requirements posed by the spinal cord for its activation. How many cortical cells are brought to firing by a local peripheral touch? Making a conservative estimate, our experience suggests that something like ten radial cell columns would become involved. How many cells is that? Experimentally we know that in any one microelectrode penetration, about 30 responding cells are encountered. Because the limit of microelectrode detection is about 0.1 mm, a restricted set of input columns of 0.5 mm total diameter would contain 25 x 30, or 750, responding cells. If only one tenth of these were pyramidal tract cells, this would suffice to set muscles into motion. But we are dealing surely with at least ten columns, or thousands of cells. In theory then, the corticospinal loop gain is adequate for dominant operation of tight coupling from periphery to motor cortex to cord.

## Activity of Central Neurons during
## Eye Movements and Limb Movements

The main result of the study of eye movements in man has been the distinction between saccadic and smooth-pursuit systems, as well as the discovery of the physical characteristics of the two systems, such as the velocities, latencies, and types of adequate inputs. These data have provided a very useful conceptual framework for the investigator interested in defining the physiological properties of the oculomotor system. Accordingly, the question whether the distinction between the

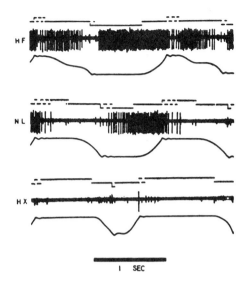

Figure 4. This figure illustrates the activity of a unit at each of three load conditions. The three loads employed are 400 g opposing flexion (top), 400 g opposing extension (bottom), and no load opposing the movement (middle). It may be seen that the unit became much more active with a load that opposed flexion and therefore required increased flexor force. Conversely, the unit was almost totally silent during periods when the movement was carried out with a heavy load opposing extension. [Evarts, 1968]

saccadic and smooth-pursuit systems can be observed at the level of single cortical cells has recently been investigated. It was found that, indeed, there are two types of cells in the frontal eye field of intact unanesthetized monkeys. One type (type I) fired each time a saccade in a given direction occurred, but remained silent during smooth-pursuit movement of the eyes elicited by a target moving slowly in front of the animal. In contrast, a second type of frontal eye field cell (type II) was found to discharge during smooth-pursuit eye movements, but not to fire during saccadic movements. These results indicate that in the frontal eye field there are neurons with patterns of discharge distinctly related either to saccadic or to pursuit movements, in line with the view that these two different types of eye movements are generated by distinct neuronal mechanisms (Bizzi, 1968; Bizzi and Schiller, 1970).

The results obtained for activity of single cortical pyramidal tract neurons of the precentral hand area in relation to flexion-extension movements of the hand by a monkey are quite different from those described above for cortical cells related to eye movement (Figure 4). When the hand movement is carried out with differing load conditions, it is found that the activity of the neuron is related to the pattern of muscular contraction rather than to the positional aspects of the movement. In a different study it was shown that for many units in the precentral hand area, activity of nerve cells precedes the first peripheral muscular contraction.

Thus, there are sharp contrasts between the way in which precentral and frontal eye field units are related to movements of the hand and eye, respectively. Precentral cells precede hand movement and are related to pattern of muscular contraction, whereas frontal eye field units follow movement and are related to positional aspects of eye movement. It remains for further investigation to discover whether or not the cerebral cortex contains units that have the same relation to hand movements as the frontal eye field units have to eye movements. Likewise, the loci of neuronal activity occurring prior to eye movement remain to be determined. In any event, these beginning studies on activity of cerebral neurons in relation to movement of limb and eye indicate that there will probably be a number of fundamentally different modes of relationship between movement and the activity of cerebral neurons in different areas of the brain.

## Summary and Conclusions

In this chapter we have, first of all, attempted to present a conceptual background for current work on reflex movements, and have tried to point out some of the differences between "reflex," "triggered," and "centrally patterned" movements. Second, we have discussed two sorts of eye movements, one of which (smooth-pursuit) is under continuous control by sensory input and the other of which (saccadic) is controlled by intermittent sampling of the input. Third, we have discussed studies concerned with the input-output properties of the cerebral cortex in relation to limb movement, and in agreement with Sperry (1969) have concluded that schemes in which sensory impulses are routed through a central network into motor responses are apt to be misleading. Fourth, we considered some studies of activity of single cortical neurons in relation to limb and to eye movements of intact monkeys—studies that illustrated basic differences in the ways in which cortical activity is related to movements of hand and eye, respectively.

From the diversity of points of view presented in this chapter, it should be clear that no amount of discussion or analysis at this point can resolve the differences; for they involve basic disagreements as to cerebral organization, research strategy, and philosophy. One may hope that the results of future experiments may someday solve these as-yet-unresolved problems. As Sperry stated in the article referred to above:

> To determine precisely how the more elemental physiological aspects of brain activity are used to build the emergent qualities of awareness becomes the central challenge for the future. At present even the general principles by which cerebral circuits produce conscious effects remain obscure.

## IV. CONTROL SYSTEMS OPERATING ON SPINAL REFLEX MECHANISMS

### by Robert E. Burke

### Introduction

Scientific study of the way in which the nervous system produces and controls movement began with examination of reflexes, which have the advantage of being stereotyped, repeatable, and controllable by the observer applying stimuli. As a result of his prodigious study of reflex mechanisms, Sherrington (1906) concluded that

> The unit reaction in nervous integration is the reflex....
> Co-ordination, therefore, is in part the compounding of reflexes.

However, as also discussed in the previous chapter of this report, there is considerable debate as to what part of, and to what extent, coordinated movements can be explained in terms of amalgamated reflexes. It seems clear on the basis of currently available evidence that physiologists must also consider "central patterning" mechanisms in addition to reflex control of movement.

The essential difference between central patterning and reflex mechanisms, as they are envisioned to operate in the control of coordinated movements, lies in the storage and readout of information specifying the spatiotemporal pattern of activation of the alpha motoneurons that produce the movement. With regard to reflex arcs, spatial factors in the motor output are specified by the built-in connections, or circuit diagram, of the reflex system, while many of the temporal features of the output pattern are not contained or specified within the nervous system. Rather, these temporal features arise as the consequence of continuous feedback of afferent information which is sampled by the CNS during execution of a movement, modifying the output by virtue of the changing sensory input. In contrast, the notion of central patterning mechanisms for movement control requires that *both* spatial and temporal information be specified within the CNS, and be capable of functioning without necessary participation of the time sequence of afferent inflow during movement execution. This notion of combined spatiotemporal factors has led to the expression "motor tape," which includes the idea of temporal specification as well as an implied system for reading out in proper sequence information stored in the motor tape. As discussed in the second chapter, some progress has been made in attempts to understand the neural mechanisms that might underlie the storage and readout of temporally sequenced information in invertebrate nervous systems, particularly those of arthropods, both crustacea and insects. In the vertebrate CNS, such questions are largely matters of speculation at present.

Much of the progress that has been made in understanding how the vertebrate CNS may control motor activity has been made along the classical line of reflex physiology—the study of input-output relations in neural systems where the input is accessible and produces a measurable and more or less repeatable output. In studying reflex systems, the time domain is reduced to short epochs during which the activity within the system is stereotyped, permitting reliable recording of such activity and enabling the observer to assimilate the information in a manageable form. Experimental study of such reflex systems is relatively simple, since the time domain is under the observer's control. The feature of prime interest is the spatial organization of the neural system mediating the observed response. The intellectual step between investigation of such pure reflex systems and the study of autochthonous rhythmically operating mechanisms, such as swimming, stepping, and breathing, is short. Similar experimental techniques can be, and have been, used in both types of endeavors.

This chapter will concentrate on a subject that is now in the process of rapid development. After almost 100 years of intensive study, neurophysiologists are now at the point of understanding certain mammalian reflex systems in considerable detail, and, perhaps more important, find that such reflex systems are not independent entities functioning in isolation. Rather, it appears that a great many reflex pathways are interconnected, with at least some other pathways, in ways that can now be more or less defined. In addition, reflex pathways present in the spinal cord segments are also under very effective, and often complex, control by motor centers in supraspinal structures. A picture emerges from this work that may enable us to propose a solution to the dichotomy between the notions of reflex control and central-patterning control. Indeed, it appears that central-patterning systems, whatever their nature, very likely operate by utilizing some of the same neuronal elements that are integral parts of reflex arcs. Afferent information, while undoubtedly controlling some types of movement patterns, can also be utilized in the production of movements that are specified by central-patterning mechanisms. The information that will be partially reviewed in this chapter suggests that the neuronal mechanisms which have been studied as reflex arcs can be utilized in a variety of different ways by virtue of the interaction between reflex pathways and by the action of control systems that are present, even at the level of the spinal cord segment. The dichotomy between reflex control and central-patterning control of movement may, in this sense, be artificial.

This chapter will be confined to a brief examination of some current ideas about the nature and function of neuronal mechanisms present in the spinal cord of mammals. The review of currently available experimental data will be by no means complete, but evidence will be reviewed in enough detail to permit some evaluation of the above idea.

## Organization of Reflex Mechanisms

For Sherrington, the term "reflex" involved the idea that afferent information from sensory receptors was processed in more or less definable pathways and delivered by these pathways to the "final common path," the motoneurons, to produce or prevent movement. Reflex loops of varying length could be viewed as nested within one another, with the motoneuron as the key focus, or nodal point, for integration of information. Sherrington (1906) expressed this as follows:

> The edifice of the whole central nervous system is reared upon two neurons—the afferent root-cell and the efferent root-cell. These form the pillars of a fundamental reflex arc. And on the junction between these two are superimposed and functionally set, mediately or immediately, all the other neural arcs, even those of the cortex of the cerebrum itself.

Such a schema tends to suggest that the major point of integration of the information in the various reflex lines is the motoneuron, such integration taking place in the form of summation of postsynaptic potentials. Such "parallel processing" of afferent information in pathways of different length and complexity, including both segmental and supraspinal loops, all focused on the segmental motoneurons, does appear to exist with respect to several muscle afferent systems. However, it is now clear from the work of Lundberg and his co-workers that the inputs to many reflex systems can be viewed as focused primarily on segmental neurons other than motoneurons, including interneurons in pathways to motoneurons as well as neurons of ascending tracts carrying information to supraspinal centers. According to this view, the motoneurons ultimately responsible for muscle activation remain the "final common path," but the route taken by information destined for motoneurons can be followed more clearly for many particular reflex systems by attending to other neurons as key nodal points for integration of the information.

This chapter will be organized by taking up classes of neurons and fibers present in spinal cord segments, regarding elements in the particular class in question as nodal points on which are focused certain other reflex or descending control systems. Such a schema is artificial, because in fact these neuronal elements are interconnected; but it seems useful to focus attention onto the different parts of the spinal motor apparatus in order to discuss experimental data. This highlighting of parts of the segmental apparatus in no way implies functional independence of the parts. Indeed, present evidence suggests just the opposite—that the segmental motor apparatus is a functional entity with all its subsystems interconnected through a variety of mechanisms, some of which can now be specified.

As a *modus operandi,* four broad categories of neuronal elements present in the spinal segments will be dealt with as nodal points onto which certain control mechanisms converge. These are the following:

*1. Motoneurons,* including the *alpha motoneurons,* which innervate extrafusal muscle fibers, which produce muscle tension and movement, and *gamma motoneurons,* which innervate the specialized intrafusal muscle fibers contained in muscle spindles.

*2. Primary afferent fibers* entering the dorsal roots, which have cell bodies in the dorsal root ganglia.

*3. Interneurons,* with cell bodies in the spinal segments and axons projecting within the same and/or other spinal segments.

*4. Tract neurons* with cell bodies in the spinal segments and axons projecting to supraspinal centers.

Elements in all four classes have been shown to receive control input from reflex and descending systems. A complete classification of segmental neuronal elements would also include the axons and terminals of supraspinal descending fibers; but as far as is known at present, such descending axons are not nodal points for the receipt of control influence.

## The Final Common Path – Alpha Motoneurons as Nodal Points

All motor mechanisms, as a matter of definition, can be said to project "mediately or immediately" onto alpha motoneurons, and a discussion of these key cells as nodal points could be stretched to include the entire nervous system. A review even of the more immediate links would encompass an enormous amount of material. This section will be limited to a discussion of some recent evidence with regard to the systems that project directly, or monosynaptically, to segmental alpha motoneurons. In order to make some functional correlations, it will be useful to preface the discussion of synaptic organization with a word about some organizational features of motoneurons themselves.

### Motor Unit Organization

Sherrington defined the term "motor unit" to include what is now known as an alpha motoneuron plus the bundle of muscle fibers innervated by that cell (see Creed et al., 1932). It appears unlikely that extrafusal muscle fibers innervated by a particular motoneuron are to any significant extent doubly innervated by another alpha cell (Brown and Matthews, 1960). Furthermore, all of the muscle fibers innervated by a given motoneuron are, for all practical purposes, of the same histochemical type (Edström and Kugelberg, 1968). Under normal conditions, all of the muscle fibers innervated by a motoneuron contract when that cell fires (see Krnjevic and Miledi, 1958). Thus, the entire motor unit can be considered as the functional final common path, and this notion has proved useful in interpreting some of the features of synaptic organization discussed below.

Motor units in a given animal species vary over a wide range in a number of characteristics. The motoneuron can vary in size and membrane properties (Burke, 1967a; Henneman et al., 1965a,b; Kernell, 1966), and such variation is correlated with the speed of contraction and tension output of the muscle fibers innervated by the particular cell (the "muscle unit" portion; see Burke, 1967a). The range of variation in both motoneuron and muscle-unit properties may be quite broad in some large limb muscles of the cat, which contain muscle units with (1) fast contraction times and large tension outputs, innervated by large motoneurons, and (2) other units with slow contraction times and small tension output, innervated by smaller alpha motoneurons. The motor unit profile of other cat muscles may be quite different. For example, in the lumbrical muscles of the paws, no separation into fast- and slow-twitch groups has been found, although the contraction speed of different units in the same muscle may vary over a rather wide range (15 to 45 msec; see Appelberg and Emonet-Denand, 1967). In the long extensor of the baboon fingers, most muscle units are rapidly contracting and generate uniformly rather small tensions (Eccles et al., 1968a). It seems useful to consider the overall characteristics of motor units in any discussion of the functional implications of particular patterns of motoneuronal synaptic organization, as has been done in a recent review by Phillips (1969).

## Synaptic Organization

Recent studies of the organization of monosynaptic inputs to alpha motoneurons have dealt with two different, but interrelated, aspects: (1) the distribution of synaptic input over the somatodendritic receptive surface of a given cell (Burke, 1967b; Porter and Hore, 1969; Rall et al., 1967; Terzuolo and Llinas, 1966), and (2) the variation in synaptic organization to different motoneurons, both those innervating a single muscle (Burke, 1968a,b) and between motoneurons innervating different, functionally related muscle groups (Eccles et al., 1957; Eccles and Lundberg, 1958a; Phillips, 1969).

With regard to the monosynaptic projection of muscle-spindle primary afferents (group Ia) to triceps surae motoneurons in the cat, there is evidence of a greater synaptic density and greater dendritic dominance of such input to motoneurons innervating slowly contracting muscle units than that to motoneurons innervating fast contracting units (Burke, 1968b). This pattern of synaptic organization correlates well with the relative ease of exciting various motor units in the decerebrate cat via the group Ia afferent system; slow-twitch motor units are more readily brought to firing than fast-twitch (Burke, 1968a). This pattern of relative excitability gradient fits with the notion of the "size principle" suggested by Henneman and his col-

leagues (Henneman et al., 1965b; Work Session), because the motoneurons innervating slow-twitch muscle units appear to be smaller than
those innervating the fast units (Burke, 1967a, 1968a,b). Because
motoneuron size and muscle unit properties are correlated, at least in
the large proximal limb muscles, the "size principle" permits some
useful generalizations about the usage of different motor units during
movement (Henneman et al., 1965b). As these authors point out, the
"size principle" hypothesis makes several implicit assumptions about
the distribution of synaptic input to motoneurons in a given motor
pool. With regard to the group Ia synaptic input, these assumptions are
supported by experimental evidence where directly tested (Burke,
1968a,b). However, there is now additional experimental data that
makes it appear that synaptic input to motoneurons from several
polysynaptic systems, both afferent and supraspinal, may be distributed
to large and small cells in a pattern that may give excitability gradients
opposite to those observed for the group Ia input (Burke et al., 1970).
It would seem that a broad functional generalization, based only on the
index of the size of neurons within a particular functional group, may
be unwarranted without additional supportive data.

Alpha motoneurons have monosynaptic connections with several descending systems. In the primate, corticospinal fibers project
monosynaptically to hindlimb (Stewart and Preston, 1967) and forelimb (Phillips, 1969) motoneurons. Observations in the baboon show
that the corticospinal monosynaptic connection with forelimb motoneurons is more powerful (presumably by virtue of synaptic density) to
neurons innervating distal muscles than proximal muscles (see Phillips,
1969). Porter and Hore (1969) have found the same pattern in
hindlimb motoneurons and further suggest that the synaptic terminals
of corticomotoneuronal fibers are situated primarily on motoneuron
dendrites. This descending monosynaptic connection in the primate,
not present in "lower" mammals, correlates with the progressive
encephalization of motor function in the higher species and probably
represents a pathway by which the distal musculature, so much
involved in environmental exploration and manipulation under control
of distance receptors, can be used efficiently via cortical loops not
directly dependent on proprioceptive and exteroceptive information.

In contrast, motoneurons in the cat do not receive direct
corticospinal projections but do receive monosynaptic excitation from
fibers in the vestibulospinal tract (from Deiters' nucleus; Lund and
Pompeiano, 1968; Wilson and Yoshida, 1969), and from fibers in the

reticulospinal system presumably originating in the region of the medial longitudinal fasciculus (MLF) (Grillner et al., 1968; Grillner and Lund, 1968; Wilson and Yoshida, 1969). Hindlimb motoneurons receive monosynaptic input from one or the other of these systems but not from both. In general, these descending connections appear to be organized along flexor-extensor lines in the knee and ankle muscles, and there is evidence that a reciprocal inhibitory circuit to antagonists may also be involved (Grillner et al., 1968). The descending monosynaptic systems thus are distributed to flexor and extensor motoneuron pools in a manner analogous to the more familiar reciprocal innervation circuit established by the group Ia afferent system (see Eccles and Lundberg, 1958b), as shown diagrammatically in Figure 5. This recip-rocal organization suggests that these direct descending links to moto-neurons may represent the output side of supraspinal loops, very likely involving the cerebellum, which may be more directly dependent on proprioceptive and exteroceptive afferent information than on input from distance receptors.

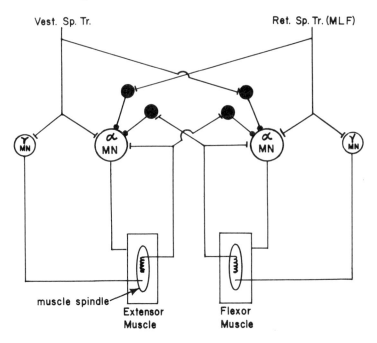

Figure 5. Descending systems projecting monosynaptically to flexor and extensor motoneuron pools in the hindlimb of the cat, compared with the group Ia input pattern. Note convention of denoting excitatory synaptic junctions with a bar and inhibitory interneurons and synapses with filled circles. [Modified from Grillner, 1969b]

As noted above, consideration of the systems that impinge directly on motoneurons only scratches the surface of the complex problem of the organization of synaptic input to alpha motoneurons. Other systems, both afferent and descending, that project to motoneurons through intercalated interneurons will be treated, at least partially, below, because many such interneurons can be regarded as nodal points in their own right. It must be emphasized that, in terms of number of synaptic terminals and strength of action (which might be termed the "gain" of a particular input line), the systems projecting monosynaptically to the alpha motoneurons represent a small part of the total input to these cells. Even the relatively powerful group Ia input probably cannot bring about motoneuron firing without concomitant excitatory action in other, polysynaptic input systems (see Burke, 1968a). This perspective must be kept in mind in making functional interpretations of synaptic organization patterns.

Discussion of primary afferent systems projecting directly to motoneurons raises the question of how the information carried by such a system can be controlled. Clearly, there are only two possibilities. Either the transmission of synaptic information can be modulated by mechanisms operating on the afferent terminals themselves (through presynaptic inhibition), or the generation of afferent impulses can be controlled by modulating the sensitivity of the transducer of peripheral information. Both mechanisms are used by the CNS in controlling the flow of afferent information in the group Ia afferent system. The mechanisms of presynaptic inhibition will be dealt with following some discussion of the modulation capacity that the CNS has with regard to muscle spindle sensitivity.

### Alpha-Gamma Linkage – Gamma Motoneurons as Nodal Points

The general principles involved in the operation of the gamma motoneuron-muscle spindle system (or "gamma loop") are so well known that they need no elaboration in a short review of this sort. Current thinking about this system has been reviewed recently by Matthews (1964), Granit (1955, 1968) and Grillner (1969b). Only a few points will be taken up to put the subject into the context of this report.

The muscle spindle is a transducer sensitive to muscle length, transmitting information to the CNS over two afferent lines, the spindle primary, or group Ia, afferents, and the spindle secondary, or group II, afferents. The sensitivity of these afferents to muscle length, and changes in it, can be modulated by the CNS through activity of the gamma motoneurons, which cause contraction of the small intrafusal muscle fibers present within the capsule of the muscle spindles. Gamma motoneurons can be divided into two functional classes: (1) static, which modulate the sensitivity of both group Ia and group II afferents, and (2) dynamic, which affect only the group Ia afferents (Matthews, 1964; Appelberg et al., 1966). The difference between the effects of activity in static and dynamic gamma motoneurons on spindle afferent sensitivity is further reflected in the fact that increased activity of static gamma motoneurons (or static spindle bias) increases the discharge rate of group Ia fibers to maintained stretch and *reduces* their responsiveness to varying stretch, while concomitantly increasing the discharge of group II afferents at given muscle lengths (the group II afferent is primarily responsive to maintained length in any case). Thus, static gamma motoneuron bias makes the output of both afferent types responsive primarily to muscle length, and there is relatively little decrease in group Ia activity during muscle shortening (or spindle unloading) (see Lennerstrand and Thoden, 1968b). On the other hand, increased activity of dynamic gamma motoneurons (or dynamic gamma bias) affects only the group Ia endings, making them more responsive to *changes* in muscle length, but it does *not* eliminate the cessation of firing of group Ia endings during shortening of the whole muscle (see Lennerstrand and Thoden, 1968a).

The importance of the differential effect that is exerted by static and dynamic gamma motoneurons is underscored by the fact that these two groups of gamma cells are apparently under independent control of synaptic systems converging upon them. While it has been clear for some time that gamma motoneurons were effectively excited or inhibited by a variety of afferent and supraspinal mechanisms (see Matthews, 1964; Jansen, 1966), recent experimental evidence has suggested that the CNS can exert very- specific patterns of differential control on static and dynamic gamma cells. Bergmans and Grillner (1969) have recently demonstrated that the balance of activity in flexor muscle static and dynamic gamma motoneurons in the spinal cat can be reversed after injection of L-dopa, a drug that mimics activation of a descending noradrenergic control system. This particular differential control of static-dynamic gamma bias is not exerted on gamma cells innervating extensor muscle spindles (Grillner, 1969a). Two other descending systems project exclusively to static gamma motoneurons: (1) the vestibulospinal tract, which connects monosynaptically with static gamma cells of extensor muscles, and (2) fibers in the reticulospinal tract (from the MLF region), which project directly to flexor static gamma cells (Grillner et al., 1969). These connections appear in the diagram in Figure 5. It should be noted that these descending systems connect in parallel ways with both alpha and gamma motoneurons, suggesting that alpha and gamma coactivation may result from descending activity in these systems, involving only the static gamma cells.

Control systems operating upon dynamic gamma motoneurons are not well defined at present. However, there is recent evidence that a descending system, unrelated to either corticospinal or rubrospinal tracts, exerts excitatory action preferentially on dynamic gamma cells (Appelberg and Jeneskog, 1969).

**Alpha-Gamma Linkage**

The problem posed by the coactivation of alpha and gamma motoneurons in many types of movements has been succinctly put by Granit (1955):

> In every instance hitherto analyzed the $\alpha$ and $\gamma$ reflexes have proved to be linked, co-excited, and co-inhibited, often with the $\gamma$ reflexes leading. All experimenters have been struck by this fact. The significance of $\alpha$-$\gamma$ linkage may now be preliminarily assessed in general terms: excitation of the $\alpha$ motoneurones over the $\gamma$-loop through nuclear bag [group Ia] afferents is sufficiently important to be organized for cooperation with direct $\alpha$ excitation. Is it important enough to have a decisive influence on motor performance?

The question of "decisive influence" has taken two forms: (1) can the gamma loop serve as the primary drive for alpha motoneurons, initiating and controlling activity of alpha cells indirectly through the gamma motoneurons as proposed by Merton (1953), or (2) does the gamma loop serve a subsidiary role, acting only in concert with other (direct) influences on alpha motoneurons, but providing feedback information about the execution of movement? These questions have been discussed at length by Matthews (1964) and by Phillips (1969). The available evidence, which includes the fact that coordinated movements are possible in totally deafferented limbs (see the second chapter, and Eldred, 1960), favors the view that gamma loop participation in centrally directed movements is additive to other systems projecting to alpha motoneurons, perhaps providing load-compensation excitatory feedback (see Phillips, 1969).

If one role for the gamma loop is the provision of load-compensatory feedback during alpha-gamma linked movements, then the static gamma motoneurons must be involved, because increase in static gamma bias provides the crucial effect—maintaining or even increasing group Ia (as well as group II) afferent firing during muscle shortening. There is indirect evidence that alpha-static gamma coactivation takes place in coordinated stepping in the mesencephalic cat (Severin et al., 1967) and in respiratory movements (von Euler, 1966, 1970).

The work of von Euler and his colleagues has provided an extensive test of the role played by the gamma loop in natural movements, in this case, intercostal muscle activity during respiration. During respiratory movements, coactivation of alpha and some gamma motoneurons (the "specifically respiratory" gammas) occurs, and the level of activity in both alpha and gamma groups fluctuates with the chemical drive for ventilatory excursion (see von Euler, 1966, 1970). The fact that primary spindle afferent discharges increase during intercostal contraction suggests that the respiratory gamma motoneurons are analogous to the static gamma cells of limb muscles. The other group of gamma motoneurons innervating intercostal muscle spindles discharge steadily during respiration (and are thus termed "tonic" gamma cells). The discharge of such tonic gamma motoneurons can be modified by stimulation of supraspinal structures, notably the cerebellum; and this group appears to contain both static and dynamic gamma cells (Corda et al., 1966). It has been suggested that the function of the *respiratory* gamma motoneuron complement is to provide load compensation and length information to control intercostal excursion, and consequently ventilatory volume. The *tonic* gamma motoneuron complement may be involved primarily in postural adjustments, a function performed by the intercostal muscles in addition to their role in respiration.

It is not clear at present whether there is differential control of static and dynamic gamma motoneurons from the motor cortex, although this has been suggested (Fidone and Preston, 1969; Vedel, 1966). There is evidence that gamma motoneurons are coactivated with alpha cells upon electrical stimulation of the motor cortex in the baboon (Phillips, 1969), although the alpha and gamma effects can be elicited independently (Koeze, 1968; Phillips, 1969). This alpha-gamma linkage is enough to overcome spindle unloading when the whole muscle length is held close to constant, and there is indirect evidence suggesting that static gamma cells are activated from the cortex (Phillips, 1969).

As noted above, alpha-gamma coactivation, or linkage, has been found in many sorts of movement in experimental animals. There is indirect evidence for similar alpha-gamma linkage in the human (see Matthews, 1964), and it has been argued that the gamma loop is responsible for the tremor present in human voluntary movements (see Lippold, 1970). More direct evidence for alpha-gamma linkage in the human has come from recent experiments of Vallbo, in which it has been observed that primary spindle afferent discharge can increase during a voluntary contraction (Vallbo, cited during the Work Session by Phillips).

If, as suggested by the above evidence, static gamma moto-neurons participate importantly in alpha-gamma linked movements, what is the functional role of the dynamic gamma cells? These efferents modulate the sensitivity of group Ia afferents to changing muscle length. With dynamic spindle bias, small changes in muscle length can result in very large changes in the rate of firing of group Ia afferents, conferring on the system a high gain for length variation. This would be desirable, particularly in situations calling for maintenance of constant muscle length (Grillner, 1969b), as occur in the control of postural attitude. It would be premature to say that the dynamic gamma cells are primarily responsive to postural control mechanisms, because it is becoming clear that such mechanisms are inextricably intertwined with the systems controlling other, phasic movements.

A point of interest in this connection is that both the static and dynamic gamma motoneurons modulate the sensitivity of group Ia afferents, albeit in different ways. These group Ia endings are most effectively connected with the motoneurons that innervate small, slowly contracting motor units, at least in the large limb muscles (Burke, 1968b). It may be that the constancy of recruitment order of motor units in human voluntary contraction (for example, Ashworth et al., 1967), as occurs also in the decerebrate cat (Henneman et al.,1965), may depend to some extent on participation of the gamma loop in an alpha-gamma linked movement.

Lurking in any discussion of the function of the gamma loop is the problem of the group II afferents. The response of these afferents to changes in muscle length and static spindle bias are well known (see above), but their activities within the CNS are rather a mystery. In the spinal cat, they cause effects on motoneurons in the pattern of the generalized flexion reflex (Eccles and Lundberg, 1959; Lloyd, 1960), but this may be variable (see Wilson and Kato, 1965) and sometimes weak (Lloyd, 1960). In any case, the effect of group II spindle afferents reaches alpha motoneurons (and both static and dynamic gamma motoneurons; Bergmans and Grillner, 1969) via segmental interneurons. Information from group II afferents also projects to a variety of supraspinal centers, including cerebellar cortex (Eccles et al., 1968b; Lundberg, 1964a; Oscarsson, 1965), the brainstem reticular formation (Lundberg, 1964a), and the sensorimotor cortex (Landgren and Silfvenius, 1969). While the group II afferents may under some circumstances contribute to excitation or inhibition of alpha moto-neurons through purely segmental mechanisms, it seems also likely, according to Grillner (1969b), that

> ... the significance of the secondary [group II] endings in the cat might not be related to a delicate spinal reflex pattern but in mediating information related to muscle length and static gamma-bias to supra-spinal structures.

This suggestion seems particularly important in relation to the question of alpha-gamma linkage in supraspinally controlled movements, which seem to involve primarily the static gamma motoneurons, which, as pointed out above, modulate the sensitivity of group II afferents, an effect not shared by the dynamic gamma cells.

## Presynaptic Modulation of Afferent Transmission
## Primary Afferents as Nodal Points

The notion that transmission of synaptic information in a particular afferent line can be modulated by mechanisms operating on the presynaptic element is now familiar to most neurophysiologists, and evidence for this has been extensively reviewed (see Eccles, 1964; Wall, 1964). It is known that afferent fibers of group Ia, group Ib (from Golgi tendon organs), and large-diameter cutaneous afferents are targets for such presynaptic modulation; and a good deal of evidence exists delineating the patterns of input to the subsystems modulating trans-mission in particular afferent channels (Eccles, 1964; Wall, 1964). With regard to proprioceptive afferent terminals, it is known that presynaptic modulation of transmission can be effected by both peripheral afferent systems and by descending systems (see Lundberg, 1964b, 1966). There are definite patterns in these effects as to which fiber systems are affected by which input systems, but there is little evidence regarding detailed, muscle-by-muscle organization of presynaptic effects. For this reason, and because the magnitude of the effect that the presynaptic modulation mechanisms may have is difficult to determine, rather little progress has been made in integrating the presynaptic mechanisms into general hypotheses of motor control, even at the segmental level.

The evidence available at present suggests that the magnitude of presynaptic modulation may be relatively small. Although such effects can be significant when taken together with other influences on the postsynaptic cell receiving information from the modulated line, it seems fair to say that the primary afferent line itself cannot be interdicted, or cut off, entirely by presynaptic effects. The remaining discussion will omit further consideration of presynaptic effects, but it should be kept in mind that they may well enter into the total motor control system in a significant way.

## Gating of Afferent Information Transmission – Interneurons of Spinal Reflex Arcs as Nodal Points

In the preceding section, the notion was discussed that transmission of information in a primary afferent line to postsynaptic neurons can be modulated, but not interdicted, by mechanisms operating presynaptically. This means that the information in the afferent line is continuously sampled by the postsynaptic neuron and, notwithstanding any modulating effects, always enters into the summation process that determines the postsynaptic cell output. The basic situation is different if an interneuron is interposed between the primary afferent fiber and the target postsynaptic neuron, for example, an alpha motoneuron. The interposed interneuron confers two additional possibilities to the system: (1) the sign of synaptic action can be changed from excitation to inhibition (see Eccles, 1964), and (2) depending on the strength of synaptic transmission from afferent line to interneuron, and upon the convergence of other synaptic inputs to the same interneuron, transmission in the link between primary afferent and target neuron, e.g., motoneuron, may be not only modulated but entirely interrupted by virtue of prevention of any of the interneurons in a specific link from firing. Transmission of information in a reflex line containing one or more interneurons is accomplished only by firing of the interneuron(s), and this may depend not only on excitatory input from primary afferents but also on other inputs converging on elements in the line, notably on the interneuron(s) but also including possible presynaptic effects on primary afferent terminals. This scheme offers the possibility that a reflex line can be opened or closed, that is, gated on and off, by control systems converging on elements in the reflex line, provided that one of the elements is an interneuron intercalated between the input line and the target neuron.

There is now considerable experimental evidence, largely from the work of Lundberg and his colleagues, that interneurons in well-known spinal reflex arcs receive input from more than one primary afferent path and also from descending systems. It seems safe to say that virtually every segmental pathway between primary afferent fibers and motoneurons receives convergent input from both descending and other afferent systems (Lundberg, 1959, 1964b, 1966, 1967). Among the many possibilities for detailed discussion, mention will be made here of only one example.

The pathway that mediates inhibition from group Ia afferents to motoneurons of antagonist muscles contains an inhibitory interneuron between the group Ia terminals and the target motoneurons (Eccles and Lundberg, 1958b). This interneuron, the "group Ia inhibitory neuron," is the nodal point for convergence of input from a number of identified systems other than group Ia afferents. Lundberg (1970) has recently reviewed this convergence pattern and has offered some provocative ideas as to the role played by this system in the control of movement.

The convergence patterns of synaptic input to group Ia inhibitory interneurons for which there is evidence are shown in Figure 6. The types of input that converge on group Ia inhibitory interneurons are clearly different for interneurons that project to flexor motoneuron pools versus those that project to extensor motoneuron pools. The Ia inhibitory interneurons projecting to flexor motoneurons are excited monosynaptically by axons originating in Deiters' nucleus and descending in the vestibulospinal tract (Vest.Sp.Tr.) (Grillner et al., 1966), and through a polysynaptic path from high threshold muscle, joint, and skin afferents (denoted FRA, or flexor reflex afferents) from the contralateral side of the spinal cord (Bruggencate et al., 1969). These input

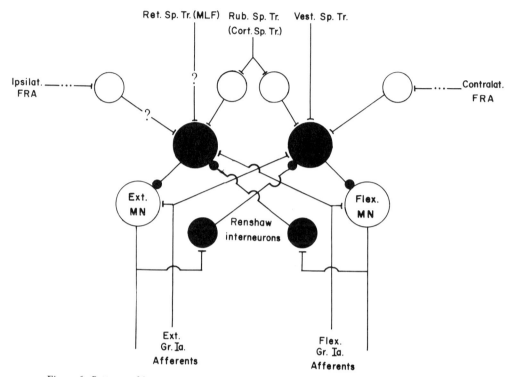

Figure 6. Patterns of input convergence onto group Ia inhibitory interneurons that project to extensor and to flexor motoneuron pools. Pathways for which there is some evidence but which have not been conclusively demonstrated are indicated with question marks. Polysynaptic paths are shown with dotted lines. [Burke: Adapted in part from Lundberg, 1970]

systems do not seem to project to Ia interneurons in the pathway to extensor motoneurons. At the latter interneuron, there is some evidence that axons in a reticulospinal tract originating in or near the medial longitudinal fasciculus (Ret.Sp.Tr. (MLF)) (Grillner et al., 1968), as well as the ipsilateral FRA (Lundberg, 1970), converge with excitatory connections; but this evidence is scanty, hence the question marks. The group Ia inhibitory interneurons projecting to both flexor and extensor motoneurons are excited by axons in the rubrospinal tract (Rub.Sp.Tr.) with disynaptic latency (Hongo et al., 1969b), and similar effects with longer latency have been seen on stimulation of the sensorimotor cortex (corticospinal tract, Cor.Sp.Tr.) (Lundberg and Voorhoeve, 1962). It has recently been shown that both types of group Ia inhibitory interneuron are also subject to inhibitory control through the Renshaw circuit from motor axon recurrent collaterals (Hultborn et al., 1968b,c). This inhibitory mechanism appears to be organized along roughly reciprocal lines, as shown in the diagram; but this is not strictly so in all cases, and the actually observed pattern is somewhat asymmetric (Hultborn et al., 1968a).

Thus the convergence of synaptic inputs onto group Ia inhibitory interneurons appears to present a very complex picture in its details—involving inputs from segmental afferent systems (group Ia and FRA effects), from segmental mechanisms not directly involved with primary afferent input (the Renshaw loop), and from descending systems (vestibulo-, rubro-, reticulo- and corticospinal systems). Some general patterns are evident in it, however, which seem to make some sense. For example, the convergence pattern to group Ia inhibitory interneurons from group Ia afferents displays a clear reciprocal organization. Note that in Figure 5 a reciprocal inhibitory link from the descending vestibulo- and reticulospinal systems to antagonist motoneurons was indicated. It seems rather likely that the disynaptic inhibition from, for example, the vestibulospinal fibers to flexor motoneurons may, at least partly, involve the action of the group Ia inhibitory interneurons (although this possibility was omitted from Figure 5 for the sake of clarity), so these descending influences on group Ia interneurons may also involve a reciprocal organization.

The existence of such interconnections as those dealt with above illustrates the point that synaptic inputs to alpha motoneurons from several different input systems may utilize common interneuronal elements, providing economy of connectivity at the segmental level and at the same time the possibility for cooperative, or interdictive, action of several control systems by virtue of the common nodal elements located in their transmission pathways. A pathway transmitting primary afferent information to motoneurons (that is, a reflex system such as the group Ia inhibitory link) may be gated on by virtue of summation of excitatory input from segmental and/or descending systems, or reduced in gain to the point of being gated off by removal of excitatory summation or the introduction of active inhibition to the interneurons. The flexibility of such a system is obvious.

Lundberg and his colleagues have recently discussed several likely functional possibilities for the group Ia inhibitory system outlined above and have proposed the notion of "alpha-gamma linked reciprocal inhibition" as a unifying concept in thinking about how the system may be used in the control of movements that are based on a flexion-extension pattern of muscle action (Hongo et al., 1969b; Lundberg, 1970). This hypothesis also includes consideration of how the flexion-extension pattern may be broken in movements requiring co-contraction of antagonists. It should be clear from the discussion to this point that the reflex influence of the gamma loop on motoneurons can be weakened or even removed by several mechanisms: (1) decrease in gamma motoneuron spindle bias, (2) presynaptic inhibition of group Ia terminals, and (3) disfacilitation or active inhibition of the reciprocal Ia inhibitory interneurons. Each of these mechanisms may be brought in by both primary afferent and descending systems. When and how each mechanism may be used in controlling a variety of diverse movements remains to be determined.

It seems likely on the basis of available evidence that a good deal of descending motor control ultimately exerted on alpha motoneurons operates through segmental interneurons, some of which may represent "private" channels, while others represent elements shared by several systems, both afferent and descending. Transmission of information in a "private" pathway would be determined only by the input signal and the strength of synaptic connection. However, transmission of descending signals through interneurons shared with other pathways would be conditioned to a great extent by these other inputs. The same descending signal might at different times encounter quite different "states" in the segmental interneurons, resulting in a variety of output signals to motoneurons. It can be envisioned that such a scheme might be advantageous, given a very specific set of interconnections, in that it might permit automatic compensation for initial limb position in the task of attaining a final limb position. However, it seems reasonable to suspect that there would also be an advantage in keeping supraspinal centers in some way "informed" about the "state of affairs" in the interneuronal circuits at the segmental level with respect to the afferent information constantly being processed there.

Two possible schemes, which are not mutually exclusive, might be proposed to accomplish the above task. Information from primary afferents may be transmitted to higher centers rather accurately through more or less "private relay paths," entering the equation determining descending control signals at or near the latter's point of origin. Such a scheme might be cumbersome and time consuming if *all* movement-control feedback occurred in this way. It would seem efficient if the CNS also included mechanisms by which descending control signals could be compared with the internal "state of affairs" at many levels of the motor control hierarchy, taking into account both the initial conditions prior to movement and the later results generated by the execution of the movement.

## Internal Feedback – Neurons of Ascending Spinal Pathways as Nodal Points

Under some conditions, certain of the ascending spinal pathways appear to carry relatively undigested information from primary afferents to higher centers, with little information processing taking place along the line. For example, the pathway carrying information from fast-adapting mechanoreceptors in the palmar skin to the postcentral cortex of the monkey functions such that, given vibratory stimuli within a certain range of frequency and amplitude, "cyclic entrainment" of cortical cell discharge with the stimulus can be observed (Mountcastle et al., 1969). This suggests that, within specified limits, transmission in this ascending pathway must be quite secure, resulting in little change in the pulse code observable at the sending and receiving stations. However, this is not to say that integrative processing of afferent information cannot take place in such a line.

Among the ascending systems that carry proprioceptive and exteroceptive information from the hindlimb region of the cat, there is growing evidence that some integrative information processing takes place even in the most securely transmitting "relay" systems. As noted before, presynaptic modulation must be kept in mind in such considerations, but there are in addition several schemes involving postsynaptic integration at the level of the spinal segment for which there is experimental evidence. Two of these are considered in the hypothetical neuronal diagrams in Figure 7.

The diagram in Figure 7A shows a hypothetical ascending tract neuron receiving synaptic convergence from three sources, one monosynaptic and two operating through interneurons (one excitatory and one inhibitory). Such a scheme appears to be representative of the situation in some cells giving rise to the dorsal spinocerebellar tract (DSCT). Many DSCT cells receive a very powerful excitatory monosynaptic connection from group Ia afferents (see Eide et al., 1969; Kuno and Miyahara, 1968). The receptive field for a given DSCT neuron carrying group Ia information is usually discrete (one or a few related muscles), and the discharge of these cells is often a linear function of muscle stretch (Jansen et al., 1966). However, excitatory influence from unspecified segmental mechanisms also converges on these cells (Jansen et al., 1966), as does polysynaptic input from a variety of primary afferent sources, including both excitatory and inhibitory effects (see Lundberg, 1964a; Oscarsson, 1965). It has been

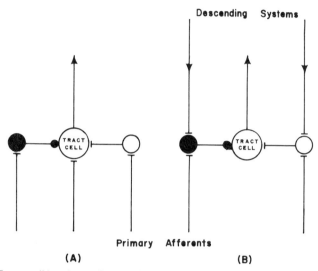

Figure 7. Two possible schemes for organization of input to neurons giving rise to ascending spinal tracts, illustrating increasing complexity of integrative function. (A) Input consists of primary afferent information, conveyed directly and through interneurons. (B) Input consists of results of convergent action of primary afferents and descending signals in segmental interneurons. [Burke: Adapted in part from Oscarsson, 1965]

demonstrated that DSCT cells firing to group Ia input from a given muscle can be inhibited by group Ia and Ib input from other functionally related muscles in the same limb (Jansen et al., 1967). This evidence suggests that the cells of origin of the DSCT, while basically subserving a relay function for rather discrete information from specific proprioceptive and exteroceptive inputs (see Jansen and Rudjord, 1965), can also perform some integrative processing at the tract cell level due to the convergence of afferent information from other sources. There is no evidence at present suggesting that descending systems converge directly on DSCT neurons (Oscarsson, 1965).

As discussed in the preceding section, interneurons of a variety of reflex arcs receive convergent input from other afferent and descending systems. If it is assumed that some of the same interneurons carrying information to motoneurons, and played upon by descending systems, also project to ascending tract cells, one has a system in which the information carried by the tract neurons to supraspinal centers reflects the results of integration of *both* afferent and descending input to segmental reflex mechanisms. This notion was advanced some time ago by Lundberg (1959), and it has received considerable attention in his laboratory and in that of Oscarsson.

The scheme illustrated in Figure 7B shows a hypothetical tract neuron converged upon by excitatory and inhibitory interneurons, which in turn receive input from both afferents and from descending systems. Depending on the relative gain at the various synaptic junctions, information relayed by such tract cells would represent the result of integration of four inputs at three nodal points. The probability that such ascending information would accurately reflect primary afferent input alone seems small (Lundberg, 1959, 1964a; Oscarsson, 1965, 1967). This organizational scheme (Figure 7B) appears to represent the situation found with respect to several ascending pathways transmitting input to the cerebellar cortex. The details of experimental observations on these ascending tracts have been described in reviews by Lundberg (1964a) and by Oscarsson (1965, 1967). Briefly summarized, it appears that several ascending pathways are activated by the FRA, often from wide receptive fields and sometimes from several limbs (Oscarsson, 1967). There is evidence that transmission from the FRA to the tract cells is influenced by descending systems. Oscarsson's current view of this matter is included in the following chapter.

At the Work Session, Oscarsson discussed a provocative elaboration of his ideas as to the function of the ascending pathways that convey information to the cerebellar cortex by way of a relay in the inferior olivary nucleus. The spino-olivocerebellar path seems to have special significance, because olivary neurons project to specific cerebellar Purkinje cells through very powerful excitatory synapses established by climbing fibers (see Eccles et al., 1967). The topographical features of the climbing fiber pathway, plus the peculiarly low rate of discharge of climbing fibers under a wide variety of conditions (Oscarsson, 1967; Thach, 1968), suggest that the climbing fiber input may play a key role in cerebellar integration (see Marr, 1969).

It has been demonstrated that neurons in the inferior olive can be activated by stimulation of the ipsilateral sensorimotor cortex (Armstrong and Harvey, 1966), and Oscarsson and co-workers have recently shown that olivary neurons can receive convergent input from both ascending spino-olivary and descending cortico-olivary systems (Miller et al., 1969). Oscarsson has suggested that such olivary neurons may function in making comparisons between the activity in descending control systems and the result of such descending activity on the interneurons of the spinal cord segments. Several of the ascending spino-olivocerebellar tracts appear to be organized in a way similar to that discussed regarding Figure 7B. If it is assumed that a given olivary neuron receives the same descending control signals that also converge on segmental interneurons projecting to spino-olivary neurons and ultimately returning to that olivary cell, an internal feedback loop would be closed that might monitor not only the spinal "state of affairs" before and during movement but would also compare this with the command signals descending to the spinal level before any result of the command signal became apparent through changed afferent input. The low discharge rate of olivary cells might signal either "match" or "mismatch" between command signals and the state of segmental motor centers. Such a system might carry with it the advantage of internal feedback lines—detection and correction of signal errors with minimum delay, that is, before the final output is produced and evaluated.

The above picture of the function of FRA spino-olivocerebellar pathways depends on the assumption of a very highly organized system of synaptic interconnections between spinal and olivary neurons. This and other assumptions alluded to above are discussed in Oscarsson's note in the next chapter.

## Reflex Arcs and the Control of Movement

This chapter began with a brief consideration of the contrast between the notions of reflex control and central-patterning control of movement. The bulk of the discussion so far has dealt essentially with some current ideas about the spatial organization of spinal mechanisms commonly termed reflex, which convey information from primary afferents to motoneurons through relatively simple paths. However, it should be clear that such "reflex" paths contain elements that can also be controlled by, perhaps even preempted by, descending systems. In the total picture, descending systems projecting directly, and perhaps through "private" interneurons, to alpha motoneurons must be added.

An example of a complex motor control system that apparently operates in a reflex fashion, dependent on certain kinds of afferent information, is the maintenance of stable posture in standing dogs as studied by Brookhart and colleagues (Brookhart et al., 1965; Mori and Brookhart, 1968a). The motor responses of different animals to the test situation are remarkably stereotyped even though the animals are awake and alert. Brookhart views his training procedure as persuading the animal *not to move,* presumably allowing automatic postural mechanisms to operate without input to them from systems not immediately concerned with postural maintenance. The similarity of motor responses from animal to animal in this test suggests that the output results from operation of a built-in, or "hard-wired," set of neuronal interconnections that are the same for all members of a given species. Whether this postulated circuitry provides any temporal, as well as spatial, specification of output remains to be determined; but the evidence at hand tends to suggest that temporal specification in this system resides to significant degree in the afferent inflow.

While the above example appears to provide an illustration of a purely reflex system operating in the intact, behaving animal, recent work with regard to an analysis of stepping gives some insight into a possible relation between reflex arcs and central-patterning mechanisms in the control of a complex movement. There is evidence that a neuronal mechanism exists in cat spinal cord segments that can initiate and maintain flexion-extension movements of the limbs without afferent participation (see Eldred, 1960; Engberg and Lundberg, 1969; Lundberg, 1969). However, in the cat and in higher mammals, afferent information appears to be necessary for precise performance of stepping (see Eldred, 1960). Recently, Engberg and Lundberg (1969) have suggested that the flexion-extension activation pattern that is basic to the step is driven by a built-in segmental mechanism, perhaps a mutually inhibiting reciprocal network of interneurons, which itself contains little detailed specification for either spatial or temporal features of the step. Because there is an apparently specialized pattern of group Ia interconnections among hip and knee muscles (Eccles and Lundberg, 1958a), these authors have proposed that such reflex connections, as well as the connections established by other afferent systems, may well provide the details for both spatial and temporal sequencing of muscle activation. It has been shown that gamma motoneurons (probably static gammas) are coactivated with alpha cells during some forms of stepping (Severin et al., 1967), and it may be that in stepping the reciprocal group Ia inhibitory circuits are also gated "on." In sum, it seems likely that normal stepping in the intact animal may involve cooperative activity of a relatively simple internal pattern-generating network, modulated by afferent information in reflex arcs that supply the spatiotemporal specification of details, giving the movement finesse and responsiveness to changing environmental conditions. The whole segmental mechanism may well be set into motion by descending command signals, which are in themselves simple, containing no information about the sequence of muscle activations to be accomplished. This notion is a clear analogy to the "command inter-neuron" scheme present in crustaceans and discussed in the second chapter of this *Bulletin*. However, there are probably no single neurons in the mammalian CNS that can function as command cells, but rather sets of neurons might be regarded as the proper analogy.

A functional interpretation of the motor system in the control of movement requires the critical element of time. Movement results from a sequence of signals that is ordered in both time and space. Having spent the past 100 years arriving at our present admittedly incomplete understanding of the spatial connections involved in move-ment control, physiologists are only beginning to find the experimental tools for investigating the temporal organization involved in control of even the simplest movements. Considering the reflex versus central-patterning notions of movement control, it would seem that the static spatial organization of connections within the CNS that can be demonstrated with present techniques provides no real clue as to which notion works "best." The crucial question seems to be "Where does the temporal specification originate?" With reflex control, the answer is relatively simple—temporal specification arises largely through con-tinuous sampling of afferent information that reports the progress of the movement.

However, we know that coordinated movements take place without afferent information, albeit often with only partial success. It would seem reasonable to conclude that, on the basis of present evidence, coordinated movement in mammals is very likely the result of a combination of two mechanisms that are not mutually exclusive: (1) internally stored spatiotemporal command sequences that can initiate, and to some extent direct, alpha motoneuron activity without imme-diate benefit of afferent information regarding either initial conditions or progress of the movement, and (2) spatially organized afferent information that imparts precision and responsiveness to changing or unanticipated conditions in the environment, and that is in all like-lihood essential to the formation of at least a part of the stored spatiotemporal sequences which can later direct movements that are "learned." Whether or not some of the complex spatiotemporal sequences stored in the mammalian CNS are present as a result of genetically controlled building-in is at present unknown, but it seems likely by analogy with evidence obtained from other animals such as amphibians and birds (see other chapters of this report).

Further progress in understanding how the CNS controls motor output would seem to depend on a broadening of experimental vision. Additional information is of course necessary to define the spatial organization of neuronal connections, since these will be utilized by the motor control mechanisms in any case. However, real progress in defining the temporal features of motor control is now necessary. This problem requires new experimental approaches in applying the techniques of single-unit analysis, so successful in the spatial organization problem, to the situation of ongoing motor activity.

## V. FEEDBACK AND COROLLARY DISCHARGE: A MERGING OF THE CONCEPTS

### by Edward V. Evarts

### Introduction

In almost any symposium on motor control, feedback is inevitably a conspicuous topic, and there are often lengthy arguments as to its importance in this or that type of motor performance. One commonly hears assertions that the "loop delay" is too long and movement duration in skilled movement (e.g., piano playing) is too short to allow feedback to have a role: the movement is all over before it has a chance to be corrected. Alternatively, it can be pointed out that massive feedback pathways exist, and that their interruption causes gross impairment of movement; it is self-evident that no skilled pianist would be able to get along without his dorsal roots. It often seems to be the case that such discussions generate more heat than light. The Work Session was no exception in respect to the extent to which feedback was discussed, but fortunately, most of the usual unproductive debates did not occur. Moreover, this Work Session may have been somewhat different from other conferences in that the discussion of feedback in the usual meaning of the term (i.e., feedback from the motor response) was expanded to include discussion of *internal feedback* occurring prior to the response (i.e., feedback arising from elements in the neural chain leading to the response). Internal feedback in turn was discussed in relation to corollary discharge. Participants agreed that some new thoughts were expressed in these general areas, and it will be the purpose of this chapter to summarize some of these ideas.

#### Feedback and Corollary Discharge: Origin of the Concepts

James Watt's introduction of a governor for control of the steam engine in 1769 (see Dorf, 1967, p. 3) is commonly cited as the initial breakthrough of negative feedback into the field of control technology, although there were many earlier devices employing the same principle. The article by Otto Mayr (1970) on "The Origins of Feedback Control" points out that the earliest known construction of a device for feedback control was a water clock invented in the 3rd century B.C. by the Greek mathematician Ktesibios, and the significance of feedback in control of movement was recognized by Magendie early in the 19th century. Regardless of when and by whom the concept of feedback was originated, it achieved formal scientific treatment in 1868 when J. C. Maxwell published a mathematical model describing the behavior of the governor (see Dorf, 1967, p. 4). Just one year before Maxwell's paper on negative feedback, Helmholtz (see Sperry, 1950) published ideas that are commonly said to represent the first scientific treatment of the concept which we now refer to as "corollary discharge."

Everyone already has his own definition of feedback, and so no attempt was made to define the term at the outset of this chapter (various different definitions of feedback will actually be taken up later). The meaning of the term "corollary discharge" is not quite so obvious, however, and it therefore seems advisable to go back to the papers of Sperry (1950), von Holst and Mittelstaedt (1950), Mittelstaedt and von Holst (1953), von Holst (1954), and Mittelstaedt (1958) to review the concepts of "corollary discharge" and the related idea of "efference copy." In 1950 Sperry proposed "that a corollary discharge of motor patterns into the sensorium may play an important adjustor role in the visual perception of movement along with non-retinal kinesthetic and postural influences from the periphery." The experimental results that led Sperry to introduce the term "corollary discharge" came out of a study on the neural basis of the spontaneous optokinetic response produced by eye rotation in fish. Sperry found that the spontaneous optokinetic response persisted after removal of forebrain, cerebellum, and labyrinth, but that the optic lobe was essential for preservation of the circling. In view of the persistence of the circling following the extirpations mentioned above, it seemed unlikely that kinesthetic influx from the periphery was of vital significance in maintaining the circling. In speculating on alternative, nonkinesthetic origins for the circling, Sperry wrote as follows:

> Another possibility must be considered, namely, that the kinetic component may arise centrally as part of the excitation pattern of the overt movement. Thus, any excitation pattern that normally results in a movement that will cause a displacement of the visual image on the retina may have a corollary discharge into the visual centers to compensate for the retinal displacement. This implies an anticipatory adjustment in the visual centers specific for each movement with regard

to its direction and speed. A central adjustor factor of this kind would aid in maintaining stability of the visual field under normal conditions during the onset of sudden eye, head, and body movement. With the retinal field rotated 180 degrees, any such anticipatory adjustment would be in diametric disharmony with the retinal input, and would therefore cause accentuation rather than cancellation of the illusory outside movement.

The postulation of such a central kinetic factor would help account also for certain other phenomena in movement perception such as the illusory displacement of the visual field in man when the eyeball is moved passively as when tapped with the finger tips and the lack of such displacement when the same movement of the eye in space is brought about through an active response. It would provide a neural basis for what Helmholtz called the sensation of the "intensity of the effort of will." The need for some kind of central mechanism to eliminate blurring of vision between fixations in eye movement has long been recognized.

Quite independently, von Holst and Mittelstaedt (1950) proposed a related theory, in which the term "efference copy" is used in a way reminiscent of the way Sperry used the term "corollary discharge." Whereas Sperry's ideas arose from observations on the circling behavior of fish whose eyes had been rotated 180°, Mittelstaedt turned the head of the insect 180° so that the two eyes were, in effect, interchanged. When such insects once began to move, they would spin rapidly to right or left until exhausted. In considering these observations, it was proposed that

> ... the efference leaves an "image" of itself somewhere in the CNS, to which the re-afference of this movement compares as the negative of a photograph compares to its print; so that, when superimposed, the image disappears. A motor impulse, a "command," from a higher center causes a specific activation in a lower center, which is the stimulus-situation giving rise to a specific efference to the effector (that is, a muscle, a joint, or the whole organism). This central stimulus situation, the "image" of the efference, may be called "efference copy." The effector, activated by the efference, produces a re-afference, which returns to the lower center, nullifying the efference copy by super-position.*

*Von Holst, 1954.

The insects whose heads had been rotated received a reafference that, instead of nullifying the efference copy, added to it in such a way as to create an error and generate endless movement, because each movement would generate a reafference, which generated additional movement, and so ad infinitum.

The cancellation of efference copy by reafference is further discussed by von Holst (1954) as follows:

> Now, we make a simple experiment and turn the paralysed eye mechanically to the right. In this case both the motor intention and also the efference-copy are lacking, but the image moves across the retina and afference is transmitted, unmatched by an efference-copy, to higher centres and produces, as is known, the perception that "the surroundings move to the left." This is also a false perception. If now we combine the first case with the second, that is, if my eye is moved mechanically at the same time I intend this movement—which is the same as *voluntarily* moving a *normal* eye—then in fact these two complementary effects just mentioned are produced: firstly, the perception of the returning "command" causing a jump of the surroundings to the right and, secondly, an image-motion on the retina producing a jump of the surroundings in the opposite direction. These two phenomena, the efference-copy and the re-afference, now compensate each other, and as a result *no moving* of the surroundings is perceived. The surroundings appear stationary during this normal eye movement, and *this* perception is *physically correct*. As we have already seen, the correct perception results from two opposite and false perceptions which cancel each other.

A somewhat different interpretation of the processes underlying stability of the environment during eye movement has been proposed by MacKay, and because MacKay's ideas lead into the concept of "feedforward control," they are introduced here. MacKay (1966) took the same example used by von Holst.

> When the direction of gaze changes, the optical image moves over the retina. If the change is imposed on the eyeball from without—say, by gentle pressure on the corner of the open eyelid—the visual world is seen to move. When the same change results from voluntary use of the eye muscles, however, no movement of the visual world is seen. What is it about voluntary control, we may ask, which makes this difference?

If we were to think of perception as an internal witnessing of incoming signals, we might be tempted to invoke the idea that the signals resulting from voluntary eye movement must be "canceled" in some way, presumably by signals derived from the oculomotor system. If so, however, the accuracy of "cancellation" is remarkably high, and it is difficult to imagine how this could be achieved on the basis of either the "outflow" signals to the eye muscles or any "inflow" signals from proprioceptors, unless visual acuity for field movement is much lower than it seems.

The problem takes on a different complexion, however, if we ask what part a consciously controlled eye movement is calculated to play within the *total* information system. Informationally its function, like that of exploratory hand movements in the dark, is to steer the receptor surface to a fresh sample of the sensory world. If we are right in supposing that the perception of *change* results only from a significant *mismatch* between the signals received and those which the action was calculated to bring about, then the changes resulting from voluntary movement in a stable environment should generate no such mismatch. They are, in fact, precisely what the action was calculated to bring about on the basis of the *existing* internal world map (MacKay, 1962).

As the case of tactile exploration makes abundantly clear, there is no question of having to *cancel* the resulting motion signals. What is wanted is not cancellation, but *evaluation,* in the light of the objective of the action. The sense of rubbing friction as the hand explores a stationary surface is part of the object of the exercise, and gives positive evidence that the surface is stationary. If it were absent or abnormally weak, this would be evidence that the surface was moving with the hand. Because the eye normally explores in saccadic jumps, smooth motion of the image over the retina is *not* part of the calculated evidence of stability; but progressive displacement of the image certainly is, and in its absence, as with a stabilized retinal image (Ditchburn and Ginsborg, 1952), or with paralysis of eye muscles (Kornmüller, 1931), the visual world seems to move.

Whatever the source of its information, however, it will be seen that the accuracy required by the evaluator in our present model need be no greater than the accuracy with which the *action* responsible can be executed. It is in this quantitative respect that the model most clearly differs from the cancellation theory of visual stability in the form advanced by von Holst (1957). His feedback (*reafferenz*) model (von Holst and Mittelstaedt, 1950) has almost the same gross flow map as our Figure 8, but the process of evaluation is replaced by one of subtraction. In its original context of sensorimotor coordination in fish and insects, this was unexceptionable, but in his later application of the model to the problem of *raumkonstanz* (von Holst, 1957), it seems to have been presupposed that what is perceived is the residue of the subtractive process; and this implies that perceptual stability should depend to a much higher degree on the accuracy of the *efferenzkopie* supposed to be subtracted. The issue still waits on physiological evidence.

Figure 8, reproduced from MacKay's paper, refers to the concept of "feedforward" control, and because this idea is to be considered by Ito (p. 132), it would seem useful to present MacKay's definition on this process, so as to distinguish clearly between "feedback" and "feedforward." In referring to this figure, MacKay points out that

It is often possible to increase speed and accuracy by arranging that in addition to the "feedback" signals from $C$, the selector system $S$ receives an input computed directly from $I_G$ itself, together with any other relevant advance indications obtainable via auxiliary sensors. This "feedforward" need only roughly approximate to the required form, but will leave $C$ more free for the task of fine adjustment, on which it still has the last word (Figure 8). It seems not unlikely that feedforward of this sort (by "alpha innervation") plays a part in the control of muscle movement (Matthews, 1964).

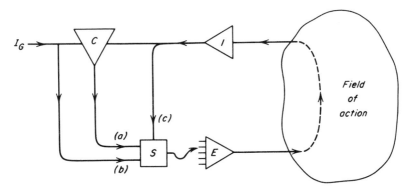

Figure 8. Simple loop with "feedforward." (*a*) Feedback, (*b*) Feedforward from indication of goal. (*c*) Feedforward from indication of field state. (See text for further explanation.) [MacKay, 1966]

With this introduction to the concepts of feedback and feed-forward to efference copy and corollary discharge, attention will now be focused on some current notions as to their role in movement and to how the neural pathways mediating certain forms of feedback may, in fact, be the same pathways traversed by corollary discharge. We will begin by considering feedback.

## Some Contemporary Views on the Role of Feedback in Motor Control

In this attempt to sketch out briefly some of the more recent ideas on the operation of feedback in movement, we will omit description of contributions in the 80 years following Maxwell's classical work and jump ahead to the late 1940's. By this time, the role of feedback in control of movement was the active concern of scientists in a number of different disciplines—cyberneticists, psychologists, neurophysiologists, and "human engineers," to name four. The range of ideas existing at this time may be indicated in two quotations, one from Wiener's *Cybernetics* (1948), and the other from a 1948 article by Taylor and Birmingham.

The first quotation, from Wiener (1948), stresses the vital role of feedback in movement:

> Now, suppose that I pick up a lead-pencil. To do this I have to move certain muscles. However, for all of us but a few expert anatomists, we do not know what these muscles are; and even among the anatomists, there are few if any who can perform the act by a conscious willing in succession of the contraction of each muscle concerned. On the contrary, what we will is *to pick the pencil up.* Once we have determined on this, our motion proceeds in such a way that we may say roughly that the amount by which the pencil is not yet picked up is decreased at each stage. This part of the action is not in full consciousness.

> To perform an action in such a manner, there must be a report to the nervous system, conscious or unconscious, of the amount by which we have failed to pick the pencil up at each instant. If we have our eye on the pencil, this report may be visual, at least in part, but it is more generally kinaesthetic, or to use a term now in vogue, proprioceptive. If the proprioceptive sensations are wanting, and we do not

replace them by a visual or other substitute, we are unable to perform the act of picking up the pencil, and find ourselves in a state of what is known as *ataxia.*

Note that in Wiener's formulation it is proposed that the feedback controls the movement at each instant.

A very different view as to the role of feedback in control of movement appears in the paper of Taylor and Birmingham (1948) on the acceleration pattern of quick manual corrective responses. In contrast to Wiener, these authors express the view that movement can occur in the absence of feedback:

> The fact that the movements observed in this study are "continuously controlled" movements in contrast to "ballistic" motions, immediately raises the question as to the nature of the control mechanism. It might, at first, be assumed that the sequence of applied forces observed in this experiment arises in response to the changing size of the perceived error as the target is moved toward the hairline. Thus, it might be assumed that the S becomes aware of the target displacement, starts to move the joy stick, watches the target move toward the hairline and guides his movement in terms of what he sees.
>
> However, it becomes apparent that this can hardly be the case when the time factor is considered. The average reaction time found in this study was 0.435 sec. Nevertheless, once the arm started to move the force pattern varied continuously. For approximately the first 0.07 sec force in the direction of motion was applied at an increasing rate, then it was applied at an increasing rate during the next 0.10 sec, and finally the negative force was applied at a decreasing rate for the last 0.10 sec of the response. Thus, the arm had passed through all phases of acceleration and had come to rest in a length of time no greater than the reaction time.
>
> Even if it is assumed that the obtained reaction time which is being used as the standard is abnormally long, it is still true that the early phases of acceleration are completed in less time than it would take to perceive and to react selectively to a visual stimulus. This must mean that at least the early phases of the response are not guided continuously by information obtained through vision. The S starts to stop before he can react to visual information which tells him how far he has already moved.
>
> It is doubtful if this effect can be accounted for by proprioception or "muscle feedback." Though it is conceivable that the

reaction time to proprioceptive cues may be shorter than that to visual stimuli, the proprioceptive reaction time would have to be infinitely small to permit kinesthetic control of the continuous variation in applied force found in this study.

It should be noted that the opinion expressed by Taylor and Birmingham was an outgrowth of thinking that had been developing for at least half a century. Thus, as early as 1899 Woodworth's studies showed that a number of different sorts of movement were not controlled by sensory inputs *during their execution.* Visual or kinesthetic inputs provided information that was essential to the subject in deciding what sort of a movement to emit, but once this decision had been made the movement was not under the guidance of external sensory inputs. Thus, Woodworth's (1899) experiments, carried out at Harvard, provided data pointing to certain limitations of Wiener's ideas developed 50 years later at a point only a few miles down the Charles River. Until recently, in fact, neurophysiologists, neurologists, cyberneticists, psychologists, and human engineers have carried out their studies in relative isolation from each other.

If we now take another jump in time, from the 1940's to the 1960's, we see that the communication barrier between disciplines is beginning to break down. One of the clearest examples of the merging of the approaches of cyberneticist, human engineer, and neurophysiologist is seen in the work of Stark et al. (1961, 1962a,b,d, 1969), Young and Stark (1963a,b), and Stark (1966, 1968). Of particular relevance to the type of problem addressed by the work of Taylor and Birmingham (1948) is the paper of Navas and Stark (1968) on the control of hand movement. In this study, Navas and Stark used a subject (man) and a form of motor output (manual tracking of a visual display) that had been extensively used by human engineers. To this human engineering approach was added the theoretical and mathematical approach of cybernetics. The results obtained help to settle some of the problems raised by the differences between Wiener's statement and the one made by Taylor and Birmingham. In addition, the paper of Navas and Stark (1968) provides some general guidelines for conduct of studies on the role of feedback in motor control.

On the basis of the results of the experiment in question, Navas and Stark postulated an alternation between voluntary and reflex control of the alpha efferent pathway, with a brief uncoupling of the motoneuron from spindle-afferent feedback at the time a preprogrammed command from higher centers took over control. In this model, the higher center preparing the program samples the visual and kinesthetic input only intermittently: once the program has been set, sensory information is no longer admitted. Wiener's notion of continuous feedback does not apply at this level of the system—and the model of Taylor and Birmingham seems more appropriate. On the other hand, although kinesthetic inputs are only intermittently sampled at the level of the system that programs the abrupt, saccadic shifts of hand position, these same kinesthetic inputs may actually be in almost continuous operation at lower levels of the system. Thus, for a tracking task in which the manual "saccades" may occur at intervals of 500 msec or more, it appears that "the system is operating proprioceptively open loop ... during a short period of time, about 100 msec." It is during this brief 100 msec period that "spindle afferent control [of the motoneuron] is reduced and the higher control of the alpha efferents is fully turned on."

Thus, for a large part of the time, the operation of the motoneuron *is* under continuous proprioceptive guidance. Additional evidence for this model comes from experiments (Stark, 1966) in which the gamma input was selectively blocked with procaine. Following such block, the initial part of skilled movements (alpha controlled) was still correct, but the termination was severely affected; the subject lost control of his final arm position, apparently because the proprioceptive feedback path was inoperative for "clamping and damping" of the movement.

These results of Navas and Stark (1968) indicate that in thinking about feedback, the following are necessary:

1. It is necessary to specify the modality over which feedback is delivered to the system.
2. For a given modality, it is necessary to specify the particular feedback loop and the phase of the movement in which this particular loop is postulated to operate.

With reference to point (1), the experiment of Navas and Stark makes it clear that in speaking of feedback, one must distinguish between visual and kinesthetic modalities, because information fed back via these two modalities was handled quite differently. The visual input was sampled intermittently, confirming the view of Taylor and Birmingham that this type of tracking behavior is not under feedback control during its occurrence. However, the situation was different for the nonvisual feedback path, because this modality *does* modify the movement during the terminal phases of its execution.

With reference to point (2), it is clear that within the kinesthetic-proprioceptive modality it is essential to consider the pathway over which feedback takes place. Thus, spindle afferents feeding back at a low level of the system (e.g., to motoneuron) were providing continuous feedback control during the terminal phases of the movement at a time when these inputs were not being admitted to the "higher level programmer."

In summary, then, it is apparent that a discussion of the role of feedback in control of movement requires specification of *modality* (e.g., visual versus proprioceptive), *type of data sampling* (e.g., continuous versus intermittent), *pathway* (e.g., direct monosynaptic onto motoneuron versus multisynaptic up to cerebrum or cerebellum and down again), *type of movement* (e.g., tracking predictable versus unpredictable targets), *phase of movement* (e.g., initial saccade or terminal slowdown), and other things as well.

Substituting the word "feedback" for "sensory input" in Weiss's (1941a) statement on the role of sensory input in movement, one may say,

> Nobody in his senses would think of questioning the importance of [feedback] control of movement. But just what is the precise scope of that control? Is the [feedback] influx a constructive agent, instrumental in building up the motor patterns, or is it a regulative agent, merely controlling the expression of autonomous patterns without contributing to their differentiation?

### Internal Feedback and Knowledge of Results

Thus far in this discussion the term "feedback" has been used in reference to information arising as a direct consequence of muscular contraction. This narrow definition of feedback does not include two important classes of information: (1) feedback arising from structures within the nervous system—let us refer to this as *internal feedback;* (2) feedback arising from the external environment as an indirect consequence of muscular contraction—let us refer to this as *knowledge of results.*

The importance of internal feedback in electromechanical

control systems hardly needs emphasis. We are well aware that in a long chain of elements, it is necessary to monitor performance and correct errors well before the final output of the system has even begun to emerge. For the nervous system, there is abundant anatomical and electrophysiological evidence that the all-or-none output (i.e., the action potential of the nerve cell) is directed not only to a target at the next step along a sequential chain, but that it is also fed back onto the neuronal systems that initiated the action potential to begin with, just as the all-or-none contraction of the muscle generates feedback onto its controller (the motoneuron) as well as exerting an effect on its target (the bone on which it inserts).

Thus, when it is said that a movement (e.g., visually guided tracking) is "reeled off on the basis of a predetermined central program without feedback" and when it is reported that "once set, the program cannot be modified by a feedback," the term "feedback" is being used in a narrow sense. In such a statement, the term "feedback" refers to signals generated by the response rather than to internal neuronal activity. But internal feedback (e.g., from the motoneuron onto its input sources) is no less significant than response feedback (e.g., from muscle onto its input sources).

Experiments carried out thus far have only just begun to investigate the effects of opening internal feedback loops, although pharmacological and neurosurgical treatments of movement disorders in man (treatments that act by altering internal feedback loops) point to the great importance of internal feedback in initiation and control of movement. Indeed, the prominence of these loops appears to grow ever greater as the nervous system phylogenetically increases in size and complexity, and subdivisions of the nervous system that were originally concerned with transmitting feedback information in the narrow sense (i.e., generated by muscular contraction) gradually acquire more and more neurons concerned with internal feedback functions. This point is clearly made in Hassler's (1966) comments on the evolutionary changes in thalamic function, in which it is pointed out that only one-eighth of the human thalamus serves as a part of somatosensory systems, with the remaining seven-eighths serving as relay stations for other systems. It seems clear that much of the thalamus is, in fact, involved in internal feedback loops—the "reverberating circuits" long ago investigated by Chang (1950).

Recent Developments on Internal Feedback: O. Oscarsson

The current work of Oscarsson is directly concerned with internal feedback loops and the role of such loops in motor control, and it was Oscarsson's remarks that were instrumental in leading to the discussion of internal feedback at the Work Session. Because Oscarsson's work has been so directly addressed to the problem of internal feedback loops, we have asked him to submit a brief note concerning these ideas, and this note follows:

It is usually assumed that ascending paths forward information about peripheral events. This is presumably the case with the dorsal spinocerebellar and cuneocerebellar tracts projecting to the cerebellum and with many components in the medial lemniscal system projecting to the cerebral sensorimotor cortex. These paths carry modality- and space-specific information from exteroceptors and proprioceptors. However, the organization of many ascending paths suggests that they monitor activity in lower motor centers rather than peripheral events. These paths include those receiving their main peripheral input from the flexor reflex afferents (FRA).* The information carried by these paths cannot be classified as either proprioceptive or exteroceptive. It lacks modality specificity and permits only very crude spatial discrimination. The importance of these paths is

*The term FRA, for flexor reflex afferents, has been defined as follows by Wall (1970):

Motoneurones which supply axons to flexor muscles are driven into activity by a variety of peripheral stimuli. A class of these stimuli sets off the flexor reflex, which seems designed to move the limb away from the stimulus point. Which muscles contract or which motoneurones fire depends therefore not only on the type of afferent fibres stimulated but also on their spatial origin. In other words, each motoneurone can be said to have a receptive field with respect to the flexor reflex stimulus. A wide range of afferents can evoke the reflex, including cutaneous, joint and smaller-diameter muscle afferents. As a stimulus at one point is increased in strength, more and more muscles take part in the response. In terms of the single motoneurone, this means that threshold varies within its receptive field. Increase in stimulus strength not only recruits more neurones but also produces repetitive response. The motoneurones are therefore subject to both spatial and temporal summation. Continuous tetanic stimulation leads to a muscular response which is initially large and slowly declines over a period of seconds to a lower sustained level of contraction. The threshold for the reflex is affected by the existence of other peripheral stimuli and by central activity. The threshold is low in the spinal animal and higher in the decerebrate. These then are the classical and defining properties of the flexor reflex which had been so fully explored by Sherrington (1906) and by Creed et al. (1932).

indicated by the fact that they include not less than six of the nine spinocerebellar paths so far investigated and listed in Table 1 (Oscarsson, 1967,1970). These ascending paths also project to other structures in the brain, including the cerebral sensorimotor cortex (Grampp and Oscarsson, 1968).

The FRA are defined as those myelinated afferents which evoke the flexion reflex in the spinal preparation. They include low- and high-threshold cutaneous afferents, groups II and III muscle afferents, and high-threshold joint afferents (Eccles and Lundberg, 1959; Holmqvist and Lundberg, 1961). It is not certain if all the afferents in these

TABLE 1

Classification of Spinocerebellar Paths and Main Afferent Input from Periphery

| Paths | Main Input |
|---|---|
| *Direct mossy fiber paths* | |
| Proprioceptive components of dorsal spinocerebellar and cuneocerebellar tracts, DSCT and CCT | Group I |
| Exteroceptive components of dorsal spinocerebellar and cuneocerebellar tracts, DSCT and CCT | Cutaneous |
| Ventral and rostral spinocerebellar tracts, VSCT and RSCT | FRA |
| *Indirect mossy fiber path* | |
| Spinoreticulocerebellar path relayed through the lateral reticular nucleus, LRN-SRCP | FRA |
| *Climbing fiber paths* | |
| Dorsal spino-olivocerebellar path, DF-SOCP | FRA |
| Dorsolateral spino-olivocerebellar path, DLF-SOCP | Cutaneous |
| Ventral spino-olivocerebellar path, VF-SOCP | FRA |
| Lateral climbing fiber—spinocerebellar path, LF-CF-SCP | FRA |
| Ventral climbing fiber—spinocerebellar path, VF-CF-SCP | FRA |

groups contribute to the FRA, and it is possible that unmyelinated afferents should be included (Franz and Iggo, 1968).

The ascending FRA paths appear to have the following characteristics in common (Oscarsson, 1970):

1. The information from the periphery is without modality specificity, because excitation and/or inhibition is evoked by all the components of the FRA.

2. The receptive fields are large and may include one or several limbs. They sometimes consist of excitatory and inhibitory areas, but they permit only very crude spatial discrimination.

3. The FRA effects to the ascending paths are mediated by pools of interneurons in the spinal cord and/or brainstem.

4. These interneurons are strongly excited and inhibited by descending tracts.

There are remarkable similarities in the organization of the FRA paths and the reflexes evoked by the FRA (Lundberg, 1959, 1964a; Oscarsson, 1967, 1970). In both cases the same combination of afferents is involved, the receptive fields are large, and the synaptic actions are mediated by interneurons. Furthermore, the interneurons are in both cases under similar supraspinal control; for example, they are excited by the pyramidal tract and inhibited by reticulospinal paths.

The diffuse afferent organization of the ascending FRA paths would seem to make them unsuitable as channels for specific information about peripheral events. On the other hand, the characteristics of these paths can be explained if they are assumed to carry information about activity in pools of interneurons that are simultaneously reflex centers and links in descending motor paths, as suggested in Figure 9 (see Lundberg, 1959, 1964a,b; Oscarsson, 1967, 1968, 1970; Larson et al., 1969a,b; Miller and Oscarsson, 1969). We do not know if the interneurons should be regarded first as reflex centers controlled by descending paths, or primarily as links in descending motor paths that are modulated by the peripheral input. However, such a

distinction might be meaningless; there is evidence that
motor performance depends partly on mobilization and
inhibition of reflex arcs by higher centers (Lundberg, 1966;
Hongo et al., 1969a,b). It is possible that the activity of the
interneurons in many cases is determined by signals in the
descending tracts, because many ascending FRA paths are
only weakly influenced by natural stimulation of receptors.

The above hypothesis would be corroborated if it
could be shown that the input from the motor cortex to the
ascending FRA paths is highly specific, for example, related
to single cortical cell columns that are known from Asanuma's investigations with intracortical microstimulation
technique to control individual muscles (Asanuma and
Sakata, 1967; Asanuma et al., 1968). Asanuma, Stoney and
Thompson* have recently described spinal interneurons
that receive a diffuse peripheral input from the FRA
and a specific input from single cortical cell columns. These
interneurons are presumably links in motor paths and are
possible candidates for the interneurons hypothesized in
Figure 9.

Figure 10 shows a hypothetical diagram of some
paths between the cerebral motor cortex, cerebellar anterior

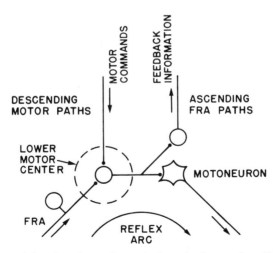

Figure 9. Suggested function of ascending paths from the flexor reflex afferents (FRA). It is
assumed that these paths monitor activity in lower motor centers whose pools of interneurons
are both reflex arcs and links in descending motor paths. [Oscarsson, 1970]

*Unpublished data.

lobe, and lower motor centers, and suggests possible
functions for these paths (Oscarsson, 1970). It is postulated
that the anterior lobe is important for correcting errors in
motor activity elicited from the cerebral cortex and carried
out by command signals through pyramidal and extrapyramidal paths. It is suggested that the anterior lobe receives
information (a) about command signals from the motor

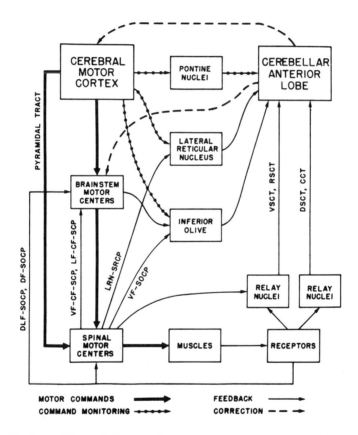

Figure 10. Some of the paths between the cerebral motor cortex, anterior lobe, and lower
motor centers with an interpretation of the function of these paths. The anterior lobe is
assumed to correct errors in motor activity elicited from the cerebral cortex 'and carried out by
command signals through pyramidal and extrapyramidal paths. The command signals are
assumed to be monitored by the anterior lobe through paths relayed in the inferior olive and
pontine and reticular nuclei. The spinocerebellar paths are assumed to serve as feedback
channels that monitor the activity in lower motor centers and the evolving movement.
Abbreviations for paths are given in Table 1. [Oscarsson, 1970]

cortex (representing "intended movement"), (b) about the effects these signals evoke in lower motor centers that are also influenced from the periphery, and (c) about the movement evolving from signals on the pathways that are more directly activated by exteroceptors and proprioceptors (the spino- and cuneocerebellar tracts). One function of the anterior lobe might be to compare these three classes of information and on this basis perform necessary corrections through the well-known efferent paths to the motor cortex and brainstem. Figure 10 suggests that the comparison might already have occurred in part in the precerebellar nuclei, for example, in the inferior olive (Miller and Oscarsson, 1970) and the lateral reticular nucleus (Bruckmoser et al., 1970). It has been shown that the olivary neurons projecting to the anterior lobe and activated from the spino-olivary paths are also activated from the motor cortex (Miller et al., 1969; Oscarsson, 1969). A comparator function of the inferior olive would be corroborated if it could be shown that the information from the motor cortex received directly through cortico-olivary paths and indirectly through corticospino-olivary paths is related to the same specific motor actions.

The point that we wish to emphasize from Oscarsson's note is that certain afferent paths "monitor activity in lower motor centers and are only partly and indirectly concerned with peripheral events." These paths are not so much concerned with feedback information in the usual sense (i.e., information consequent upon muscular contraction). Here, then, we have a case of internal feedback. The notion of internal feedback is also apparent in Ito's ideas as to the function of the massive interconnections between cerebral cortex and cerebellum. Ito proposes that these interconnections allow the cerebellum to (1) monitor cortical output and (2) feed back corrective signals to cortex so as to correct errors of cortical output long before this output has given rise to motoneuronal discharge. Many additional examples of internal feed-back loops might be cited—but for the purposes of this chapter two should suffice for the moment. The point we wish to emphasize is that in seeking to unravel the functional significance of various parts of the nervous system and to understand central control of movement, internal feedback pathways of the type proposed by Oscarsson and Ito would seem to be of very great significance.

### Knowledge of Results

The third type of feedback to be taken up in this essay is at the opposite extreme from internal feedback, having to do with events taking place *after* the motor response rather than before it, and depending on the behavior of the environment rather than the behavior of the organism. This type of feedback has been given a number of different names (see Bilodeau, 1966). Among these names are extrinsic feedback, learning feedback, reinforcement, information feedback, reward, indirect feedback, nontopographic feedback, and knowledge of results. For the purposes of this paper, the last term, knowledge of results, will be used. Any overall consideration of the role of feedback in movement should thus deal with three types of feedback: (1) internal feedback, (2) response feedback, and (3) knowledge of results.

When response feedback is eliminated by deafferentation, it is still possible that feedback generated internally or by knowledge of results may be of critical importance in motor control. Indeed, some very interesting notions concerning the role of feedback in movement have come from work of two groups of investigators studying acquisition of conditioned limb movement following dorsal root section. In such a situation, it is possible to examine the extent to which new movements can be learned on the basis of knowledge of results and internal feedback, but without response feedback. Because of the implications of these experiments for ideas of internal feedback, a separate section of this chapter will be devoted to these studies.

### Conditioned Movements in Deafferented Limbs

The first group of experiments to be reported on are those of Jankowska and Gorska, as summarized by Konorski (1967). In these experiments (performed on cats, dogs, and rats) it was found that learned operant responses (Type II conditioned responses in Konorski's terminology) could be established in fully deafferented limbs. Furthermore, it was found that conditioned movements established prior to deafferentation were not eliminated by that procedure. It is interesting to consider the behavior of the animals before and after deafferentation, for even though the conditioned reflex was preserved following deafferentation, its character was grossly altered. Thus, Konorski states,

After this operation the unconditioned scratch reflex was preserved, although its character was drastically changed. The animal failed to reach the ear with his hindleg, performing the scratch movements "in the air" with a stiffly extended leg. Exactly in the same manner the instrumental pseudoscratch CR was changed. The movement was clumsy and was performed with the extended leg, but there was no doubt about its full preservation. The same was true when deafferentation affected both hindlegs.

The point to be emphasized here is that although this motor learning took place in the absence of response feedback, it did not take place in the absence of all feedback: both reinforcement (i.e., knowledge of results) and internal feedback were present.

Thus, Konorski points out,

Repeated non-reinforcement led to the extinction of the trained movement, which was restored when the CS was again followed by food. In other words, the properties of the instrumental CR involving the deafferentated limb did not differ in any way from those involving the unimpaired leg, except in the clumsiness of the movement owing to the rigidity of the distal portions of the leg.

Konorski goes on to comment on these results, as follows:

All these data leave no doubt that, for both the formation of the type II CR and its preservation, the sensory feedback from the limb involved in the trained movement is not necessary. The generally held misconception concerning this point, wide-spread among neurologists and neurophysiologists, probably has its source in the observation that the deafferentated limb is of no use in any skillful, manipulatory movements, such as reaching a goal with the forelimb, pulling or pushing objects, or seizing them. Since such activities are composed of successive motor acts in which a feedback from the fulfillment of one segment provides a stimulus for eliciting the next one, they obviously cannot be performed by the deafferentated limb. As a matter of fact, the deafferentated limb does not take part even in locomotion which involves at least proprioceptive information about its position and tactile information concerning its touching the ground. In consequence, the non-employment of the deafferentated limb because of its inutility is easily mistaken for the inability of this limb to perform *any* instrumental movements. Thus, precise evidence showing the preservation of the instrumental performance in the deafferentated limb can be provided only by the "artificial" experimental condition in which the animal is required to perform a movement which obviously plays no practical role in his life, since it does not involve any precision connected with the sensory feedback. In other words, a movement which in normal life is completely useless in artificially simplified conditions remains fully instrumental for achieving a goal consisting in obtaining the food or avoiding an aversive US. Therefore, its utility is totally preserved, and in consequence its performance is not inhibited.

A second set of experiments on conditioned limb movements following deafferentation has been carried out by Taub and Berman (1968). Their results, obtained in monkeys, were for the most part in agreement with those described by Konorski. In considering mechanisms that might have been responsible for learning in the absence of response feedback, Taub and Berman suggested a possible role for central efferent monitoring—a term that they use in the same sense as the term internal feedback has been used in this chapter. Taub and Berman point out that one issue raised by their results relates to the question of how an animal could learn to use a deafferented limb in the absence of vision. In considering this issue, Taub and Berman state that

Instrumental conditioning involves the ability to repeat consistently certain movements; but how can animals with a deafferented limb that they cannot see learn to repeat movements, when by all classical considerations they should not know where the limb is, whether it has moved, and if moved, in what way? Since the required information concerning the topography of their movements could not have been conveyed over peripheral pathways, it must have been provided by some central mechanism that does not involve the participation of the peripheral nervous system. Such a mechanism could be one of two general types: either it would involve feedback, but of wholly central origin, or it would involve no feedback whatever.

The former mechanism requires the existence of a purely central feedback system that could, in effect, return information concerning future movements to the CNS before the impulses that will produce these movements have reached the periphery. An animal could thus determine the general position of its limb in the absence of peripheral sensation. Indeed, just such a mechanism has been demonstrated electrophysiologically, first by Chang (1955) and Li (1958), subsequently by a large number of investigators (for a partial summary see Levitt, Carreras, Liu and Chambers, 1964) and anatomically by

Kuypers (1960). It would seem to involve afferent collaterals from the medullary pyramidal tracts to the nuclei gracilis and cuneatus, thence back to the cerebral cortex through ventralis lateralis. In fact, in the last 15 years it has been found, in contrast to the classical view, that most of the thalamic nuclei are really two-way streets. Indeed, if central feedback is of significance for behavior following deafferentation, it seems reasonable to assume that not one but several "loop" pathways would be involved. That is, if one were to set out a priori to construct a servomechanism that was maximally effective and sensitive to control, one would certainly establish a feedback loop at *each* level of the system from command center to output.

## Corollary Discharge

The studies on movement in deafferented limbs led Taub and Berman to consider the role of "central efferent monitoring" in learning. As one reads their proposal that collaterals from medullary pyramidal fibers to dorsal column nuclei mediate this process, and thinks back on Sperry's definition of corollary discharge as "discharge of motor patterns into the sensorium," one cannot avoid the impression that the internal feedback or central efferent monitoring is another name for corollary discharge. In order to clarify the relation between corollary discharge and internal feedback (or central feedback or central efferent monitoring), we will cite some of the points made by Teuber (1966) concerning corollary discharge:

> We assume that each voluntary movement, or change of posture, involves not only the downward discharge to the peripheral effectors, but a simultaneous central discharge from motor to sensory systems preparing the latter for those changes that will occur as a result of the intended movement. A voluntary eye movement which transports contours across the retina would thus leave the spatial order of perception undisturbed, because the impulses to the eye muscles are accompanied by appropriate corollary discharges which preset the visual system for all anticipated shifts in the spatial order of visual inputs. By contrast, when we push against our eyeball, moving it passively, the visual scene jumps, and the same apparent shift of scene is perceived whenever we intend to move our eye but there is inability to move, either by mechanically restricting the globe in the orbit, or by virtue of an acute extraocular palsy.

Such instances of paralysis are particularly revealing, since there is no displacement of contours on the retina; yet any unsuccessful intention to move is experienced as an illusory displacement of the scene in a direction opposite to the intended motion (Kornmüller, 1931). These illusory motions of contours in the presence of ocular palsy provide the most direct evidence for the continual operation of the compensatory mechanisms which counteract the normally inevitable shifts of input that result from voluntary movement. In paralysis, these counteractive signals turn into illusions, since they are not annulled by the result of the motion. Correspondingly, if one wears spectacles that invert the visual scene, as has been done by Stratton (1897), Kohler (1951), and Held (1961), one will inevitably perceive at first extensive illusory shifts of the scene with every voluntary movement of the eye. This is due to the fact that the compensatory signal is not subtracted from those shifts in contours that result from the eye movements, but is added to these shifts, as a consequence of the optical reversal induced by the spectacles.

All of these considerations go back at least as far as Helmholtz (1867), who included them in an outflow theory about the role of eye movements in perception. He was severely criticized for attributing the compensatory effects to what he called "feelings of innervation"; his critics claimed that they could not discover these feelings among their own introspections. Modern versions of the outflow theory have tried to preserve its essential features while eliminating the postulated feelings of innervation. Such versions were formulated simultaneously and independently by the team of von Holst and Mittelstaedt (1950) and by Sperry (1950) about 15 years ago.

In speaking of discharge from motor to sensory systems, one wonders whether the terms "motor" and "sensory" should be interpreted in their very broad, functional sense or in a narrow, anatomical sense. Should "motor" suggest precentral motor cortex (area 4) and "sensory" suggest postcentral somatosensory receiving area (areas 3, 1, and 2)? Certainly the original use of these terms by Sperry was in the broad, functional sense; he spoke of "motor patterns" and "sensorium," and it appeared that both the "motor patterns" and the "sensorium" that Sperry referred to in this case might have resided in the optic lobe of the fish.

Teuber (1967) has postulated that

> ... a voluntary movement (e.g., of the eyes) is always charac-
> terized by a twofold process: an efferent discharge to the effectors (in
> the case of eye movements, to the extraocular musculature), and a
> simultaneous central discharge (the "corollary discharge") to the
> appropriate sensory systems (here, the visual system) which forewarns
> them, so-to-speak, of the impending change.

Again, in this statement, should the term "visual system" be interpreted
in a narrow, anatomical sense—or should we include within the visual
system structures such as the caudate, putamen, and globus pallidus?

It seems most appropriate to assume for the moment that no
definite anatomical loci are implied by the terms "motor" and "sensory," but that by "motor" we should mean regions of the brain that
play a role in initiating movement, and that by "sensory" we should
mean regions that have some role in perception.

As one seeks out the structures that may be significant in
carrying "motor patterns" into the "sensorium," it is of interest to
consider Teuber's (1966) observations on the occurrence of certain
visual perceptual defects in patients with lesions of the frontal lobes,
basal ganglia, and ventralis lateralis. Teuber suggested that the perceptual defects in these patients resulted from interruption of pathways
or centers essential for the occurrence of corollary discharge, but these
structures (frontal lobe, ventralis lateralis, globus pallidus) are ones that
loom prominently on any list of links in the chain which may
constitute the internal feedback loops by which the activity of the
cerebral cortex feeds back on itself. Thus, corollary discharge would
seem to have a variety of internal feedback loops as its neurophysiological and neuroanatomical substrates. For example, it is known
that fibers travel:

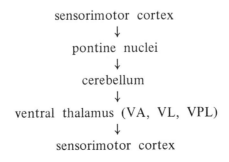

sensorimotor cortex
↓
pontine nuclei
↓
cerebellum
↓
ventral thalamus (VA, VL, VPL)
↓
sensorimotor cortex

This cortico-ponto-cerebello-thalamo-cortical loop might constitute one of many possible anatomical substrates for the occurrence of corollary discharge.

Another set of experiments demonstrating pathways that might mediate corollary discharge are those carried out by Towe and associates (Chapter II), showing that cortical output via the pyramidal tract could cause enhancement of cortical input by means of facilitatory actions of pyramidal tract neurons on transmission through the dorsal column nuclei—"constitut[ing] one route by which the cerebral cortex can modify its own afferent input" (Adkins et al., 1966).

A case in which "motor patterns" pass into the "sensorium" at even lower levels of the system is seen in Oscarsson's model of the role of FRA pathways in monitoring the central inputs to the motoneuronal pool. Thus, inputs from motor areas can return to sensory systems via the pons and the cerebellum, via the dorsal column nuclei, via the interneuronal pools of the spinal cord—and via many other pathways that have been delineated earlier.

Another instance of the operation of internal feedback loops is seen in the studies of Thach (1970a,b) on activity of cerebellar neurons prior to arm movements in the monkey, and Evarts's (1970) observations on discharge of neurons in the VL nucleus prior to movement. These studies show cerebellar and VL neurons (which are links in one of the internal feedback loops from motor cortex back to itself) do in fact discharge prior to learned movements and well in advance of any response feedback.

Neuroanatomical and electrophysiological studies have now shown internal feedback loops that transmit information from motor to sensory, from motor to motor, and from sensory to motor areas. These pathways would seem appropriate candidates for mediating "corollary discharge" or "feedforward control," and we should now seek to design definitive experiments concerning the way in which internal feedback loops operate during movement, for it seems likely that an understanding of the activities of these internal feedback loops will greatly help us to understand the neurophysiological basis of central motor control.

Before leaving this topic, it would seem useful to summarize briefly one additional point made in the Work Session discussion. This relates to the hypothesis that corollary discharge has a special relation to "voluntary" as contrasted to "reflex or passive" movement. Of course, as the terms "corollary discharge" or "efference copy" were originally used, they explained a highly nonvoluntary movement in a fish or insect. Indeed, what could be less voluntary and more reflex than lifelong useless circling? Certainly there is reason to believe that the internal feedback loops we have spoken of *do* operate in a variety of reflex as well as in voluntary movements. We may conclude that if the definitions of corollary discharge or efference copy include the requirement that they be peculiar to voluntary movement, then what we have said about their neurophysiological substrate must be taken back. Alternatively, one might use these terms without attempting to invoke them as criteria in deciding whether a movement is reflex or voluntary.

In concluding this section, it would seem worth while to mention a recent study on movement in man showing that certain types of movement errors start to be corrected without response feedback. This finding appeared in a study by Higgins and Angel (1970) on correction of tracking errors. It was found that error correction time was less than proprioceptive reaction time, and it was proposed that subjects

> ... are able to monitor their own behavior internally, comparing the actual motor commands with some reference value, which is the "right" or "proper" command under the given circumstances. If there is a discrepancy between the actual command and the "right" command, the response can be arrested prematurely, before any information has been fed back from the periphery.

Here again, the notion of internal feedback is raised. It seems probable that this "modern" view of internal feedback, and the lack of a sharp distinction between feedback from the periphery and feedback from the nervous system to itself would have been applauded by Magendie, who is commonly thought of as emphasizing distinctions between dorsal and ventral roots—but who also recognized the unity of the neuromuscular apparatus in normal motor function. Indeed, it would seem appropriate to end this chapter with a quotation from Magendie (1824), giving the last word to the man who said one of the first words on this topic:

> The organs which concur in muscular contraction are the brain, the nerves, and the muscles. We have no means of distinguishing in the

brain those parts which are employed exclusively in sensibility, and in intelligence, from those that are employed alone in muscular contraction. The separation of the nerves into nerves of feeling and nerves of motion is of no use: this distinction is quite arbitrary.

## VI. STRATEGIES AND TACTICS IN RESEARCH ON CENTRAL CONTROL OF MOVEMENT
### Edward V. Evarts and W. Thomas Thatch, Jr.

### Introduction

The preceding chapters have dealt with some of the conceptual issues that emerged as focal points of discussion at the NRP Work Session on "Central Control of Movement." But in addition to discussing the generalities of how one should think about the problems of central motor control, we also tried to come up with experimental approaches that might actually help to solve some of these problems. In this chapter, we shall consider several experimental approaches that were discussed.

To begin with, there will be reports by two Work Session participants concerning new techniques currently being applied in their laboratories. Frank will discuss methods whereby the information outflow from the nervous system might be used to control external devices (e.g., prosthetic limbs), and Brookhart will describe a method for investigating central postural control mechanisms.

Following the reports of Frank and of Brookhart, we will consider ideas as to how certain current theories of motor control might be tested in moving animals. For obvious technical reasons, hypotheses as to neurophysiological mechanisms of motor control do not often arise from experiments carried out in moving animals, and most current theories of central motor control processes actually grew out of experiments on immobilized preparations. It is clearly desirable that these theories ultimately be tested in moving animals, however, and one aim of the Work Session was to generate ideas for experiments that could provide such tests. This chapter of the *Bulletin* will present summaries by three Work Session participants—Henneman, Ito, and Phillips—as to how some of their ideas might be tested in normally moving animals. Each of these participants was asked a question designed to elicit these views, and, following the Work Session, wrote a summary of his answer.

### Use of Neural Signals to Control External Devices: K. Frank

In this consideration of prospects and problems that seem to lie ahead as newer techniques are applied to neurophysiological analysis of central motor control mechanisms, we are especially grateful to Karl Frank—who was an absentee at the Work Session because of the flu—for preparing the following statement:

> The aim of this statement is to put forth the view that, despite the huge gaps in our understanding of the nervous system, it is now practical to make a concerted attack on the problem of harnessing the information outflow from the nervous system to the direct control of external devices, such as artificial limbs, teleoperators, and communications equipment. Similarly, it should be possible to develop arrays of chronically implanted electrodes and appropriate electronics so that time-space patterns of electrical stimuli could be used for direct communication with the nervous system (Frank, 1968).

> Dealing first with the question of outward information transfer, signals may be tapped from the muscles, peripheral nerves, motoneurons, cerebellar cells, pyramidal cells of the motor cortex, or a number of brainstem nuclei. The signals may be trains of spikes from single units or gross potentials averaging the spike and synaptic activity of a number of different cells. There are two basic approaches that apply to all of these routes for sampling outward information from the nervous system: (1) information may be sampled as it is generated naturally, during the attempted performance of a particular pattern of motor activity, and used to control movement normally associated with that pattern; or (2) the nervous system can be modified through learned behavior so that signals are delivered to implanted electrodes as required for the operation of some external device; with this approach a code must be learned as in typewriting or in the operation of a crane.

> The former approach is well illustrated by the development of an EMG-operated upper arm prosthesis at the Moss Rehabilitation Hospital in Philadelphia (1968). Surface electromyogram (EMG) signals are detected from 10 pairs of electrodes recording the activity of neck, shoulder, and pectoral muscles secondarily related to the prime movements of the lower arm. A multifactor analysis of such signals from a normal individual, recorded together with components of

movement of his forearm, permits the derivation of a set of time-potential templates by which decisions can be made from related muscle potentials as to which movement components are called for. While the original analysis requires the use of a digital computer, a simple electronic device has now been built that performs these decisions and controls the movement of an artificial arm. The importance of this principle of utilizing naturally occurring related muscle information is illustrated by observing that newly fitted amputees can operate such an arm immediately without a learning period.

A related approach is being applied with monkeys by Humphrey's group in the Laboratory of Neural Control of NINDS (Humphrey et al., 1970). Instead of sampling EMG potentials, this group is recording simultaneously the spike trains from up to 10 pyramidal cells of the motor cortex, selected on the basis of their association with one or more recorded components of arm movement in monkeys trained to operate a lever against an adjustable force. The arm movement represents a steplike transition from one maintained state to another. Information sufficient to identify the state (in terms of force and position) may or may not be contained in the signals from the selected pyramidal cells. While the state is being maintained, these cells fire irregularly at frequencies that may differ for different maintained states. The cells may be individually selected because they show a more or less easily recognizable change in firing frequency, associated with a transition between selected force-position states. As shown by Evarts, some cells (the tonic cells) maintain their altered firing rates as long as a new state is maintained, and they may do so for a variety of new positions so that their firing patterns are not generally diagnostic of position. More often, however, cells (the phasic cells) show a transient burst of activity associated with and generally preceding a change in force-position state.

It is attractive to suppose that these phasic cells are representative of the kind of control exerted by the brain, so no new cortical information is required for the maintenance of a steady state, therefore arguing from the standpoint of efficiency that cortical output channels need not be wasted in duplicating signals already carried by the motoneurons; series redundancy in a control system adds only noise to the final output. One can easily imagine regenerative circuits capable of maintaining a given pattern of muscular contraction until they receive a new command signal to change patterns. However, there are a great many muscles that are involved in a supportive role, bracing the body as it were for backing up any transition in states for even so simple a movement as a wrist flexion. Statistically, therefore, one might expect to find pyramidal cells associated with these supportive, phasic muscle contractions more often than those restricted to the actual movement observed. Also, nature is rife with examples of apparent inefficiencies and redundancies of all kinds.

Whether or not any group of cells lies on a causal pathway for a given motor act is, in a sense, not relevant to the possible role such cells might play in control of an external device. Fairly discrete movements are possible even after section of the pyramidal tracts, but this only shows that there are other control pathways that can serve a similar function. There is a real possibility, of course, that pyramidal cells of the motor cortex represent a side branch in the control pathway, despite the fact that they may show a burst of activity 100 msec before the movement starts and that electrical stimulation may evoke a closely related movement. In this case, the firing patterns of even a large number of such cells may not contain the information necessary to specify the motor state or even the particular transition of states observed.

In contrast to the sampling of spike trains from single nerve cells, gross potentials from the brain can be used to indicate intended changes in motor state. Techniques are already available for recording such averages of cell behavior chronically, with little damage to the nervous system. However, this approach has certain basic limitations. The experiments of Woody (1970) suggest that cortical cells subserving rather different motor functions appear to be distributed through the same cortical space. Thus, potentials recorded from single gross electrodes would be expected to be an ambiguous composite of several intended motor patterns.

(This is not fundamentally different from the participation of a particular neuron in a variety of muscle contractions, as described above.) The simultaneously recorded time courses of potentials or signatures from a number of such gross electrodes might, however, permit discrimination between various patterns of motor behavior. It seems probable that this method of sampling neuronal output will lend itself more readily to the detection of abrupt motor changes, such as an eye blink, than to discrimination between a number of motor steady states.

So far there has been no mention of the role of feedback mechanisms. The enormous advantages to be gained by feedback loops in any control system make it likely that in addition to known sensory feedback systems there exist large numbers of internal feedback loops within the nervous system. For example, if the nuclei of spinal motoneurons are under supervisory control from the motor cortex, one would certainly expect that information about the firing state of the spinal neurons and even the distribution of excitatory and inhibitory background over their somata and dendrites would be fed back to the brain, thus providing a mechanism for modulation of motor cortical outflow of an error-correcting nature. Perhaps this concept could be tested experimentally by locally modifying the firing state of the spinal cord (with ventral roots cut) while recording the outflow from appropriate pyramidal cells.

In developing practical direct control mechanisms, it will be important to investigate the effects of feedback loops of all sizes. The longest loops are perhaps the visual and auditory assessment of the final movement or other output to be controlled. In the case of limb prostheses, important shorter loops will be the "reactionary" forces, which feed back through joint, tendon, stretch, and tactile receptors. In addition, there will certainly be a need for feedback loops within the external artificial mechanisms under control. The Moss myographically controlled arm uses such external feedback of force and position information, and the group is investigating the advantages of feeding back velocity information.

However, the most intriguing possibility is the feedback of externally sensed information through an array of
implanted electrodes on the sensory side (Brindley and
Lewin, 1968). This approach presents very difficult problems, but through the exploration of such hybrid feedback
loops may come a profoundly greater understanding of
control mechanisms operating in the unassisted central
nervous system.

I believe it is especially in this area of feedback
control that a systems engineering approach will be valuable.
Models of parts of both natural and artificial control systems
may help to describe their behavior and at the same time will
show up limitations and suggest alterations that might form
the goals for future experiments.

## A Technique for Investigating Central Control of Posture: J. M. Brookhart

Several recent publications by Brookhart and co-workers (Mori
and Brookhart, 1968b; Petersen et al., 1965; Brookhart et al., 1965;
Mori et al., 1970; Brookhart et al., 1970) have described a technique
for studying the input-output relations of the postural control system
in the dog. On the basis of these publications, the following statement-
question was put to Brookhart at the Work Session:

Brookhart's experiments with intact standing dogs indicate that
under particular conditions of experimental design, it is possible to
obtain meaningful analysis of the input-output relations of the motor
system during the course of a behavioral motor response to an
"adequate" stimulus. The critical delays in the postural response to
horizontal displacement and the more or less stereotyped nature of the
response in different animals suggests the existence in the CNS of a
uniform pattern of neural changes that underlies the external manifestations of response. The predictable behavior of the system implies that
it will be possible to inquire more deeply into the manner in which the
CNS functions while carrying out the generation of such complex but
stable reactions. One may ask how this test system can be utilized to
investigate the role of known neuronal mechanisms, both spinal and
supraspinal, in the production of the response. Brookhart has already

begun to isolate the sources of afferent information that are necessary for the complete response. When this is done, how can the problem of the central processing of this information be attacked?

In response to this question, Brookhart prepared an outline of the objectives of the technique, the strategy of achieving these objectives, and the means of implementing this strategy. In what follows, Brookhart's outline will be reproduced almost verbatim:

The *Objectives* of the approach are to understand more about the manner in which the CNS operates in accepting a variety of inputs, processing them, and issuing decision orders to effectors in such a way as to bring about adaptive reactions.

*Strategy* for achieving these goals involves attention to the following points:

1. It is imperative to avoid unnatural or artificial stimuli, because these present nonsensical and meaningless problems to the CNS, which it was not designed to handle.

2. Some pattern of behavior should be selected, preferably requiring only reflex operations, that can be depended upon for replication within measurable, small limits of variability.

3. Because many CNS functions appear to be based on probabilities, the experiment should be designed so that the necessary sample sizes for statistical treatments can be acquired by repeated presentations of uniform natural stimuli.

4. These patterns of afferent, efferent, and central activity that are repetitive and time-locked to some aspect of behavior must be determined in the normal state.

5. Knowing the normal system operations, alterations in function of subunits within the system can then be produced, and the changes in outputs and in the behavior of other subunits within the system can be examined.

6. With this information, it will then be possible to deduce the nature of the interactions among subunits from the alterations in the patterns of behavior.

On the question of *Implementation,* Brookhart had the following comments:

      1. Postural control operation has been selected as the preferred system for examination because

            a. It appears to function as an automatic feedback-operated control system.

            b. As such, it should be free from the requirement of conscious perception and interpretation.

            c. In the adult animal it appears to be "hard wired," the learning process having been completed.

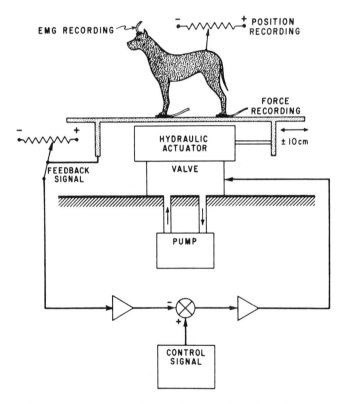

Figure 11. Experimental arrangements for postural control study. The trained animals stand on force recorders supported by a platform which can be moved in the longitudinal axis of the animal. The movement of the table constitutes a disturbance of the quiet standing posture which the dog is trained to maintain. The reactions to this disturbance can be detected in a number of ways, some of which are indicated. [Mori and Brookhart, 1968b]

2. Animals can be trained to adapt and to minimize deviations from a uniform quiet stance in the face of postural disturbances. This amounts to training animals to allow their hard-wired system to operate without interference from volitional or other types of activity. (See Figure 11.)

3. The next step is to establish the range of variability of response characteristics and the degree of replicability of elements of response (Figure 12). This has been accomplished using step disturbances; the reactions to sinusoidal disturbance are now undergoing examination (Figures 13 and 14).

4. Because the output behavior is measured quantitatively, it can be characterized using several informative analytic techniques (Figures 15 and 16).

5. Time and magnitude relationships, between behavior and CNS activity, can be established.

In considering future applications, it was suggested that the technique might someday be used in exploration of mechanisms underlying compensatory functional changes after lesions, and in exploration of mechanisms underlying learning.

Several points in Brookhart's outline will now be singled out for discussion. First of all, there is his warning to avoid unnatural stimuli. At first glance, such a warning might seem quite superfluous; after all, much the same dictum was put forward by Sherrington long ago. Also, almost everyone recognizes the importance of "adequate" stimuli. Upon inspection of current practices employed in research on central control of movement, however, it is apparent that the majority of investigators seeking to elicit a given motor output do so by means of an electric shock. This unnatural stimulus, while of great value in tracing out pathways and providing electroanatomical data, is of considerably less value in revealing normal patterns of organization. At times, the use of electric shocks in awake, moving animals may obscure the artificiality of such stimuli, especially when the evoked motor response is one that appears to be a part of the animal's natural repertoire. Thus it might be a good idea to have Brookhart's warning on display in every neurophysiology laboratory—rather like the ubiquitous signs saying "THINK."

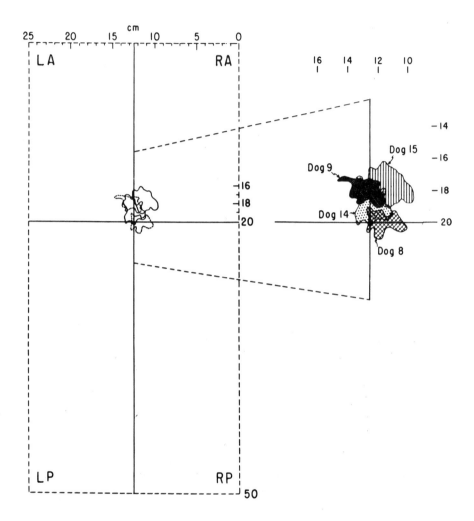

Figure 12. Representation of the replicability of quiet stance for four dogs trained to stand quietly erect. The corners of the large rectangle to the left indicate the positions of the feet of the animals. The vertical forces at each foot have been used to compute the position of the center of mass of the animals, and this position projected to the plane of the supporting surface. The irregular areas enclose the extreme positions of the center of mass registered during 10 consecutive trial periods, each lasting 5 min. The central portion of the figure has been scaled up and is shown on the right. Only the extremes have been used to delimit the borders; time spent in any given position is not represented. [Brookhart et al., 1965]

Following the Work Session, Brookhart provided some elaboration of the view (expressed in strategic point 2) that the movement to be investigated should involve "only reflex operations."

The desirability of a reflexly controlled reaction is an important consideration. The processes underlying consciousness, perception, and decision-making are entirely outside the realm of our understanding in neurophysiological terms. Therefore, it is impossible for us to manipulate the underlying operations in any closely controllable manner. On the other hand, reflexly controlled behavior has been defined historically as behavior that is stereotyped in its manifestations and independent of conscious perception and volitionally controlled decision-making. These qualities imply the existence of neuronal assemblies that are activated in a predictable fashion by the proper constellation of events at the inputs to the assem-

A                                                    N = 10

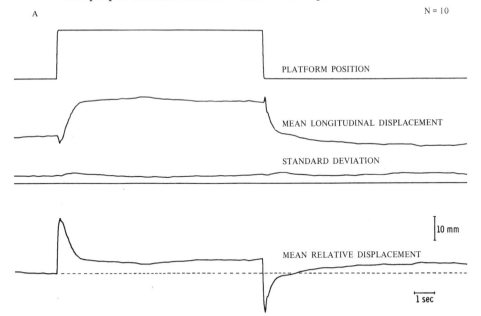

Figure 13. An example of average responses to step displacements of the platform. The platform moved in the headward direction, remained stable for 10 sec, and moved in the tailward direction. Both movements were 20 mm in 60 msec. The mean longitudinal displacement reflects the movements of a point over the pelvis with reference to a point on the floor. Below that is the SD relative to its zero value. The lowest record is the difference between the other two, and shows the mean relative displacement with respect to the platform. The analog voltage records were sampled, converted to digital form, and averaged with the aid of a computer. [Brookhart et al., 1970]

bly. These qualities do not imply anything about the anatomical distribution of the elements of this assembly, or about the processes by which the assembly was put together. They could have been built into the system by genetic programming or acquired by processes akin to those underlying learning. The result is a mode of response generation that operates automatically without the requirement for the participation of the uncontrollable processes of consciousness and volition in the implementation of the details of the response.

Whereas Brookhart's strategic points 1, 2, and 3 deal with selection and evocation of the motor behavior, strategic points 4, 5, and 6 deal with analyzing the afferent channels whereby this motor behavior is initiated and with determining the central activity controlling output units that generate the response.

The method can provide high-precision data and should be quite effective in detecting effects of lesions. Previous studies of the effects

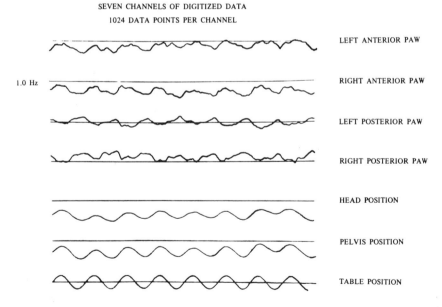

Figure 14. Averaged records from 30 trials each consisting of 8 cycles of sinusoidal displacement (lowest record) at 1.0 Hz and 80 mm. The record from each of the paw force plates and from the head and pelvis position indicators is shown with its reference line. The records were measured at a sampling rate of 168 samples per cycle, digitized, averaged, and displayed on a computer graphics terminal from which the photograph was taken. Over 7,000 measurements are represented in this one figure. [Brookhart]

of damage to components of the postural control system have often relied on relatively imprecise "clinical" observations, and results of such investigations have often yielded surprisingly negative results. In the past, the high degree of precision available at the level of neurophysiology in the immobilized animal has not been matched by comparable precision at the level of evaluation of functional role of components when the system is actually operating. Brookhart's technique would seem to provide a remedy for this mismatch.

Brookhart's points on implementation again raise the question of volitional control, and in elaboration of the brief statement given in the outline, Brookhart submitted the following statement:

The stance of these animals reflects completely normal behavior and represents a capability acquired early in postnatal life and maintained by frequent use. The training of an animal for food reward is regarded as having the effect of establishing a goal to be adhered to during the observation period. When animals are naive, the reaction patterns accompanying disturbance are erratic and unpredictable; this is due

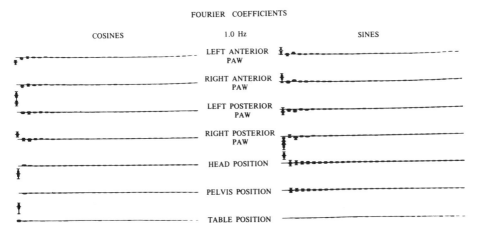

Figure 15. Data such as that shown in Figure 12 can be summarized through the use of the fast Fourier transform. The figure represents the first 32 cosine and sine coefficients using the forcing frequency of 1.0 Hz as the fundamental. The extremes of the vertical symbols indicate the range; the thick portion of the vertical line delimits the standard deviation, and the mean value of each coefficient is indicated by the middle crossbar. Again, the values were computed and displayed on a graphic terminal, which was then photographed. It would appear that the majority of useful information about the patterns of the waves shown in Figure 12 is incorporated in the first eight coefficients. [Brookhart]

to variable initial conditions, because the animals have not
learned to stand quietly. Only after they learn to stand quietly
do the response patterns become predictable. The reactions
to step displacement are initiated in muscles so rapidly that
Brookhart believes there is not enough time for the perception-
decision-execution processes that occur during "reaction time"
($< 100$ msec versus 200 to 500 msec). This fact, coupled with
older observations that many disintegrated elements of postural control reactions can be elicited in decerebrate, thalamic,
and decorticate animals, led Brookhart to the following proposition: In this experimental situation, the role played by
conscious volition is goal definition. The training establishes
the desirability of maintaining a stable posture and the most

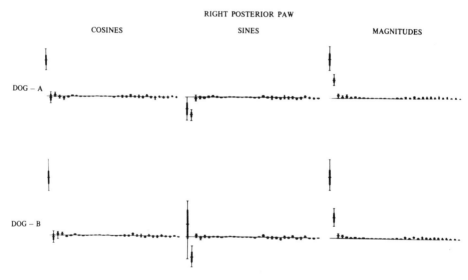

Figure 16. Comparison of patterns of force change on the right posterior foot exhibited by two
dogs. Data summarizations from 30 trials using 2.0 Hz at 80 mm. are shown in two forms. The
cosine and sine Fourier coefficients are shown to the left and in the center. These values have
been used to compute the magnitude of the power function shown to the right. Note that the
patterns of the cosine coefficients and the power magnitudes are essentially the same for the
two animals. Except for the fundamental, where the difference in sign is related to a small
phase difference, the patterns of the sine coefficients also are similar. These kinds of measures
can thus be used to characterize, by a relatively small number of values, the qualities of the
responses to postural disturbance and the degree of uniformity in time and between animals.
[Brookhart]

successful way to achieve that goal in the face of familiar disturbances. Once this is achieved, the conscious volitional mechanisms operate only to generate the analog of a reference, and the reflex mechanisms are programmed to minimize deviations from the reference. Thus, conscious volition is involved in the operation of the system once the training is acquired, but it is operative in a constant and uniform way. It need not be involved in the structuring of the detailed and specific commands that are required for the genesis of the complex behavioral reaction. The problems of pattern recognition and response are now resolved at a subconscious, reflex level of organization.

## Over What Range of Normal Movement May We Expect Henneman's "Size Principle" to Operate?

Henneman's experiments concerning the recruitment of motoneurons have demonstrated that in decerebrate preparations a variety of excitatory inputs to a pool of extensor motoneurons cause recruitment of cells in an orderly sequence beginning with the smallest alpha motoneurons. In similar preparations the order of susceptibility to inhibition also appears to be related to neuron size, in that discharge of the smallest cells is more resistant to inhibition from several afferent sources than discharge in the larger motoneurons. The data are consistent with the hypothesis that the excitability of neurons is determined by their input resistance, which is a function of cell size. In discussing these results, Henneman (1968) has stated,

> It might be supposed that the central nervous system could select out of the available supply any group of motor units it required for use according to the needs of the moment. This is not the case ... the excitabilities of motor neurons depend upon their dimensions, which range very widely. As a consequence the motor units of a muscle can be fired in only one particular order, as determined by the sizes of their motor neurons.

In the questions we directed to Henneman, we sought to get at his ideas as to the generality of this "size principle" both (1) for different sorts of muscle, and (2) for different sorts of movement (e.g., voluntary movements in intact animals). The first question was the following: What picture of functional organization of muscle emerges when one examines motoneuronal and motor unit properties over a wide range of muscles? To clarify this question, we will repeat that within a given muscle (e.g., gastrocnemius), motor units with large axons generate greater twitch tensions than motor units with small axons. When we compare motor units *between* different muscles, e.g., extraocular versus soleus, however, we find that the tiny motor units of lateral rectus are innervated by axons as large or larger than the axons innervating much larger motor units of soleus. Looking at motor units between muscles, therefore, how does Henneman view the relation between axonal and motor unit size? His reply follows:

> As studies on four different limb muscles show quite clearly, the diameters of the axons innervating a given muscle bear a relation to a number of different properties of motor units. These include factors other than the size of the unit, i.e., number of muscle fibers innervated, speed of contraction, fatigability, number of mitochondria, richness of capillary supply, enzymatic constitution, and myoglobin content. Owing to the differing sizes of the somas of the motoneurons, also correlated with axonal diameter are susceptibility to discharge (i.e., excitability), maximal frequency of discharge, and mean daily frequency of discharge. The relation between axon diameter and motor unit size is obviously one of the most important of these relations, but it is clearly not the only one, as an examination of motor units in eye muscles suggests. The very high rate of firing that occurs in motoneurons of extraocular muscles probably requires bigger channels for axoplasmic flow and larger neuromuscular junctions for secure transmission at high frequencies. I would suggest that the individual terminals of these motor axons are larger than those to other muscles and therefore that the parent axon must also be larger. As I said at the meeting, I would expect that the high rate of discharge of these neurons would shift the whole distribution pattern for these muscles to the right—i.e., to the large end of the scale.

The second question addressed to Henneman stated that his finding that motor units of a muscle are fired in a particular order raises the question as to whether this order is fixed regardless of the nature of the movement, or whether there may be some movements in which there is simultaneous, parallel activation of large and small motoneurons—or possibly even prior activation of large motoneurons. We would like to know how Henneman thinks one might go about studying the pattern of motor-unit activation during voluntary movement. In response to this question, Henneman replied,

Our observations are that recruitment follows the same small-to-large pattern for all types of gradually increasing tension, whether the movement is elicited by a stretch, an electrical stimulation of the muscle nerves, a crossed extension reflex, a flexor reflex, or an electrical stimulation of the cerebral cortex, brainstem, or cerebellum. Sudden, large inputs of any kind simultaneously discharge a large number of cells of different sizes, but the cutoff point between cells fired and those not discharged is determined by the sizes of the motoneurons. This is clear in studies of single units responding to large synchronous volleys.

Studies of the pattern of motor unit activation during voluntary movement can be carried out by means of EMG recordings. Even in recordings from muscles, the order of recruitment is surprisingly clear. However, only the smallest 10 to 20 percent of the motor-unit population can be studied with current techniques before the total amount of EMG activity becomes too great for analysis of the individual units. New techniques, perhaps involving data processing with computers, may be necessary to overcome the difficulties.

### How Might Neurophysiological Studies Demonstrate the Special Functional Role of the Cerebellum in Normal Movement?

Comparative anatomical studies show that the cerebellum originally developed in relation to the vestibular apparatus and that primary vestibular afferents project in parallel to motoneurons and to the cerebellum. Thus, vestibular inputs control motoneurons by at least two parallel pathways—one direct and (in the fish) monosynaptic or (in mammals) disynaptic via an interneuron, the other indirect and poly-synaptic via the cerebellum. Ito has been especially interested in the functional meaning of these two parallel pathways, and we therefore asked Ito to respond to the following two questions:

1. Do these two pathways simply reveal redundancy, designed in to provide functional security? Is the motoneuronal input identical from the two sources? Or is it different? If different, in what way, through what means, and to what end? We hoped that this question would lead Ito to give his ideas on the special functional increment provided by the cerebellum.

2. What specific experiments could be carried out to demonstrate the special functional role of the cerebellum in those normal movements for which it may seem to be most important?

Ito's answer to these questions follows:

The two pathways, one direct and the other indirect through the cerebellum, certainly provide much more than

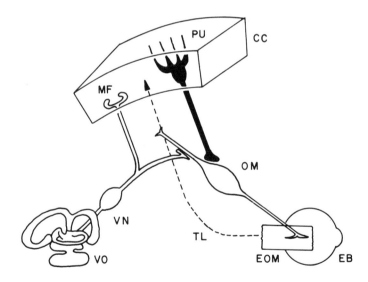

Figure 17. Vestibulo-cerebellar-ocular pathway in fish. [Based on Kidokoro, 1968, 1970]

| | |
|---|---|
| VO  vestibular organ | CC  cerebellar cortex |
| VN  primary vestibular neuron | OM  oculomotor neuron |
| MF  mossy fiber terminal | EOM  extraocular muscle |
| PU  Purkinje cell | EB  eye ball |
| TL  postulated checking line | FL  flocculo-vestibular projection |
| | SV  secondary vestibular neurons |

redundancy for functional security. These pathways differ from one another in the following characteristic way, as demonstrated in the vestibulo-ocular reflex arc. The primary vestibular impulses have an excitatory action upon their target neurons in the vestibular nuclear complex, in the oculomotor neurons of fish (but not of mammals), and in the cerebellum (via the granule cells to the Purkinje cells). After the primary vestibular impulses pass through the cerebellar pathway, however, they are inverted into inhibitory impulses by the Purkinje cells. In fish, these excitatory vestibular and inhibitory cerebellar impulses converge onto oculomotor neurons (Figure 17), and in rabbit and cat, onto vestibular nucleus cells that are involved in the vestibulo-ocular reflex (Figure 18). These cells thus receive the sum of the effects through the two pathways that should be adjusted by means of cerebellar cortical information processing to produce optimum activities in the reflex.

The special meaning of the presence of the cerebellar pathway is apparent when the vestibulo-ocular reflex is compared with the spinal reflex arc, which is not related directly to the cerebellum. In the latter, for example, the follow-up length servo composed of muscle spindle group Ia afferents and motoneurons, or the head-holding reflex

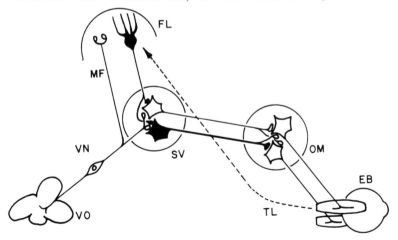

Figure 18. Vestibulo-cerebellar-ocular pathway in rabbit and cat. Notation follows Figure 17. [Based on work by Ito, Highstein, and Tsuchiya, 1970; Ito, Highstein, anf Fukuda, 1970.]

effected by vestibular organ, ventral Deiters' neurons, and neck motoneurons, are characterized by straightforward negative feedback. In contrast, the vestibulo-ocular reflex arc does not appear to have such a straightforward negative feedback loop from the position or movement of the eyeball back to the vestibular organ. Even if the vestibular efferent system provides a kind of feedback, this may not be a simple one like the one utilized in the usual negative feedback-operated control system. The construction of the vestibulo-ocular reflex arc instead appears to be a feedforward control* system, which, as discovered recently by engineers, exhibits much more flexible and subtle performance than a negative feedback-operated system. A feedforward control system can be constructed only with the aid of a computer having a learning capacity. Thus, the author's major point is to regard the cerebellum as a kind of computer having a learning capacity, as in the feedforward control system.

The experimental design now needed is to develop a precise measurement of the performance of the rabbit vestibulo-ocular reflex; roles of certain afferent or efferent pathways relevant to this reflex would be evaluated by their destruction and stimulation. An important logical deduction from the above consideration is that the cerebellar cortex should have a plasticity for learning (Ito, 1970). This learning should be carried out through a certain checking line related to the visual pathways that criticize inadequate performance of the vestibulo-ocular reflex by signaling any blurring in the visual field (see T - - L, Figures 17 and 18). The entity of this checking line will also be revealed in behavioral tests of the performance of the vestibulo-ocular reflex in combination with destruction and stimulation techniques.

Of course, I still strongly believe that neuronal circuit analysis with microelectrodes has to be advanced further in parallel with these systems-analysis types of experiments.

Ito's notion that the cerebellum may have some role in learning has been suggested by several other investigators as well. Thus, Brindley (1964) advanced the hypothesis that the role of the cerebellum is to learn and to initiate movements, and more recently Marr (1969) has suggested a mechanism whereby this learning might take place.

*For definition of feedforward control, see MacKay's statement on p. 91.

In brief, Brindley (1964) proposed the following model:

> It is therefore suggested that the message sent down from the forebrain in initiating a voluntary movement is often insufficient as instructions for all the anterior horn cells that take part in the movement even if elaborated as far as it can be according to fixed rules; it needs to be further elaborated in the cerebellum in a manner that the cerebellum learns with practise, and this further elaboration makes use of information from sense organs. The cerebellum is thus a principal agent in the learning of motor skills.

Marr (1969), using much of the information that is available on cerebellar circuitry, proposed a model that has two principal features: (1) a system of "pattern detectors" that can recognize the many possible "contexts" in which a particular movement might occur, and (2) a system for selecting which of the pattern detectors (and thus, which of the contexts) are to be used to trigger the particular movement. The pattern recognition, he suggests, is handled by the mossy fiber-Golgi cell-granule cell apparatus. Information about the context (conditions in the environment, body, brain, etc.) in which a movement might occur is brought to the cerebellum over mossy fibers. The mossy fibers that are active in relation to any one context may be many; the role of the mossy fiber-granule cell relay is to reduce the number of neurons (granule cells) active in relation to that event. A granule cell (driven by several mossy fibers) becomes active in relation to complex events; the Golgi cells, driven both by mossy fibers and by granule cells, inhibit and thus reduce the number of granule cells active in relation to a complex event. The granule cell projects to the Purkinje cell; because relatively few granule cells are active in relation to any one context, and because one Purkinje cell receives inputs from over 200,000 granule cells, the range of contexts that may affect the Purkinje cells, though limited, is very large. Marr suggests that in the naive state the granule cells are incapable of driving the Purkinje cell, but when *active* granule cell-Purkinje cell synapses are accompanied by concurrent discharge in the climbing fiber, they become capable (in the future) of driving the Purkinje cell, and remain so.

Each climbing fiber (like the Purkinje cell to which it projects) is viewed as representing an individual movement (or element of movement), and may itself be triggered in either or both of two ways:

1. The command for the movement that begins in the motor cortex and goes to the motoneurons is postulated to go also to the

inferior olive (and therefore to the climbing fiber). In activating the climbing fiber concurrent with (or just prior to) the movement, the cerebral command not only causes a movement, but also facilitates the input to the Purkinje cell of those granule cells that are active at the time and thus "teaches" the Purkinje cell to "learn" the context in which the movement has been called for. In the future, when the context has been learned, occurrence of the context alone is sufficient to fire the Purkinje cell, which then causes the next element of movement. In time, a complex series of elements of movement may be pieced together by the cerebellum, one element triggering the next in a chain reflex. The entire complex could then be initiated with a much abbreviated set of instructions from the cerebral cortex, and the function of generating a complex movement would have then been largely transferred from the cerebrum to the cerebellum.

2. The climbing fiber may be triggered by peripheral inputs (indirectly) from sensory receptors during maintained postures: a receptor activates a climbing fiber, which instructs a Purkinje cell to become active under that condition also. Discharge of the Purkinje cell causes postural change that decreases discharge of the receptor, thus stabilizing a posture by a negative feedback mechanism.

At the Work Session, Ito applied Marr's model to the vestibular oculomotor responses in the rabbit. The vestibular organs project to the vestibular nuclei, which then project to the oculomotor nuclei; this comprises a "basic" circuit whereby the eyes use vestibular information to maintain their visual fixation in space despite movements of the head. But as any fixed system may become "untuned," a "fine tuning" circuit is superimposed; the Purkinje cell, receiving from vestibular nuclear inputs, inhibits the vestibular nuclear cells. If the climbing fibers were operated by information indicating inappropriate eye movement during head movement (e.g., by visual feedback), then the Purkinje cell might learn to "tune" the basic circuit to appropriate performance by changing the vestibular input-oculomotor output relations within the vestibular nuclei.

In Ito's model the climbing fiber might serve as a comparator, detecting the discrepancies between what was intended (ocular fixation despite head turning) and what was achieved. Miller and Oscarsson (1970) have specifically suggested that the inferior olive (and thus the climbing fiber) might serve as a comparator between a command for movement and the resulting movement or the effect of the command on cells within the spinal cord. If this is so, then an "error signal" based on the comparison of command and response might be used to "teach" the cerebellum to "tune" the command-response relationship as suggested by Ito.

The model as proposed by these authors may or may not be correct; the main assumptions and conclusions to be tested are as follows: (1) whether the granule cell does serve as a complex pattern detector, firing to complex but not simple input patterns; (2) whether granule cell inputs to Purkinje cells are facilitated by conjunction with climbing fiber discharge; (3) whether climbing fiber discharge can simulate the "motor command," or be triggered solely by receptor discharges, or serve as a comparator between command and response; (4) whether movement-related neuronal activity does shift from cerebrum to cerebellum during motor learning; and (5) whether Purkinje cells do in fact play a role in initiating movement. The value of the hypothesis lies in its proposal of new ideas on function that are consistent with what is currently known of circuitry, and in giving clear-cut criteria by which it may be experimentally tested.

### How Can One Discover the Relative Functional Roles of Pyramidal Tract Inputs to Alpha as Compared to Gamma Motoneurons? C. G. Phillips

The experiments of Phillips and associates have shown the presence of a direct monosynaptic projection from pyramidal cells of the precentral gyrus of the baboon to alpha motoneurons. This direct component is especially powerful in its action on the motoneurons innervating muscles of the forearm and hand. It has also been shown that the same cortical area from which the projection to alpha motoneurons originates also gives rise to a projection to fusimotor neurons supplying the muscle spindles of the hand. Thus, motor and fusimotor projections are potentially capable of providing for a controlled alpha-gamma coexcitation during movement. Given this parallel projection from cortex to alpha and gamma motoneurons, we asked Phillips to give us his ideas on: (1) the sorts of movement (if any) in which one or the other pathway might be of especial functional significance, and (2) experimental approaches, in man or animal, to the demonstration of the role of the cerebral cortex in control of alpha as compared to gamma motoneurons during normal, voluntary movement. Phillips's response to these questions follows:

I will begin by disclaiming belief in any concept of "The Pyramidal System" as a single functional entity in any mammal, and by deliberately abstracting, from the complex totality of pyramidal tract projections in a primate,

1. The monosynaptic corticomotoneuronal (CM) projection (Bernhard et al., 1953; Preston and Whitlock, 1960, 1961), which is especially well-developed in relation to the hand (Kuypers, 1964; Landgren et al., 1962; Phillips and Porter, 1964; Clough et al., 1968); and

2. A corticofusimotor (CF) projection, which can offset the unloading of muscle spindles in cortically evoked "isometric" tetani of the finger dorsiflexor, extensor digitorum communis (EDC) (Koeze et al., 1968), but whose "offsetting" power still awaits testing when the muscle is allowed to shorten freely. The excitatory effect of the spindle feedback to EDC motoneurons is probably only marginal (Phillips, 1969), but would reinforce the CM action on the motoneuron membrane (Clough et al., 1968).

The projections thus abstracted are *structurally* the closest and densest of those from the cortex to the segmental apparatus, and *functionally* the lowest in the hierarchy of cortical levels. Evarts (1967) found that when a monkey lets go of a telegraph key in response to a flash of light and is rewarded if it does so as rapidly as possible, the reaction time of movement-related pyramidal tract cells is 100 msec. There is thus abundant time for prior activity in cortex, basal ganglia, cerebellum (Evarts and Thach, 1969), thalamus, etc., upstream of this lowest cortical level.

In residual hemiparesis following strokes, the movements most severely impaired are those of the fingers (Walshe, 1963), and in monkeys the fingers are especially vulnerable to pyramidal tract lesion (Lawrence and Kuypers, 1965, 1968). Electrical activity in the perirolandic region precedes voluntary finger movement in man (Deecke et al., 1969). These considerations lead to the proposal that (see Question (2)) "experimental approaches to the demonstration of the role of the cerebral cortex in control of alpha as compared to gamma motoneurons during normal, voluntary movement" should begin with voluntary movements of individual fingers. These could be trained under visual guidance, but tested without vision.

In *isometric experiments* the rate of development, steady level, and rate of decline of torque would be prescribed and recorded. In other experiments *free movements* would be made over specified angular distances, with the rate of acceleration, steady velocity, and rate of deceleration being prescribed and recorded. Attempts could then be made to assess the roles of CM and CF activity in these performances by the following methods that have already been introduced by other workers:

A. Vallbo (1970, Figure 6) found that in isometric flexion of the human index finger, a single spindle afferent, which was silent at rest, began to discharge at the onset of contraction and ceased at the relaxation. CM and CF were therefore coactivated in this performance—"α-γ linkage" (Granit, 1968). Would such fusimotor-induced spindle feedback be found also in free shortening; i.e., would the fusimotor activity be sufficient to offset the unloading of the spindle when the muscle shortened freely? (The experiment might be possible without dislodging the microwire in the median nerve.) If so, will the spindle accelerate suddenly if the prescribed movement is unexpectedly resisted, as suggested by Matthews (1964) and found in respiratory movements by Corda et al. (1965)?

B. Movements of the fingers are severely impaired by cutting the dorsal roots (Foerster, 1927). It is desirable that whenever the opportunity presents itself, patients should be trained to execute prescribed finger movements before rhizotomy and tested again afterwards. (If, as seems probable, the movements are then impossible, the relative importance of spindle as compared to skin and joint afferents will remain unknown, see C.)

C. Reversible denervation of skin and joints of fingers is possible, leaving intact the spindle innervation of the long flexors and extensors (Merton, 1964). Under these conditions a subject was still able to move his thumb through a prescribed angle. He believed he had moved when, in fact, his thumb had been restrained by the examiner. Responses of single spindle afferents could be investigated under these conditions (see A, above). Is the output of force increased (load-compensation reflex) when the movement is resisted? Is there acceleration of spindles? (The illusion of movement might be associated with CF-induced spindle feedback. Eklund and Hagbarth (1966) reported that there was sometimes, during tonic vibration reflexes, an illusion of the movement that would have taken place if the muscle had been allowed to shorten.)

Regarding Question (1), "the sorts of movement (if any) in which one or the other pathway might be of especial functional significance," CF activation might be important in performances requiring steady isometric force, in performances requiring steady velocities, or in performances of both kinds.

An experiment that would discover the role of fusimotor activity during normal voluntary movement is obvious in principle but impossible of execution. Human subjects would be trained to extend a finger through a prescribed angle in a prescribed time, under varied conditions of load, and, ultimately, without visual aid. Graphic recording would measure the onset, acceleration, and arrest of the movement. One would then block the gamma axons reversibly during alternate performances of the movement. The alternate performances of the (unblocked) controls should show by their identity that identical command signals were being projected to the alpha and gamma motoneurons by the brain. The records would be searched for quantitative differences between the normal and the blocked condition.

The ideal comparison being unattainable, something might be done with more prolonged blockade by injection of procaine (Rushworth, 1960). Strict controls (Matthews and Rushworth, 1957) would be needed to prove that the alpha motor fibers and the spindle afferents were unaffected, and such controls would be difficult to apply in man. A single spindle afferent could, however, be recorded by the method of Hagbarth and Vallbo (Vallbo, 1970) and used to test the degree of nerve block in the manner of Matthews and Rushworth (1958).

## Comment

In the five questions and answers that have been dealt with in this chapter, we get some idea of the range of conceptual and technological approaches which will be applied to research on central control of movement in coming years. The proposals of these five Work Session participants do indeed span a great range—from Frank's plans for use of CNS activity to control prostheses all the way to Ito's ideas on the possibility that the cerebellar cortex may have a role in learning. As the Work Session ended, all participants could at least agree that many different approaches could be fruitfully and simultaneously applied to the problems in which they shared a common interest.

## List of Abbreviations

| | |
|---|---|
| CF | corticofusimotor |
| CM | corticomotoneuronal |
| CNS | central nervous system |
| Cort. Sp. Tr. | corticospinal tract |
| DSCT | dorsal spinocerebellar tract |
| EDC | extensor digitorum communis |
| EMG | electromyogram |
| EPSP | excitatory postsynaptic potential |
| FRA | flexor reflex afferents |
| MLF | medial longitudinal fasciculus |
| MN | motoneuron |
| Ret. Sp. Tr. | reticular spinal tract |
| Rub. Sp. Tr. | rubrospinal tract |
| VA | anteroventralis thalamus |
| Vest. Sp. Tr. | vestibulospinal tract |
| VL | ventralis lateralis thalamus |
| VPL | ventralis lateralis posterior thalamus |

# BIBLIOGRAPHY

This bibliography contains two types of entries: (1) citations given or work alluded to in the report, and (2) additional references to pertinent literature by conference participants and others. Citations in group (1) may be found in the text on the pages listed in the right-hand column.

Page

Adkins, R.J., Morse, R.W., and Towe, A.L. (1966): Control of somatosensory input     41,110
by cerebral cortex. *Science* 153:1020-1022.

Adrian, E.D. (1931): Potential changes in the isolated nervous system of *Dytiscus*     14
*marginalis. J. Physiol.* 72:132-151.

Amassian, V.E., Rosenblum, M., and Weiner, H. (1969): Thalamocortical systems     53,56
related to contact placing of forelimb of cat. *Fed. Proc.* 28:455.

Amassian, V.E., Rosenblum, M., and Weiner, H. (1970): Role of thalamic n.     53,56
ventralis lateralis and its cerebellar input in contact placing. *Fed. Proc.* 29:792.

Andén, N.E., Jukes, M.G., Lundberg, A., and Vyklický, L. (1964): A new spinal
flexor reflex. *Nature* 202:1344-1345.

Appelberg, B., Bessou, P., and Laporte, Y. (1966): Action of static and dynamic     69
fusimotor fibres on secondary endings of cat's spindles. *J. Physiol.* 185:160-171.

Appelberg, B. and Emonet-Dénand, F. (1967): Motor units of the first superficial     65
lumbrical muscle of the cat. *J. Neurophysiol.* 30:154-160.

Appelberg, B. and Jeneskog, T. (1969): A dorso-lateral spinal pathway mediating     70
information from the mesencephalon to dynamic fusimotor neurones. *Acta
Physiol. Scand.* 77:159-171.

Armstrong, D.M. and Harvey, R.J. (1966): Responses in the inferior olive to     81
stimulation of the cerebellar and cerebral cortices in the cat. *J. Physiol.*
187:553-574.

Asanuma, H. and Sakata, H. (1967): Functional organization of a cortical efferent     101
system examined with focal depth stimulation in cats. *J. Neurophysiol.*
30:35-54.

Asanuma, H., Stoney, S.D., Jr., and Abzug, C. (1968): Relationship between     51,52,
afferent input and motor outflow in cat motorsensory cortex. *J. Neurophysiol.*     53,101
31:670-681.

Ashworth, B., Grimby, L., and Kugelharg, E. (1967): Comparison of voluntary and     72
reflex activation of motor units. Functional organization of motor neurones. *J.
Neurol. Neurosurg. Psychiat.* 30:91-98.

Page

Bard, P. (1933): Studies on the cerebral cortex. I. Localized control of placing and          53
hopping reactions in the cat and their normal management by small cortical
remnants. *Arch. Neurol. Psychiat.* 30:40-74.

Barron, D.H. (1934): The results of unilateral pyramidal section in the rat. *J.*             42
*Comp. Neurol.* 60:45-56.

Becker, W. and Fuchs, A.F. (1969): Further properties of the human saccadic                   37
system: eye movements and correction saccades with and without visual
fixation points. *Vision Res.* 9:1247-1259.

Bergmans, J., Burke, R., and Lundberg, A. (1969): Inhibition of transmission in the
recurrent inhibitory pathway to motoneurones. *Brain Res.* 13:600-602.

Bergmans, J. and Grillner, S. (1969): Reciprocal control of spontaneous activity            69,
and reflex effects in static and dynamic flexor $\gamma$-motoneurones revealed by an          73
injection of DOPA. *Acta Physiol. Scand.* 74:629-636.

Bernhard, C.G., Bohm, E., and Petersén, I. (1953): Investigations on the organiza-           136
tion of the cortico-spinal system in monkeys *(Macaca Mulatta)*. *Acta Physiol.*
*Scand.* 29:79-105.

Bickel, A. (1897): Uber den Einfluss der sensibelen Nerven und der Labyrinthe auf            24,
die Bewegung der Tiere. *Pflügers Arch. Ges. Physiol.* 67:299-344.                            27

Bilodeau, I. (1966): Information feedback. *In: Acquisition of Skill,* Bilodeau, E.A.,        104
ed. New York: Academic Press, pp. 255-296.

Bizzi, E. (1966): Changes in the orthodromic and antidromic response of optic                 37
tract during the eye movements of sleep. *J. Neurophysiol.* 29:861-870.

Bizzi, E. (1968): Discharge of frontal eye field neurons during saccadic and                  57
following key movements in unanesthetized monkeys. *Exp. Brain Res.* 6:69-80.

Bizzi, E. and Schiller, P.H. (1970): Single unit activity in the frontal eye fields of        57
unanesthetized monkeys during eye and head movement. *Exp. Brain Res.*
10:151-158.

Bossom, J. and Ommaya, A.K. (1968): Visuo-motor adaptation (to prismatic                      28
transformation of the retinal image) in monkeys with bilateral dorsal rhizotomy.
*Brain* 91:161-172.

Brindley, G.S. (1964): The use made by the cerebellum of the information that it             132
receives from sense organs. *IBRO Bull.* 3(3):80.

Brindley, G.S. and Lewin, W.S. (1968): The sensations produced by electrical                 118
stimulation of the visual cortex. *J. Physiol.* 196:479-493.

Bromiley, R.B. and Brooks, C.M. (1940): Role of neocortex in regulating postural              42
reactions of the opossum *(Didelphis virginiana)*. *J. Neurophysiol.* 3:339-346.

Page

Bronk, D.W. and Bullock, T.H. (1970): Reliability in neurons. *J. Gen. Physiol.* 55:563-584.

Brookhart, J.M., Parmeggiani, P.L., Petersen, W.A., and Stone, S.A. (1965): Postural stability in the dog. *Amer. J. Physiol.* 208:1047-1057.
83,118,
122

Brookhart, J.M., Mori, S., and Reynolds, P.J. (1969): Digital afferent contributions to postural reactions. *Fed. Proc.* 28:713. (Abstr.)

Brookhart, J.M., Mori, S., and Reynolds, P. (1970): Postural reactions to two directions of displacement in dogs. *Amer. J. Physiol.* 218:719-725.
118,
123

Brooks, V.B. (1969): Information processing in the motorsensory cortex. *In: Information Processing in the Nervous System.* Leibovic, K.N., ed. New York: Springer-Verlag, pp. 231-243.
53

Brooks, V.B., Horvath, F., Atkin, A., Kozlovskaya-Avdeyeva, I., and Uno, M. (1969): Reversible changes in voluntary movement during cooling of a sub-cerebellar nucleus. *Fed. Proc.* 28:780. (Abstr.)
52

Brooks, V.B., Jasper, H.H., Patton, H.D., Purpura, D.P., and Brookhart, J.M. (1970): Symposium on cerebral and cerebellar motor control. *Brain Res.* 17:539-552.

Brooks, V.B. and Stoney, S.D., Jr. (1971): Motor mechanisms: The role of the pyramidal system in motor control. *Ann. Rev. Physiol.* 33:337-392.

Brown, M.C. and Matthews, P.B.C. (1960): An investigation into the possible existence of polyneuronal innervation of individual skeletal muscle fibres in certain hind-limb muscles of the cat. *J. Physiol.* 151:436-457.
64

Brown, T.G. (1911a): The intrinsic factors in the act of progression in the mammal. *Proc. Roy. Soc. B.* 84:308-319.
27

Brown, T.G. (1911b): Studies in the physiology of the nervous system. IX. Reflex terminal phenomena—rebound—rhythmic rebound and movements of progression. *Q. J. Exp. Physiol.* 4:331-397.
27

Brown, T.G. (1914): On the nature of the fundamental activity of the nervous centres; together with an analysis of the conditioning of rhythmic activity in progression, and a theory of the evolution of function in the nervous system. *J. Physiol.* 48:18-46.
27

Bruckmoser, P., Hepp-Reymond, M.-C., and Wiesendanger, M. (1970): Cortical influence on single neurons of the lateral reticular nucleus of the cat. *Exp. Neurol.* 26:239-252.
103

Bruggencate, G. ten, Burke, R., Lundberg, A., and Udo, M. (1969): Interaction between the vestibulospinal tract, contralateral flexor reflex afferents and Ia afferents. *Brain Res.* 14:529-532.
75

Page

Bucy, P.C. (1957): Is there a pyramidal tract? *Brain* 80:376-392.                        43

Bucy, P.C., Keplinger, J.E., and Siqueira, E.B. (1964): Destruction of the "pyrami-    43
dal tract" in man. *J. Neurosurg.* 21:385-398.

Bucy, P.C., Ladpli, R., and Ehrlich, A. (1966): Destruction of the pyramidal tract     42
in the monkey. The effects of bilateral section of the cerebral peduncles. *J.
Neurosurg.* 25:1-23.

Buechner, H.K. (1963): Territoriality as a behavioral adaption to environment in       45
the Uganda kob. *Proc. XVI Int. Cong. Zool.* 3:59-62.

Bullock, T.H. (1961): The origins of patterned nervous discharge. *Behaviour*        10,12,
17:48-59.                                                                            14,16,31

Bullock, T.H. (1962): Integration and rhythmicity in neural systems. *Amer. Zool.*
2:97-104.

Bullock, T.H. (1965): Functional integration in nervous systems. Presented at the
International Symposium in the Life Sciences; Massachusetts Institute of
Technology, December 2-3.

Bullock, T.H. (1965): Physiological bases of behavior. *In: Ideas in Modern Biology,
Vol. 6 (XVI International Congress of Zoology).* Moore, J.A., ed. New York:
National History Press, pp. 452-482.

Bullock, T.H. (1967): Signals and neuronal coding. *In: The Neurosciences: A Study
Program.* Quarton, G.C., Melnechuk, T., and Schmitt, F.O., eds. New York: The
Rockefeller University Press, pp. 347-352.

Bullock, T.H. (1969): Species differences in effect of electroreceptor input on
electric organ pacemakers and other aspects of behavior in electric fish. *Brain
Behav. Evol.* 2:85-118.

Bullock, T.H. (1970): Operations analysis of nervous functions. *In: The Neurosciences: Second Study Program.* Schmitt, F.O., editor-in-chief. New York:
Rockefeller University Press, pp. 375-383.

Bullock, T.H. and Horridge, G.A. (1965): *Structure and Function in the Nervous*      19
*Systems of Invertebrates, Vol. I.* San Francisco: W.H. Freeman and Co.

Burke, R.E. (1967a): The composite nature of the monosynaptic excitatory            64,65,
postsynaptic potential. *J. Neurophysiol.* 30:1114-1137.                              66

Burke, R.E. (1967b): Motor unit types of cat triceps surae muscle. *J. Physiol.*        65
193:141-160.

Burke, R.E. (1968a): Firing patterns of gastrocnemius motor units in the decere-       65,
brate cat. *J. Physiol.* 196:631-654.                                                  66

Page

Burke, R.E. (1968b): Group Ia synaptic input to fast and slow twitch motor units          65,66,
of cat triceps surae. *J. Physiol.* 196:605-630.          68,72

Burke, R.E., Jankowska, E., and Bruggencate, G. ten, (1970): A comparison of          66
peripheral and rubrospinal synaptic input to slow and fast twitch motor units of
triceps surae. *J. Physiol.* 207:709-732.

Buxton, D.F. and Goodman, D.C. (1967): Motor function and the corticospinal          42
tracts in the dog and raccoon. *J. Comp. Neurol.* 129:341-360.

Carpenter, D.O. and Henneman, E. (1966): A relation between the threshold of
stretch receptors in skeletal muscle and the diameter of their axons. *J.
Neurophysiol.* 29:353-368.

Casey, K.L. and Towe, A.L. (1961): Cerebellar influence on pyramidal tract
neurones. *J. Physiol.* 158:399-410.

Chang, H.-T. (1950): The repetitive discharges of corticothalamic reverberating          97
circuit. *J. Neurophysiol.* 13:235-257.

Chang, H.-T. (1955): Activation of internuncial neurons through collaterals of          106
pyramidal fibers at cortical level. *J. Neurophysiol.* 18:452-471.

Chow, K.L. and Leiman, A.L. (1970): The structural and functional organization of
the neocortex. *Neurosciences Res. Prog. Bull.* 8(2):153-220.

Clough, J.F.M., Kernell, D., and Phillips, C.G. (1968): The distribution of          53,
monosynaptic excitation from the pyramidal tract and from primary spindle          54,
afferents to motoneurones of the baboon's hand and forearm. *J. Physiol.*          136
198:145-166.

Coghill, E.G. (1924): Correlated anatomical and physiological studies of the growth          15
of the nervous system of amphibia. IV. Rates of proliferation and differentiation in the central nervous system of Amblystoma. *J. Comp. Neurol.* 37:71-109.

Corda, M., Eklund, G., and von Euler, C. (1965): External intercostal and phrenic          137
alpha-motor responses to changes in respiratory load. *Acta Physiol. Scand.*
63:391-400.

Corda, M., von Euler, C., and Lennerstrand, G. (1966): Reflex and cerebellar          71
influences on alpha and on 'rhythmic' and 'tonic' gamma activity in the
intercostal muscle. *J. Physiol.* 184:898-923.

Creed, R.S., Denny-Brown, D., Eccles, J.C., Liddell, E.G.T., and Sherrington, C.S.          64,
(1932): *Reflex Activity of the Spinal Cord.* Oxford: Clarendon Press.          98

Davis, W.J. (1968): Quantitative analysis of swimmeret beating in the lobster. *J.*          23
*Exp. Biol.* 48:643-662.

Page

Davis, W.J. (1969a): The neural control of swimmeret beating in the lobster. *J. Exp. Biol.* 50:99-117.       23

Davis, W.J. (1969b): Reflex organization in the swimmeret system of the lobster. I. Intrasegmental reflexes. *J. Exp. Biol.* 51:547-563.       23

Davis, W.J. (1969c): Reflex organization in the swimmeret system of the lobster. II. Reflex dynamics. *J. Exp. Biol.* 51:565-573.       23

Deecke, L., Scheid, P., and Kornhuber, H.H. (1969): Distribution of readiness potential, pre-motion positivity, and motor potential of the human cerebral cortex preceding voluntary finger movements. *Exp. Brain Res.* 7:158-168.       136

Denny-Brown, D. (1950): Disintegration of motor function resulting from cerebral lesions. *J. Nerv. Ment. Dis.* 112:1-45.       53

Denny-Brown, D. (1956): Positive and negative aspects of cerebral functions. *North Carolina Med. J.* 17:295-303.       54

Denny-Brown, D. (1960): Motor mechanisms–introduction: the general principles of motor integration. *In: Handbook of Physiology: Neurophysiology, Vol. II.* Magoun, H.W., ed. Washington, D.C.: American Physiological Society, pp. 781-796.       53, 54

Ditchburn, R.W. and Ginsborg, B.L. (1952): Vision with a stabilized retinal image. *Nature* 170:36-37.       90

Dorf, R.C. (1967): *Modern Control Systems.* Reading, Mass.: Addison-Wesley Publishing Co.       86, 87

Doty, R.W. (1967): Neural organization of deglutition. *In: Handbook of Physiology, Alimentary Canal, Vol. IV.* Code, C.F. and Prosser, C.L. eds. Washington, D.C.: American Physiological Society, pp. 1861-1902.       18, 19

Doty, R.W. and Bosma, J.F. (1956): An electromyographic analysis of reflex deglutition. *J. Neurophysiol.* 19:44-60.       18

Doty, R.W., Richmond, W.H., and Storey, A.T. (1967): Effect of medullary lesions on coordination of deglutition. *Exp. Neurol.* 17:91-106.       18

Eccles, J.C. (1964): *The Physiology of Synapses.* New York: Academic Press.       73,74

Eccles, J.C., Eccles, R.M., and Lundberg, A. (1957): The convergence of mono-synaptic excitatory afferents on to many different species of alpha moto-neurones. *J. Physiol.* 137:22-50.       65

Eccles, R.M. and Lundberg, A. (1958a): Integrative patterns of Ia synaptic actions on motoneurones of hip and knee muscles. *J. Physiol.* 144:271-298.       65, 83

Eccles, R.M. and Lundberg, A. (1958b): The synaptic linkage of 'direct' inhibition. *Acta Physiol. Scand.* 43:204-215.       67, 75

Page

Eccles, R.M. and Lundberg, A. (1959): Synaptic actions in motoneurones by          73
afferents which may evoke the flexion reflex. *Arch. Ital. Biol.* 97:199-221.

Eccles, J.C., Ito, M., and Szentágothai, J. (1967): *The Cerebellum as a Neuronal*   81
*Machine.* Berlin: Springer-Verlag.

Eccles, R.M., Phillips, C.G. and Wu, C.-P. (1968a): Motor innervation, motor unit    65
organization and afferent innervation of M. extensor digitorum communis of the
baboon's forearm. *J. Physiol.* 198:179-192.

Eccles, J.C., Provini, L., Strata, P., and Táboríková H. (1968b): Analysis of        73
electrical potentials evoked in the cerebellar anterior lobe by stimulation of
hindlimb and forelimb nerves. *Exp. Brain Res.* 6:171-194.

Edström, L. and Kugelberg, E. (1968): Histochemical composition, distribution of     64
fibres and fatiguability of single motor units, anterior tibial muscle of the rat. *J.*
*Neurol. Neurosurg. Psychiat.* 31:424-433.

Egger, M.D. and Wyman, R.J. (1969): A reappraisal of reflex stepping in the cat. *J.*   27
*Physiol.* 202:501-516.

Eide, E., Fedina, L., Jansen, J., Lundberg, A., and Vyklický, L. (1969): Unitary     79
components in the activation of Clarke's column neurones. *Acta Physiol. Scand.*
77:145-158.

Ekblom, B. and Lundberg, A. (1968): Effect of physical training on adolescents
with severe motor handicaps. *Acta Paediat. Scand.* 57:17-23.

Eklund, G. and Hagbarth, K.E. (1966): Normal variability of tonic vibration
reflexes in man. *Exp. Neurol.* 16:80-92.

Eldred, E. (1960): Posture and locomotion: *In: Handbook of Physiology: Neuro-*      71,
*physiology, Vol. II.* Magoun, H.W., ed. Washington, D.C.: American Physiologi-       83
cal Society, pp. 1067-1088.

Engberg, I. and Lundberg, A. (1969): An electromyographic analysis of muscular       28,
activity in the hindlimb of the cat during unrestrained locomotion. *Acta*           83
*Physiol. Scand.* 75:614-630.

Evarts, E.V. (1966): Pyramidal tract activity associated with a conditioned hand     136
movement in the monkey. *J. Neurophysiol.* 29:1011-1027.

Evarts, E.V. (1967): Representation of movements and muscles by pyramidal tract
neurons of the precentral motor cortex. *In: Neurophysiological Basis of Normal*
*and Abnormal Motor Activities.* Yahr, M.D. and Purpura, D.P., eds. New York:
Raven Press, pp. 215-251.

Evarts, E.V. (1968): Relation of pyramidal tract activity to force exerted during    49
voluntary movement. *J. Neurophysiol.* 31:14-27.

Evarts, E.V. (1969): Activity of pyramidal tract neurons during postural fixation. *J.*  49
*Neurophysiol.* 32:375-385.

Page

Evarts, E.V. (1970): Activity of ventralis lateralis neurons prior to movement in the monkey. *Physiologist* 13:191.

Evarts, E.V. and Thach, W.T. (1969): Motor mechanisms of the CNS: cerebrocere-     136
bellar interrelations. *Ann. Rev. Physiol.* 31:451-498.

Evarts, E.V. and Thach, W.T. (1969): Motor mechanisms of the CNS: cerebrocere-     136
bellar interrelations. *In: Annual Review of Physiology.* Hall, V.E., Giese, A.C.,
and Sonnenschein, R.R., eds. Palo Alto: Annual Reviews, Inc., pp. 451-498.

Evoy, W.H. and Kennedy, D. (1967): The central nervous organization underlying control of antagonistic muscles in the crayfish. I. Types of command fibers. *J. Exp. Zool.* 165:223-238.

Evoy, W.H., Kennedy, D., and Wilson, D.M. (1967): Discharge patterns of neurones supplying tonic abdominal flexor muscles in the crayfish. *J. Exp. Biol.* 46:393-411.

Fidone, S.J. and Preston, J.B. (1969): Patterns of motor cortex control of flexor     71
and extensor cat fusimotor neurons. *J. Neurophysiol.* 32:103-115.

Fields, H.L., Evoy, W.H., and Kennedy, D. (1967): Reflex role played by efferent     18
control of an invertebrate stretch receptor. *J. Neurophysiol.* 30:859-874.

Fillenz, M. (1955): Responses in the brainstem of the cat to stretch of extrinsic     38
ocular muscles. *J. Physiol.* 128:182-199.

Foerster, O. (1927): Schlaffe und spastische Lähmung. *In: Handbuch der normalen*     137
*und pathologischen Physiologie.* Bethe, A., von Bergmann, G., Embden, G., and
Ellinger, A., eds. Berlin: Springer-Verlag, pp. 900-901.

Frank, K. (1968): Some approaches to the technical problem of chronic excitation     114
of peripheral nerve. *Ann. Otol.* 77:761-771.

Franz, D.N. and Iggo, A. (1968): Dorsal root potentials and ventral root reflexes     100
evoked by nonmyelinated fibers. *Science* 162:1140-1142.

Fuchs, A.F. and Kornhuber, H.H. (1969): Extraocular muscle afferents to the     38
cerebellum of the cat. *J. Physiol.* 200:713-722.

Goodman, L.J. (1965): The role of certain optomotor reactions in regulating     21
stability in the rolling plane during flight in the desert locust, *Schistocerca*
*gregaria. J. Exp. Biol.* 42:385-407.

Grampp, W. and Oscarsson, O. (1968): Inhibitory neurons in the group I projection area of the cat's cerebral cortex. *In: Structure and Function of Inhibitory Neuronal Mechanisms.* von Euler, C., Skoglund, S., and Söderberg, U., eds. Oxford: Pergamon Press, pp. 351-356.

Granit, R. (1955): *Receptors and Sensory Perception.* New Haven: Yale University     68
Press.

Page

Granit, R. (1968): The functional role of the muscle spindle's primary end organs.          68,
*Proc. Roy. Soc. Med.* 61:69-78.          137

Granit, R. (1970): *The Basis of Motor Control.* New York: Academic Press.

Gray, J. (1933a): Studies in animal locomotion. I. The movement of fish with          26
special reference to the eel. *J. Exp. Biol.* 10:88-104.

Gray, J. (1933b): Studies in animal locomotion. II. The relationship between waves          26
of muscular contraction and the propulsive mechanism of the eel. *J. Exp. Biol.*
10:386-390.

Gray, J. (1933c): Studies in animal locomotion. III. The propulsive mechanism of          26
the whiting (*Gadus merlangus*). *J. Exp. Biol.* 10:391-400.

Gray, J. (1939): Aspects of animal locomotion. *Proc. Roy. Soc. B.* 128:28-61.          24

Gray, J. and Lissmann, H.W. (1940): The effect of deafferentation upon the          24
locomotor activity of amphibian limbs. *J. Exp. Biol.* 17:227-236.

Gray, J. and Lissmann, H.W. (1946a): The co-ordination of limb movements in the          24
amphibia. *J. Exp. Biol.* 23:133-142.

Gray, J. and Lissmann, H.W. (1946b): Further observations on the effect of          24,
deafferentation on the locomotory activity of amphibian limbs. *J. Exp. Biol.*          45
23:121-132.

Gray, J., and Lissmann, H.W. (1947): The effect of labyrinthectomy on the          45
co-ordination of limb movements in the toad. *J. Exp. Biol.* 24:36-40.

Grillner, S. (1969a): The influence of DOPA on the static and the dynamic          69
fusimotor activity to the triceps surae of the spinal cat. *Acta Physiol. Scand.*
77:490-509.

Grillner, S. (1969b): Supraspinal and segmental control of static and dynamic          67,68,
γ-motoneurones in the cat. *Acta Physiol. Scand. Suppl.* 327, pp. 1-34.          72,73

Grillner, S., Hongo, T., and Lund, S. (1966): Interaction between the inhibitory          75
pathways from the Deiters nucleus and Ia afferents to flexor motoneurones.
*Acta Physiol. Scand. Suppl.* 277, p. 61.

Grillner, S., Hongo, T., and Lund, S. (1968): Reciprocal effects between two          67,
descending bulbospinal systems with monosynaptic connections to spinal          76
motoneurones. *Brain Res.* 10:477-480.

Grillner, S., Hongo, T., and Lund, S. (1969): Descending monosynaptic and reflex          70
control of γ-motoneurones. *Acta Physiol. Scand.* 75:592-613.

Grillner, S. and Lund, S. (1968): The origin of a descending pathway with          67
monosynaptic action on flexor motoneurones. *Acta Physiol. Scand.*
74:274-284.

Page

Guiot, G. and Pecker, J. (1949): Traitement du tremblement parkinsonien par la          43
pyramidotomie pedonculaire. *Sem. Hop. Paris* 25:2620-2624.

Hagiwara, S. (1961): Nervous activities of the heart in Crustacea. *Ergebn. Biol.*      11,
24:287-311.                                                                             14

Hagiwara, S. and Watanabe, A. (1956): Discharges in motoneurons of cicada. *J.*         17
*Cell. Comp. Physiol.* 47:415-428.

Hall-Craggs, J. (1962): The development of song in the blackbird, *Turdus merula.*
*Ibis* 104:277-300.

Harris, F., Jabbur, S.J., Morse, R.W., and Towe, A.L. (1965): The influence of the
cerebral cortex on the cuneate nucleus of the monkey. *Nature* 208:1215-1216.

Hassler, R. (1966): Thalamic regulation of muscle tone and the speed of move-           97
ments. *In: The Thalamus.* Purpura, D.P. and Yahr, M.D., eds. New York:
Columbia University Press, pp. 419-438.

Heimer, L., Ebner, F.F., and Nauta, W.J.H. (1967): A note on the termination of
commissural fibers in the neocortex. *Brain Res.* 5:171-177.

Held, R. (1961): Exposure-history as a factor in maintaining stability of perception    108
and coordination. *J. Nerv. Ment. Dis.* 132:26-32.

Heinroth, O. (1910): Beiträge zur Biologie, insbesonders Psychologie und Ethologie      44
der Anatiden. *Verh. V. Int. Ornithol. Kongr.*

Helmholtz, H.L.F. von (1867): *Handbuch der physiologischen Optik. Vol. III.*           108
Leipzig: Leopold Voss.

Henneman, E. (1968): Peripheral mechanisms involved in the control of muscle. *In:*     127
*Medical Physiology, Vol. II.* Mountcastle, V.B., ed. St. Louis: C.V. Mosby Co.,
pp. 1697-1716.

Henneman, E., Somjen, G., and Carpenter, D.O. (1965a): Excitability and inhibit-        64
ability of motoneurons of different sizes. *J. Neurophysiol.* 28:599-620.

Henneman, E., Somjen, G., and Carpenter, D.O. (1965b): Functional significance          64,65,
of cell size in spinal motoneurons. *J. Neurophysiol.* 28:560-580.                      72

Hering, H.W. (1893): Ueber die nach Durchschneidung der hinteren Wurzeln               24
auftretende Bewegungslosigkeit des Ruckenmarkfrosches. *Pflügers Arch. Ges.*
*Physiol.* 54:614-640.

Higgins, J.R. and Angel, R.W. (1970): Correction of tracking errors without sensory     111
feedback. *J. Exp. Psychol.* 84:412-416.

Hinde, R.A. (1958): Alternative motor patterns in chaffinch song. *Anim. Behav.*
6:211-218.

Page

Hodos, W. and Karten, H.J. (1966): Brightness and pattern discrimination deficits in the pigeon after lesions of nucleus rotundus. *Exp. Brain Res.* 2:151-167.

Holmqvist, B. and Lundberg, A. (1961): Differential supraspinal control of synaptic actions evoked by volleys in the flexion reflex afferents in alpha motoneurones. *Acta Physiol. Scand. Suppl.* 186, pp. 1-51.

Hongo, T., Jankowska, E., and Lundberg, A. (1969a): The rubrospinal tract. I. Effects on alpha-motoneurones innervating hindlimb muscles in cats. *Exp. Brain Res.* 7:344-364.

Hongo, T., Jankowska, E., and Lundberg, A. (1969b): The rubrospinal tract. II. Facilitation of interneuronal transmission in reflex paths to motoneurones. *Exp. Brain Res.* 7:365-391.

Horridge, G.A. and Burrows, M. (1968): Efferent copy and voluntary eyecup movement in the crab, *Carcinus. J. Exp. Biol.* 49:315-324.

Houk, J. and Henneman, E. (1967): Responses of Golgi tendon organs to active contractions of the soleus muscle of the cat. *J. Neurophysiol.* 30:466-481.

Hoyle, G. (1964): Exploration of neuronal mechanisms underlying behavior in insects. *In: Neural Theory and Modeling.* Reiss, R.F., ed. Stanford: Stanford University Press, pp. 346-376.

Hughes, G.M. and Wiersma, C.A.G. (1960): The co-ordination of swimmeret movements in the crayfish, *Procambarus clarkii* (Girard). *J. Exp. Biol.* 37:657-670.

Hultborn, H., Jankowska, E., and Lindström, S. (1968a). Distribution of recurrent inhibition of Ia IPSPs in motoneurones. *Acta Physiol. Scand.* 74:17A-18A.

Hultborn, H., Jankowska, E., and Lindström, S. (1968b): Recurrent inhibition from motor axon collaterals in interneurones monosynaptically activated from Ia afferents. *Brain Res.* 9:367-369.

Hultborn, H., Jankowska, E., and Lindström, S. (1968c): Recurrent inhibition of reflex transmission to motoneurones. *Acta Physiol. Scand.* 73:41A.

Humphrey, D.R., Schmidt, E., and Thompson, W.D. (1970): Patterns of activity across simultaneously observed cortical units during simple arm movements. *Fed. Proc.* 29:791. (Abstr.)

Ito, M. (1970): Neurophysiological aspects of the cerebellar motor control system. *J. Internat. Neurol.* (In press)

Ito, M., Highstein, S.M., and Fukuda, J. (1970): Cerebellar inhibition of the vestibulo-ocular reflex in rabbit and cat and its blockage by picrotoxin. *Brain Res.* 18:524-526.

Page

Ito, M., Highstein, S.M., and Tsuchiya, T. (1970): The postsynaptic inhibition of        131
rabbit oculomotor neurones by secondary vestibular impulses and its blockage
by picrotoxin. *Brain Res.* 17:520-523.

Ito, M., Hongo, T., and Okada, Y. (1969): Vestibular-evoked postsynaptic potentials in Deiters neurones. *Exp. Brain Res.* 7:214-230.

Ito, M., Hongo, M., Okada, Y., and Obata, K. (1964): Antidromic and transsynaptic activation of Deiters' neurones induced from the spinal cord. *Jap. J.
Physiol.* 14:638-658.

Ito, M., Kawai, N., Udo, M., and Mano, N. (1969): Axon reflex activation of
Deiters neurones from the cerebellar cortex through collaterals of the cerebellar
afferents. *Exp. Brain Res.* 8:249-268.

Ito, M., Kawai, N., and Udo, M. (1968): The origin of cerebellar-induced inhibition
of Deiters neurones. III. Localization of the inhibitory zone. *Exp. Brain Res.*
4:310-320.

Ito, M., Kawai, N., Udo, M., and Sato, N. (1968): Cerebellar-evoked disinhibition in
dorsal Deiters neurones. *Exp. Brain Res.* 6:247-264.

Ito, M., Obata, K., and Ochi, R. (1966): The origin of cerebellar-induced inhibition
of Deiters neurones. II. Temporal correlation between the trans-synaptic
activation of Purkinje cells and the inhibition of Deiters neurones. *Exp. Brain
Res.* 2:350-364.

Ito, M. and Yoshida, M. (1966): The origin of cerebellar-induced inhibition of
Deiters neurones. I. Monosynaptic initiation of the inhibitory postsynaptic
potentials. *Exp. Brain Res.* 2:330-349.

Ito, M., Yoshida, M., Obata, K., Kawai, N., and Udo, M. (1970): Inhibitory control
of intracerebellar nuclei by the Purkinje cell axons. *Exp. Brain Res.* 10:64-80.

Jabbur, S.J. and Towe, A.L. (1961): Cortical excitation of neurons in dorsal
column nuclei of cat, including an analysis of pathways. *J. Neurophysiol.*
24:499-509.

Jacobsen, C.F. (1932): Influence of motor and premotor area lesions upon the          42
retention of skilled movements in monkeys and chimpanzees. *Proc. Ass. Res.
Nerv. Ment. Dis.* 13:225-247.

Jansen, J.K.S. (1966): On fusimotor reflex activity. *In: Muscular Afferents and
Motor Control. (Nobel Symposium I).* Granit, R., ed. Stockholm: Almqvist and
Wiksell, pp. 91-105.

Jansen, J.K.S., Nicolaysen, K., and Rudjord, T. (1966): Discharge pattern of
neurons of the dorsal spinocerebellar tract activated by static extension of
primary endings of muscle spindles. *J. Neurophysiol.* 29:1061-1086.

Page

Jansen, J.K.S., Nicolaysen, K., and Walloe, L. (1967): On the inhibition of          80
transmission to the dorsal spinocerebellar tract by stretch of various ankle
muscles of the cat. *Acta Physiol. Scand.* 70:362-368.

Jansen, J.K.S. and Rudjord, T. (1965): Dorsal spinocerebellar tract: response        80
pattern of nerve fibers to muscle stretch. *Science* 149:1109-1111.

Johnston, J.B. (1913): The morphology of the septum, hippocampus, and pallial        42
commissures in reptiles and mammals. *J. Comp. Neurol.* 23:371-478.

Johnstone, J.R. and Mark, R.F. (1969): Evidence for efference copy for eye
movements in fish. *Comp. Biochem. Physiol.* 30:931-939.

Karten, H.J. (1967): The organization of the ascending auditory pathway in the
pigeon (*Columba livia*) I. Diencephalic projections of the inferior colliculus
(nucleus mesencephali lateralis, pars dorsalis) *Brain Res.* 6:409-427.

Karten, H.J. (1969): The organization of the avian telencephalon and some
speculations on the phylogeny of the amniote telencephalon. *Ann. N.Y. Acad.
Sci.* 167:164-179.

Kawai, N., Ito, M., and Nozue, M. (1969): Postsynaptic influences on the vestibular
non-Deiters nuclei from primary vestibular nerve. *Exp. Brain Res.* 8:190-200.

Kendig, J.J. (1968): Motor neurone coupling in locust flight. *J. Exp. Biol.*        22
48:389-404.

Kennard, M.A. (1940): Studies of motor performance after parietal ablations in       42
monkeys. *J. Neurophysiol.* 3:248-257.

Kennard, M.A. (1944): Reactions of monkeys of various ages to partial and            42
complete decortication. *J. Neuropath. Exp. Neurol.* 3:289-310.

Kennedy, D. (1968): Input and output connections of single arthropod neurons.        17
*In: Physiological and Biochemical Aspects of Nervous Integration.* Carlson,
F.D., ed. Englewood Cliffs, N.J.: Prentice-Hall, pp. 285-306.

Kennedy, D., Evoy, W.H., and Hanawalt, J.T. (1966a): Release of coordinated          43
behavior in crayfish by single central neurons. *Science* 154:917-919.

Kennedy, D., Evoy, W.H., and Fields, H.L. (1966b): The unit basis of some            17
crustacean reflexes. *Symp. Soc. Exp. Biol.* 20:75-109.

Kennedy, D., Evoy, W.H., Dane, B., and Hanawalt, J.T. (1967): The central nervous
organization underlying control of antagonistic muscles in the crayfish. II.
Coding of position by command fibers. *J. Exp. Zool.* 165:239-248.

Kennedy, D., Selverston, A.I., and Remler, M.P. (1969): Analysis of restricted
neural networks. *Science* 164:1488-1496.

Page

Kennedy, T.T., Grimm, R.J., and Towe, A.L. (1966): The role of cerebral cortex in evoked somatosensory activity in cat cerebellum. *Exp. Neurol.* 14:13-32.

Kernell, D. (1966): Input resistance, electrical excitability and size of ventral horn cells in cat spinal cord. *Science* 152:1637-1640.                                            64

Kidokoro, Y. (1968): Direct inhibitory innervation of teleost oculomotor neurones by cerebellar Purkinje cells. *Brain Res.* 10:453-456.                                       130

Kidokoro, Y. (1970): Cerebellar and vestibular control of fish oculomotor neurons. *In: Neurobiology of Developmental and Evolutional Aspect of Cerebellum.* Llinas, R., ed. Chicago: Amer. Med. Assn., pp. 257-276.                                130

Koeze, T.H. (1968): The independence of corticomotoneuronal and fusimotor pathways in the production of muscle contraction by motor cortex stimulation. *J. Physiol.* 197:87-105.                                                                       72

Koeze, T.H., Phillips, C.G., and Sheridan, J.D. (1968): Thresholds of cortical activation of muscle spindles and $\alpha$ motoneurones of the baboon's hand. *J. Physiol.* 195:419-449.                                                                  136

Kohler, I. (1951): *Ueber Aufbau und Wandlungen der Wahrnehmungswelt; insbesondere über "bedingte Empfindungen."* Wien: Rudolph M. Rohrer.                                   108

Konishi, M. (1963): The role of auditory feedback in the vocal behaviour of the domestic fowl. *Z. Tierpsychol.* 20:349-367.                                                    15

Konishi, M. (1965a): Effects of deafening on song development in American robins and black-headed grosbeaks. *Z. Tierpsychol.* 22:584-599.                                      43

Konishi, M. (1965b): The role of auditory feedback in the control of vocalization in the white-crowned sparrow. *Z. Tierpsychol.* 22:770-783.                                 15,43

Konishi, M. and Nottebohm, F. (1969): Experimental studies in the ontogeny of avian vocalizations. *In: Bird Vocalizations in Relation to Current Problems in Biology and Psychology.* Hinde, R.A., ed. Cambridge: Cambridge University Press, pp. 29-48.                                                                            43

Konorski, J. (1967): *Integrative Activity of the Brain.* Chicago: University of Chicago Press.                                                                         104,105

Kornmüller, A.E. (1931): Eine experimentelle Anästhesie der äusseren Augenmuskeln am Menschen und ihre Auswirkungen. *J. Psychol. Neurol.* 41:354-366.                       90,108

Kozlovskaya, I., Atkin, A., Horvath, F., Uno, M., and Brooks, V.B. (1969): Reversible movement disorders during cooling of the dentate nucleus. (Fourth International Congress of Neurological Surgery and Ninth International Congress of Neurology). *Excerpta Medica*, International Congress Series No. 193, p. 241. (Abstr.)

Page

Krnjević, K. and Miledi, R. (1958): Failure of neuromuscular propagation in rats. *J. Physiol.* 140:440-461.
64

Kuno, M. and Miyahara, J.T. (1968): Factors responsible for multiple discharge of neurons in Clarke's column. *J. Neurophysiol.* 31:624-638.
79

Kupfermann, I. and Kandel, E.R. (1969): Neuronal controls of a behavioral response mediated by the abdominal ganglion of *Aplysia. Science* 164:847-850.
20

Kuypers, H.G.J.M. (1960): Central cortical projections to motor and somato-sensory cell groups. An experimental study in the rhesus monkey. *Brain* 83:161-184.
49,107, 136

Kuypers, H.G.J.M. (1962): Corticospinal connections: postnatal development in the Rhesus monkey. *Science* 138:678-680.
49

Kuypers, H.G.J.M. (1963): The organization of the "motor system." *Int. J. Neurol.* 4:78-91.
49

Kuypers, H.G.J.M. (1964): The descending pathways to the spinal cord, their anatomy and function. *Progr. Brain Res.* 11:178-202.
49

Lade, B.I. and Thorpe, W.H. (1964): Dove songs as innately coded patterns of specific behaviour. *Nature* 202:366-368.

Landgren, S., Phillips, C.G., and Porter, R. (1962): Minimal synaptic actions of pyramidal impulses on some alpha motoneurones of the baboon's hand and forearm. *J. Physiol.* 161:91-111.
136

Landgren, S. and Silfvenius, H. (1969): Projection to cerebral cortex of group I muscle afferents from the cat's hind limb. *J. Physiol.* 200:353-372. ·
73

Lanyon, W.E. (1960): The ontogeny of vocalizations in birds. *In: Animal Sounds and Communication.* Lanyon, W.E. and Tavolga, W.N., eds. Washington, D.C.: American Institute of Biological Sciences, pp. 321-347.
43

Larimer, J.L. and Kennedy, D. (1969a): The central nervous control of complex movements in the uropods of crayfish. *J. Exp. Biol.* 51:135-150.
18

Larimer, J.L. and Kennedy, D. (1969b): Innervation patterns of fast and slow muscle in the uropods of crayfish. *J. Exp. Biol.* 51:119-133.
18

Larson, B., Miller, S., and Oscarsson, O. (1969a): A spinocerebellar climbing fibre path activated by the flexor reflex afferents from all four limbs. *J. Physiol.* 203:641-649.
100

Larson, B., Miller, S., and Oscarsson, O. (1969b): Termination and functional organization of the dorsolateral spino-olivocerebellar path. *J. Physiol.* 203:611-640.
100

Page

Lashley, K.S. (1924): Studies of cerebral function in learning. V. The retention of     42
motor habits after destruction of the so-called motor areas in primates. *Arch.
Neurol. Psychiatr.* 12:249-276.

Lassek, A.M. (1953): Inactivation of voluntary motor function following rhizot-     27
omy. *J. Neuropath. Exp. Neurol.* 12:83-87.

Latour, P.L. (1962): Visual threshold during eye movements. *Vision Res.*     37
2:261-262.

Laursen, A.M. (1966): Motion speed and reaction time after section of the     42
pyramidal tract in cats. *Bull. Schweiz. Akad. Med. Wiss.* 22:336-340.

Laursen, A.M. and Wiesendanger, M. (1966): Motor deficits after transsection of a     42
bulbar pyramid in the cat. *Acta Physiol. Scand.* 68:118-126.

Lawrence, D.G. and Kuypers, H.G.J.M. (1965): Pyramidal and non-pyramidal     136
pathways in monkeys: anatomical and functional correlation. *Science*
148:973-975.

Lawrence, D.G. and Kuypers, H.G.J.M. (1968): The functional organization of the     42,49,
motor system in the monkey. I. The effects of bilateral pyramidal lesions. *Brain*     136
91:1-14.

Lennerstrand, G. and Thoden, U. (1968a): Position and velocity sensitivity of     69
muscle spindles in the cat. II. Dynamic fusimotor single-fibre activation of
primary endings. *Acta Physiol. Scand.* 74:16-29.

Lennerstrand, G. and Thoden, U. (1968b): Position and velocity sensitivity of     69
muscle spindles in the cat. III. Static fusimotor single-fibre activation of primary
and secondary endings. *Acta Physiol. Scand.* 74:30-49.

Leopold, A. (1949): *A Sand County Almanac and Sketches Here and There.* New     45
York: Oxford University Press.

Levitt, M., Carreras, M., Liu, C.N., and Chambers, W.W. (1964): Pyramidal and     106
extrapyramidal modulation of somatosensory activity in gracile and cuneate
nuclei. *Arch. Ital. Biol.* 102:197-229.

Lewis, R.P. and Brindley, G.S. (1965): The extrapyramidal cortical motor map.     42
*Brain* 88:397-406.

Li, C.-L. (1958): Activity of interneurons in the motor cortex. *In: Reticular     100
Formation of the Brain.* Jasper, H.H., Proctor, L.D., Knighton, R.S., Noshay,
W.C., and Costello, R.T., eds. Boston: Little, Brown, pp. 459-472.

Lippold, O.J.C. (1970): Oscillation in the stretch reflex arc and the origin of the     72
rhythmical, 8-12 c/s component of physiological tremor. *J. Physiol.*
206:359-382.

Page

Lissmann, H.W. (1946a): The neurological basis of the locomotory rhythm in the        26
spinal dogfish (*Scyllium canicula, Acanthias vulgaris*). I. Reflex behaviour. *J. Exp. Biol.* 23:143-161.

Lissmann, H.W. (1964b): The neurological basis of the locomotory rhythm in the        26
spinal dogfish (*Scyllium canicula, Acanthias vulgaris*). II. De-afferentation. *J. Exp. Biol.* 23:162-176.

Lloyd, D.P.C. (1960): Spinal mechanisms involved in somatic activities. *In:*          73
*Handbook of Physiology: Neurophysiology, Vol. II.* Magoun, H.W., ed. Washington, D.C.: American Physiological Society, pp. 929-949.

Lund, S. and Pompeiano, O. (1968): Monosynaptic excitation of alpha moto-            66
neurones from supraspinal structures in the cat. *Acta Physiol. Scand.* 73:1-21.

Lundberg, A. (1959): Integrative significance of patterns of connections made by    75,80,81,
muscle afferents in the spinal cord. *XXI International Congress of Physiological*    100
*Sciences Symposia Vol. 2.* Buenos Aires, pp. 100-105.

Lundberg, A. (1964a): Ascending spinal hindlimb pathways in the cat. *Prog. Brain*   73,79,81,
*Res.* 12:135-163.                                                                   100

Lundberg, A. (1964b): Supraspinal control of transmission in reflex paths to         74,75
motoneurones and primary afferents. *Prog. Brain Res.* 12:197-221.

Lundberg, A. (1966): Integration in the reflex pathway. *In: Muscular Afferents and*  74,75,
*Motor Control.* (Nobel Symposium). Granit, R., ed. Stockholm: Almqvist and           101
Wiksell, pp. 275-305.

Lundberg, A. (1967): The supraspinal control of transmission in spinal reflex         75
pathways. *Electroenceph. Clin. Neurophysiol. (Suppl.)* 25:35-46.

Lundberg, A. (1969): Reflex control of stepping. *Proc. Norwegian Acad. Sci. Lettr.*  83
Oslo: Universitetsforlaget.

Lundberg, A. (1970): The excitatory control of the Ia inhibitory pathway. *In:*      75,76,
*Excitatory Synaptic Mechanisms.* (Proceedings of the Fifth International              77
Meeting of Neurobiologists). (In press)

Lundberg, A. and Voorhoeve, P. (1962): Effects from the pyramidal tract on spinal     76
reflex arcs. *Acta Physiol. Scand.* 56:201-219.

MacKay, D.M. (1962): Theoretical models of space perception. *In: Aspects of the*     90
*Theory of Artificial Intelligence.* Muses, C.A., ed. New York: Plenum Press, pp.
83-104.

MacKay, D.M. (1966): Cerebral organization and the conscious control of action.      89,91
*In: Brain and Conscious Experience.* Eccles, J.C., ed. New York: Springer-
Verlag, pp. 422-445.

Page

Magendie, F. (1824): *An Elementary Compendium of Physiology.* Philadelphia:        111
James Webster. (Transl. by E. Milligan)

Marler, P. (1964): Inheritance and learning in the development of animal vocalizations. *In: Acoustic Behavior of Animals.* Busnel, R.G., ed. Amsterdam: Elsevier,
p. 228.

Marler, P. (1967): Comparative study of song development in sparrows. *In:
Proceedings of the 14th International Ornithological Congress.* Snow, D.W., ed.
Oxford: Blackwell Scientific Publication, pp. 231-244.

Marler, P. (1970): A comparative approach to vocal learning: song development in        15,43
the white-crowned sparrows. *J. Comp. Physiol. Psychol.* (Monogr.) 71(2):1-25.

Marler, P. and Hamilton, W.J., III (1966): *Mechanisms of Animal Behavior.* New
York: John Wiley and Sons.

Marler, P. and Mundinger, P. (1971): Vocal learning in birds. *In: The Ontogeny of*        43
*Vertebrate Behavior.* Moltz, H., ed. New York: Academic Press. (In press)

Marler, P. and Tamura, M. (1964): Culturally transmitted patterns of vocal behavior        15,43
in sparrows. *Science* 146:1483-1486.

Marr, D. (1969): A theory of cerebellar cortex. *J. Physiol.* 202:437-470.        81,132,
                                                                                                 133
Matin, L. and Pearce, D.G. (1965): Visual perception of direction for stimuli
flashed during voluntary saccadic eye movements. *Science* 148:1485-1488.

Matthews, P.B.C. (1964): Muscle spindles and their motor control. *Physiol. Rev.*        68,69,70,
44:219-288.                                                                                       72,91,137

Matthews, P.B.C., and Rushworth, G. (1957): The relative sensitivity of muscle        138
nerve fibers to procaine. *J. Physiol.* 135:263-269.

Matthews, P.B.C., and Rushworth, G. (1958): The discharge from muscle spindles        138
as an indicator of $\gamma$ efferent paralysis by procaine. *J. Physiol.* 140:421-426.

Mayr, O. (1970): The origins of feedback control. *Sci. Amer.* 223:110-118.        86

Merton, P.A. (1953): Speculations on the servo-control of movement. *In: The*        70
*Spinal Cord.* Wolstenholme, G.E.W., ed. London: Churchill.

Merton, P.A. (1964): Human position sense and sense of effort. *Symp. Soc. Exp.*        137
*Biol.* 18:387-400.

Miller, S., Nezlina, N., and Oscarsson, O. (1969): Projection and convergence        81,103
patterns in climbing fibre paths to cerebellar anterior lobe activated from
cerebral cortex and spinal cord. *Brain Res.* 14:230-233.

Page

Miller, S. and Oscarsson, O. (1969): Termination and functional organization of     100,103
spino-olivocerebellar paths. *In: The Cerebellum in Health and Disease.* Fields,
W.S. and Willis, W.D., eds. St. Louis: Warren H. Green, Inc., pp. 172-200.

Mittelstaedt, H. (1958): The analysis of behavior in terms of control systems. *In:*     87
*Group Processes,* Transactions of the Fifth Conference, Princeton, N.J.
Schaffner, B., ed., pp. 45-84.

Mittelstaedt, H. and von Holst, E. (1953): Reafferenzprinzip und Optomotorik.     87
*Zool. Anz.* 151:253.

Mori, S. and Brookhart, J.M. (1967): Delay times in postural control. *Physiologist*
10(3):255.

Mori, S. and Brookhart, J.M. (1968a): Postural control system delays. *Fed. Proc.*     82
27:1318 (Abstr.)

Mori, S. and Brookhart, J.M. (1968b): Characteristics of the postural reactions of     118,120
the dog to a controlled disturbance. *Amer. J. Physiol.* 215:339-348.

Mori, S., Reynolds, P.J., and Brookhart, J.M. (1970): The contribution of pedal     118
afferents to postural control in the dog. *Amer. J. Physiol.* 218:726-734.

Morse, R.W., Adkins, R.J., and Towe, A.L. (1965): Population and modality
characteristics of neurons in the coronal region of somatosensory area I of the
cat. *Exp. Neurol.* 11:419-440.

Moss Rehabilitation Hospital (1968): Quarterly Progress Report, May, June, July.     114
Social and Rehabilitation Service, Krusen Center for Research and Engineering,
Philadelphia, Pennsylvania.

Mott, F.W. and Sherrington, C.S. (1895): Experiments on the influence of sensory     27
nerves upon movement and nutrition of the limbs. *Proc. Roy. Soc. B*
57:481-488.

Mountcastle, V.B. (1957): Modality and topographic properties of single neurons of
cat's somatic sensory cortex. *J. Neurophysiol.* 20:408-434.

Mountcastle, V.B. (1966): The neural replication of sensory events in the somatic
afferent system. *In:' Brain and Conscious Experience.* Eccles, J.C., ed. New
York: Springer-Verlag, pp. 85-115.

Mountcastle, V.B. (1967): The problem of sensing and the neural coding of sensory
events. *In: The Neurosciences: A Study Program.* Quarton, G.C., Melnechuk, T.,
and Schmitt, F.O., eds. New York: Rockefeller University Press, pp. 393-408.

Mountcastle, V.B., Poggio, G.F., and Werner, G. (1963): The relation of thalamic     50
cell response to peripheral stimuli varied over an intensive continuum. *J.*
*Neurophysiol.* 26:807-834.

Mountcastle, V.B., Talbot, W.H., and Kornhuber, H.H. (1966): The neural transformation of mechanical stimuli delivered to the monkey's hand. *In: Touch, Heat and Pain.* (Ciba Symposium) de Reuck, A.V.S. and Knight, J., eds. Boston: Little, Brown, pp. 325-358.

Mountcastle, V.B., Talbot, W.H., Darian-Smith, I., and Kornhuber, H.H. (1967): A neural base for the sense of flutter-vibration. *Science* 155:597-600.

Mountcastle, V.B., Talbot, W.H., Sakata, H., and Hyvarinen, J. (1969): Cortical              79
neuronal mechanisms in flutter-vibration studied in unanesthetized monkeys. Neuronal periodicity and frequency discrimination. *J. Neurophysiol.* 32:452-484.

Mulligan, J.A. (1966): Singing behavior and its development in the song sparrow,          15,43
*Melospiza melodia. Univ. Calif. Publ. Zool.,* 81:1-76.

Myers, R.E., Sperry, R.W., and McCurdy, N.M. (1962): Neural mechanisms in visual           42
guidance of limb movement. *Arch. Neurol.* 7:195-202.

Nakao, C. and Brookhart, J.M. (1967): Effects of labyrinthine and visual deprivation on postural stability. *Physiologist* 10(3):259.

Nauta, W.J.H. and Ebbesson, S.O., eds. (1970): *Contemporary Research Methods in Neuroanatomy.* New York: Springer-Verlag.

Nauta, W.J.H. and Karten, H.J. (1968): Organization of retino-thalamic projections in the pigeon and owl. *Anat. Rec.* 160:373.

Nauta, W.J.H. and Karten, H.J. (1970): A general profile of the vertebrate brain,          16
with sidelights on the ancestry of cerebral cortex. *In: The Neurosciences: Second Study Program,* Schmitt, F.O., editor-in-chief. New York: Rockefeller University Press, pp. 7-26.

Nauta, W.J.H. and Mehler, W.R. (1966): Projections of the lentiform nucleus in the monkey. *Brain Res.* 1:3-42.

Navas, F. and Stark, L. (1968): Sampling or intermittency in hand control system dynamics. *Biophys. J.* 8:252-302.

Nelson, B. (1968): *Galápagos: Islands of Birds.* New York: William Morrow and              45
Co., Inc.

Nottebohm, F. (1967): The role of sensory feedback in the development of avian              15
vocalizations. *In: Proceedings of the 14th International Ornithological Congress.* Snow, D.W., ed. Oxford: Blackwell Scientific Publication, pp. 265-280.

Nottebohm, F. (1968): Auditory experience and song development in the chaffinch, *Fringilla coelebs. Ibis* 110:549-568.

Nottebohm, F. (1969): The "critical period" for song learning. *Ibis* 111:386-387.

Page

Nottebohm, F. (1969): The song of the chingolo, *Zonotrichia capensis*, in
Argentina: description and evaluation of a system of dialects. *Condor*
71:299-315.

Nottebohm, F. (1970): Ontogeny of bird song. *Science* 167:950-956.          16

Nottebohm, F. and Nottebohm, M. (1971): Vocalizations and breeding behaviour          15
of surgically deafened ring doves. *Anim. Behav.* (In press)

Obersteiner, H. (1896): *Anleitung beim Studium des Baues der nervösen Central-*          42
*organe im gesunden und kranken Zustande.* Leipzig: Deutike.

Olson, C.B., Carpenter, D.O., and Henneman, E. (1968): Orderly recruitment of
muscle action potentials. *Arch. Neurol.* 19:591-597.

Oscarsson, O. (1965): Functional organization of the spino- and cuneocerebellar          73,79,
tracts. *Physiol. Rev.* 45:495-522.          80,81

Oscarsson, O. (1966): The projection of group I muscle afferents to the cat cerebral
cortex. *In: Muscular Afferents and Motor Control.* (Nobel Symposium I).
Granit, R., ed. Stockholm: Almqvist and Wiksell, pp. 307-316.

Oscarsson, O. (1967): Functional significance of information channels from the          81,99,
spinal cord to the cerebellum. *In: Neurophysiological Basis of Normal and*          100
*Abnormal Motor Activities.* Yahr, M.D. and Purpura, D.P., eds. New York:
Raven Press, pp. 93-117.

Oscarsson, O. (1968): Termination and functional organization of the ventral          100
spino-olivocerebellar path. *J. Physiol.* 196:453-478.

Oscarsson, O. (1969): The sagittal organization of the cerebellar anterior lobe as
revealed by the projection patterns of the climbing fibre system. *In: Neuro-*
*biology of Cerebellar Evolution and Development.* Llinas, R., ed. Chicago:
Amer. Med. Assn., pp. 525-537.

Oscarsson, O. (1969): Termination and functional organization of the dorsal          103
spino-olivocerebellar path. *J. Physiol.* 200:129-149.

Oscarsson, O. (1970): Functional organization of spinocerebellar paths. *In: Hand-*          99,100,
*book of Sensory Physiology, Vol. II. Somatosensory System.* Iggo, A., ed.          101,102
Berlin: Springer-Verlag, pp. 121-127.

Oscarsson, O. and Rosen, I. (1966): Response characteristics of reticulocerebellar
neurones activated from spinal afferents. *Exp. Brain Res.* 1:320-328.

Oscarsson, O. and Uddenberg, N. (1966): Somatotopic termination of spino-
olivocerebellar path. *Brain Res.* 3:204-207.

Pabst, H. and Kennedy, D. (1967): Cutaneous mechanoreceptors influencing motor
output in the crayfish abdomen. *Z. Vergl. Physiol.* 57:190-208.

Page

Page, C.H. and Wilson, D.M. (1970): Unit responses in the metathoracic ganglion of            22
the flying locust. *Fed. Proc.* 29:590. (Abstr.)

Patton, H.D., Towe, A.L., and Kennedy, T.T. (1962): Activation of pyramidal tract
neurons by ipsilateral cutaneous stimuli. *J. Neurophysiol.* 25:501-514.

Perkel, D.H. and Bullock. T.H. (1968): Neural coding. *Neurosciences Res. Prog.
Bull.* 6(3):221-348. Also *In: Neurosciences Research Symposium Summaries,
Vol. 3.* Schmitt, F.O., Melnechuk, T., Quarton, G.C., and Adelman, G., eds.
Cambridge, Mass.: M.I.T. Press. 1969. pp. 405-527.

Petersen, W.A., Brookhart, J.M., and Stone, S.A. (1965): A strain-gage platform for          118
force measurements. *J. Appl. Physiol.* 20:1095-1097.

Phillips, C.G. (1969): Motor apparatus of the baboon's hand. *Proc. Roy. Soc. B*   65,66,70,
173:141-174.                                                                      71,72,136

Phillips, C.G. and Porter, R. (1964): The pyramidal projection to motoneurones of            136
some muscle groups of the baboon's forelimb. *Progr. Brain Res.* 12:222-245.

Pitres, A. (1884): Recherches anatomo-cliniques sur les scléroses bilatérales de la
moelle épinière consécutives à des lésions unilatérales du cerveau. *Arch.
Physiol. Norm. Path.* 3:142-185.

Porter, R. and Hore, J. (1969): Time course of minimal corticomotoneuronal               65,66
excitatory postsynaptic potentials in lumbar motoneurons of the monkey. *J.
Neurophysiol.* 32:443-451.

Poulsen, H. (1951): Inheritance and learning in the song of the chaffinch, *Fringilla*      43
*coelebs. Behaviour* 3:216-227.

Preston, J.B. and Whitlock, D.G. (1960): Precentral facilitation and inhibition of         136
spinal motoneurons. *J. Neurophysiol.* 23:154-170.

Preston, J.B. and Whitlock, D.G. (1961): Intracellular potentials recorded from            136
motoneurons following precentral gyrus stimulation in primate. *J. Neurophysiol.* 24:91-100.

Rall, W., Burke, R.E., Smith, T.G., Nelson, P.G., and Frank, K. (1967): Dendritic           65
location of synapses and possible mechanisms for the monosynaptic EPSP in
motoneurons. *J. Neurophysiol.* 30:1169-1193.

Rao, K.P., Babu, K.S., Ishibko, N., and Bullock, T.H. (1970): Effectiveness of
temporal pattern in the input to a ganglion: Inhibition in the cardiac ganglion of
spiny lobsters. *J. Neurobiol.* 1:233-245.

Rashbass, C. (1961): The relationship between saccadic and smooth-tracking eye              39
movements. *J. Physiol.* 159:326-338.

Reed, D.J. and Reynolds, P.J. (1969): A joint angle detector. *J. Appl. Physiol.*
27:745-748.

Page

Revzin, A.M. and Karten, H.J. (1967): Rostral projections of the optic tectum and
the nucleus rotundus in the pigeon. *Brain Res.* 3:264-276.

Roberts, A. (1968): Some features of the central co-ordination of a fast movement    17
in the crayfish. *J. Exp. Biol.* 49:645-656.

Roberts, B.L. (1969): Spontaneous rhythms in the motoneurons of spinal dogfish    26
(*Scyliorhinus canicula*). *J. Mar. Biol. Ass. (U.K.)* 49:33-49.

Robinson, D.A. (1968): The oculomotor control system: a review. *Proc. IEEE*    36,38
56:1032-1049.

Romer, A.S. (1966): *Vertebrate Paleontology,* 3rd Ed. Chicago: University of    43,44
Chicago Press.

Rushworth, G. (1960): Spasticity and rigidity: an experimental study and review. *J.*    138
*Neurol. Neurosurg. Psychiat.* 23:99-118.

Salmoiraghi, G.C. and von Baumgarten, R. (1961): Intracellular potentials from    15
respiratory neurones in brain-stem of cat and mechanism of rhythmic respiration. *J. Neurophysiol.* 24:203-218.

Schafer, E.A. (1910): Experiments on the paths taken by volitional impulses    42
passing from the cerebral cortex to the cord: the pyramids and the ventrolateral
descending tracts. *Quart. J. Exp. Physiol.* 3:355-373.

Schiller, P.H. (1970): The discharge characterisitics of single units in the oculomotor and abducens nuclei of the unanesthetized monkey. *Exp. Brain Res.*
10:347-362.

Semmes, J. and Chow, K.L. (1955): Motor effects of lesions of precentral gyrus and    42
of lesions sparing this area in monkey. *Arch. Neurol. Psychiat.* 73:546-556.

Severin, F.V., Orlovsky, G.N., and Shick, M.L. (1967): Work of the muscle    38,71,
receptors during controlled locomotion. *Biofizika* 12:502-511.    83

Seyffarth, H. and Denny-Brown, D. (1948): The grasp reflex and the instinctive    54,55
grasp reaction. *Brain* 71:109-183.

Sherrington, C.S. (1906): *The Integrative Action of the Nervous System.* (1st Ed.)    7,60,62,
New Haven: Yale University Press.    98

Sherrington, C.S. (1947): *The Integrative Action of the Nervous System.* (2nd Ed.)    34
New Haven: Yale University Press.

Sherrington, C.S. (1913): Further observations on the production of reflex stepping    27
by combination of reflex excitation with reflex inhibition. *J. Physiol.*
47:196-214.

Page

Somjen, G., Carpenter, D.O., and Henneman, E. (1965): Responses of motoneurons
of different sizes to graded stimulation of supraspinal centers of the brain. *J.
Neurophysiol.* 28:958-965.

Sperry, R.W. (1947): Cerebral regulation of motor coordination in monkeys        43
following multiple transection of sensorimotor cortex. *J. Neurophysiol.*
10:275-294.

Sperry, R.W. (1950): Neural basis of the spontaneous optokinetic response       87,108
produced by visual inversion. *J. Comp. Physiol. Psychol.* 43:482-489.

Sperry, R.W. (1969): A modified concept of consciousness. *Psychol. Rev.*        51,58
76:532-536.

Stark, L. (1959): Stability, oscillations, and noise in the human pupil servo-     31
mechanism. *Proc. IRE* 47:1925-1939.

Stark, L. (1966): Neurological feedback control systems. *In: Advances in Bio-*   94,95
*engineering and Instrumentation.* Alt, F., ed. New York: Plenum Press, pp.
289-385.

Stark, L. (1968): *Neurological Control Systems.* New York: Plenum Press.          94

Stark, L., Iida, M., and Willis, P.A. (1961): Dynamic characteristics of the motor  94
coordination system in man. *Biophys. J.* 1:279-300.

Stark, L., Michael, J.A., and Zuber, B.L. (1969): Saccadic suppression: a product of  94
the saccadic anticipatory signal. *In: Attention in Neurophysiology.* Evans, C.R.
and Mulholland, T.B., eds. London: Butterworths, pp. 281-303.

Stark, L., Okabe, Y., and Willis, P.A. (1962a): Sampled-data properties of the      94
human motor coordination system. *Res. Lab. Electron. (M.I.T.)* QPR
67:220-223.

Stark, L., Okabe, Y., Willis, P.A., and Rhodes, H.E. (1962b): Simultaneous hand    94
and eye tracking movements. Also: A transient response of the human motor
coordination system. *Res. Lab. Electron. (M.I.T.)* QPR 66:389-401.

Stark, L., Van der Tweel, H., and Redhead, J. (1962c): Pulse response of the pupil.  31
*Acta Physiol. Pharmacol. Neerl.* 11:235-239.

Stark, L., Vossius, G., and Young, L.R. (1962d): Predictive control of eye tracking  94
movements. *IRE Trans. Human Factors in Electronics.* HFE-3:52-57.

Stensiö, E.A. (1958): Les cyclostomes fossibles, ou ostracodermes. In: *Traité de*   43
*Zoologie,* Grassé, P.N., ed. Paris: Masson.

Stewart, D.H. and Preston, J.B. (1967): Functional coupling between the pyramidal  66
tract and segmental motoneurons in cat and primate. *J. Neurophysiol.*
30:453-465.

Page

Stratton, G.M. (1897): Vision without inversion of the retinal image. *Psychol. Rev.*          108
4:463-481.

Straznicky, K. (1963): Function of heterotopic spinal cord segments investigation
in the chick. *Acta Biol. Acad. Sci. Hung.* 14:143-153.

Székely, G. (1968): Development of limb movements: embryological, physiological
and model studies. *In: Growth of the Nervous System.* Wolstenholme, G.E.W.
and O'Connor, M., eds. Boston: Little, Brown, pp. 77-93.

Székely, G., Czéh, G., and Vörös, G. (1969): The activity pattern of limb muscles          25
in freely moving normal and deafferented newts. *Exp. Brain Res.* 9:53-72.

Taub, E. and Berman, A.J. (1968): Movement and learning in the absence of          27,106
sensory feedback. *In: The Neuropsychology of Spatially Oriented Behavior.*
Freedman, S.J., ed. Homewood, Ill.: Dorsey Press, pp. 173-192.

Taylor, F.V. and Birmingham, H.P. (1948): Studies of tracking behavior. II. The          92,93,
acceleration pattern of quick manual corrective responses. *J. Exp. Psychol.*          94
38:783-795.

Terzuolo, C. and Llinas, R. (1966): Distribution of synaptic inputs in the spinal          65
motoneurone and its functional significance. *In: Muscular Afferents and Motor
Control.* Granit, R., ed. Stockholm: Almquist and Wiksell, pp. 373-384.

Teuber, H.-L. (1964): The riddle of frontal lobe function in man. *In: The Frontal
Granular Cortex and Behavior.* Warren, J.M. and Akert, K., eds. New York:
McGraw-Hill, pp. 410-444.

Teuber, H.-L. (1966): Alterations of perception after brain injury. *In: Brain and*          107,109
*Conscious Experience.* Eccles, J.C., ed. New York: Springer-Verlag, pp.
182-216.

Teuber, H.-L. (1967): Lacunae and research approaches to them. I. *In: Brain*          108
*Mechanisms Underlying Speech and Language.* Darley, F.L. and Millikan, C.H.,
eds. New York: Grune and Stratton, pp. 204-216.

Teuber, H.-L. (1968): Disorders of memory following penetrating missile wounds
of the brain. *Neurology.* 18:287-288.

Teuber, H.-L. (1969): Speech as a motor skill with special reference to monaphasic
disorders. *Monogr. Soc. Res. Child Develop.* 29:131-138.

Thach, W.T. (1968): Discharge of Purkinje and cerebellar nuclear neurons during          81
rapidly alternating arm movements in the monkey. *J. Neurophysiol.*
31:785-797.

Thach, W.T. (1970a): Discharge of cerebellar neurons related to two maintained
postures and two prompt movements. I. Nuclear cell output. *J. Neurophysiol.*
33:527-536.

Page

Thach, W.T. (1970b): Discharge of cerebellar neurons related to two maintained postures and two prompt movements. II. Purkinje cell output and input. *J. Neurophysiol.* 33:537-547.

Thorpe, W.H. (1958): The learning of song patterns of birds, with especial reference to the song of the chaffinch, *Fringilla coelebs*. *Ibis*. 100:535-570.                              43

Thorpe, W.H. (1961): *Bird Song: The Biology of Vocal Communication and Expression in Birds*. Cambridge: Cambridge University Press.                              43

Towe, A.L. and Jabbur, S.J. (1961): Cortical inhibition of neurons in dorsal column nuclei of cat. *J. Neurophysiol.* 24:488-498.

Towe, A.L., Nyquist, J.K., and Tyner, C.F. (1969): Composition of the corticofugal reflex. *Brain Res.* 16:530-534.

Towe, A.L., Patton, H.D., and Kennedy, T.T. (1963): Properties of the pyramidal system in the cat. *Exp. Neurol.* 8:220-238.

Towe, A.L., Patton, H.D., and Kennedy, T.T. (1964): Response properties of neurons in the pericruciate cortex of the cat following electrical stimulation of the appendages. *Exp. Neurol.* 10:325-344.

Towe, A.L., Whitehorn, D., and Nyquist, J.K. (1968): Differential activity among wide-field neurons of the cat postcruciate cerebral cortex. *Exp. Neurol.* 20:497-521.

Towe, A.L. and Zimmerman, I.D. (1962): Peripherally evoked cortical reflex in cuneate nucleus. *Nature* 194:1250-1251.

Tower, S.S. (1936): Extrapyramidal action from the cat's cerebral cortex: motor and inhibitory. *Brain* 59:408-444.                              42

Tower, S.S. (1940): Pyramidal lesion in the monkey. *Brain* 63:36-90.                              42

Travis, A.M. and Woolsey, C.N. (1956): Motor performance of monkeys after bilateral partial and total cerebral decortications. *Amer. J. Phys. Med.* 35:273-310.                              42

Twitchell, T.E. (1954): Sensory factors in purposive movement. *J. Neurophysiol.* 17:239-252.                              27

Twitchell, T.E. (1965): The automatic grasping responses of infants. *Neuropsychologia* 3:247-259.                              55

Tyner, C.F., and Towe, A.L. (1970): Interhemispheric influences on sensorimotor neurons. *Exp. Neurol.* 28:88-105.

Vallbo, A.B. (1970): Slowly adapting muscle receptors in man. *Acta Physiol. Scand.* 78:315-333.                              137,138

Page

Vedel, J.P. (1966): Mise en évidence d'un controle cortical de l'activité des fibres    71
fusimotrices dynamiques chez le chat par la voie pyramidale. *C.R. Acad. Sci. D.*
262:908-911.

von Euler, C. (1966): Proprioceptive control in respiration. *In: Muscular Afferents*    71
*and Motor Control.* Granit, R., ed. Stockholm: Almquist and Wiksell, pp.
197-207.

von Euler, C. (1970): The role of fusimotor activity in spindle control of    71
movements with special reference to respiration. *In: Excitatory Synaptic*
*Mechanisms.* Proceedings of the Fifth International Meeting of Neurobiologists.
(In press)

von Holst, E. (1935a): Alles oder Nichts, Block, Alternans, Bigemini und verwandte    26
Phänomene als Eigenschaften des Rückenmarks. *Pflüger's Arch. Ges. Physiol.*
236:515-532.

von Holst, E. (1935b): Uber den Prozess der zentralnervösen Koordination.    26
*Pflüger's Arch. Ges. Physiol.* 236:149-158.

von Holst, E. (1936a): Versuche zur Theorie der relativen Koordination. *Pflüger's*    26
*Arch. Ges. Physiol.* 237:92-121.

von Holst, E. (1936b): Vom Dualismus der motorischen und der automatisch-    26
rhythmischen Funktion im Rückenmark und vom Wesen des automatischen
Rhythmus. *Pflüger's Arch. Ges. Physiol.* 237:356-378.

von Holst, E. (1954): Relations between the central nervous system and the    87,88,
peripheral organs. *Brit. J. Anim. Behav.* 2:89-94.    89

von Holst, E. (1957): Aktive Leistungen der menschlichen Gesichtswahrnehmung.    89,91
*Studium Generale* 10:234.

von Holst, E. (1968): Relations between the central nervous system and the
peripheral organs. *In: Contemporary Theory and Research in Visual Perception.*
Haber, R.N., ed. New York: Holt, Rinehart and Winston, pp. 497-503.

von Holst, E. and Mittelstaedt, H. (1950): Das Reafferenzprinzip (Wechsel-    87,88,
wirkungen zwischen Zentralnervensystem und Peripherie). *Naturwiss.*    91,108
37:464-476.

Vossius, G. (1960): Das system der augenbewegung. *Z. Biol.* 112:27-57.    38

Walker, A.E. (1949): Cerebral pedunculotomy for the relief of involuntary    43
movements. I. Hemiballismus. *Acta Psychiat. Neurol.* 24:723-729.

Walker, A.E. (1952): Cerebral pedunculotomy for the relief of involuntary    43
movements. II. Parkinsonian tremor. *J. Nerv. Ment. Dis.* 116:766-775.

Walker, A.E. and Richter, H. (1966): Section of the cerebral peduncle in the    42
monkey. *Arch. Neurol.* 14:231-240.

Page

Wall, P.D. (1964): Presynaptic control of impulses at the first central synapse in the          73
cutaneous pathway. *Progr. Brain Res.* 12:92-118.

Wall, P.D. (1970): Habituation and post-tetanic potentiation in the spinal cord. *In:*          98
*Short-term Changes in Neural Activity and Behaviour,* Horn, G. and Hinde,
R.A., eds. Cambridge: Cambridge University Press, pp. 181-210.

Walshe, F.M.R. (1963): *Diseases of the Nervous System,* 10th Ed. Baltimore:          136
Williams and Wilkins.

Weiss, P. (1934): Function of de-afferented amphibian limbs. *Proc. Soc. Exp. Biol.*
*Med.* 32:436-438.

Weiss, P. (1935): Homologous function in supernumerary limbs after elimination of
sensory control. *Proc. Soc. Exp. Biol. Med.* 33:30-32.

Weiss, P. (1935): Homologous (resonance-like) function in supernumerary fingers in
a human case. *Proc. Soc. Exp. Biol. Med.* 33:426-430.

Weiss, P. (1935); Unmodifiability of locomotor coordination in amphibia demonstrated by the reverse functioning of mutually exchanged right and left limbs.
*Proc. Soc. Exp. Biol. Med.* 33:241-242.

Weiss, P. (1936): A study of motor coordination and tonus in deafferented limbs of          24
amphibia. *Amer. J. Physiol.* 115:461-475.

Weiss, P. (1941a): Does sensory control play a constructive role in the development          7,96
of motor coordination? *Schweiz. Med. Wschr.* 71:406-407.

Weiss, P. (1941b): Self-differentiation of the basic patterns of coordination. *Comp.*          25
*Psychol Monogr.* 17(4):1-96.

Weiss, P. (1950): Experimental analysis of coordination by the disarrangement of          25
central-peripheral relations. *Symp. Soc. Exp. Biol.* 4:92-111.

Weiss, P. and Brown, P.F. (1941): Electromyographic studies on recoordination of
leg movements in poliomyelitis patients with transposed tendons. *Proc. Soc.*
*Exp. Biol. Med.* 48:284-287.

Weiss, P. and Ruch, T.C. (1936): Further observations on the function of
supernumerary fingers in man. *Proc. Soc. Exp. Biol. Med.* 34:569-570.

Welt, C., Aschoff, J.C., Kameda, K., and Brooks, V.B. (1967): Intracortical          51,52
organization of cat's sensorimotor neurons. *In: The Neurophysiological Basis of*
*Normal and Abnormal Motor Activities,* Purpura, D.P. and Yahr, M.D., eds. New
York: Raven Press, pp. 255-293.

Westheimer, G. and Conover, D.N. (1954): Smooth eye movements in the absence          39
of a moving visual stimulus. *J. Exp. Psychol.* 47:283-284.

Page

Wheeles, L.L., Jr., Boynton, R.M., and Cohen, G.H. (1960): Eye movement    37
responses to step and pulse-step stimuli. *J. Opt. Soc. Amer.* 56:956-960.

Whitehorn, D. and Towe, A.L. (1968): Postsynaptic potential patterns evoked upon
cells in sensorimotor cortex of cat by stimulation at the periphery. *Exp. Neurol.*
22:222-224.

Whitman, C.O. (1899): Animal behavior. *In: Biological Lectures* (Marine Biol. Lab.,    44
Woods Hole, 1898). Boston: Ginn and Co., pp. 285-338.

Wiener, N. (1948): *Cybernetics: or Control and Communication in the Animal and*    92
*the Machine.* New York: John Wiley and Sons, Inc.

Wiersma, C.A.G. (1947): Giant nerve fiber system of the crayfish. A contribution    17
to comparative physiology of synapse. *J. Neurophysiol.* 10:23-38.

Wiesendanger, M. (1969): The pyramidal tract. Recent investigations on its    42
morphology and function. *Ergebn. Physiol.* 61:71-136.

Wiesendanger, M. and Tarnecki, R. (1966): Die Rolle des pyramidalen Systems bie    42
der sensomotorischen Integration. *Bull. Schweiz. Akad. Med. Wiss.* 22:306-328.

Willows, A.O. and Hoyle, G. (1969): Neuronal network triggering a fixed action    19
pattern. *Science* 166:1549-1551.

Wilson, D.M. (1961): The central nervous control of flight in a locust. *J. Exp. Biol.*    10,11,
38:471-490.    20

Wilson, D.M. (1964a): The origin of the flight-motor command in grasshoppers. *In:*    10,14,
*Neural Theory and Modeling.* Reiss, R., ed. Stanford, California: Stanford    20,22
University Press, pp. 331-345.

Wilson, D.M. (1964b): Relative refractoriness and patterned discharge of locust    22
flight motor neurons. *J. Exp. Biol.* 41:191-205.

Wilson, D.M. (1968): Inherent asymmetry and reflex modulation of the locust    21
flight motor pattern. *J. Exp. Biol.* 48:631-641.

Wilson, D.M. and Gettrup, E. (1963): A stretch reflex controlling wing beat
frequency in grasshoppers. *J. Exp. Biol.* 40:171-185

Wilson, D.M. and Waldron, I. (1968): Models for the generation of the motor out-    22
put pattern in flying locusts. *Proc. IEEE* 56:1058-1064.

Wilson, V.J. and Kato, M. (1965): Excitation of extensor motoneurons by group II    73
afferent fibers in ipsilateral muscle nerves. *J. Neurophysiol.* 28:545-554.

Wilson, V.J. and Yoshida, M. (1969): Comparison of effects of stimulation of    66,67
Deiters' nucleus and medial longitudinal fasciculus on neck, forelimb and
hindlimb motoneurons. *J. Neurophysiol.* 32:743-758.

Page

Woodworth, R.S. (1899): The Accuracy of Voluntary Movement. *Psychol. Monogr.*     94
Vol. III, 114 pp.

Woody, C.D. (1970): Conditioned eye blink: gross potential activity at coronal-
precruciate cortex of the cat. *J. Neurophysiol.* 33:838-850.

Young, L.R. and Stark, L. (1963a): Variable feedback experiments testing a     36,94
sampled data model for eye tracking movements. *IEEE Trans. Human Factors in
Electronics* HFE-4:38-51.

Young, L.R. and Stark, L. (1963b): A discrete model for eye tracking movements.     94
*IEEE Trans. Military Electronics* MIL-7(2-3):113-115.

Zeigler, H.P., Green, H.L., and Karten, H.J. (1969): Neural control of feeding
behavior in the pigeon. *Psychon. Sci.* 15:156-157.

# INDEX

# Brain Monoamines and Endocrine Function

A Report Based on an NRP Work Session
held May 5-7, 1968

by

**Richard J. Wurtman**
Department of Nutrition and Food Science
Massachusetts Institute of Technology
Cambridge, Massachusetts

Catherine M. LeBlanc
NRP Writer-Editor

CONTENTS

LIST OF PARTICIPANTS

Dr. Adelbert Ames, III
Massachusetts General Hospital
Boston, Massachusetts 02114

Dr. Fernando Anton-Tay
Department of Physiology
Institute of Biomedical Research
U. N. A. M.
Mexico 20, D.F., Mexico

Dr. Julius Axelrod
Laboratory of Clinical Science
National Institute of Mental Health
National Institutes of Health
Bethesda, Maryland 20014

Dr. Floyd E. Bloom
Laboratory of Neuropharmacology
National Institute of Mental Health
Saint Elizabeth's Hospital
Washington, D. C. 20032

Dr. John A. Coppola
Department of Endocrine Research
Lederle Laboratories
Pearl River, New York 10965

Dr. Annica Dahlström
Institute of Neurobiology
Medicinaregatan 5
University of Göteborg
Göteborg 33, Sweden

Dr. Alfredo O. Donoso
Instituto de Histologia y Embriologia
Facultad de Ciencias Medicas U. N. C.
Mendoza, Argentina

Dr. Kjell Fuxe
Department of Histology
Karolinska Institutet
Stockholm 60, Sweden

Dr. Robert Galambos
Department of Neurosciences
University of California, San Diego
La Jolla, California 92037

Dr. William F. Ganong
Department of Physiology
University of California Medical Center
San Francisco, California 94122

Dr. Roger A. Gorski
Department of Anatomy
University of California
    School of Medicine
Los Angeles, California 90024

Dr. Harvey J. Karten
Department of Psychology
Massachusetts Institute of Technology
Cambridge, Massachusetts 02139

Dr. Seymour S. Kety
Psychiatric Research Laboratories
Massachusetts General Hospital
Boston, Massachusetts 02114

Dr. Seymour Levine
Department of Psychiatry
Stanford University
    School of Medicine
Palo Alto, California 94304

Dr. Samuel M. McCann
Department of Physiology
Southwestern Medical School
Dallas, Texas 75235

Dr. Bruce S. McEwen
The Rockefeller University
New York, New York 10021

Dr. Donald M. MacKay
Department of Communication
University of Keele
Keele, Staffordshire ST5 5BG, England

Dr. Roger P. Maickel
Department of Psychology
Indiana University
Bloomington, Indiana 47401

Dr. Bernard H. Marks
Department of Pharmacology
The Ohio State University
Columbus, Ohio 43210

Dr. Luciano Martini
Department of Pharmacology
School of Medicine
University of Milan
20129 Milano, Italy

Dr. Bengt Meyerson
Department of Pharmacology
University of Uppsala
Uppsala, Sweden

Dr. Robert Y. Moore
Division of Neurology
University of Chicago
Chicago, Illinois 60637

Dr. Eugenio E. Müller
Department of Pharmacology
School of Medicine
University of Milan
20129 Milano, Italy

Dr. Walle J. H. Nauta
Department of Psychology
Massachusetts Institute of Technology
Cambridge, Massachusetts 02139

Dr. Larissa Pohorecky
Department of Nutrition and
 Food Science
Massachusetts Institute of Technology
Cambridge, Massachusetts 02139

Dr. Gardner C. Quarton
Mental Health Research Institute
University of Michigan
Ann Arbor, Michigan 48104

Dr. Charles H. Sawyer
Department of Anatomy
University of California
 School of Medicine
Los Angeles, California 90024

Dr. Berta V. Scharrer
Department of Anatomy
Albert Einstein College of Medicine
Bronx, New York 10461

Dr. Francis O. Schmitt
Neurosciences Research Program
Brookline, Massachusetts 02146

Dr. John Urquhart
Biomedical Engineering
University of Southern California
Los Angeles, California 90007

Dr. Richard J. Wurtman
Department of Nutrition and
 Food Science
Massachusetts Institute of Technology
Cambridge, Massachusetts 02139

Dr. Michael J. Zigmond
Department of Biology
University of Pittsburgh
Pittsburgh, Pennsylvania 15213

Note: NRP Work Session summaries are reviewed and revised by participants prior to publication.

# I. INTRODUCTION

In general, one thinks of the output of the brain as the transmission of neuronal signals to other cells by release of neurotransmitter substances at synapses. The synaptic signals may be transmitted to other neurons, skeletal or smooth muscle cells, myocardium, adipose tissue cells, or a variety of other cells. However, over the past few decades it has become clear that the brain has additional output channels that allow some cells to transmit regulatory signals by humoral means. Although their input is synaptic, the output channels of these cells utilize hormones, which are delivered to target cells by the circulation. The translation, or transduction, of synaptic input signals to humoral output signals is performed by a type of cell that has been termed the "neuroendocrine transducer cell"; this concept will be discussed in detail in Section II of this report.

The Neurosciences Research Program convened meetings dedicated to neuroendocrinology on October 29-30, 1967, and May 5-7, 1968. The first meeting, a preliminary conference, "New Directions in Neuroendocrine Research," considered neural and humoral information transfer in a variety of neuroendocrine systems, and the concept of the neuroendocrine transducer cell within that broad context. Most of the neuroendocrine transducer cells for which information is available either release a monoamine (adrenomedullary chromaffin cells), or receive a monoamine as their input neurotransmitter (pineal). Accordingly, the second meeting, and the subject of this report, was a full Work Session on "Brain Monoamines and Endocrine Function." The Work Session considered information transfer involving a particular family of neuroendocrine transducer cell in the mammalian brain: cells of the hypothalamus that are thought to liberate "releasing factors" and "inhibiting factors," or "hypophysiotropic hormones." These hormones are secreted into a special portal vascular system, the hypothalamohypophyseal portal system, which carries them directly to the anterior pituitary gland, or adenohypophysis (Figure 1). In response, the adenohypophysis releases tropic hormones that regulate peripheral hormone release and other functioning by target endocrine glands (Table 1, Figure 2). Studies on hypophysiotropic hormones will be described in Section IV.

The over-all purpose of the Work Session was to examine the role of brain monoamines in controlling anterior pituitary function. As

TABLE 1

Hypophysiotropic and Tropic Hormones [Wurtman]

| Hypophysiotropic | Tropic |
|---|---|
| Corticotropin releasing factor (CRF) | Adrenocorticotropin (ACTH) |
| Follicle-stimulating hormone releasing factor (FRF) | Follicle-stimulating hormone (FSH) |
| Growth hormone releasing factor (GRF) | |
| Growth hormone inhibiting factor (GIF) | Growth hormone (GH) |
| Luteinizing hormone releasing factor (LRF) | Luteinizing hormone (LH) or interstitial-cell-stimulating hormone (ICSH)* |
| Prolactin inhibitory factor (PIF) | Prolactin or luteotropic hormone (LtH)† |
| Thyroid-stimulating hormone releasing factor (TRF) | Thyroid-stimulating hormone (TSH) |
| Melanocyte-stimulating hormone releasing factor (MRF) | |
| Melanocyte-stimulating hormone inhibitory factor (MIF) | Melanocyte-stimulating hormone (MSH) |

*Termed LH in the female because it induces the formation of the corpus luteum in the ovary; termed ICSH in the male because it stimulates the testicular cells to secrete testosterone.
†Termed prolactin in species in which it stimulates the development of milk-producing cells during and after pregnancy; termed LtH in those relatively few species in which it acts physiologically to prolong the life of a corpus luteum.

is well known clinically, if brain monoamine metabolism is modified pharmacologically (e.g., by treatment with reserpine or chlorpromazine), human patients not infrequently display clinical evidence of altered pituitary function (e.g., they may fail to ovulate, or may begin to lactate at an inappropriate time). Evidence described in this *Bulletin* indicates that the effect of the drugs is not on the pituitary itself, but on the "hypophysiotropic cells" in the hypothalamus, or on neurons providing an input to these cells. It is generally accepted that the monoamines norepinephrine (NE), dopamine (DA), and serotonin (5-HT) function as neurotransmitters within the brain; but it seems clear that these compounds are released by only a tiny fraction of the central neurons. However, in spite of their poor total representation in

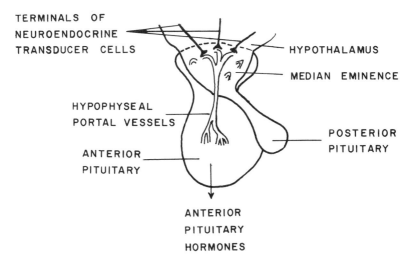

Figure 1. Diagram of the final common pathway regulating anterior pituitary secretion. Terminals of hypothalamic hypophysiotropic cells converge on the median eminence, where they liberate releasing factors into the capillary loops that give rise to the portal hypophyseal vessels. These vessels transport the factors directly to the anterior pituitary. [Modified from Harris, 1964]

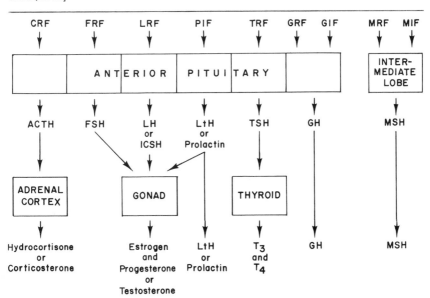

Figure 2. Diagram showing stages of hormone release from the hypothalamus to the periphery: hypothalamic hypophysiotropic hormones, pituitary tropic hormones, and target organ hormones. $T_3$, triiodothyronine; $T_4$, thyroxine. [Wurtman]

brain, the neurons that release monoamines do appear to be of especial significance in certain important brain functions, such as mood, or affect, and sleep (earlier *NRP Bulletins* in these areas: "Neural Properties of the Biogenic Amines," by Kety and Samson, 1967; "Sleep, Wakefulness, Dreams and Memory," by Nauta et al., 1966), as well as in endocrine processes such as the control of reproduction and the regulation of the extracellular fluid. Biochemical and cytological characteristics of brain monoamines will be reviewed in Section III.

One goal of the Work Session was to evaluate research strategies that might be applied to the study of brain monoamine control of endocrine function. Section V of this report describes research relating brain monoamines and endocrine function; but however fruitful have been the correlations resulting from this research, it is clear that unknowns often interpose themselves when the investigator suggests mechanisms for his observations. As described in Section IV, the neuroendocrine-transducer level of function has been and continues to be difficult to study, partly because of its anatomical inaccessibility and partly because of the multiplicity of input-output-feedback influences operative at that site. Also, as described in Section III, brain monoamines could exert their influence on and/or in the hypothalamus at a variety of sites and by a variety of mechanisms. Thus, needs for advances in the state of the art are suggested at all levels of brain-endocrine relationships.

Another goal of the Work Session was to acquaint neurobiologists and endocrinologists with one another's fields. Until recently, the two communities have had little interaction, and very few scientists have been trained in both fields. Moreover, the approach to knowledge-gathering has often differed strikingly between neurobiologists and endocrinologists. For example, historically, endocrinologists have had great success in understanding and treating human disease states; and so their experimental paradigms have generally involved treating whole animals with vast amounts of hormones, or ablating the source of the hormone entirely. Structure is usually of secondary importance in endocrinology; hence, the endocrinologist usually thinks nothing of homogenizing a gland before he tries to identify its active principle. In contrast, structure is paramount in neurobiology, and the perturbations that neurobiologists utilize to examine the systems that they are studying are more often within the functional range. However, their formulations have too infrequently passed the ultimate test of generating therapies that are really effective in diseased humans.

In the time that has elapsed since the Work Session, almost all of the participants who were not already there have crossed the bridge between neurobiology and endocrinology, and have published original studies dealing with brain monoamines and endocrine function. This period of scientific history coincides with a striking increase in interest in neuroendocrinology in general, and in neuropharmacological approaches to the control of pituitary function in particular. This report will include some of these recent advances. It is hoped that publication of this *Bulletin* may tempt additional life scientists to apply themselves to the two disciplines that seek to describe this exciting and important field.

## II. NEUROENDOCRINE TRANSDUCTION*

## An Essay by R. J. Wurtman

In general, the brain sends signals to other organs via synaptically connected chains of neurons, with peripheral nerves constituting the last link in these chains. The peripheral axons terminate directly on skeletal muscle, adipose tissue, and exocrine glands, and transmit instructions to these cells by releasing acetylcholine (ACh) or norepinephrine (NE). However, among the many tasks the brain must perform, at least two cannot be mediated solely through synaptic neuronal channels, because the organs with which the brain must communicate do not respond to neurotransmitters released in the immediate vicinity of the organs but to other chemical signals that are delivered to the organs by the blood stream. These tasks include regulation of concentrations of certain substances in the extracellular fluid (e.g., homeostasis of glucose, cortisol, osmolarity), and control of the mechanisms of reproduction and lactation.

For example, when plasma glucose concentrations fall below an allowable minimum, the brain activates an effector mechanism that causes the rate of glycogenolysis in the liver to be accelerated. The immediate signal that instructs the hepatic cells to initiate the depolymerization of glycogen is a hormone, epinephrine, which reaches the hepatocytes, not by diffusing across a synapse but by being taken up from the circulation. Similarly, the brain of the female mammal instructs the ovary to develop a mature follicle and subsequently directs it to ovulate not by sending nerve impulses to individual ovarian cells but by causing the anterior pituitary gland to secrete specific hormones, the gonadotropins. These substances reach the ovary via the blood stream and activate biochemical processes leading to the maturation of the ovum, ovulation, and the synthesis of ovarian hormones.

The ability of the mammalian brain to regulate substances in the extracellular fluid and to control gonadal function requires that the central and peripheral nervous systems contain groups of specialized communications cells capable of transducing neuronal inputs to hormonal outputs. These cells have been termed "neuroendocrine transducer

*This topic appears in more detail in Wurtman, R.J. (1970c): Neuroendocrine transducer cells in mammals. *In: The Neurosciences: Second Study Program.* Schmitt, F.O., editor-in-chief. New York: Rockefeller University Press, pp. 530-538.

cells" (Wurtman and Axelrod, 1965). Their input signal is that of a neuron; they respond to a neurotransmitter diffusing across a short distance, typically at a synapse. Their output signal is that of a true endocrine cell; they emit coded chemical messages that are delivered by the circulation to many or all cells in the body.

### Some Differences Between Neural and Endocrine Communication

The transmission of signals from one neuron to another, or from a neuron to a cell that it innervates, is mediated by a familiar process: the neurotransmitter, a specific substance stored within characteristic subcellular vesicles, is released and liberated near the receptor cell. Most commonly, this release occurs at an identifiable anatomical locus, the synapse. The neurotransmitter then diffuses across a short distance to reach a specialized zone (the postsynaptic membrane) on the receptor cell where it alters the flux of specific ions. This causes a change in electrical potential within the postsynaptic cell and, if the cell is a neuron, alters the probability that an action potential will be generated. Nearly all of the compounds thought to function as neurotransmitters have similar chemical characteristics; i.e., they are low-molecular-weight, water-soluble amines and, possibly, amino acids. Moreover, they are rapidly inactivated by physical and chemical processes, such as reuptake into their cell of origin or enzymatic transformation. Consequently, their concentrations in the blood tend to be very low.

In contrast, the transmission of hormonal signals between organs via the blood stream utilizes a variety of chemicals far broader than those of the current list of putative neurotransmitters. The hormones seem to lack common chemical characteristics. Thus, insulin is water soluble, but progesterone is highly nonpolar; thyroxine is a low-molecular-weight amino acid, but TSH appears to be a large glycoprotein. Epinephrine is rapidly cleared from the circulation (by enzymatic transformation or uptake into sympathetic nerve endings), but cortisol persists in the blood for relatively long periods. The specific anatomic loci on receptor cells at which hormones act have yet to be identified, but almost certainly lack the well-defined structural features of the postsynaptic membrane. In addition, no characteristic response has been identified in hormone-responsive cells that is analogous to the ion fluxes and potential changes observed in the postsynaptic neuron or at the neuromuscular junction. An experimental problem in this regard is that one can usually tell whether a given neuron has received and responded to a neurotransmitter within seconds, whereas considerably more time is required to determine, for example, whether a thyroid cell has responded to circulating TSH.

Also in contrast to neurotransmitters, the tendency for hormones to be localized in organelles seems to be correlated with their chemical nature rather than with the communications properties of their cells of origin (see Fawcett et al., 1969). Neuroendocrine transducer cells in the adrenal medulla and the supraoptic nucleus contain characteristic granular vesicles that can be identified by electron microscopy, segregated by ultracentrifugation, and shown by chemical or biological assays to store the secretory products of their respective cells. On the other hand, pineal parenchymal cells do not seem to store their secretions in a characteristic organelle, probably because these secretions are lipid soluble. Water-soluble hormones are often present in relatively high concentrations within their cells of origin; their partial confinement within vesicles allows these concentrations to exist without unduly raising the cytoplasmic solute loads. The binding of epinephrine and insulin to subcellular particles also serves to protect these hormones from catabolic intracellular enzymes (i.e., mitochondrial monoamine oxidase (MAO) and lysosomal proteases), which would destroy them if they floated freely in the cytoplasm. Lipid-soluble hormones tend not to be localized within vesicles or similar subcellular particles. For example, estrogens and thyroxine are stored in reservoirs outside the cell, that is, in the follicular fluid of the ovary or the colloid of the thyroid; corticosterone is probably synthesized *de novo* as it is needed for secretion. Hence, the tendency for the secretory product of a cell to be localized within a visible particle cannot be used to identify the physiological function of that cell as neuroendocrine transduction.

Perhaps the most characteristic difference between the transmission of signals by neurotransmitters and that by hormones lies in the techniques used by these communications systems to attain "privacy." Nervous systems obtain privacy by *anatomical* means: a given neuron apparently transmits signals only to the small number of cells with which it makes synapses, or to cells lying within a few hundred Angstroms of its terminal boutons. Thus, even though the particular chemical signal emitted when a particular neuron fires might be capable of stimulating $10^8$ other neurons within the brain, only $10^3$ or $10^4$

cells respond, because only this smaller number of neurons is anatomically accessible to the neurotransmitter.

Hormonal systems obtain privacy by *biochemical* means: a given signal may be distributed by the blood to every cell in the body; the signal is, however, coded, and only a relatively small number of cells are able to perform the necessary decoding operation to obtain information. The high degree of specificity attainable by hormonal communications systems is well illustrated by the physiological operation of the thyroid gland. The input to this organ, TSH, is carried by the circulation to every organ in the body; its output, thyroxine, is distributed in essentially the same volume. However, only the thyroid gland appears capable of responding to the circulating TSH, while biochemical responses to circulating thyroxine are shown by the heart, the liver, and most other organs.

## Neuroendocrine Transducer Cells

The conversion of neural to hormonal signals is accomplished by the neuroendocrine transducer cells. Information currently available about the input signals to several specific neuroendocrine transducers suggests that these cells respond to typical neurotransmitter substances (ACh and NE), which reach them by diffusing across synapses or from nearby terminal boutons (i.e., in the pineal). Their output signals exhibit all the variety typical of hormones: epinephrine is water soluble, but melatonin is nonpolar; renin is a high-molecular-weight protein, but epinephrine and melatonin are low-molecular-weight derivatives of single amino acids. The output signals (the hypophysiotropic hormones) emitted by the hypothalamic transducer cells that mediate the neural control of the anterior pituitary apparently each act on only one group of cells in this single target organ. In contrast, oxytocin, a hormonal signal emitted by the hypothalamic paraventricular nucleus via the posterior pituitary, carries instructions to both the uterus and the myoepithelium of the mammary glands.

The demonstration that a given cell functions as a neuroendocrine transducer requires two types of evidence: (1) it must be shown by anatomical methods that the cell receives a direct innervation, and (2) it must be demonstrated that the ability of the cell to secrete its hormone under appropriate physiological conditions is impaired on interruption of this innervation. According to these criteria, at least five types of cells can now be termed neuroendocrine transducers, some peripherally innervated (1-3), others centrally innervated (4-5):

1. The chromaffin cells of the adrenal medulla, which respond to a sympathetic cholinergic input by releasing the monoamine hormone epinephrine (Wurtman, 1966).

2. The juxtaglomerular cells of the mammalian kidney, which respond to a sympathetic noradrenergic input by releasing renin into the blood stream (Bunag et al., 1966).

3. The parenchymal cells of the mammalian pineal organ, which respond to a sympathetic noradrenergic input by synthesizing and releasing the hormone melatonin (Wurtman, Axelrod, and Kelly, 1968a).

4. Hypothalamic cells of the supraoptic and paraventricular nuclei, which respond to noradrenergic or cholinergic inputs, or both, by releasing the hormones vasopressin and oxytocin via the neurohypophysis (Scharrer and Scharrer, 1963). These as well as the fifth type are representatives of neurosecretory neurons. For discussions of the functional interpretation of the phenomenon of neurosecretion, see Scharrer and Scharrer, 1963; E. Scharrer, 1965, 1966; B. Scharrer, 1967, 1969a-c, 1970; Scharrer and Weitzman, 1970. Specific reference to the concept of neuroendocrine transducer cells is made in B. Scharrer, 1970.

5. Hypothalamic cells that reside within the arcuate and other nuclei, and that appear to secrete releasing factors and inhibiting factors, or hypophysiotropic hormones, into the pituitary portal circulation, possibly in response to monoaminergic inputs (evidence given in this *Bulletin*).

It seems likely that this list will continue to expand. Whenever it can be demonstrated that the brain influences the secretion of a hormone from a peripheral organ (e.g., insulin from the pancreas), a *prima facie* case has been made for the participation of a neuroendocrine transducer in the secretory process.

In several cases, it has been possible to show that neuroendocrine transducer cells also respond to hormonal inputs. Thus, the cells of the hypothalamus that secrete CRF may be directly inhibited by plasma cortisol; estrogens inhibit melatonin synthesis in the pineal; and cortisol stimulates epinephrine synthesis in the adrenal medulla. These "feedback" effects are undoubtedly significant, but as yet (see Section IV) to an undetermined degree. The ability of these cells to respond to hormonal inputs is not surprising, inasmuch as all of the neuroendo-crine transducer cells are bathed by hormone-containing extracellular fluid. What is surprising is their basic functional ability, the ability to secrete hormones in direct response to neuronal signals. This ability allows them to serve as a major link between the brain and the viscera in the control of reproduction and lactation and in the regulation of key substances in the internal milieu.

## III. BRAIN MONOAMINES

The brains of mammals are able to synthesize at least three monoamines that appear to be neurotransmitters and to have a special role in the control of anterior pituitary function. These compounds are norepinephrine (NE), dopamine (DA), and serotonin (5-hydroxy-tryptamine, 5-HT). The first two are catecholamines and have their origin in circulating tyrosine; 5-HT is an indoleamine and is synthesized from circulating tryptophan.

It is important to recognize that tyrosine and tryptophan have several additional fates in the body that are unrelated to their ability to serve as precursors for brain monoamines. Both amino acids are incorporated into protein by brain and other tissues. Both are also metabolized by specific enzymes in the liver, i.e., tyrosine transaminase and tryptophan pyrrolase. Circulating tyrosine serves as the precursor for the hormones of the thyroid gland (thyroxine and 3,5,3'-triiodo-thyronine) and for cutaneous melanin. It has recently been suggested (Curzon and Green, 1968) that such hormones as hydrocortisone can influence the rate at which the brain synthesizes 5-HT by inducing hepatic tryptophan pyrrolase, which metabolizes its amino acid precursor. The brain probably synthesizes other monoamines in addition to NE, DA, and 5-HT. Epinephrine has been found in the hypothalamus of several species, and has been shown to be synthesized in vivo from NE within the olfactory bulbs of rats. This reaction is catalyzed by the enzyme phenylethanolamine-N-methyl transferase (PNMT). Histamine, octopamine, and tryptamine are also present in the brains of several species. The Work Session limited its consideration of brain mono-amines to DA, NE, and 5-HT, because the evidence that those compounds function as central neurotransmitters is most nearly ade-quate, and because indirect evidence, described in detail below, suggests strongly that these compounds are involved in the control of tropic hormone secretion from the anterior pituitary gland.

### Biosynthesis and Fate: J. Axelrod

The biochemistry and pharmacology of brain monoamines was discussed by Axelrod at the Work Session. These topics have also been summarized in a previous *NRP Bulletin* ("Neural Properties of the Biogenic Amines," by Kety and Samson, 1967).

Catecholamines

   The biosynthesis of brain catecholamines (Figure 3) is initiated by the uptake of the amino acid tyrosine from the circulation. Although the concentrations of amino acid in human plasma and rat brain exhibit characteristic daily fluctuations (tyrosine levels in human blood are almost twice as high late in the morning as they are at 2-4 A.M., Wurtman et al., 1968c), it is not known whether the availability of circulating tyrosine influences the rate of brain catecholamine synthesis under normal conditions. However, perinatal protein malnutrition has recently been shown to suppress the accumulation of NE in the brains of weanling rats (Shoemaker and Wurtman, 1970, 1971). Since the brains contained elevated amounts of tyrosine hydroxylase, the enzyme that normally rate-limits catecholamine biosynthesis (Nagatsu et al., 1964), the most likely basis for this phenomenon was inadequate precursor tyrosine. There is some evidence that uptake of tyrosine into brain cells is mediated by an active transport process (Chirigos et al., 1960).

   The first biochemical transformation in the synthesis of brain catecholamines involves the *meta*-hydroxylation of tyrosine. This process is catalyzed by the enzyme L-tyrosine hydroxylase, a protein that appears to be characteristic of cells that synthesize catecholamines. There is considerable debate about the location of tyrosine hydroxylase

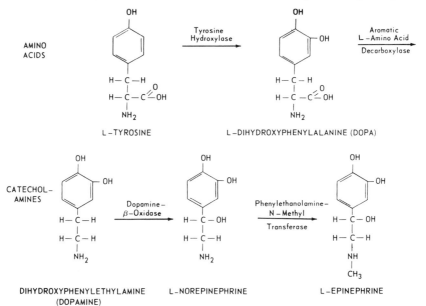

Figure 3. Biosynthesis of catecholamines. [Wurtman, 1966]

within the cell; it has been variously ascribed to the mitochondria, the synaptic vesicles, and the cytoplasm. As indicated above, it is generally believed that tyrosine hydroxylase activity normally controls the rate at which catecholamines are synthesized in vivo. The activity of this enzyme appears to be regulated by end-product inhibition. It has been postulated that a small fraction of the NE molecules present in nerve terminals constitute an "active" pool that inhibits tyrosine hydroxylase activity, perhaps by binding with a cofactor. When this pool of NE is released, for example, following nerve stimulation, the inhibition of tyrosine hydroxylase is removed, and more NE is synthesized as a result. This hypothesis would seem ·to explain why the level of NE within neurons tends to show such little variability from time to time despite great variations in the rate at which the catecholamine is being released into the synapse.

The rate at which neurons hydroxylate tyrosine has been shown experimentally to bear direct relations to their bioelectric activity (reviewed by Weiner, 1970). Weiner and Rabadjija (1968) examined the effect of electrically stimulating postganglionic sympathetic neurons on their ability to convert $^3$H-tyrosine to catecholamines, using the isolated vas deferens preparation. While the postganglionic nerve was being stimulated, within a short period of time (minutes) there was a sizable increase in net catecholamine synthesis. This increase could be blocked by adding NE to the medium, but not by adding such inhibitors of protein synthesis as cycloheximide. After electrical stimulation of the nerve was stopped, the rate of $^3$H-catecholamine synthesis continued to increase. However, the mechanism of this increase is unclear; it could be blocked by inhibitors of protein synthesis, but did not reflect an increase in the amount of newly synthesized tyrosine hydroxylase. Recent studies in Axelrod's laboratory on neurons subjected to a chronic (24-48 hour) increase in presynaptic input have demonstrated an increase in tyrosine hydroxylase activity within postganglionic sympathetic neurons. Animals treated with reserpine (which enhances the activity of preganglionic sympathetic neurons, Iggo and Vogt, 1960) displayed increased tyrosine hydroxylase activity in the superior cervical ganglia (Mueller et al., 1969a). This increase could be blocked by decentralization of the ganglia (Thoenen et al., 1969a) or by treatment with cycloheximide or actinomycin D (Mueller et al., 1969b). These observations taken together indicate that neuronal activity enhances NE synthesis via at least two distinct mechanisms: an initial release of tyrosine hydroxylase from inhibition by NE, and a long-term transsynaptic induction of the hydroxylating enzyme.

The hydroxylation of tyrosine results in the formation of a catechol amino acid, L-dihydroxyphenylalanine (L-dopa). Little, if any, of this amino acid can normally be detected in brain or blood, suggesting that it is metabolized to DA (by the enzyme aromatic L-amino acid decarboxylase, or dopa decarboxylase) almost immediately after it is formed. L-Dopa is now being used as a drug in the therapy of Parkinson's disease and certain other neurological disorders. It has been presumed that the amino acid produces its beneficial effect by being taken up within "catecholaminergic" neurons in the brain, and being decarboxylated to DA, and perhaps further transformed to NE. However, recent studies suggest that the large doses of L-dopa that are utilized may produce their therapeutic effects by other metabolic mechanisms, e.g., by lowering the concentration of S-adenosylmethionine in the brain (Wurtman et al., 1970).

DA (3,4-dihydroxyphenylethylamine) appears to be a neurotransmitter in some brain neurons (see below), and to be the precursor of a different neurotransmitter, NE, within other neurons. The conversion of DA to NE is catalyzed by an enzyme, dopamine-β-oxidase, which, like tyrosine hydroxylase but unlike dopa decarboxylase, appears to be restricted to catecholamine-producing cells and to be present in relatively small amounts. Dopamine-β-oxidase is localized within NE storage granules, i.e., neural synaptic vesicles (Potter and Axelrod, 1963) and adrenal chromaffin granules (Kirshner, 1959).

The fates of brain DA and NE are complex. Within any neuron, the catecholamines probably exist in more than one metabolic compartment. This complicates attempts to measure catecholamine synthesis and turnover, or to relate catecholamine turnover (i.e., the sum of the processes causing catecholamine molecules to disappear from the neuron) to the physiological activity of the neuron. The bulk of the catecholamine within any neuron is concentrated within synaptic vesicles. In sympathetic nerve terminals and in some, if not all, central noradrenergic neurons, these vesicles contain a central core, or granule, which is visible after tissues have been fixed with osmium tetroxide (see this Section, Granular Vesicles as Criteria for Intracellular Localization of Monoamines). The "bound" catecholamine is presumed to be in equilibrium with a much smaller pool, possibly identical with the pool that inhibits tyrosine hydroxylase, which is in the cytoplasm or perhaps in the plasma membrane of the terminal bouton. Nerve stimulation causes catecholamine molecules to be released into the synaptic cleft. A small fraction of these molecules presumably interacts with "receptors" on the postsynaptic membrane.

The physiological actions of NE within the synaptic cleft are terminated largely by the process of reuptake into the presynaptic terminal. The nerve terminal has been likened to a wet sponge: when the sponge is squeezed (i.e., by a nerve impulse), NE molecules drip out; these reenter the metaphoric sponge when it is allowed to resume its normal shape (i.e., when the nerve impulse has terminated). Within the synaptic cleft NE and DA can be inactivated by enzymatic mechanisms (Figure 4); they are O-methylated through the action of the enzyme catechol-O-methyl transferase (COMT). They can also be destroyed enzymatically through the action of monoamine oxidase (MAO). Catecholamines can be metabolized by MAO at two different loci: within their neuron of origin and, if circulating, within the liver or other viscera. In fact, the major fraction of the NE molecules within the nerve terminal never have the opportunity to do physiological work; instead, they are destroyed within the nerve terminal by oxidative

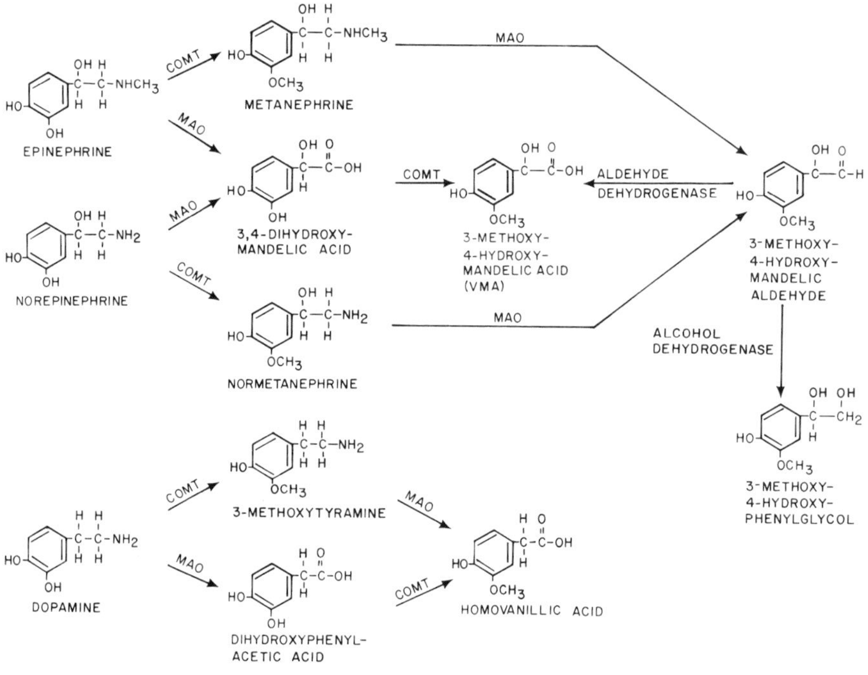

Figure 4. Metabolism of brain catecholamines. Some minor metabolites are omitted from this diagram. These include the alcohols formed from the reduction of aldehydes generated by the oxidative deamination of DA, NE, and epinephrine. The aldehydes have not been demonstrated to be present in tissues, and are presumed to be reactive intermediates. [Modified from Axelrod]

deamination, and leave the neuron as physiologically inert metabolites (Kopin et al., 1962; Wurtman, 1966). When the investigator measures the turnover of brain NE by usual methods, his measurements largely describe the behavior of these physiologically inert molecules.

*Quantitative Studies*

No method has yet been described that allows a quantitative estimation of the turnover of catecholamines in the human brain, much less a measurement of the number of molecules that leave the neuron as a result of nerve stimulation. At least four techniques have been proposed for estimating catecholamine synthesis or turnover in the brains of experimental animals. All of these techniques are based on assumptions that are only partly true; hence, the data that they generate, while useful for comparative purposes, cannot be taken as truly quantitative.

In the first approach, animals receive isotopically labeled tyrosine systemically, and the rate at which $^3$H-catecholamines accumulate in brain is monitored as an index of catecholamine synthesis. This approach has several disadvantages. $^3$H-Tyrosine levels in brain vary in a complex manner after the systemic administration of the amino acid (Zigmond and Wurtman, 1970). Moreover, the $^3$H-amino acid molecules present within catecholaminergic neurons cannot be distinguished from those in other neurons or in glia. As a consequence, it is almost impossible to estimate the specific activity of the $^3$H-tyrosine precursor pool in the few cells that actually synthesize $^3$H-NE or $^3$H-DA. Furthermore, the release and intraneuronal metabolism of radioactive NE commence within minutes after the administration of $^3$H-tyrosine, and the synthesis of $^3$H-catecholamines from $^3$H-tyrosine continues for a long time after the administration of the amino acid. Hence, measurements of brain $^3$H-catecholamine levels at almost any interval after the administration of the amino acid reflect both synthesis and turnover, and tend to lead to underestimation of both rates.

In the second approach, $^3$H-DA or $^3$H-NE is injected into the lateral cerebral ventricle or the cisterna magna, and subsequently taken up within catecholaminergic neurons in the brain. The rate at which brain $^3$H-catecholamine levels decline is followed and taken as an index of catecholamine turnover. However, exogenous $^3$H-NE or $^3$H-DA injected into the cerebrospinal fluid may not become localized uniformly and uniquely within catecholamine-containing cells. There is evidence that the likelihood that a given noradrenergic cell will take up exogenous catecholamine depends in part on the dose administered (see this Section, Autoradiographic Studies) and on the amount of time that has passed subsequent to the administration of the catecholamine (Iversen and Glowinski, 1966). Furthermore, as described above, at the cellular level, catecholamine turnover reflects more than only the physiological release of NE or DA into the synaptic cleft; it also varies with the efficiency of catecholamine reuptake and with the rate of intraneuronal metabolism by oxidative deamination. Hence, even a foolproof estimate of catecholamine turnover need not necessarily provide useful information about the physiological activity of noradrenergic or dopaminergic neurons.

In the third approach, animals are given a synthetic amino acid, $\alpha$-methyl-$p$-tyrosine ($\alpha$-MPT), which inhibits the enzyme tyrosine hydroxylase and thus blocks catecholamine biosynthesis. The rate at which brain catecholamine levels fall is then monitored. It is sometimes assumed that this decrease is proportional to the amount of catecholamine that would normally have been synthesized in the animal had it not received the tyrosine hydroxylase inhibitor. However, there is no physiological basis for the assumption that a marked decrease in the synthesis of brain NE will not also modify the turnover of the catecholamine. Indeed, as discussed above (Weiner and Rabadjija, 1968), experimental evidence has recently been presented that turnover and synthesis are physiologically coupled. The administration of $\alpha$-MPT disturbs the steady-state relationships that control brain NE levels. The subsequent decline in brain NE reflects the excess of NE turnover over NE synthesis (Wurtman, Anton-Tay, and Anton, 1969; see also Section V, Analysis of Norepinephrine Metabolism in Nonsteady States).

In the fourth approach, animals are treated with drugs that inhibit the enzyme MAO, and the rate at which brain catecholamine levels increase is monitored. It is often assumed that this increase is proportional to what catecholamine turnover would be in brains of untreated animals. This method is subject to the same general objections as the third method above; i.e., it induces a disturbance of the steady state, without taking into account that changes in catecholamine turnover might also influence the rate of catecholamine synthesis. Moreover, as described above, the portion of catecholamine turnover blocked by inhibitors of MAO is precisely the portion that is *not* related to the physiological activity of the catecholaminergic neurons.

Serotonin

The biosynthesis of brain 5-HT (Figure 5) is initiated by the uptake of the amino acid tryptophan from the plasma. The concentration of tryptophan exhibits diurnal rhythms in plasma (Wurtman et al., 1968c) and in brain*; tryptophan concentrations vary as much as twofold during each 24-hour period. Tryptophan appears to have a role that is unique among amino acids in controlling protein metabolism. In addition to its special role as the precursor for a putative brain neurotransmitter, i.e., 5-HT, tryptophan is the limiting amino acid in most dietary proteins; moreover the entry of tryptophan into the body is specifically limited by the hepatic enzyme tryptophan pyrrolase, which is rapidly induced when the mammal starts to consume protein. Hence total body tryptophan stores are lowest of all the amino acids, and tryptophan can serve a regulatory function. Pronczuk et al. (1968) and others have shown that the availability of tryptophan controls protein synthesis in the liver, and perhaps elsewhere. In the absence of dietary tryptophan, most of the RNA in the rat liver cell is disaggregated and sediments on sucrose density gradients with the monosomal and disomal fractions. After the animal is given access to tryptophan (by eating protein), these subunits aggregate to form polysomes, and protein synthesis is reinitiated.

Tryptophan levels in the brain are relatively low, and there is abundant evidence that changes in these levels induced by tryptophan administration or tryptophan-free diets produce parallel changes in brain 5-HT content (Moir and Eccleston, 1968; Wurtman and Fernstrom, 1971). The first biochemical transformation in 5-HT biosynthesis involves the 5-hydroxylation of tryptophan by tryptophan hydroxylase to form 5-hydroxytryptophan (5-HTP). This amino acid, like L-dopa, is rapidly decarboxylated, through the action of the enzyme aromatic-L-amino acid decarboxylase, to form the biogenic

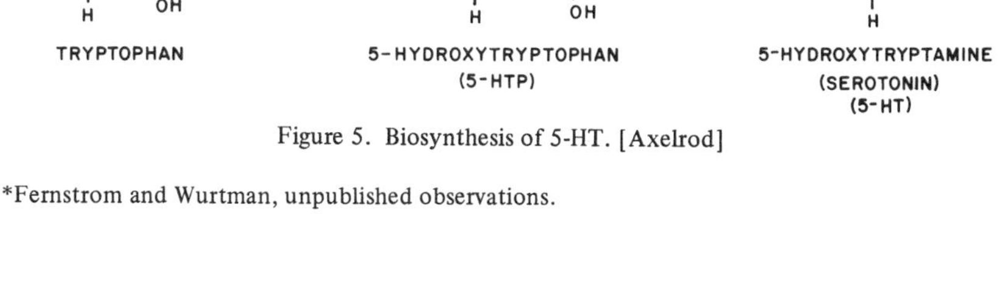

Figure 5. Biosynthesis of 5-HT. [Axelrod]

*Fernstrom and Wurtman, unpublished observations.

amine 5-HT. It is not known whether tryptophan hydroxylase exercises the same role in the regulation of 5-HT biosynthesis that tyrosine hydroxylase exercises in the control of catecholamine synthesis. This seems unlikely. Brain tryptophan levels, as described above, are low; and the $K_m$ for tryptophan hydroxylase is much greater than that for tyrosine hydroxylase. Hence, the extent to which tryptophan hydroxylase is saturated with substrate, rather than the amount of the enzyme in the cell, is probably of primary importance in controlling 5-HT biosynthesis. Apparently, 5-HT does not exert feedback control over its own synthesis similar to the end-product inhibition of tyrosine hydroxylase ascribed to NE (Jequier et al., 1969).

Much less is known about the fate of 5-HT within the brain than about NE. It has proved considerably more difficult to introduce a radioactive label into the brain 5-HT stores than to label brain catecholamines. The main pathway for 5-HT metabolism appears to involve oxidative deamination (catalyzed by MAO); the resultant product, an aldehyde, can then be oxidized to 5-hydroxyindole acetic acid, or reduced to 5-hydroxytryptophol (Figure 6). It is not known whether 5-HT released from nerve endings is also inactivated by reuptake.

The mammalian pineal organ has the unique capacity to transform 5-HT and related compounds (i.e., N-acetylserotonin) to methoxyindoles such as melatonin (Axelrod and Weissbach, 1961; Wurtman, Axelrod, and Kelly, 1968a). These compounds are secreted, possibly into the cerebrospinal fluid, and may produce their physiological effects by acting on serotoninergic neurons within the brain. The administration of exogenous melatonin to rats causes a rapid rise in brainstem 5-HT levels (Anton-Tay, Chou, Anton, and Wurtman, 1968).

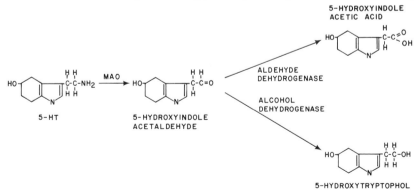

Figure 6. Metabolism of 5-HT. [Wurtman]

Regional Distribution of Brain
Catecholamines and Serotonin: A. Dahlström

Dahlström described the distribution of catecholamines and
5-HT in the various regions of the mammalian CNS. She compared
observations made by the methods of chemical assay, histochemical
fluorescence, and autoradiography.

Using bioassays, Vogt (1954) first demonstrated regional differences in the distribution of NE in the brain beyond those known to
correlate with the sympathetic innervation of blood vessels. On this
basis, she first suggested that NE might function as a central neurotransmitter. Subsequently, several experimenters assayed the concentrations
of NE and DA in macrodissected brain regions by chemical methods
(reviewed by Iversen, 1967; see Table 2). Although the exact concentrations in any particular brain region differ slightly among mammals, the
percent of total brain NE present in a given region is surprisingly
constant from species to species. NE is widely distributed in the CNS,
but in small amounts, although the concentrations in hypothalamus and
brainstem are noticeably higher than those in other regions. Strikingly
high concentrations of DA are present in the neostriatum (caudate-
putamen).

**Histochemical Fluorescence Studies**

Chemical assays yielded important data on the regional distribution of brain amines, but they did not allow these compounds to be
localized at a cellular level. Dahlström and her colleagues in Stockholm
have been studying the localization of monoamines in individual
neurons by the precise histochemical fluorescence method of Falck and
Hillarp (e.g., Falck et al., 1962), in which sections of nervous tissue are
rapidly freeze-dried and exposed to formaldehyde vapor. The tissue
monoamines condense with the formaldehyde to produce characteristic
highly fluorescent compounds. Catecholamines form green to yellow-
green isoquinolines; 5-HT, a yellow carboline. Although once a
problem, small 5-HT-containing axons can now be visualized among
larger catecholamine-containing axons; treatment with the monoamine
depletor reserpine plus the MAO inhibitor nialamide causes a temporary
increase in 5-HT fluorescence without inducing catecholamine fluorescence. The catecholamines can sometimes be distinguished from one
another. Epinephrine requires more severe reaction conditions to form

TABLE 2

Distribution of Catecholamines in the Central Nervous System [Iversen, 1967]

| | | Catecholamine content ($\mu$g/g) | | |
|---|---|---|---|---|
| Brain region | Species | NE | DA | Reference |
| Whole brain | Dog | 0.16 | 0.19 | |
| Whole brain | Rabbit | 0.29 | 0.32 | |
| Whole brain | Rat | 0.49 | 0.60 | Bertler & Rosengren, 1959 |
| Whole brain | Cat | 0.22 | 0.28 | |
| Medula oblongata | Dog | 0.37 | 0.13 | |
| Medulla oblongata | Rat | 0.72 | —— | Glowinski & Iversen, 1966 |
| Mesencephalon | Dog | 0.33 | 0.20 | Bertler & Rosengren, 1959 |
| Pons | Dog | 0.41 | 0.10 | Bertler & Rosengren, 1959 |
| Cerebellum | Rat | 0.17 | —— | Glowinski & Iversen, 1966 |
| Cerebellar cortex | Dog | 0.07 | —— | Vogt, 1954 |
| Cerebellar cortex | Dog | 0.06 | 0.03 | Bertler & Rosengren, 1959 |
| Hypothalamus | Rat | 1.79 | —— | Glowinski & Iversen, 1966 |
| Hypothalamus | Rat | 1.29 | 0.14 | Laverty & Sharman, 1965 |
| Hypothalamus | Dog | 1.00 | —— | Vogt, 1954 |
| Hypothalamus | Cat | 1.40 | —— | Vogt, 1954 |
| Striatum | Rat | 0.25 | 7.50 | Glowinski & Iversen, 1966 |
| Caudate nucleus | Rat | 0.27 | 6.39 | Laverty & Sharman, 1965 |
| Caudate nucleus | Cat | 0.10 | 9.90 | Laverty & Sharman, 1965 |
| Lentiform nucleus | Dog | 0.08 | 1.63 | Bertler & Rosengren, 1959 |
| Caudate nucleus | Dog | 0.10 | 5.90 | Bertler & Rosengren, 1959 |
| Cerebral cortex | Dog | 0.05 | —— | Vogt, 1954 |
| Cerebral cortex | Rat | 0.18 | <0.01 | Laverty & Sharman, 1965 |
| Cerebral cortex | Rat | 0.24 | —— | Glowinski & Iversen, 1966 |

a fluorescent product than do DA and NE (Falck and Owman, 1965). The Stockholm investigators can sometimes distinguish DA from NE with the aid of pharmacological pretreatment with the monoamine depletor $\alpha$-methyl-$m$-tyrosine ($\alpha$-MMT); the DA stores are replenished more rapidly and can be seen in the absence of NE (Fuxe, 1965a). (As will be described later, since the time of the Work Session Björklund et al. (1968) have developed a microspectrofluorimetrical method for distinguishing DA from NE.) Precursors of the monoamines (i.e., dopa, 5-HTP) could also form fluorescent products, but their levels in the brain are normally too low to interfere with the monoamine fluorescence.

In catecholaminergic neurons, the catecholamines are more concentrated (i.e., their fluorescence intensity is greater) in varicosities at axonal endings than in perikarya and in nonterminal portions of the axon, although 5-HT appears to be more homogeneously distributed within serotoninergic neurons. Electron microscopy shows that vesicles thought to store monoamine neurotransmitters also are more concentrated in the varicosities. The Stockholm group found that after several pharmacological treatments that deplete the monoamine stores in the whole neurons (e.g., reserpine) the recovery of fluorescence starts at the soma and then proceeds progressively outward (Dahlström and Fuxe, 1964); so they suggest that the monoamine-storing vesicles are synthesized in the soma and transported to axonal terminals by axonal flow. Their method for mapping monoaminergic neuronal pathways (some of which are apparently quite long) utilizes this sequence of events. They make brain lesions and then follow anterograde and retrograde changes in neuronal fluorescence after axotomy. Distal to the lesion the monoamine disappears; proximally the fluorescence increases and the axon swells, suggesting a damming effect. As shown in Figure 7,

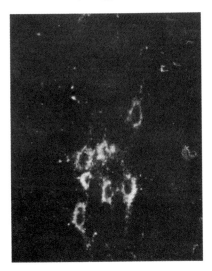

Figure 7. Left: Noradrenergic nerve cell bodies in the nucleus reticularis lateralis of the medulla oblongata of normal rat. Right: Noradrenergic nerve cell bodies in the same area following a 4-day-old lesion placed in the tegmental reticular midbrain at the level of the inferior colliculus. The figures demonstrate the retrograde cell body changes (increased fluorescence intensities, apparently increased cell volume, and displaced nucleus) observed in neurons of which the axons have been lesioned. [Andén, Dahlström, Fuxe, et al., 1966]

retrograde increase in fluorescence also occurs in the soma; this is accompanied by a displacement of the nucleus. To visualize serotoninergic neurons, pretreatment with an MAO inhibitor is useful to increase the fluorescence intensity (Dahlström and Fuxe, ,1964; Fuxe, 1965a).

Dahlström claimed that the relative intensity of the fluorescence in a given brain region provides a semiqualitative measure of its content of monoamines. Within a limited range, i.e., in tissues partially depleted of the compounds, the correlations are reasonably good between estimates of regional catecholamine content based upon chemical assays and estimates based upon histochemical studies. However, according to chemical assays done by the Stockholm group (Fuxe et al., 1968), the correlation between 5-HT content and apparent 5-HT fluorescence (Andén, 1967, and unpublished; Andén, Fuxe, and Ungerstedt, 1967) is somewhat variable, particularly in thalamus, amygdala, and the septal region. This may reflect the fact that the fluorescence method for 5-HT is less sensitive than that for catecholamines; the 5-HT fluorescent product is much more labile to ultraviolet light, with an apparent half-life of only a few seconds in the microscope.

*Cellular Localizations*

Brain neurons whose cell bodies contain monoamines have been identified in only a relatively small portion of the brainstem (Dahlström and Fuxe, 1964) and in parts of the hypothalamus (Fuxe, 1964). Brainstem noradrenergic neurons occur mostly in the tegmentum of the mesencephalon, with fairly large numbers in the pons and medulla oblongata*; serotoninergic neurons are localized to the raphe nuclei; brainstem dopaminergic neurons are concentrated within the substantia nigra, and may also be found surrounding the interpeduncular nucleus. The main monoaminergic neuron systems mapped out by fluorescence histochemical techniques are shown in Figure 8.

*Aminergic Tracts from the Brainstem:* The brainstem neurons give rise to several descending and ascending tracts:

1. NE and 5-HT bulbospinal tracts (Dahlström and Fuxe, 1964).

2. Large NE and 5-HT systems, most of the fibers ascending in the medial forebrain bundle, to the limbic system, as well as to the

---

*Using immunohistochemical techniques and inhibitors of dopamine-β-hydroxylase, Corrodi, Fuxe, et al. (1970) have recently obtained evidence that noradrenergic neuron somata are strictly localized to the pons and medulla oblongata.

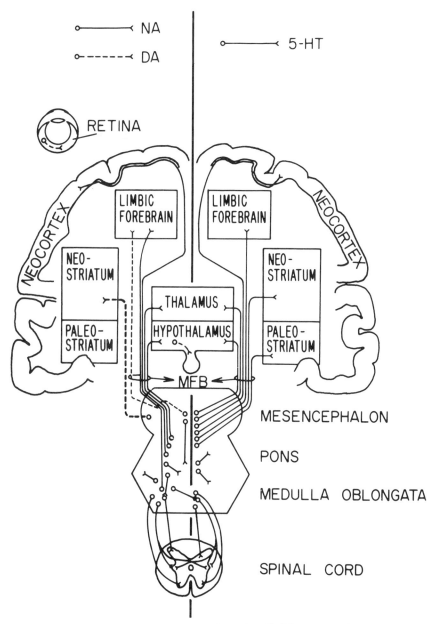

Figure 8. Schematic drawing illustrating, in highly simplified form, the main monoaminergic neuron systems in the CNS, as mapped out with the use of the Hillarp-Falck fluorescence method in combination with lesion techniques and biochemical estimations of monoamines. [Andén, Dahlström, Fuxe, et al., 1966]

hypothalamus and preoptic area (Dahlström, Fuxe, et al., 1964; Andén, Dahlström, Fuxe, and Larsson, 1965; Andén, Dahlström, Fuxe, et al., 1966; Fuxe et al., 1970), and to cerebellum and cerebral cortex (Andén, Fuxe, and Ungerstedt, 1967). As will be discussed later, however, the existence of monosynaptic aminergic tracts to the telencephalon within the medial forebrain bundle is not universally accepted. Moore (see Section V, Interpretation of Effects of Lesions on Localization of Brain Monoamines) and others fail to observe typical axonal degeneration in the limbic telencephalon following medial forebrain bundle lesions that cause telencephalic NE levels to decline. On this basis, Moore suggests that the fall in NE levels results from a transsynaptic effect, i.e., a decrease in input to noradrenergic cell bodies distal to the lesion, and that the noradrenergic fibers in the medial forebrain bundle represent the terminals and short axons of a multisynaptic pathway. This difference in interpretation remains unresolved.

   3. A large DA nigro-neostriatal system, which ascends from cell bodies in the substantia nigra to the caudate nucleus and putamen, via the internal capsule (Andén, Carlsson, Dahlström, Fuxe, et al., 1964).

   4. A DA system innervating the limbic forebrain, especially the nucleus accumbens and the tuberculum olfactorium, probably arising from the neurons surrounding the interpeduncular nucleus (Andén, Dahlström, Fuxe, et al., 1966).

   *Aminergic Tracts from the Hypothalamus:* The hypothalamic catecholaminergic neurons send fibers to the median eminence and to the proximal infundibular stem of the neurohypophysis (Fuxe, 1964). Fuxe suggested, on the basis of treatment with α-MMT, that the terminals were mainly dopaminergic, but that noradrenergic terminals were also present. The terminals arise from nonsympathetic nerve fibers and are densely packed in all neurohypophyseal regions where the primary plexus of the hypophyseal portal system arises. This led Fuxe to propose that the terminals were part of the tuberoinfundibular system (Szentágothai et al., 1962), with the cells of origin being in the arcuate nucleus and the ventral part of the anterior periventricular nucleus of the hypothalamus; and subsequent lesion studies by Fuxe and Hökfelt (1966) confirmed this proposal.

   Since the time of the Work Session, Björklund et al. (1968) described a modification of the histochemical fluorescence method for catecholamines that allows direct differentiation of NE-containing from DA-containing structures. This microspectrofluorimetric method involves exposing tissue sections, previously treated with formaldehyde, to vapors of hydrochloric acid. The acidification causes a marked decrease in the intensity of the NE fluorophore; in contrast, the fluorescence intensity of the DA fluorophore is retained, although its excitation maximum shifts.

Using this modification, Björklund et al. (1970) have mapped aminergic fiber systems running between the rat hypothalamus and pituitary, and have identified the dominant monoamine in several of these systems. They observed the following five groups of monoamine-containing axons entering the median eminence:

1. A large group of DA-containing axons that originate in the cells of the arcuate nuclei and the ventral parts of the anterior periventricular nuclei.

2. A large group of NE-containing axons that probably originate beyond the mediobasal hypothalamus. These fibers intermingle with the DA-containing fibers in the internal layer and the deeper part of the external layer of the median eminence.

3,4. Two minor groups of axons that contain a primary catecholamine,* one that reaches the median eminence (probably mingled with the supraopticohypophyseal tract), and another passing at least as far as the arcuate nucleus. The nature of the catecholamine and the site of origin of these fibers have not yet been identified.

5. A small group of scattered axons containing an unidentified fluorigenic substance differing from the catecholamines or 5-HT, but suspected by the authors to be an indole derivative.

The arcuatohypophyseal DA neurons appear to give rise to some of the terminals in the external zone of the median eminence (i.e., the terminals first observed by Fuxe, 1964), and to most, if not all, of the terminals in the internal zone. The tuberohypophyseal NE fibers terminate in the external zone, and also contribute to the catecholamine innervation of the neural and intermediate lobes of the pituitary.†

*Primary catecholamine: A compound with fluorescence characteristics similar to those of DA or NE but which has not yet been demonstrated to be one of these compounds by rigid chemical criteria.

†Fuxe has unpublished observations in disagreement with some of Björklund's conclusions. Using the Halász knife technique (described by Gorski in Section V), Fuxe and Hökfelt found that nearly all NE nerve terminals disappeared in the hypothalamus; so they suggest that no short-axon NE neurons are present in the hypothalamus. Also, from pharmacological studies involving dopamine-β-oxidase inhibitors, Fuxe concludes that practically all of the catecholaminergic nerve terminals in the external layer contain DA and practically all of the terminals in the internal layer contain NE. He further suggests that the NE terminals derive from ascending NE axons from the pons and medulla oblongata.

These authors comment on the fact that their histochemical findings are consistent with chemical evidence that the median eminence contains relatively large amounts of NE as well as of DA (cf. Rinne and Sonninen, 1968), and they suggest that studies on the functional role of aminergic mechanisms in the hypothalamic regulation of pituitary function should consider the possible significance of NE as well as that of DA. Dahlström added that since the time of the Work Session Ungerstedt has performed a detailed study of monoaminergic systems, which will be published in *Brain Research* in 1971.

**Autoradiographic Studies**

Dahlström also compared the Stockholm group's data on the distribution of brain monoamines with that obtained by Reivich and Glowinski (1967) using autoradiographic methods. These authors followed the uptake of $^{14}$C-NE in various regions of rat brain after its intraventricular injection. In general, the correlation between the results of the two methods was good. The relative concentration of the label was similar to the relative fluorescence intensity in spinal cord, the limbic system, substantia nigra (DA neurons also took up the NE label), medial forebrain bundle fibers, preoptic area, hypothalamus, and neostriatum, among others.

The only discrepancies were found in the trapezoid body and the oculomotor nuclei; these structures took up label, but by fluorescence were free of monoamines. However, fluorescence studies of intraventricularly injected nonradioactive monoamines indicated potential sources of error in the isotopic methods used by Reivich and Glowinski (Fuxe and Ungerstedt, 1968); i.e., the injected monoamines were not distributed in a pattern similar to that of endogenous amines, but could be detected intraneuronally only at distances from the ventricles and subarachnoid spaces that were less than 300 $\mu$. These results suggested that the uptake of labeled monoamines from the cerebrospinal fluid does not provide an adequate map of monoaminergic neurons. Moreover, higher doses of NE tended to be taken up nonspecifically in pericytes and endothelial cells. However, the discrepancy between the findings of Reivich and Glowinski and of Fuxe and Ungerstedt may be related to the considerably greater sensitivity of the autoradiographic method: the density of NE molecules required to be visualized by histochemical fluorescence is much greater than that necessary for demonstration by autoradiography. On the other hand,

Dahlström and Fuxe pointed out that the autoradiographic method does not permit differentiation between the monoamines and their metabolites; this possible source of error does not apply to the fluorescence method because the metabolites do not fluoresce.

It should also be noted that the studies both of Reivich and Glowinski and of Fuxe and Ungerstedt utilized amounts of NE (2-8 μg) that were considerably greater than normally present in brain. Other investigators who study NE turnover in brain inject smaller quantitites of labeled NE, which would be more likely to label NE neurons specifically.

### Granular Vesicles as Criteria for Intracellular Localization of Monoamines: F. E. Bloom

Bloom discussed the support that fine structural analyses give to the idea that electron-opaque synaptic vesicles are the storage organelles for monoamines. For peripheral sympathetic nerves, the evidence is quite conclusive that granular synaptic vesicles, 400-600 A in diameter, are the main storage site of NE. However, the evidence remains less conclusive for the CNS, where larger 800-1200 A granular vesicles (LGV) are under consideration, as well as 400-600 A vesicles.

In the peripheral nervous system, the small granular vesicles (SGV) are observed in every sympathetically innervated structure (Grillo, 1966; Richardson, 1966); and when sympathetic nerves are destroyed, the vesicles disappear (Pellegrino de Iraldi et al., 1965). Extensive data also relate the granular vesicles to measures of mono-amine content. First, after labeling with radioactive catecholamines (Wolfe et al., 1962) or precursors (Taxi and Droz, 1966), the autoradio-graphic grains are seen by electron microscopy to concentrate over nerve endings or axons that contain the vesicles. Second, the electron opacity of the SGV varies with the NE content of the nerve, as judged by the intensity of histochemical fluorescence and by biochemical assays for NE, in studies utilizing depleting drugs to reduce the levels of catecholamines stored (Van Orden, Bloom, et al., 1966; Van Orden et al., 1967; Gillis et al., 1966; Bloom and Barrnett, 1966). Third, the granular contents of the vesicles can be restored after depletion by application of NE in vitro (Van Orden, Bloom, et al., 1966) and in vivo (Bondareff and Gordon, 1966; Tranzer and Thoenen, 1967). Thus, the data strongly indicate that the granular material itself represents monoamine content. This idea is strengthened by the results of electron microscopic histochemical experiments on adrenal medullary chromaffin granules (Coupland and Hopwood, 1966) that show that NE-containing granules are made electron opaque by the fixatives glutaraldehyde and osmium tetroxide.

Small granular vesicles (400-600 A) are present in the CNS, but until recently granularity had been observed only in the larger 800-1200 A vesicles. Experimental approaches similar to those used in the periphery resulted in some correlations between the LGV and nerve monoamine content. First, the vesicles were shown to be present in almost all of the nerve endings and axons over which autoradiographic grains appeared after labeling with $^3$H-NE, $^3$H-dopa, and $^3$H-5-HT (Aghajanian and Bloom, 1966, 1967a,b; Lenn, 1967). Second, the incidence of the vesicles in several regions of rat brain correlates with biochemical and fluorescent-histochemical measurements of NE and 5-HT levels, although not of DA (Bloom and Aghajanian, 1968a; Fuxe et al., 1965). Also, degeneration studies showed that after lesions in the raphe nucleus, which is rich in 5-HT, the degenerating nerve fiber endings in the suprachiasmatic nucleus contain large granular vesicles (Aghajanian, Bloom, and Sheard, 1969).

As pointed out by Fuxe at the Work Session, granularity in small CNS vesicles was obtained by Hökfelt (1967) by fixing slices of rat median eminence and locus coeruleus, areas known to be rich in monoamines, with concentrated $KMnO_4$. Hökfelt used Richardson's method (Richardson, 1966), which in the sympathetic nervous system resulted in the highest frequency of staining of SGV. A second larger granular vesicle also occurs in the terminals containing stained small vesicles. Fuxe suggested that the large vesicles represent a second storage pool, perhaps having to do with monoamine metabolism. Bloom and Aghajanian (1968b) also found a method for making small vesicles electron opaque. Small granular vesicles, similar in appearance to the $KMnO_4$-fixed vesicles, result from gluteraldehyde-$OsO_4$ double fixation, if the $OsO_4$ fixation is done at an elevated temperature (60°C). Presumably the high temperature increases the potency of $OsO_4$ as an oxidizing agent, so that its effects are similar to those of cold $KMnO_4$.

Bloom's studies, however, led him to question whether granularity indicates the presence of the monoamine itself, in either large or small vesicles in the CNS (Figures 9 and 10). Pharmacological depletion studies on large vesicles (Bloom and Aghajanian, 1968a) and on small vesicles (Bloom and Aghajanian, 1968b) made granular with hot $OsO_4$

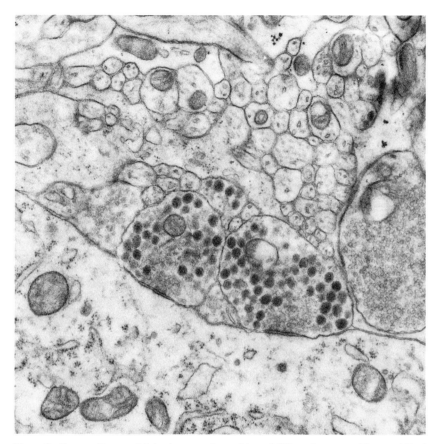

Figure 9. Nerve endings containing large numbers of dense LGV can be seen making specialized axosomatic contacts in the paraventricular hypothalamus of a reserpine-treated rabbit. The electron opacity of the granular vesicles is equal to or greater than the matrix of the nerve ending mitochondrion. Note that the LGV do not enter the vicinity of the specialized contact. [Bloom and Aghajanian, 1968a]

failed to relate granularity to monoamine levels. Furthermore, Bloom and Aghajanian found that the elevated temperature caused a loss of half of the $^3$H-NE grains in labeled tissue, indicating that the process that permits visualization of the granular material extracts the catecholamine. Therefore, Bloom suggests that, while granular vesicles may be an index to monoamine storage, the granular material is not exclusively the monoamine itself, but includes a proteinaceous matrix substance or other intravesicular component. These results have recently been analyzed in detail (Bloom, 1970).

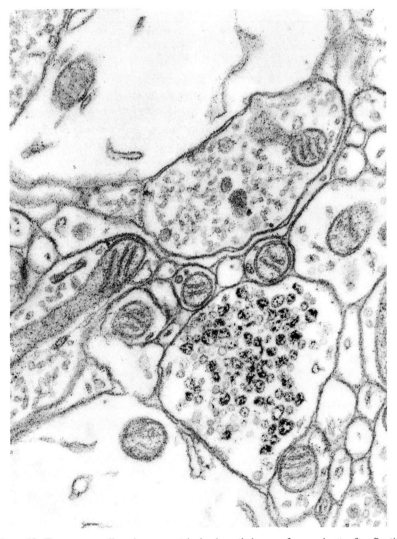

Figure 10. Two nerve endings in paraventricular hypothalamus of normal rat, after fixation with glutaraldehyde and exposure to 1 percent OsO$_4$ at 60°C for 30 min. Synaptic vesicles in lower ending are filled with electron-opaque osmiophilic precipitates, while vesicles in upper nerve ending are electron lucent. [Bloom and Aghajanian, 1968b]

## Possible Function of Monoamines as Neurotransmitters:
## F. E. Bloom

Bloom reviewed evidence that the biogenic monoamines, especially NE, function as synaptic neurotransmitters in the CNS. Although studies such as those given earlier show that NE, DA, and 5-HT are present in the brain, Bloom stressed that the mere presence of the substance is not sufficient to identify it as a transmitter; putative neurotransmitters are also present in nonneural tissues—5-HT in platelets, ACh in placenta. Of four criteria generally used to identify neurotransmitters at peripheral synapses, two or three have been satisfied for several central pathways; but all four criteria have not yet been satisfied for any synapse in mammalian brain. Nevertheless, most neurophysiologists grant grudging acceptance to the hypothesis that one or more monoamines (plus, of course, ACh) do fill this function.

The four criteria usually suggested for identifying substances as central neurotransmitters are the following:

1. That the substance be released during stimulation of the nerve.

2. That postsynaptic application of the substance mimic the effect of stimulating the presynaptic nerve.

3. That drugs which specifically block or potentiate natural transmission have parallel effects on the response to the applied substance.

4. That the substance, and its associated synthetic and catabolic enzymes, be present in the region of the synapse.

These criteria are testable in the periphery by macromethods (for example, the classic ACh studies of Loewi, 1921), but not in the CNS. Accordingly, Bloom also discussed newer electrophysiological and pharmacological micromethods now available for the characterization of substances as CNS neurotransmitters.

### Microelectrophoretic Studies

The only method currently available for testing the electrophysiological criterion (2) in the CNS is microelectrophoresis. This technique permits simultaneous application of test transmitter substances and measurement of neuronal potentials. The method is also useful in testing the pharmacological criterion (3); i.e., drugs that reach the synaptic region with difficulty following systemic administration may be microelectrophoretically applied to individual neurons. Using a fused multibarrel assembly of glass microcapillary electrodes, neuronal potentials are recorded from the central electrode while ionized solutions of drugs or transmitter substances are ejected from the surrounding micropipet barrels by "microelectrophoretic" currents. The drugs are held in the electrodes by a "holding" current, and the ejecting microelectrophoretic current is simply a reversal in polarity of the holding current. Thus, the drug application is localized and time-controlled, and it circumvents blood-brain permeability problems. In the experiments described by Bloom the recording electrode measured extracellular activity.

Although the microelectrophoretic method is more direct than application of drugs via the blood stream or by a gross syringe, Bloom pointed out that the interpretation of the observed response does not permit certain knowledge of the exact molecular site of action (Figure 11). For example, if the drug has an excitatory effect by purely synaptic mechanisms, the excitation mechanism could be either a direct excitatory action on one synapse or an inhibitory effect on an inhibitor of the synapse. Even direct effects of drugs on a specific transmitter at a specific synapse could act on a variety of stages of transmitter activity: storage, release, diffusion across the cleft, inactivation, and receptor site activity. In addition, the drugs diffuse away from the pipet in several directions, so that the effects could be nonsynaptically mediated—as direct membrane permeability alterations on nonsynaptic neuronal membranes, general metabolic effects, or effects on glial cells.

*Microelectrophoretic Responses*

Despite the interpretative difficulties with the method, it yields reproducible and interesting results. Bloom tests for general responsiveness of neurons to applied chemicals by the change in firing rate—an increase indicating excitation; a decrease, inhibition. Characteristically, the responses to monoamines begin seconds after the application and outlast it by several seconds. In contrast, the response to ACh is of the same duration as its period of application. Any one neuron tends to respond to more than one monoamine and ACh as well; in spinal cord and hypothalamus the ability of a neuron to respond to one substance was found not to be a predictor either of ability to respond or type of response to any other substance (NE, 5-HT, ACh).

In some regions of the CNS, gross correlations have been made between microelectrophoretic responsiveness of neurons to NE and

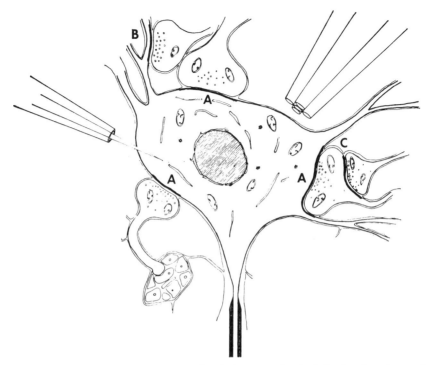

Figure 11. Diagrammatic representation of a single neuronal unit, indicating three types of synaptic arrangements [axosomatic (A), axodendritic (B), and axoaxonic (C)] at which postsynaptic (A and B) and presynaptic (C) drug effects could occur. Nonsynaptic portions of the membrane are covered by glia. Subsynaptic patches are emphasized by thickening of the postsynaptic membrane. Tips of the two types of electrode assemblies (concentric and five-barreled) used in microelectrophoretic studies are drawn to approximate scale. [Salmoiraghi and Bloom, 1964]

5-HT and their composition determined cytochemically. Units found responsive to NE in hypothalamus are located in areas containing the largest concentration of NE- and 5-HT-containing fibers (Bloom et al., 1963). Cells responsive to NE and to dexamethasone (both of which most often inhibit single unit activity) occur in similar regions of the midline hypothalamus and the midbrain (Steiner et al., 1969). In spinal cord, there is a general regional correlation between unit responses to NE and 5-HT and their monoaminergic inputs as identified by fluorescence nerve-ending histochemical data (Engberg and Ryall, 1966).

Identification of substances as specific transmitters according to the four criteria requires knowledge of specific pathways. In cat lateral geniculate, the effects of 5-HT have been studied on cells projecting to visual cortex, identified by their antidromic response to cortical stimulation and by an orthodromic response to visual stimulation. The orthodromic response is abolished by microelectrophoretic application of 5-HT (Curtis and Davis, 1962). However, drug effects cannot be tested because no specific inhibitor of 5-HT is available that is releasable microelectrophoretically.

### Norepinephrine as a CNS Transmitter

For NE, the pathway best documented by the microelectrophoretic technique at the time of the Work Session* was the mitral cell-lateral olfactory tract (LOT, primarily efferent mitral cell axons) system in rabbit olfactory bulb. (Lesion studies on rats have provided evidence that the olfactory bulb contains noradrenergic neurons, which project to the telencephalon (Pohorecky, Larin, and Wurtman, 1969), but such neurons have not yet been identified using histochemical fluorescence methods.) The rabbit olfactory bulb is the only system in which all four neurotransmitter criteria have been tested (microelectrophoretic studies by Salmoiraghi, Bloom, and Costa, 1964):

*1. Release of NE with Stimulation:* In vitro, rabbit olfactory bulb slices take up $^3$H-NE and release it with electrical stimulation (Baldessarini and Kopin, 1967). In vivo, functional olfactory stimulation leads to $^3$H-NE release, but labeled metabolically inert compounds such as $^{14}$C-urea are also released (Chase and Kopin, 1968), implying that this test system is not specific for neurotransmitters.

*2. Applied NE Mimicking Stimulation:* Stimulation of the LOT causes a 50 to 150 msec inhibition of extracellular mitral cell activity and an intracellular IPSP of similar duration. Electrophoretically applied NE inhibits mitral cell activity, but 5-HT and ACh do also, reemphasizing the additional necessity for pharmacological tests.

*3. Effects of Applied Drugs:* Dibenamine, known to be a peripheral α-adrenergic blocker, blocks the inhibition of mitral cell activity by NE; it also shortens the period of inhibition induced by LOT stimulation, but not the responses induced by ACh or 5-HT. Phentolamine and LSD, the only other blockers of synaptic transmission in a long series of drugs tested, also blocked both LOT- and NE-induced inhibition of mitral cell activity (Figure 12).

*More recent work has established the NE projection to rat cerebellar Purkinje cells as an even more prominent system (see Rall, T.W. and Gilman, A.G. (1970): The role of cyclic AMP in the nervous system. *Neurosciences Res. Prog. Bull.* 8:221-323).

OLFACTORY NEURON

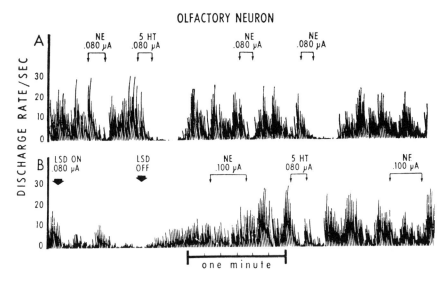

Figure 12. Continuous polygraph record of discharge rate of one olfactory neuron responding to both NE and 5-HT with reduction in rate of firing. NE response was blocked after LSD (0.10 μamp) electrophoresis, while 5-HT response was not. [Bloom et al., 1964]

Metaraminol and tyramine, false transmitters, increased the duration of the inhibitory period following LOT stimulation but did not affect the NE-induced inhibition. Acute intracellular depletion of monoamines with reserpine prevented the response to LOT stimulation from taking place, but not the response to applied NE. Chronic selective depletion of NE by α-MMT shortened the duration of the LOT-induced inhibition, although did not completely block it (averaged over many cells, the inhibitory period was decreased from 91 to 40 msec).

*4. Presence of NE:* In the LOT-mitral cell system of the rabbit, rather few catecholamine-containing (probably NE-containing) nerve endings have been observed on mitral cells (Dahlström, Fuxe, et al., 1965). However, studies by Rall and Shepherd indicate that this may be a peculiar system—the input to mitral cells not being recurrent axon collaterals. On the basis of theoretical analyses of field potentials and electron micrographic observations of synapses with vesicles on both sides of the cleft, Rall and Shepherd suggest that the synaptic transmission is effected by "two-way" dendrodendritic synapses between mitral cells and granule cells (Figure 13; Rall, Shepherd, et al., 1966; Rall and Shepherd, 1968; Rall, 1970; Shepherd, 1970).

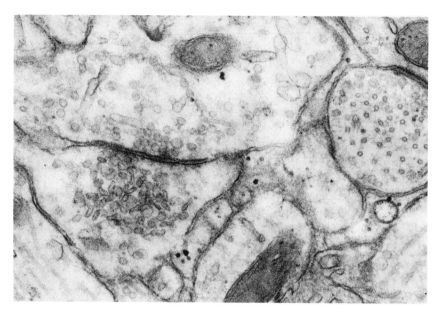

Figure 13. Micrograph of rat olfactory bulb illustrating the reciprocal synapses between the mitral cell dendrite (upper left) and the granule cell gemmule (lower left); note that the synaptic vesicles in the mitral cell are spherical and that those in the granule cell are flattened. A conventional synapse between an axon en passage containing spherical vesicles is seen at upper right. [Bloom]

## Possible Sites of Action of Monoamines in Hypothalamohypophyseal Function: R. J. Wurtman

Given the hypothesis that specialized neuroendocrine transducer cells in the hypothalamus release specific hypophysiotropic hormones into the pituitary portal circulation at a rate depending in part on their neurotransmitter inputs, several possibilities exist for distinct sites of action at which monoamines could participate in the activity of such cells. These sites may include not only synaptic influences upon the cells but also activity within them, since all three monoamines have been found to be present in the median eminence. Therefore, when relating monoamines to pituitary function, it should be realized that experimental variations in monoamine synthesis or release could modify the neuronal-hormonal link at a variety of sites (Figure 14).

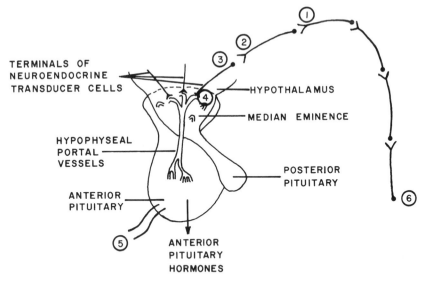

Figure 14. Possible loci at which monoamines might participate in the control of secretion from the anterior pituitary gland: (1) as neurotransmitters in a tract that provides an input to hypophysiotropic cells in the hypothalamus (e.g., the medial forebrain bundle), (2) as the neurotransmitters that cause the hypothalamic cell to secrete the releasing factor, (3) within the hypophysiotropic cell to liberate the releasing factor, (4) as neurohormones acting directly on the anterior pituitary, i.e., either as releasing factors or by modifying pituitary response to a "true" releasing factor. Monoamines might also affect anterior pituitary function from extracranial sources–administered systemically, or secreted from the adrenal medulla or sympathetic nerve endings: (5) they might be delivered to the pituitary by the general circulation and modify its secretion of releasing factors; (6) they might alter the sensory input to the brain; this might activate reflexes, secondarily altering the secretion of the releasing factor. [Wurtman]

## Brain Monoamines

Four possible CNS sites of monoamine action, not mutually exclusive, are the following:

1. There could be various monoaminergic neurons in multi-synaptic pathways that carry inputs from the limbic system, or from other brain regions that influence hypothalamic function, to the releasing-factor cells. For example, as described earlier by Dahlström, the medial forebrain bundle, perhaps the most important input to the medial hypothalamus, contains a large proportion of noradrenergic and serotoninergic axons. Pharmacologic inhibition of the synthesis or release of the monoamines at these loci might be expected to interfere selectively with responses of the anterior pituitary that depend on

particular synaptic pathways while leaving other responses intact (e.g., for ACTH secretion, the diurnal rhythm might be altered while stress-induced increases in secretion might remain intact).

2. Monoaminergic neurotransmitters could act directly on the releasing-factor cells to stimulate or inhibit their secretory activity. In this event, inhibition of monoamine synthesis or release would free these cells (and the anterior pituitary) from neural regulation while leaving the cells susceptible to feedback inhibition by circulating target-organ hormones, such as thyroxine or corticosterone. As will be described in Section IV, McCann and co-workers have evidence that DA stimulates LRF secretion by this mechanism.

3. The monoamine could act within the hypophysiotropic cell to influence, e.g., activate, the secretion of its releasing factor. Abrahams et al. (1957) have postulated such a role for ACh in the release of hormones from the posterior pituitary. In this event, pharmacologic interference with monoamine synthesis or secretion probably would cause the secretion of the releasing factor to become uncoupled from its normal physiological control mechanisms, or to cease entirely.

4. The monoamine itself could be released into the pituitary portal system. It could be a releasing factor or an inhibiting factor; i.e., it could be released in physiologically significant quantities into the pituitary portal vascular system and delivered to the anterior pituitary, where it might liberate a tropic hormone or inhibit the secretion of that hormone. Alternatively, the liberated monoamine could modify the pituitary response to a concurrently secreted "true" releasing factor in the portal circulation. Techniques for collecting undiluted pituitary portal blood have become available only during the past few years; hence, it has not yet been determined whether this blood contains unusually high concentrations of monoamines.

It has been very difficult to design experiments that specifically test the participation of a monoamine at each of the sites listed above. Some of the problems intrinsic to this research will be described later in the report. For example, it has not been possible to inject monoamines directly into the pituitary portal vessels, or to administer drugs that block monoamine synthesis at selected sites, e.g., within the medial forebrain bundle but not within the median eminence.

As mentioned earlier, Steiner et al. (1969) have recently begun to obtain data on the responses of hypothalamic cells to microelectrophoretically applied chemicals. They have identified cells in which the electrical activity is inhibited by either NE or dexamethasone, a potent synthetic glucocorticoid. It is tempting to speculate that these cells integrate a feedback input (circulating glucocorticoid) and a neuronal input (i.e., NE neurotransmitter) that determines the set-point for the glucocorticoid plasma levels, to control the rate of CRF secretion. However, to date, it has not been possible to study the input-output relations of these cells directly by determining whether they respond to the local application of NE by secreting more or less CRF into the pituitary portal circulation.

## Peripheral Monoamines

Peripheral monoamines—catecholamines or indoleamines injected into the experimental animal, or catecholamines endogenously secreted from the adrenal medulla or from sympathetic nerve endings— could influence anterior pituitary function in several additional ways:

1. The monoamines could be delivered to the pituitary by the general circulation, where they might act directly to modify pituitary secretion. As will be described in Section V, the median eminence and adenohypophysis lie outside the blood-brain barrier, and Marks and others have related ACTH secretion to systemically administered catecholamines (Vernikos-Danellis and Marks, 1962).

2. The monoamines could modify the sensory input to the brain, for example, by raising the blood pressure; this effect might ultimately influence the secretion of hypophysiotropic hormones, perhaps by changing the set-point around which the homeostasis of a peripheral target-organ hormone (e.g., corticosterone) is maintained. Research by Ganong and by Marks to be described in Section V raises this question with regard to stress-induced ACTH secretion.

3. Peripheral catecholamines might also influence pituitary secretion by changing the levels of regulated substances (e.g., glucose) in the general circulation, thereby activating neuroendocrine reflexes. For example, drugs that block the secretion of epinephrine from the adrenal medulla in hypoglycemic rats might have the secondary effect of increasing the stimulatory effect of the hypoglycemia on the release of GH from the anterior pituitary. This possibility becomes especially important in experiments in which drugs are used that affect *both* peripheral and central catecholamines.

## IV. HYPOPHYSIOTROPIC HORMONES: HUMORAL
## SIGNALS FROM BRAIN

### State of the Art: S. M. McCann

McCann discussed evidence that the brain controls pituitary function via releasing factors, or hypophysiotropic hormones, that are secreted by hypothalamic neuroendocrine transducer cells (see McCann and Porter, 1969). He also described the status of experimental studies on releasing factors. Such studies have been hampered by the relative inaccessibility of the cells thought to secrete them, and of the portal venous blood into which they presumably are liberated.

### Evidence for Brain Control of Pituitary Function

There is general agreement that the secretion of each of the six major tropic hormones of the anterior lobe of the pituitary gland, and of MSH from the intermediate lobe of the pituitary, is under control of the CNS; however, the mechanisms of control are not well established. Evidence that the brain controls pituitary secretion includes the following:

1. Electrical stimulation at specific brain loci, e.g., within the hypothalamus and the amygdala, can modify the secretion of specific pituitary hormones (as estimated by changes in the weights or functional activities of their target organs).

2. Brain lesions involving the median eminence, or the deafferentation of this region and the basal hypothalamus, can impair pituitary secretion (e.g., suppressing ovulation, or disturbing the daily rhythms in adrenocortical secretion). An interesting exception to this general pattern is the control of the secretion of prolactin. Prolactin secretion apparently increases if the pituitary is deprived of signals from the brain (i.e., by making lesions in the median eminence), suggesting that the hypothalamus tonically inhibits prolactin secretion.

3. Extracts of the hypothalamus contain substances that stimulate or inhibit the synthesis or secretion of pituitary hormones, when administered in vivo or in vitro.

It is now generally accepted that the anterior pituitary receives insignificant innervation from the CNS. Instead, this gland receives a unique vascular channel, the hypothalamohypophyseal portal veins, which bring it chemical signals from the brain. This portal system carries venous blood that has drained the primary capillary plexus of the median eminence and is delivered to the cells of the anterior pituitary gland without prior dilution by other venous blood. It is now widely accepted that terminals of hypothalamic neuroendocrine transducer cells lie in close approximation to the primary capillary plexus, and secrete specific hormones into it that travel through the portal system to stimulate or inhibit the release of particular pituitary hormones. It has been suggested that several varieties of these hypothalamic cells exist, each producing a specific releasing factor. The anatomic loci of the somata of these cells are not known with certainty. Most investigators place them within the nuclei of the tuberoinfundibular region, some more specifically within the region occupied by the ventromedial and arcuate nuclei. Other investigators suggest that releasing factors are secreted into the third ventricle from sites outside the basal hypothalamus, and are then transported to the pituitary portal system by specialized ependymal cells that span the median eminence (see Kobayashi and Matsui, 1969). McCann, however, believes that there is no direct evidence for this latter suggestion, and it would constitute a very inefficient mechanism.

### Bioassay Techniques for Identifying Releasing Factors

McCann next described bioassay techniques for identifying the releasing factor activity in hypothalamic extracts, or in chemicals isolated from such extracts. These studies have proceeded slowly, largely because of ambiguities inherent in the techniques. A bioassay method for determining hypothalamic CRF is shown in Figure 15.

Among the available methods are in vitro assays in which the test extract is added to a cultured pituitary gland. Measures of activity are changes in the amount of either (a) tropic hormone released from the pituitary tissue or (b) tropic hormone stored in the pituitary. These studies are done in either short-term incubation systems or in long-term tissue or organ cultures. However, these assays have the problem that compounds tested for releasing factor activity might act nonspecifically on the cell membrane to evoke release of stored pituitary hormones.

In vivo tests for releasing factor activity are complex. After systemic administration of the test compound, three indicators of releasing factor activity are used, all involving measures of pituitary tropic hormone secretion induced by the releasing factor: (1) the increase in level of pituitary hormone in the circulation; (2) the change

|  | Rat number 1 | Rat number 2 | Rat number 3 |
|---|---|---|---|
| Pretreatment | None | Dexamethasone + reserpine to block release of endogenous CRF | Hypophysectomy one day earlier to remove source of endogenous ACTH |
| Assay | 1. Remove hypothalamus 2. Homogenize in Krebs-Ringer buffer. 3. Purify extract. | 1. Inject purified extract of no. 1 hypothalamus. 2. Remove pituitary after 10-60 min. 3. Homogenize pituitary. 4. Purify extract. | 1. Inject purified extract of no. 2 pituitary. 2. Take adrenals or plasma after 10-60 min. 3. Assay for corticosterone. |

Figure 15. An in vivo bioassay method for determining hypothalamic CRF in rat number 1. The pituitary in rat number 2 secretes ACTH in response to the CRF of rat number 1. The adrenals in rat number 3 synthesize and secrete corticosterone in response to the ACTH of rat number 2. If the hypothalamus of rat number 1 contains large amounts of CRF, the pituitary of rat number 2 will secrete large amounts of ACTH and will be depleted, and the adrenals of rat number 3 will synthesize and secrete relatively small amounts of corticosterone. Conversely, if the hypothalamus of rat number 1 contains small amounts of CRF, the pituitary of rat number 2 will secrete little ACTH and not be depleted, and the adrenals of rat number 3 will synthesize and secrete relatively large amounts of corticosterone. Note pretreatments to block or prevent endogenous hormone release.

Assays for ACTH or corticosterone can also be done in vitro. In the case of ACTH, its release from the pituitary in response to hypothalamic CRF is assayed by incubating rat pituitaries with the hypothalamic extract, and the amount of ACTH released into the medium or remaining in the gland is assayed in rat number 3. The secretion of corticosterone in response to pituitary ACTH can be assayed in vitro by incubating quartered adrenal glands with the sample to be assayed for ACTH. [Wurtman]

in hormone content within the pituitary gland; and (3) the change in weight or functional activity of the target organ of the tropic hormone secondary to increased pituitary secretion—i.e., adrenal cortex (ACTH), thyroid (TSH), gonad (FSH, LH, prolactin), width of tibial epiphyseal plate (GH), or melanophore index (MSH).

When increased hormone activity is found in vivo, besides testing the activity of exogenous releasing factor in the normal animal, the activity is also tested in animals treated such that endogenous release of releasing factor is eliminated. For example, animals can be treated with target-organ hormone for feedback suppression of endogenous release. Animals also are treated with CNS depressants (i.e., barbiturates, morphine) to eliminate central influences on releasing-factor secretion—especially important in assaying for CRF because ACTH is released in response to several nonspecific stresses. Another treatment is to make lesions in the median eminence, destroying the cells that secrete endogenous releasing factors. Finally, the test compound can be directly infused by inserting a cannula into the anterior pituitary.

However, these assays present problems that limit their use. Typical problems include: (1) the presence of pharmacologically active substances, e.g., NE, 5-HT, substance P, vasopressin, and oxytocin, in extracts to be tested; these substances can elicit release in certain in vivo assays by acting on the pituitary directly or on the CNS; (2) the contamination of hypothalamic extracts with pituitary hormones; (3) nonspecific activity, e.g., by acting on the cell membrane to evoke release of stored pituitary hormones.

### Chemical Isolation and Identification of Releasing Factors

Although it is not difficult to extract substances with releasing factor activity from the hypothalamus, the isolation and identification of these hormones has been exceedingly difficult. Part of this difficulty may arise from the problems with tests used in their bioassay. The major difficulty has stemmed from the fact that the quantities of releasing factors stored in the hypothalamus are minute, thus necessitating the extraction of hundreds of thousands, and even millions, of hypothalami to obtain sufficient material for isolation and determination of the structure of the factors. McCann and others have used tall columns of Sephadex G-25, which has made it possible to separate GRF from GIF and to separate both of these from CRF, TRF, and other releasing factors. All of the releasing and inhibiting factors have been separated from one another by this technique, with the possible exception of PIF, which has been difficult to separate completely from LRF. Further purification of the factors has been obtained by ion-exchange chromatography on carboxymethyl cellulose and by electrophoresis. Most of the factors have been obtained in highly purified forms that are active in eliciting pituitary hormone release at microgram or even nanogram intravenous doses.

In the laboratories of Guillemin and Schally, TRF has recently been shown to be a relatively simple tripeptide (Bowers et al., 1970; Burgus et al., 1970); this work represents an important breakthrough in releasing factor chemistry. The structure is shown in Figure 16. It has been widely held for several years that at least some of the releasing factors are small polypeptides, and McCann believes it logical to postulate that factors other than TRF are also small polypeptides. However, Wurtman suggests that releasing factors could constitute a chemically heterogeneous group; some could be peptides, others could be proteins or other substances that include a peptide bond (because trypsin inactivates them); yet others could be monoamines or polyamines, or other types of chemicals known to be present in the brain.

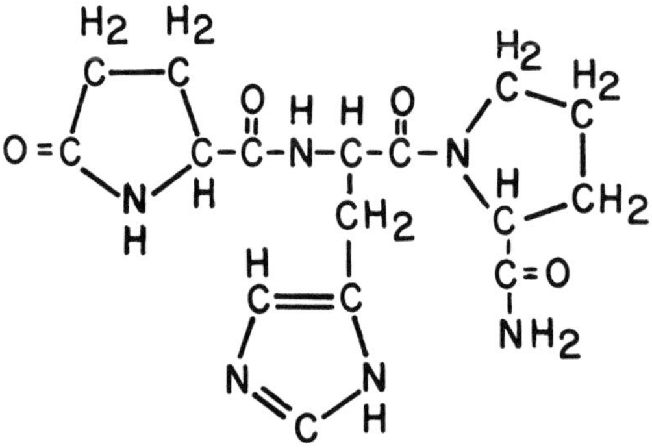

Figure 16. Structure found by Bowers et al. (1970) for porcine TRF and by Burgus et al. (1970) for ovine TRF. The structure is 2-pyrrolidone-5-carboxylyl-histidyl-proline amide.

### Localization of Cells Secreting Releasing Factors

McCann also described experiments designed to locate the anatomic sites at which specific releasing factors are made and/or stored (Watanabe and McCann, 1968; Crighton, Schneider, and McCann, 1970). With the aid of a freezing microtome, the hypothalamus was sectioned along three different planes at right angles to each other. The sections were then assayed for the presence of releasing factors. With this method, it was found that FRF and LRF are present in the median eminence and in the medial basal tuberal region; but LRF is also found in the region overlying the optic chiasm, suggesting that it is stored in neurons with long axons that project to the median eminence (Figure 17).

Another approach used for anatomic localization involved destroying the area where the releasing factor was supposed to be located, and, after degeneration was complete, assaying the basal tuberal hypothalamus for its presence. Lesions in the suprachiasmatic

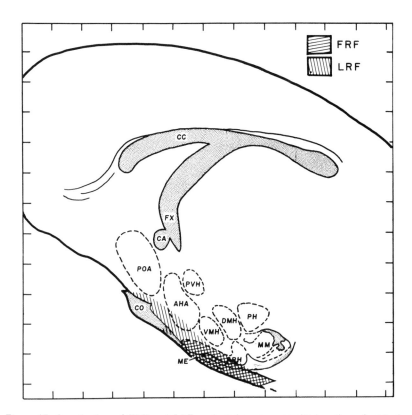

Figure 17. Localization of FRF and LRF projected on a parasagittal section of rat brain as determined by assay of frozen hypothalamic sections. Key to abbreviations: CC, corpus callosum; FX, fornix; CA, anterior commissure; POA, preoptic area; PVH, paraventricular nucleus; AHA, anterior hypothalamic area; CO, optic chiasm; VMH, ventromedial nucleus; DMH, dorsomedial nucleus; ARH, arcuate nucleus; ME, median eminence; PH, posterior hypothalamic nucleus; MM, medial mamillary nucleus. [McCann]

region in female rats were followed by a decrease in basal tuberal LRF content, but the fact that some LRF activity remained led to the hypothesis that there are two different populations of LRF neurons: one with long axons that project from the suprachiasmatic area to the median eminence, and another with short axons that originate in the basal tuberal region that was not destroyed by the lesion (Schneider, Crighton, and McCann, 1969).

### Estimating the Secretion Rates of Releasing Factors

Releasing factors are probably not secreted at a constant rate; it might be anticipated that their secretion would be enhanced when the plasma level of the substance that they regulate declines (i.e., a sudden decrease in plasma cortisol should evoke an increase in CRF secretion). A variety of techniques has been used to study their dynamics. One of the most successful is to measure changes to the content of the releasing factor within the hypothalamus at a time when the secretion of the pituitary hormone that it controls is also known to be changing. If the content of the releasing factor changes coincidentally with the rate of pituitary secretion, it is postulated that the releasing factor mediated the change. For instance, it has been reported that hypothalamic LRF decreases in proestrus or during puberty, i.e., at times when LH secretion is high. Hypothalamic LRF levels are also depressed by implants of estrogen or testosterone in the median eminence, which depress LH secretion from the pituitary. In this case, the fall in LRF content is thought to be the cause of impaired LH secretion.

The cannulation of the hypophyseal portal vessels and the demonstration of releasing factors in the portal blood would be most useful; however, this approach has great technical difficulties (for example, portal blood is easily contaminated by systemic arterial or venous blood). Several laboratories have only recently been successful in collecting portal blood. Porter first demonstrated the presence of the releasing factor CRF in this blood, and recently he and his associates have been able to detect LRF, FRF, GRF, and PIF (see this Section, Dopamine and Releasing Factor Activity).

Another technique used to estimate the secretion of releasing factors involves removing the pituitary gland and following the rate at which releasing-factor concentrations rise in peripheral blood. The concentrations of these substances are normally very low, because in the normal animal the portal venous blood containing these hormones reaches the pituitary without prior dilution of systemic blood; hence, only small amounts of releasing factors need be secreted. However, in the chronically hypophysectomized animal sufficient quantities of releasing factors are liberated to allow CRF, LRF, GRF, and FRF to be detected in peripheral blood (McCann and Porter, 1969). (This high blood level is probably the basis for a residual function found in pituitaries transplanted to a site at a distance from the hypothalamus.) Hypophysectomized animals in which the peripheral blood contains high concentrations of releasing factor can be treated with target-organ hormone (e.g., cortisol, estradiol); and the rate at which the releasing-factor concentration falls is related to its secretory rate in the untreated animal.

### Dopamine and Releasing Factor Activity

McCann, Porter, and co-workers have related DA to releasing factor activity, especially to LRF activity. Using an in vitro system, Schneider and McCann (1969) found that DA liberates LRF from hypothalamic fragments, thus enhancing the release of LH from incubated pituitaries. The catecholamine has no direct effect on pituitary secretion and does not alter the response of the pituitary to known amounts of partially purified LRF. Dopamine injected into the third ventricle is also capable of stimulating LH release in vivo by stimulating the release of LRF (Schneider and McCann, 1970a). Following intraventricular injection of DA an increase in LRF is present in peripheral blood of hypophysectomized rats (Schneider and McCann, 1970b) and in hypophyseal portal blood (Kamberi et al., 1969). Schneider and McCann (1970b) also found that the stimulatory effect of intraventricular DA was blocked by the prior intraventricular injection of estradiol (Figure 18). These results suggest that DA may be a synaptic transmitter that triggers LRF release and that a negative feedback action of estrogen may block the dopaminergic stimulation of LRF release.

Dopamine has also been demonstrated to stimulate release of FRF (Kamberi et al., 1970) and PIF (Kuhn et al., 1970). In contrast,

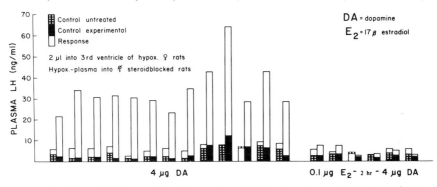

Figure 18. Effect of third ventricular DA on plasma LRF in hypophysectomized (hypox) rats, and blockade of the response to DA by prior injection into the ventricle of estradiol ($E_2$). Each bar gives the plasma LH of an ovariectomized, steroid-blocked assay rat prior to (stippled or solid portion) and 10 min after receiving 1 ml of plasma from a hypox donor (total height of column). The open portion of the bar is the response to the plasma sample and is an indication of its LRF content. Each pair of bars represents the results from a single hypox donor. The left-hand bar gives the resting plasma LH (control untreated, stippled bar) and the response to a control plasma sample from the hypox donor. The right-hand bar gives the resting plasma LH (control experimental, solid bar) and the response to an experimental plasma from a hypox donor 15 min after injection of DA into the third ventricle. [Schneider and McCann, 1970b]

NE and epinephrine are either inactive or less active in activating hormone release.


## Neuroendocrine Transducer Input-Output Relations:
## J. Urquhart

Urquhart discussed the unusual difficulties involved in studying the input-output relationships of hypothalamic neuroendocrine transducer cells. As previously discussed by McCann, problems arise from the anatomic inaccessibility of these cells and of their output channel, the hypothalamohypophyseal portal system, requiring that hypothalamic function be measured indirectly in terms of effects on the experimentally accessible peripheral components of the neuroendocrine systems. Urquhart went on to describe the complex connectivities within and among neuroendocrine systems, which include unmeasurable and nonlinear processes. These processes both distort measures of hypothalamic function and also affect hypothalamic function itself by means of feedback effects. He therefore recommended a new research strategy utilizing a systems approach and dynamic testing by computer modeling.

### Neuroendocrine Systems, Subsystems, and Processes

Urquhart first discussed examples of commonly used indirect measurements of hypothalamic function, i.e., effects on the pituitary gland or on target organs. He cautioned that measuring the change in tropic hormone content within the pituitary gland can sometimes give misleading information, since such changes reflect both synthesis and secretion of the hormone, and not simply secretion alone. Without reliable measurements both of synthesis and of secretion, observations on pituitary content of tropic hormone are "only of anecdotal value." On the other hand, measurement of changes in the target gland content of its secretory product has proved to be useful with steroidogenic target glands, because so little steroid secretory product is stored that synthesis and secretion are tightly coupled. Also, notably in the rat, measurement of changes in target-gland weight has provided much useful qualitative information about longer-term (days-weeks) changes in neuroendocrine function.

However, the complexities of neuroendocrine systems become clear when one considers measurement of short-term (minutes-hours) changes in hypothalamic function as indicated by changes in plasma concentration of target-gland hormone. Figure 19 is a general scheme of the connectivity of neuroendocrine systems, applicable to the adrenocortical, thyroid, and gonadal systems. The concentration of target-gland hormone in systemic blood (V in Figure 19) might appear to be the poorest choice for study, on the grounds that the largest number of variables stand to disturb the relation between it and the rate of

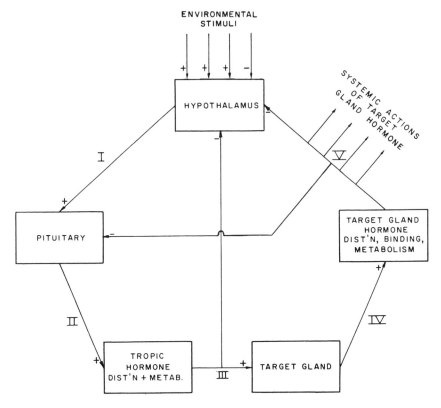

Figure 19. A general minimally complex scheme of the connectivity in the adrenocortical, thyroid, and gonadal neuroendocrine systems. The Roman numerals refer to the following variables: I, concentration of releasing factor in hypophyseal portal blood; II, rate of tropic hormone secretion; III, concentration of tropic hormone in systemic blood; IV, rate of target gland secretion; V, concentration of target gland hormone in systemic blood. Each box represents a group of processes: synthesis and secretion of hormone and their regulation in the case of the hypothalamus, pituitary, and target gland; dilution, circulatory transport, metabolic inactivation, and macromolecular binding in the case of the other two boxes. [Urquhart]

secretion of releasing factor. It would thus seem preferable to measure some variable that is functionally closer to the hypothalamus. However, this proposition is deceptively simple because it ignores the feedback effects of both tropic and target-gland hormones, illustrated in Figure 19. These feedback effects pose a vexing problem, because they render the hypothalamic secretion of a given releasing factor potentially dependent on each of the corresponding system's variables, I-V.

In general there are two classes of hypothalamic response that result in altered secretion of a releasing factor. The class usually considered is the neurosecretory part of the central nervous response to changes in either external or internal environment. But among the influential factors in the internal environment are the variables of the neuroendocrine system itself, some of which cannot currently be measured. Table 3 is a partial list of physiological factors whose constancy one must assume in associating changes in the systemic blood or plasma concentration of a target-gland hormone with changes in hypothalamic output of the corresponding releasing factor; yet through feedback effects, these variables have the potentiality of influencing hypothalamic secretory function. Urquhart considered it helpful to group these factors into a class named the "process dynamics" of the neuroendocrine systems. The extent to which changes in process dynamics would influence hypothalamic function will depend upon the strengths and dynamic ranges of the feedback actions of target-gland

TABLE 3 [Urquhart]

Variables that demonstrably, plausibly, or definitionally can influence the relation between the rate of hypothalamic secretion of releasing factor and the plasma concentration of the corresponding biologically active target-gland hormone.

Hypophyseal portal blood flow
Pituitary responsivity to releasing factor
Rate of metabolism of tropic hormone
Target-gland blood flow
Target-gland responsivity to tropic hormone
Rate of metabolism of target-gland hormone
Macromolecular binding of target-gland hormone

In the text, these variables are described collectively as the "process dynamics" of a neuroendocrine system.

and tropic hormones, but information on these important quantitative points is sparse. Even with the aid of a qualitative picture such as Figure 19, it is difficult to make directional predictions; quantitative data on the feedback actions are needed to tell which effects are negligibly small and which are not.

An illustrative example of a significant change in process dynamics is the effect of thyroid hormones on the adrenocortical system—an example also raising the problem of interactions occurring among different neuroendocrine systems. Hyperthyroidism shortens the half-life of plasma cortisol (Peterson, 1958) by enhancing the metabolic inactivation of the steroid (Hellman et al., 1961). In evident compensation for this apparently primary event, there is an increase in ACTH secretion (as judged by increased plasma concentration of ACTH, Hilton et al., 1962), and an increase in cortisol secretion rate. Presumably, too, there is an increased rate of CRF secretion. However, the net effect is little or no change in plasma cortisol concentration (Peterson, 1958). This multidimensional view of the adrenocortical system in hyperthyroidism makes it clear that the adrenal hypersecretion in hyperthyroidism is not a stress response, but is instead attributable to a change in the system's process dynamics. Furthermore, in one given physiological state, there may be multiple changes in process dynamics. The work of Labrie indicates, for example, that the thyroid hormones influence the process dynamics of the adrenocortical system at points in addition to the one discussed above (Labrie, 1967); and Kitay's work indicates that the gonadal hormones can influence the process dynamics of the adrenocortical system at multiple sites (Kitay, 1968). While the effects of a single, sustained change in process dynamics are simple to predict, it is difficult or impossible to predict the outcome of multiple and/or nonsteady-state changes.

These considerations serve to illustrate the fact that neuroendocrinology presents major systems problems. The problems are systems problems both because the measurements of hypothalamic function must be made remotely and thus are possibly distorted by changes in process dynamics, and also because hypothalamic function itself is dependent to an uncertain extent on these process dynamics because of feedback. It is a *major* systems problem because of the elaborate connectivity of the neuroendocrine systems, and also because the dynamics of each system's component parts are complexly nonlinear (see Nugent et al., 1964; Urquhart and Li, 1968). Thus, Urquhart calls for a new strategy in both experimentation and conceptualization—

"different from the approach of simplistic, one-step, cause-effect relations, which has served endocrinology well in the past but has conspicuously failed to develop neuroendocrinology beyond a sketchily qualitative beginning. The evident need for this conference testifies at least to the latter point."

## A Systems Research Strategy

Urquhart proposed a new neuroendocrinological research strategy (Urquhart, 1970b) with four parts, as follows:

1. Study the dynamic functional characteristics of accessible subsystems in each neuroendocrine system, under experimental conditions in which the subsystem is functionally isolated from the rest of the system. The definition of a subsystem is dictated in part by what it is possible to isolate, without destroying function, at one or two levels of physiological organization lower than the total system (a more complete discussion of this point can be found in Yates et al., 1968). The three extracranial processes in Figure 19 have been considered and studied separately in at least one neuroendocrine system, the adrenocortical system (for illustrative examples see Nugent et al., 1964; Urquhart and Li, 1968; Lake and Gann, 1970; Miller and Yates, 1970).

2. Formulate dynamic models of the operational characteristics of each subsystem. Whatever mathematical virtuosity the models may reveal, they rest on the data obtained in part (1). The mechanistic details of the models are irrelevant in the present context; it suffices only that a model simulate the dynamic operational characteristics of the corresponding biological subsystem with fidelity. However, there is the opportunity for a dynamic model to do double scientific duty—not merely simulating the relatively microscopic mechanisms of the subsystem, but also linking them together to show how the subsystem's relatively macroscopic dynamic properties arise (for an example, see Urquhart et al., 1968).

3. Assemble the subsystem models together with verified or verifiable dynamic descriptions of their interconnections.

4. Compare the functional properties of the biological system with those of the assembled model.

If it were possible to perform part (1) on each of the five subsystems shown in Figure 19, then the result would be a system model that could simulate the functional properties of the intact neuroendocrine system. However, given the inability to isolate and test the intracranial subsystems (i.e., the hypothalamus and pituitary), the accessible subsystems must be modeled with fidelity in order to make inferences about the dynamic functional properties of the inaccessible ones. One must ascribe functional properties to models of the inaccessible subsystems as necessary to enable the model of the total system to simulate the responses of the intact neuroendocrine system to interpretable experimental interventions (illustrative examples are Yates et al., 1968; Gann et al., 1968; Yates and Brennan, 1969). This strategy calls for biological experimentation that can be simulated with the model, and it is in that sense that the word "interpretable" was used above. If an experiment can be simulated on the model, its outcome will either confirm or contradict some aspect of the model, while an inability to simulate the experiment will reflect adversely on either or both the experiment or the scope of the model. For example, if the modeling follows the outlines of Figure 19, in which the variables are rates of hormonal secretion and concentrations of hormone in blood or plasma, an experiment is uninterpretable when exogenous hormone is administered subcutaneously, intramuscularly, or intraperitoneally. One must either change the experimental design or else provide quantitative information on hormonal absorption from the injection sites. Whichever the outcome, the result at least can be related to current concepts.

In conclusion, Urquhart commented on two aspects of this strategy. "One is that some sort of computer is required, which, together with the mathematical representation of endocrine and neuroendocrine phenomena, prompts the inference that the strategy is, in the pejorative sense, a purely theoretical one. The second is that the modeling changes at a pace that is unheralded for concepts in endocrinology, prompting the inference that, with so much jitter in the modeling, it does not warrant serious attention. However, change in the modeling reflects two things: that experiments were performed which had decisive bearing on the then current concepts and that these concepts were phrased precisely enough to be testable. Rapid change in the modeling signifies a brisk flow of decisive experimental data. And that is the name of the game in this or any other scientific endeavor."

## V. RESEARCH RELATING BRAIN MONOAMINES
## AND ENDOCRINE FUNCTION

Most of the Work Session on Brain Monoamines and Endocrine Function was devoted to consideration of research into the roles of specific brain monoamines in controlling the secretion of such anterior pituitary hormones as GH, ACTH, and the gonadotropins. In general, for the reasons given in the previous section, pituitary function was not necessarily measured directly but often by target-organ activity.

The research reported by Work Session participants followed three main strategies:

1. Varying brain monoamine levels, by pharmacological treatment or by administration of exogenous monoamine, and measuring the effects on endocrine function—effects on GH, by Müller; on ACTH, by Ganong, Marks, Maickel, and Martini; and on gonadotropins, by Sawyer, Coppola, and Meyerson.

2. Using natural or induced changes in endocrine activity and studying the effect on brain monoamine metabolism—effects of gonadotropins, by Donoso, Coppola, Fuxe, and Anton-Tay and Wurtman.

3. Studying the effects of brain lesions both on brain monoamine metabolism and on endocrine function—effects on monoamines, by Moore; and effects on endocrine function, by Gorski.

### Effects of Manipulation of Brain Monoamines
### on Pituitary Function

#### Catecholamines and Control of Growth Hormone Secretion:
#### E. E. Müller

Müller described recent studies relating brain catecholamines to the control of the secretion of GH from the anterior pituitary gland. As a model system for examining the physiological regulation of GH secretion, Müller used the GH response to insulin injections in rats. It is well established that the hypoglycemia that follows the insulin administration is a powerful stimulus in evoking GH secretion (Roth et al., 1963). GH secretion was estimated by measuring the amount of GH-like activity remaining in the pituitary glands of treated animals, by bioassays of pituitary homogenates; the assay used was the increase in width of the tibial epiphyseal plate of hypophysectomized assay rats (Greenspan et al., 1949). Müller et al. (1967b) first observed that of many neuropharmacological agents tested, CPZ and reserpine were most effective in blocking the secretion of GH in insulin-treated animals. They demonstrated that the anterior pituitary was not the site at which reserpine and CPZ acted, by showing that administration of stalk median eminence extracts from untreated animals, which contained GRF, produced almost identical depletion of pituitary GH in control animals and in rats pretreated with CPZ or reserpine.

Since CPZ and reserpine possess antiadrenergic activity, Müller et al. (1967c) did further studies to determine whether the effect on GH secretion was mediated by brain catecholamines. By testing other NE depletors they found that drugs affecting brain catecholamine levels ($\alpha$-methyldopa, $\alpha$-MMT, or tetrabenazine) were effective in blocking insulin-induced GH depletion, while peripheral catecholamine depletors (guanethidine, tyramine, or bretylium) were ineffective. Furthermore, when brain catecholamines were protected against monoamine depletion by prior administration of the MAO inhibitor iproniazid, reserpine no longer blocked the depletion of pituitary GH. (The earlier study had also shown that iproniazid given alone resulted in a more marked insulin-induced depletion of pituitary GH than normal.)

Systemically administered catecholamines have no effect on pituitary GH content and are thus thought to be unable to affect GH release (Roth et al., 1963; Müller et al., 1965). This lack of effect is probably a consequence of the blood-brain barrier to these substances (Weil-Malherbe, Whitby, and Axelrod, 1961). To circumvent this barrier, Müller et al. (1967c) injected catecholamines into the lateral ventricle. Both NE and DA (0.5 and 0.1 $\mu$g) induced a depletion of pituitary GH. Epinephrine was also effective, but only at the higher dose level. Intraventricular injection of such other hypothalamic substances as 5-HT, ACh, histamine, vasopressin, or oxytocin did not elicit depletion of pituitary GH.

The possibility existed that the catecholamines affected the release of GH via a decrease in body or hypothalamic temperature, both of which are known to cause GH release (Müller et al., 1967c; Machlin et al., 1968; Gale and Jobin, 1967). Hence, Müller et al. (1968) measured body temperature following intraventricular injection of catecholamines, and found no statistically significant difference from the temperature of rats injected with saline. The intraventricular administration of 5-HT did induce a significant decrease in body temperature, but it had no effect on pituitary GH content.

Since the majority of stimuli able to induce release of GH do so by triggering hypothalamic GRF mobilization (Katz, Dhariwal, and McCann, 1967; Müller et al., 1967a,c), in further experiments, Müller (1970) and Müller et al. (1970) studied whether catecholamines acted directly at the pituitary level or through GRF mobilization. Intraventricular administration of NE or DA at doses that induced release of pituitary GH caused disappearance of GRF activity from the hypothalamus of intact or hypophysectomized rats, while 5-HT, which does not induce release of pituitary GH, left hypothalamic GRF activity unaffected. This finding implies that the two catecholamines released pituitary GH by triggering the discharge of hypothalamic GRF and not simply by acting on the pituitary following diffusion into the portal vessels. Furthermore, as shown in Figure 20, simultaneous determination of GRF activity in the hypothalamus and plasma of intact or hypophysectomized rats, showed that only in the latter was depletion

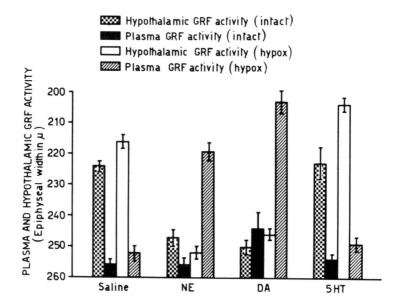

Figure 20. Effect on hypothalamic and plasma GRF activity of NE, DA, or 5-HT given by intraventricular injection to intact or hypophysectomized (hypox) rats. [Müller]

of GRF from the hypothalamus accompanied by its appearance in the plasma.

In the last series of these experiments Müller compared the GH-releasing ability of NE or DA at lower doses. It appeared that NE was active at the very small doses of 5 and 2.5 ng, while at 5 ng DA appeared to be inactive. This would suggest that, of the two amines, primarily NE is concerned with the control of GH secretion.

Finally, Müller summarized recent experiments suggesting that cyclic 3'5'-adenosine monophosphate (cyclic AMP) mediates the effect of brain catecholamines on GH secretion (Müller et al., 1969). The brain is known to contain large amounts of adenyl cyclase, the enzyme that catalyzes the formation of cyclic AMP; and Sutherland, Rall, and Menon (1962) showed that the levels of cyclic AMP in brain slices are enhanced by incubating them with NE (reviewed by Rall and Gilman, 1970). The destruction of cyclic AMP is catalyzed by the enzyme phosphodiesterase, which can be inhibited by the drug theophylline. Müller showed that intraventricular theophylline caused a slight increase in GH secretion. This effect was magnified by concurrently administering theophylline and cyclic AMP. Recently, Steiner et al. (1970) have shown that the increased release of GH from anterior pituitaries incubated in vitro which is induced by crude hypothalamic extracts is associated with an increase in the pituitary concentration of cyclic AMP.

**Monoamines and Control of Glucocorticoid Secretion**
   **from the Adrenal Cortex: W. F. Ganong, R. P. Maickel,**
   **L. Martini, B. H. Marks**

Ganong discussed the use of the adrenocortical secretory response to stress as a model system for examining the role of brain monoamines in ACTH secretion. A wide variety of stressful stimuli (e.g., ether anesthesia, burns, crowding, subcutaneous injection of formalin or histamine, laparotomy) are known to evoke a marked hypersecretion of ACTH from the pituitary gland, and a consequent increase in the secretion of glucocorticoids (i.e., hydrocortisone, corticosterone) from the adrenal cortex. Theoretically, the role of monoamines in neural pathways mediating this response could be examined by administering drugs that block their synthesis or secretion, and determining whether the animal retains the adrenocortical stress response. However, it is important to remember that there are at least three modalities of brain control of adrenocortical secretion, as follows: (1) In the unstressed animal, the brain participates in a closed feedback loop system that maintains plasma corticoid levels within a narrow homeostatic range. (2) Mechanisms within the brain also cause the homeostatic range to change as a function of time of day. There appears to be a daily rhythm in the concentration of plasma corticoids that is necessary to suppress ACTH secretion; in consequence, there is also a rhythm in adrenocortical secretion. (3) The brain seems to mediate the stress-induced increases in adrenocortical secretion. Furthermore, several studies (Yates, 1967; Yates, Brennan, and Urquhart, 1969; Urquhart, 1970a) have shown that the stress effects can be divided into two groups: one group of stressors stimulates adrenocortical secretion regardless of what the plasma cortocoid levels might be, while the effect of the other group of stressors can be blocked by administration of large doses of exogenous corticoids. The first group of stressors has been termed "open-loop." The second group is thought to operate within the closed feedback loop that normally controls plasma corticoid levels, and to produce its effect by raising the apparent set-point around which the corticoid levels are maintained.

Ganong pointed out species differences in the effects of various CNS-active drugs on ACTH secretion. In the rat, a wide variety of drugs has been reported to block the ACTH secretion response to stress; these include reserpine, CPZ, morphine, diphenylhydantoin, and meprobamate. However, epinephrine and vasopressin also block stress responses in this species; and it is possible that any stress which produces ACTH secretion can produce a temporary blockade of response to a subsequent stress (Ganong, 1963). The cause of this period of relative unresponsiveness to a second stress is unsettled; but there is some evidence that it is a neural phenomenon, perhaps relating to a depletion of ACTH from the pituitary (Kitay et al., 1959). In the meantime, Ganong believes that drug experiments on the rat must be interpreted with caution. In the dog, unresponsiveness to a second stress does not seem to occur. This species also differs from the rat in its response to CPZ, which does not block ACTH secretion but instead stimulates it and also produces adrenal hypertrophy (Betz and Ganong, 1963).

*ACTH Secretory Stress Responses in the Dog*

Tullner and Hertz (1964) showed that Monase ($\alpha$-ethyltryptamine), a drug presumed to act by inhibiting MAO, could block the release of ACTH from the pituitary of the stressed dog. It affects ACTH secretion after surgical stress and in response to insulin hypoglycemia. Ganong and associates subsequently carried out a series of experiments to find the mechanisms by which Monase exerted its inhibitory influence. ACTH release was assayed indirectly, by measuring changes in adrenal venous output of glucocorticoids. Ganong et al. (1965) first mapped areas in the brainstem where electrical stimulation produced an increase in ACTH secretion, to see where blockade of the response to such stimulation might be produced by systemically administered Monase. They found that stimulation along the dorsal longitudinal fasciculus and the mammillary peduncle produced increases in ACTH secretion that were blocked by Monase, and in subsequent unpublished studies the increase in ACTH secretion produced by stimulation of the limbic system was also found to be blocked. However, stimulation of the ventral hypothalamus produced increases in ACTH secretion that were not blocked by Monase. This suggested that Monase might act by inhibiting transmission at the synapses between afferent neurons and CRF-secreting neuroendocrine transducer cells in the hypothalamus.

The effect of Monase on ACTH secretion could also be related to the ability of this drug to raise blood pressure. In another series of experiments Lorenzen and Ganong (1967) discovered that a number of other pressor agents, including amphetamine, methamphetamine, 2-aminoheptane, and the cyclic sympathomimetic agent, clopane, similarly inhibited stress-induced ACTH secretion when given in quantitites sufficient to produce a prolonged rise in blood pressure. Monamine oxidase inhibitors tested (other than Monase), which did not have pressor effects, did not inhibit ACTH secretion. The correlation between pressor responses and inhibition was marked. If the pressor response was small in any given animal, inhibition was usually absent. When experimental hypertension was produced by raising intracranial pressure (i.e., "Cushing hypertension"), there was an associated small, but statistically significant, inhibition of ACTH secretion.* If the blood pressure rise produced by Monase was prevented by hemorrhage, inhibition failed to occur (Ganong et al., 1967). Finally, it was observed that denervation of the carotid baroreceptors (by intracranial section of the ninth to eleventh cranial nerves), which deprived the brain of information about the rise in blood pressure, also prevented Monase from inhibiting ACTH secretion.† They thus tentatively concluded that

*Wise, Boryczka, Shackelford, and Ganong, unpublished observations.

†Wise, Lorenzen, and Ganong, unpublished observations.

the effect of Monase resulted not from a chemical interaction with monoaminergic neurons but from its ability to produce an increase in blood pressure. This stimulates the carotid and aortic baroreceptors, and such stimulation activates synaptic pathways that exert an inhibitory effect on ACTH secretion.

However, Ganong went on to describe two pieces of data that do not fit this hypothesis. When given by constant systemic infusion, NE and DA failed to inhibit stress-induced ACTH secretion despite the production of marked increases in blood pressure, raising the possibility that the correlation with pressor activity was only coincidental. On the other hand, unlike the other hypertensive agents tested, NE and DA do not cross the blood-brain barrier. Dopa, which does cross the blood-brain barrier, in doses of 50 mg/kg did inhibit stress-induced ACTH secretion. Further research on the mechanism of these effects is underway. In the meantime, Ganong cautioned that effects mediated by variables such as blood pressure must be kept in mind in interpreting experiments on neuroendocrine control mechanisms.

*Monoamines and Stress Responses in the Rat*

Maickel described pharmacological experiments on the role of brain monoamines in the stress-induced increase in ACTH secretion from the rat pituitary. For indices of pituitary ACTH secretion, three parameters were selected: the levels of plasma corticosterone, adrenal ascorbic acid depletion, and liver tryptophan pyrrolase activity.

Maickel et al. (1961) found that a classical stress, cold exposure, or a single sedative dose of reserpine caused a prolonged increase in all the ACTH secretion parameters. Reserpine also produced depletion of brain NE and 5-HT, while reserpine derivatives that did not deplete brain amines produced no adrenal activation and also no sedation. This pituitary-adrenal response to reserpine treatment did not occur unless brain amine levels were depleted by at least 50 percent, either by a single large dose of reserpine or by several small doses (Westermann, Maickel, and Brodie, 1962).

A direct effect of reserpine on the adrenal cortex was ruled out, since reserpine did not produce these effects in hypophysectomized animals. Also, no direct effect on the adrenal cortex regarding its sensitivity to ACTH was caused by either cold stress or reserpine, since ACTH administration to reserpine-pretreated animals was as effective as in nonpretreated animals. The increased corticosterone levels produced by reserpine or cold stress could also not be explained as an impairment of steroid metabolism. Thus Maickel and co-workers concluded that the reserpine influence occurred at the level of the ACTH secretion from the pituitary. Furthermore, no correlation was seen between sedation and adrenal activation (Martel, Westermann, and Maickel, 1962), despite the requirement for the CNS (Westermann, Maickel, and Brodie, 1962).

Since partial depletion of brain NE with $\alpha$-MMT did not result in pituitary-adrenal activation, Westermann, Maickel, and Brodie (1962) tentatively concluded that the reserpine effect was related to the depletion of a different monoamine, 5-HT. However, studies by others (Bhattacharya and Marks, 1970) showed that the administration of doses of $p$-chlorophenylalanine (PCPA), an inhibitor of tryptophan hydroxylase, that cause a marked depression in brain 5-HT levels, had no apparent effect on adrenal stress responses.

Maickel stressed the importance of checking the effect of drugs on monoamine levels in several brain regions, because recent studies showed that 5-HT and NE levels in different parts of the brain were affected to different degrees after MAO inhibition, or treatment with reserpine, $\alpha$-MPT, or 4-chloromethamphetamine (Miller, Cox, and Maickel, 1968). The measurement of whole brain monoamine levels could obscure significant changes that may be taking place in a critical region of the brain. One conclusion that seems inescapable about the reserpine studies was that a drug with so many known biochemical actions is a poor choice for use in working on the role of specific brain monoamines in the control of anterior pituitary function.

Martini described experiments implicating hypothalamic 5-HT levels in the control of ACTH secretion.* (It has recently been shown by direct chemical assay that the bovine median eminence is very rich in 5-HT (Piezzi, Larin, and Wurtman, 1970).) Martini and co-workers found that implants of 5-HT within the median eminence blocked the increase in plasma corticoids that normally follows ether stress. At the same time pituitary ACTH levels were decreased, possibly reflecting a decrease in the synthesis of the hormone. Martini suggested that 5-HT-containing neurons may inhibit ACTH secretion, especially the hypersecretion associated with stresses. This effect of 5-HT is not specific for ACTH. In castrated rats, 5-HT implants decrease pituitary FSH levels by half, although pituitary LH levels are not affected (Fraschini, Mess, Piva, and Martini, 1968).

*Motta, Piva, and Martini, unpublished observations.

*Catecholamines and ACTH Secretions*

Marks described the results of two approaches he used to study the relationship between catecholamines and the regulation of ACTH secretion in the rat: (1) the administration of exogenous catecholamines, and (2) the alteration of the content of biological actions of endogenous catecholamines.

*Exogenous Catecholamines:* Many studies on rats and humans of the systemic administration of exogenous catecholamine have shown that these compounds, particularly those that activate alpha adrenergic receptors (i.e., those that tend to contract smooth muscle), elicit a prompt secretion of ACTH (Kitay et al., 1959; Vernikos-Danellis and Marks, 1962; Mangili, Motta, and ·Martini, 1966; Vernikos-Danellis, 1968). These responses can be blocked by adrenergic alpha-blocking drugs such as phentolamine, if given in doses that block their peripheral pressor responses (Vernikos-Danellis, 1968). Consequently, these effects have generally been considered to originate through activation of peripheral receptor mechanisms, whose loci remain somewhat obscure, but may include the carotid baroreceptors, as described earlier by Ganong. This hypothetical mechanism of action appears to be supported by abundant evidence that circulating catecholamines are kept out of most of the brain by a blood-brain barrier.

However, systemically administered catecholamines are able to penetrate into the interstitial space of at least two brain regions that are concerned with the control of anterior pituitary function. These are the median eminence of the tuber cinereum and the adenohypophysis itself (Samorajski and Marks, 1962). The rat adenohypophysis has a number of biochemical properties suggesting an ability to respond directly to administered catecholamine:

1. It demonstrates considerable uptake and binding of isotopically labeled catecholamines (Samorajski and Marks, 1962).

2. It has a very high content of the enzyme phosphodiesterase, which, as mentioned previously, inactivates cyclic AMP, whose synthesis is stimulated by catecholamine. If the phosphodiesterase is inhibited with methylxanthines, such as caffeine or theophylline, stress responses are augmented (Vernikos-Danellis and Harris, 1968). (Note Müller's finding, reported earlier, that intraventricular injections of theophylline also stimulate the secretion of GH from the anterior pituitary gland.)

3. The intravenous administration of catecholamines causes a marked increase in the concentration of adenohypophyseal ACTH, even

in doses that fail to provoke marked pressor responses or ACTH secretion.* Therefore, this response appears to be mediated by changes in the synthesis of pituitary ACTH. This increase is blocked by the adrenergic beta blocker, MJ-1999 (Vernikos-Danellis, 1968). (Similar changes in the biosynthesis of ACTH induced by stress will be described below.)

Marks and co-workers also examined the effects of exogenous catecholamine administered intraventricularly, with the expectation that any effects observed would more probably result from actions on synapses in the CNS than would result with systemic administration. They studied the effects of NE and DA, and also the cholinomimetic, carbachol, on ACTH secretion, using the adrenal content of corticosterone as the index of ACTH liberation. All three drugs caused ACTH secretion (Marks et al., 1970). Dose-response relationships were defined so that appropriate doses could be determined for antagonists and the types of pharmacologic receptor mechanisms could be identified. Marks believes that regionally implanted antagonists are generally applied in doses too high for specificity in their agonist-antagonist relationships. It should also be pointed out that neither the cholinergic nor the adrenergic receptors in the brain that are involved in ACTH secretion have yet been pharmacologically classified.

To summarize, the administration of exogenous catecholamines stimulates ACTH release. Both alpha- and beta-type adrenergic receptor mechanisms in the CNS or in the periphery may be involved in this response, part of which may be due to enhancement of pituitary biosynthetic mechanisms.

*Endogenous Catecholamines:* Effects of alteration of the content or of the biological activity of endogenous catecholamine leads to an entirely different view of the role of catecholamines in the neuroendocrine mechanisms regulating ACTH secretion. Either the depletion of catecholamines by reserpine or their antagonism by CPZ causes intense activation of ACTH secretion (Maickel et al., 1961; Bhattacharya and Marks, 1969b). Then, after a period of latency, there is a blockade of further ACTH release.

Either reserpine or CPZ also causes marked reduction of the CRF content of the hypothalamus; this may be the basis of the subsequent blockade of ACTH release in response to further stresses (Bhattacharya and Marks, 1969a). Catecholamines in the hypothalamus or elsewhere may act as inhibitory transmitters that normally prevent

*Vernikos-Danellis and Marks, unpublished observations.

the liberation of CRF. (This hypothesis is supported by the observations of Steiner et al. (1969), previously mentioned, of hypothalamic cells whose electrical activity is inhibited by local application of either dexamethasone or NE.) The depletion or the antagonism of catecholamine would allow for uncontrolled CRF secretion until the system breaks down by exhaustion of the CRF stores.

Marks suggested that inadequacy of pituitary ACTH content is not likely to be a cause of ACTH blockade following reserpine or CPZ treatment, inasmuch as there is considerable evidence that the pituitary relies for its stress-induced ACTH secretion upon rapid biosynthetic mechanisms: with stress, the pituitary ACTH content actually increases, as does the incorporation of amino acids into pituitary ACTH (Jacobowitz, Marks, and Vernikos-Danellis, 1963); drugs that interfere with the synthesis of RNA or protein also block the stress-induced increase in ACTH secretion (Vernikos-Danellis, 1965; Marks and Vernikos-Danellis, 1963).

Further evidence in favor of the hypothesis that hypothalamic catecholamines tonically inhibit CRF secretion is provided by the following observations: (1) Administration of drugs that may enhance the release of catecholamines from neurons (i.e., amphetamine or the MAO inhibitor Pargyline) shuts down stress-induced ACTH secretion. (2) Pretreatment with Pargyline prevents the effects of reserpine on both the ACTH release and the depletion of CRF that follows reserpine treatment (Bhattacharya and Marks, 1969a). (3) The intraventricular administration of small doses of NE into rats rapidly inhibits the ACTH hypersecretion produced by reserpine pretreatment (Marks et al., 1970).

## Monoamines and the Control of Gonadotropin Secretion:
### C. H. Sawyer, J. A. Coppola, B. Meyerson

*Catecholamines and the Control of Ovulation*

Sawyer reviewed both early and recent research relating CNS catecholamines to the control of ovulation. The first studies demonstrating a role of the CNS in the control of gonadotropin secretion were those performed by Harris (1937) and by Haterius and Derbyshire (1937) on the induction of ovulation by electrical stimulation of the brain. Sawyer and co-workers then found that electrical stimulation was effective in causing ovulation in rabbits when applied to the hypothalamus, but not when applied directly to the pituitary gland (Markee, Sawyer, and Hollinshead, 1946a). This suggested that the control of the pituitary was exerted by elements other than nerve fibers within the gland, and led to the idea of neurohumoral mediation of CNS effects on the pituitary.

In those days, 20 years ago, the only neurohumors known were those of the peripheral autonomic nervous system, epinephrine and ACh. If these substances were infused systemically, they were ineffective in inducing ovulation; but if epinephrine was injected directly into the pituitary gland, using a parapharyngeal approach, ovulation could be induced in the rabbit (Markee, Sawyer, and Hollinshead, 1946b). When adrenergic blocking agents of the dibenamine type became available, these blocking drugs, as well as the anti-cholinergic agent atropine, were found to block ovulation induced by coital stimulation, if administered within seconds after mating. Thus Sawyer et al. (1949) postulated a cholinergic-adrenergic sequence for the control of the release of gonadotropins, and they suggested three possible pathways for this control: (1) a direct innervation of the anterior pituitary by adrenergic nerve fibers; (2) release of an adrenergic neurotransmitter from nerve endings into the hypothalamohypophyseal portal circulation; (3) release of an adrenergic neurohumor from cells in the median eminence into the portal circulation in response to cholinergic nerve stimulation. However, none of these possibilities was demonstrated to be valid.

When stereotaxic techniques were perfected, and pure synthetic epinephrine and NE became available, Sawyer (1952) administered the drugs intraventricularly and found they induced ovulation. This ovulation could be prevented by prior administration of adrenergic blocking drugs. Other drugs thought to act only on neurons, such as atropine and pentobarbital, were also effective in blocking NE-induced ovulation, suggesting that this catecholamine acted as a central stimulant rather than as a mediator carried to the pituitary by the portal vascular system (Sawyer, 1964). This led to subsequent studies attempting to localize the site of the central action of NE. Early work (Saul and Sawyer, 1961) had indicated that NE was not involved in a limbic neuronal system related to ovulation. Ovulatory release of pituitary gonadotropins can be induced by electrical stimulation of the medial amygdala. Reserpine and adrenergic blockers were ineffective in blocking the effect, suggesting that the pathway between the medial amygdala and the hypothalamus is not adrenergic. However, the ovulatory response is an "all or none" phenomenon, and partial blockade would not have been evident in these experiments.

Sawyer later observed, by recording electrical activity from the hypothalamus and several other brain regions, that intraventricular administration of NE produced characteristic changes in the EEG. In the olfactory bulb, where these electrical changes were very prominent, the changes in electrical activity comprised an initial slowing of activity and then an increase in amplitude and frequency for about an hour. Similar EEG changes were recorded in the lateral preoptic region and in the medial forebrain bundle (a major input to the hypothalamus), suggesting that this NE activation stimulated hypothalamic function at a more central level. Concomitantly Sawyer found that the EEG changes in the olfactory bulb following NE administration would not occur in rabbits if they were pseudopregnant (thereby anestrous and nonovulating). The electrical response to intraventricular NE returned with the return of estrus, and ovulation followed. Interestingly, the electrical response to NE was elicited in one pseudopregnant animal, and it ovulated. Thus, Sawyer concluded that this NE stimulation of the CNS was related to pituitary control of ovulation.

However, the total olfactory pathway is not essential for NE control of ovulation, because intraventricular NE can induce ovulation when the olfactory bulbs are removed. It is interesting to note that Sawyer (1955) found that histamine, a monoamine which is often implicated in pituitary functions, affects the olfactory-pituitary pathway, but in a manner different from the effect of NE. Histamine, when administered intraventricularly, also produces an activation of olfactory bulb electrical activity that is associated with ovulation. Both histamine effects require pentobarbital anesthesia, and the ovulatory response is blocked by midbrain lesions and by olfactory bulb transections. The NE effects do not require anesthesia; and the ovulatory response, as well as being unaffected by olfactory bulb removal, is unaffected by midbrain lesions.

As a result of the lesion effects, Sawyer suggested that the critical site of NE action was located rostral to the midbrain and behind the olfactory bulbs, perhaps in the median eminence itself. He next investigated ovulation induced electrically at various sites. If the electrode was located in the posterior tuberal region close to the median eminence, centrally active drugs such as atropine, morphine, pentobarbital, and alcohol blocked the induced ovulation; but these drugs did not block ovulation if the electrode impinged directly on the median eminence. However, adrenergic blockers and reserpine were ineffective in blocking the ovulation induced by electrical stimulation of this basal tuberal region. Thus, the site of NE action apparently is central to the median eminence and its axonal input.

Subsequent to the Work Session, Rubinstein and Sawyer (1970) found that the intraventricular infusion of a few micrograms of epinephrine or NE would induce ovulation in the pentobarbital-blocked proestrous rat but that DA was considerably less effective. Quite recently Weiner, Rubinstein, and Sawyer (1970) have reported that intraventricular epinephrine evokes a biphasic elevation (3-6 min) and depression (30-120 min) in the multiple unit electrical activity (MUA) recorded from the median eminence of the estrogen-primed artificially proestrous rat. Again DA was quite ineffective. It was concluded that the pattern of epinephrine-induced changes in MUA in the median eminence might represent a neurophysiological correlate of activation of the release of an ovulatory surge of pituitary gonadotropin by the catecholamine.

### Catecholamines and Secretion of LH and Prolactin

Coppola described a set of experiments in which the levels of catecholamine within the brain were altered pharmacologically, and resulting changes in the secretion of prolactin (or LtH) and LH were examined (Coppola et al., 1965; 1966). Taken as evidence of prolactin release were the development and maintenance of pseudopregnancy, or of deciduomata in the traumatized horn of the rat uterus; and as an index of LH secretion, the presence of ovulation and the number of ova present in the ampulla of the oviduct. Coppola and co-workers found that drugs which depressed the levels of catecholamines in the brain (such as reserpine, tetrabenazine, or α-MPT) impaired the secretion of LH and facilitated the secretion of prolactin. The pretreatment of the animals with bretylium, a drug that protects peripheral, but not brain, catecholamine stores from reserpine depletion, did not prevent the effects of reserpine on LH or prolactin secretion. This suggested that the reserpine effect was mediated by its action on brain catecholamines. This hypothesis was supported by the further observation that pretreatment of animals with such MAO inhibitors as iproniazid, which elevate brain catecholamine levels, protected the animal against the effects of reserpine on the secretion of prolactin and LH (Lippmann, Leonardi, Ball, and Coppola, 1967).

*Dopamine and Serotonin Effects on Ovulation*

In studies reported subsequent to the Work Session, Kordon, Glowinski, and their collaborators have examined the roles of brain monoamines in the induced ovulation caused by administering the two gonadotropins Pregnant Mare's Serum (PMS), which exhibits FSH activity, and Human Chorionic Gonadotropin (HCG), which has LH activity, to immature rats. The animals receive PMS on their twenty-fifth day of life, which is nearly two weeks before their gonads might be expected to undergo spontaneous maturation; and 52 hours later they are injected with HCG. The brain enters a "critical period" of ovulation control 52 hours after PMS administration. Presumably the PMS causes the immature ovaries to secrete estrogens into the circulation; these hormones then act on the brain to stimulate the release of endogenous gonadotropins, which cause ovulation. The HCG presumably facilitates the secretion of endogenous gonadotropins but also can by itself ovulate the prepared follicle.

Kordon et al. (1968) found an apparent inhibitory effect of brain 5-HT on ovulation. Inhibition of MAO by the drug nialamide during the critical period blocks the subsequent ovulation. This blockade is prevented by prior PCPA-inhibition of brain 5-HT synthesis, but is unaffected by $\alpha$-MPT-inhibition of catecholamine synthesis. Kordon (1969) determined that the site of action of nialamide during the critical period is in the mediobasal tuberal hypothalamus (including the median eminence), after microinjecting it into various hypothalamic and hypophyseal sites; its ability to block ovulation if placed at that site persisted after $\alpha$-MPT administration but was antagonized by prior PCPA treatment.

Kordon and Glowinski (1969) also found that brain DA has an apparent stimulatory effect on ovulation. When catecholamine synthesis is blocked with $\alpha$-MPT before or during the critical period, ovulation is blocked or reduced in intensity (i.e., reduced number of ova produced). Subsequent restoration of both brain NE and DA levels, by administration of their common precursor dopa, restores ovulation (although to subnormal levels), while restoration of NE level alone, by administration of its precursor dops, is ineffective. The DA effect was localized to the median eminence region, by following the ovulation-inhibiting effect of perfusion of the false transmitter $\alpha$-methyldopa into various regions of the hypothalamus.*

*Kordon, unpublished observations.

Thus, Kordon and Glowinski (1969) suggest, "A specific dopaminergic activating mechanism seems therefore to intervene...in the course of hypothalamic ovulation control. LH release from the pituitary is thus submitted to a double [hypothalamic] aminergic control system, with a positive dopaminergic component and an inhibitory serotoninergic one, which can be shown to intervene in ovulation blockade by monoamine-oxidase inhibitors [Kordon et al., 1968]."

*Serotonin- and Progesterone-Related Behavior*

Meyerson described experiments in which he and co-workers studied the effects of monoamines and hormones on rat female copulatory behavior. The behavior observed was the lordosis response, which is displayed when an animal in heat is mounted by a male. Meyerson studied hormone-treated ovariectomized animals. It has long been known that the lordosis response is lost following castration, but can be restored by treatment with estrogen followed later by progesterone.

As a result of neuropharmacological studies on estrogen/progesterone-treated ovariectomized rats, Meyerson and co-workers suggest that copulatory behavior in the female rat is tonically inhibited by 5-HT-containing neurons. Inhibitors of MAO (Pargyline, nialamide, pheniprazine), which raise brain monoamine levels, inhibit the lordosis response (Meyerson, 1964a,b); Pargyline-induced inhibition of estrous behavior was not prevented by α-MMT depletion of catecholamines,* but the inhibition was prevented by administration of α-propyldopacetamide, a drug that Meyerson (1964a) found suppresses the Pargyline-induced increase in brain 5-HT levels. Combined MAO inhibitor/monoamine precursor treatment gave further evidence for an inhibitory effect of 5-HT on estrous behavior (Pargyline dose 50-100 mg/kg s.c., Meyerson, 1964a,b). The degree of the inhibitory effect of Pargyline on estrous behavior, increasing over a period of 4 hours as shown in Figure 21, was unaffected by coadministration of dops (Meyerson, 1964a), with similar results for dopa (Meyerson, 1964b); but 5-HTP led to a total inhibition (none of the animals responded) at 1 hour (Meyerson, 1964a,b), which continued for at least 2.5 hours (Meyerson, 1964b).

For the ovariectomized rat, Meyerson (1964c) found that, although estrogen treatment was essential for the lordosis response, monoamine depletors such as reserpine or tetrabenazine could be

*Meyerson, unpublished observations.

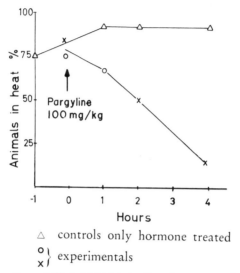

△   controls only hormone treated

○ ⎫
x ⎭  experimentals

Figure 21. The effect of MAO inhibition by Pargyline on estradiol/progesterone-activated estrus behavior in spayed rats. [Modified from Meyerson, 1964a]

substituted for progesterone to produce the response; and Meyerson and Lewander (1970) found that the 5-HT synthesis inhibitor PCPA could substitute for the progesterone. Thus, progesterone is physiologically equivalent to 5-HT depletion, raising the possibility that progesterone acts by inhibiting the activity of certain 5-HT-containing neurons that normally suppress the lordosis response.

Meyerson (1968b) showed that treatment with estrogen followed either by progesterone or by monoamine depletors (reserpine, tetrabenazine) will elicit a lordosis response in male rats castrated as adults (although the female pattern of mating behavior is difficult to activate in male rats). Estrogen alone did not produce female behavior in any of the males, but 10 percent of the animals responded when the estrogen treatment was followed by progesterone. The most striking response was obtained by monoamine-depletor treatment—40 to 50 percent of the males responded—suggesting that the male rat also has monoaminergic pathways that inhibit components of female sexual behavior.

Lindström and Meyerson (1967) also obtained evidence that increased cholinergic activity inhibits the copulatory response in estrogen/progesterone-treated ovariectomized rats. The copulatory response was decreased by the muscarinic cholinomimetic drugs pilocarpine, arecoline, and oxotremorine; and this effect was prevented by the antimuscarinic drug atropine but not by methylatropine, demonstrating that the effect is exerted on the CNS. Possible interactions of monoaminergic and cholinergic mechanisms in the production of mating behavior are under further investigation.

Regarding the question of whether progesterone might function to inhibit inhibitory neurons, Meyerson discussed the well-known characteristic of progesterone-like steroids of having "general inhibitory effects" on brain activity; e.g., intravenous administration is followed by decreased single cell activity and a general anesthetic effect. However, studies by Meyerson (1967) indicated that the anesthetic effects of progestins were not correlated with their effect on estrous behavior; the dose-response characteristics indicated no correlation between the lordosis response and the potency of the progestins as anesthetics, but did indicate a correlation between the lordosis response and the effectiveness of the progestins to maintain pregnancy in ovariectomized rats as shown in others' studies. Meyerson (1967) also showed that the duration of the anesthetic effect was of the order of minutes, while the onset of the estrous effect was of the order of hours. He therefore suggests that the anesthetic and estrous effects of progestins result from different mechanisms. However, Meyerson does not exclude as an alternative mechanism for the estrous effect of progesterone that it may act synergistically with estrogen to facilitate an excitatory system, which would then override an inhibitory system.

In discussing Meyerson's presentation, Sawyer raised questions as to the specificity of the action of progesterone. In the estrogen-primed ovariectomized rabbit, progesterone first fosters estrous behavior and secondarily produces a state of anestrus. Kawakami and Sawyer (1959) found that progesterone initially lowered the threshold of EEG arousal for a few hours and then raised the threshold to very high levels by 24 hours after treatment. Parallel effects were seen on another EEG threshold known as the EEG afterreaction threshold, which is closely correlated with pituitary-gonad function. Elevated thresholds in this system were conducive to blockade of ovulation; and steroids raising this threshold are now being used as contraceptive agents, presumably exerting at least part of their effect on the CNS (Sawyer and Kawakami, 1961).

Sawyer also discussed experiments on the female rat under urethane anesthesia. Barraclough and Cross (1963) described what they considered a specific effect of progesterone on hypothalamic neurons in such animals. The activation of hypothalamic units by vaginal stimulation was inhibited for periods up to an hour by intravenous injections of 400 $\mu$g of progesterone in propylene gylcol. These experiments were repeated in Sawyer's laboratory (Ramirez et al., 1967; Beyer et al., 1967; and Komisaruk et al., 1967); and the conclusion reached was that the progesterone exerted a generalized depressant effect on the animal, as evidenced by a sleeplike EEG and a depression of responses to other stimuli besides vaginal probing. The sleeplike EEG appeared to be an effect of elevated blood pressure induced by the progesterone on the carotid sinus, for it was blocked by denervating the carotid sinus (Beyer et al., 1967). Sawyer felt that these results did not deny that progesterone may exert specific effects on hypothalamic cells, but that such effects require further demonstration.

Subsequent to the Work Session, Terasawa and Sawyer (1970) recorded progesterone-induced changes in the multiple unit electrical activity (MUA) of the rat median eminence, changes which were independent of EEG alterations. Subcutaneous injection of progesterone late in the morning of proestrus stimulated an elevation in MUA correlatable with "advancement of the critical period" (Redmond, 1968) whereas an afternoon injection produced an initial depression of MUA.

Meyerson reported on unsuccessful attempts to relate hypothalamic 5-HT levels and treatment of rats with physiological doses of steroid hormones. No significant changes in hypothalamic 5-HT or catecholamine content were induced in ovariectomized rats by administering estradiol plus progesterone, or the two hormones given separately (Meyerson, 1966). Also, no effect was obtained by hormone treatment on either the accumulation of 5-HT after MAO inhibition with Pargyline (Meyerson, 1966) or on the decline of the 5-HT content after 5-HT synthesis inhibition with PCPA (Meyerson and Lewander, 1970). However, Meyerson cited recent studies that have demonstrated cyclic variations of catecholamine content in the hypothalamus during the estrous cycle of the female rat and after castration in the male rat (Anton-Tay and Wurtman, 1968; Lichtensteiger, 1969; Donoso et al., 1969; Fuxe and Hökfelt, 1969), and he suggested that a direct relationship between these fluctuations and hormone production with sexual behavior would appear to be possible but remains to be demonstrated.

## Effects of Varying Endocrine Function
## on Brain Monoamine Metabolism

If NE, DA, and 5-HT participate in the central regulation of anterior pituitary function, either as neurotransmitters or as compounds secreted into the pituitary portal vessels, it might be anticipated that major changes in the concentrations of circulating hormones, especially those hormones ultimately regulated by brain monoamines, would also influence the levels of the monoamines in the brain, or their rates of release, synthesis, or metabolism. During the last few years, perhaps ten laboratories in several countries have published data on the effects on brain catecholamine metabolism of experimentally modifying gonad function, i.e., by altering steroid hormone levels or by gonadectomy.

### Gonad Function and Hypothalamic Catecholamines:
#### A. O. Donoso, J. A. Coppola

Donoso summarized his laboratory's studies on this subject (see Table 4). In their procedure, the hypothalamus of male or female rats is removed and divided into anterior, middle, and posterior portions for analysis; the anterior portion includes the preoptic region. Donoso et al. (1967) found that the effect of castration was to increase the NE concentration and decrease the DA concentration in the anterior portion of the hypothalamus. This effect is typically observed 20 days after surgery but can also be demonstrated as early as 10 days. Donoso and Stefano (1967) determined that the effect can be blocked by very large doses of steroid hormones (50 $\mu$g of estradiol and 25 mg of progesterone); however it is not blocked by lower, physiological doses of estradiol (2 $\mu$g) or of estradiol plus progesterone (2 mg).

During the normal estrous cycle, cyclic variations in NE content are seen in both the anterior and the middle portions of the hypothalamus. During proestrus, the NE is high; it drops significantly after ovulation and attains its lowest concentrations during estrus (Stefano and Donoso, 1967). Recent experiments indicate that the fall of hypothalamic NE after treatment with $\alpha$-MPT is more marked in proestrus than in estrus (Donoso and Gutierrez Moyano, 1970). Donoso also observed data suggesting a diurnal rhythm in hypothalamic NE content; on the day of estrus, NE levels were higher in the morning than in the afternoon. (Daily rhythms in the NE contents of various

TABLE 4

Norepinephrine and Dopamine Levels in the
Anterior Portion of Hypothalamus  [Donoso]

| Condition | NE, $\mu$g/g* | DA, $\mu$g/g* | Reference |
|---|---|---|---|
| Normal males | 1.39 ± 0.01 | 0.42 ± 0.004 | Donoso et al., 1967 |
| Castrated males | 1.91 ± 0.02 | 0.30 ± 0.002 | Donoso et al., 1967 |
| Ovariectomized females | 2.06 ± 0.16 | 0.30 ± 0.01 | Donoso and Stefano, 1967 |
| Ovariectomized females, estradiol-progesterone-treated † | 1.57 ± 0.04 | 0.44 ± 0.01 | Donoso and Stefano, 1967 |
| Androgen-sterilized females | 1.39 ± 0.22 | – – – | Donoso and Cukier, 1968 |
| Females, estradiol-treated ‡ | 1.25 ± 0.11 | – – – | Donoso and Cukier, 1968 |
| Normal females | | | Stefano and Donoso, 1967 |
|    Proestrus | 2.36 ± 0.10 | – – – | |
|    Estrus morning | 1.64 ± 0.07 | – – – | |
|    Estrus afternoon | 1.29 ± 0.04 | – – – | |
|    Diestrus | 1.89 ± 0.12 | – – – | |

* Grams of wet tissue.

† 50 $\mu$g of estradiol plus 25 mg of progesterone.

‡ 1 $\mu$g/day/7 days of estradiol benzoate.

regions of rat and cat brain have been reported by other laboratories, Reis and Wurtman, 1968.) The concentrations of hypothalamic NE attained during estrus are similar to those found in female rats that have been treated with estrogen (2 $\mu$g/day) for a week or in androgen-sterilized rats.

Coppola (1968, 1969) also observed an increase in hypothalamic NE pool size following castration in rats. The rate of brain NE turnover, estimated by administering $\alpha$-MPT, also was elevated following castration. Both the storage and the kinetics of hypothalamic NE could be restored to normal in castrated animals by the administration of large doses of estrogens.

### Pituitary-Gonad Function and Brain Norepinephrine:
### A. O. Donoso; F. Anton-Tay and R. J. Wurtman

Donoso et al. (1969) examined the possibility that the gonad-associated changes in NE levels represented a modification in its synthesis. They administered radioactively labeled tyrosine by injection into the lateral cerebral ventricle and measured the amount of isotopically labeled NE present in brain 2 hours later. Levels of $^3$H-NE were higher in brains of castrated animals than in those of controls in some experiments; however $^3$H-normetanephrine levels were greater among castrates in all experiments. The increment in brain NE and normetanephrine observed in rats following gonadectomy is increased by MAO inhibition with nialamide. The magnitude of the increase is less in unoperated control rats. On the basis of these observations, Donoso concluded that the secretion of gonadotropins is under the control of a noradrenergic mechanism.

Anton-Tay and Wurtman (1968) examined the effect of gonadectomy on the turnover of $^3$H-NE taken up by brain after its injection into the lateral cerebral ventricle. These observers found that brains of castrated animals contained somewhat more NE, and that its turnover rate was markedly accelerated. Twenty-four hours after the administration of $^3$H-NE, less than half as much of the $^3$H-catecholamine remained in brains of castrated animals as in those of normal animals. Anton-Tay, Pelham, and Wurtman (1969) found that the acceleration of $^3$H-NE turnover was greatest in the hypothalamus, but was also observed in the brainstem and in the olfactory bulb and tubercle. Since endogenous brain NE levels did not decrease in castrated animals, Anton-Tay and Wurtman hypothesized that castration must cause an acceleration in brain NE synthesis as well as an acceleration in turnover. This hypothesis was supported by data on the accumulation of brain $^3$H-NE in animals given $^3$H-tyrosine, i.e., in experiments similar to those described by Donoso.

Anton-Tay and Wurtman found that hypophysectomy did not reproduce the effect of gonadectomy on brain $^3$H-NE turnover. Since hypophysectomy should have had the same effect on plasma estrogen levels as gonadectomy (i.e., it should block estrogen secretion), these authors examined the possibility that the mechanism by which gonadectomy modified brain $^3$H-NE turnover involved not the absence of circulating gonadal steroids but the presence of excessive amounts of pituitary gonadotropins in the blood. There is considerable evidence that gonadectomy is associated with a marked increase in the levels of FSH and LH in the plasma. Since this increase would be missing in hypophysectomized animals, Anton-Tay, Pelham, and Wurtman (1969) did experiments to determine whether either of the pituitary gonadotropins alone could reproduce the effect of gonadectomy on brain $^3$H-NE turnover in control animals or in animals subjected to hypophysectomy and gonadectomy. In both experimental preparations, it was found that FSH alone caused a marked increase in $^3$H-NE turnover. LH was without effect. Hence, these authors postulated that the mechanisms responsible for the increase in brain NE synthesis and turnover was due at least in part to a direct effect of high plasma levels of FSH upon the brain. They interpreted the ability of estradiol to block the effect of gonadectomy as resulting from its inhibiting effect on gonadotropin secretion.

There seems little debate that catecholamine-containing neurons in the hypothalamus are influenced by changes in the levels of gonadal and pituitary hormones in the circulation. Although the significance of the changes in $^3$H-catecholamine synthesis and turnover for gonadal regulation remain obscure, it should be noted that a variety of other mechanisms in which central catecholaminergic neurons are thought to participate (i.e., the control of sympathetic nervous tone) are probably also influenced by gonadal steroids. It seems possible that the changes in running activity, body temperature regulation, and blood vessel reactivity that are associated with ovulation in several mammalian species are related to the direct effects of FSH and/or gonadal steroids on central catecholaminergic neurons.

### Analysis of Norepinephrine Metabolism in Nonsteady States: R. J. Wurtman and F. Anton-Tay

Fuxe and Hökfelt (1969) reported that their group found no clear-cut difference in the rate of disappearance of hypothalamic NE between controls and castrated rats, after treatment of both with α-MPT. They contrasted this result with the observation by Anton-Tay and Wurtman (1968) that $^3$H-NE disappeared from the brains of castrated rats at a faster rate than from those of controls. Both groups refer to their findings as "turnover" of NE, but experiments by Wurtman, Anton-Tay, and Anton (1969) suggest that the method of Fuxe and Hökfelt may not provide an accurate measure of NE turnover.

Changes in NE level during any time interval $(t_1 - t_0)$ can be expressed mathematically. Defining $(Level)_{t_1} - (Level)_{t_0}$ as $(Increment)_{t_1 - t_0}$:

$$(\text{Increment})_{t_1 - t_0} = (\text{Input} - \text{Output})_{t_1 - t_0}$$
$$= (\text{Synthesis} - \text{Turnover})_{t_1 - t_0}$$
$$= (t_1 - t_0) \, [(\text{Synthesis rate})_{t_1 - t_0} - (\text{Turnover rate})_{t_1 - t_0}] \quad (1)$$
$$(\text{Increment rate})_{t_1 - t_0} = (\text{Synthesis rate})_{t_1 - t_0} - (\text{Turnover rate})_{t_1 - t_0} \quad (2)$$

where turnover reflects all of the processes by which NE is lost from the brain (oxidative deamination, O-methylation, synaptic release).

Since the action of $\alpha$-MPT is to inhibit tyrosine hydroxylase and thereby suppress NE synthesis, Fuxe and colleagues assume that the value of the "synthesis rate" term in equations (1) and (2) is decreased by the $\alpha$-MPT to be equal or nearly equal to zero, and is decreased to the same extent in control and castrated animals. Therefore, observing "increment (decrement) rate," Fuxe interprets the observation as a direct reflection of "turnover rate." Anton-Tay and Wurtman, on the other hand, using tracer amounts of $^3$H-NE, examined "turnover rate" directly, rather than "increment rate."

As described earlier, the level of NE in the brain is remarkably constant; and Wurtman, Anton-Tay, and Anton postulated that drugs such as $\alpha$-MPT that disturb this steady-state NE level may not act as simply as is often presumed. To analyze the nonsteady-state changes in NE level according to equations (1) and (2) in castrated and in control rats, they administered $\alpha$-MPT or only its diluent and (a) compared brain NE levels to obtain increment rate and (b) labeled the brain intracisternally with $^3$H-NE to measure turnover rate. During the fifth through eighth hours after $\alpha$-MPT administration,* the level of brain NE fell more rapidly in control rats than in castrated rats. However, the smaller rate of decline in castrated animals, instead of reflecting a lower turnover rate in these animals, reflected an acceleration of both turnover and synthesis, with the synthesis rate as calculated from equation (1) a surprising three times greater in castrated animals than in controls. Stated alternatively, a given dose of $\alpha$-MPT was much less effective in depressing NE synthesis in castrated animals than in control rats.

This study demonstrated two problems inherent in using the "synthesis inhibition" method for quantitative measurement of NE turnover in animals that are in different physiological states: (a) the effect of the drug cannot be attributed solely to inhibition of NE synthesis and (b) the drug may be more or less effective in the animals that the

*Fuxe studied an earlier time interval after $a$-MPT administration, and he suggests that the two sets of experiments are not comparable on that basis.

investigator wishes to study (i.e., castrated rats) than it is in normal animals. The altered sensitivity to α-MPT among castrated rats may be unrelated to the mechanism that causes the steady-state turnover of brain NE to be accelerated in these animals.

### Gonad Function and Tuberoinfundibular Dopamine: K. Fuxe

Fuxe described experiments relating changes in gonadal function of rats to the activity of the dopaminergic tuberoinfundibular neuron system. As mentioned in Section III, the tuberoinfundibular system is of special interest in that some of its terminal boutons appear to lie close to the surface of the median eminence. Hence, it is possible that catecholamines released from these terminals could enter the capillary plexus of the hypothalamohypophyseal portal system, or could modify the activity of other neurons within the median eminence. Fuxe and co-workers examined endocrine-associated changes in the DA of tuberoinfundibular neurons using semiquantitative histochemical fluorescence techniques. They examined changes in the DA level in somata in the arcuate nucleus and changes in the DA turnover in nerve endings in the median eminence, the latter according to the α-MPT synthesis inhibition method.* Fuxe considers that increases in catecholamine turnover in nerve are correlated with increases in impulse flow (see Andén, Corrodi, and Fuxe, 1969), so that he takes an increase in DA turnover rate in nerve endings to be an indication of increased bioelectric activity.

Fuxe and Hökfelt (1967) concluded that these neurons are probably not involved in ACTH secretion, since their DA turnover was unaffected by adrenalectomy or by administration of hydrocortisone or dexamethosome. Fuxe and co-workers then correlated changes in DA turnover with gonad function (Fuxe et al., 1969b) and with sex hormones (Fuxe et al., 1969a).

Pregnancy was associated with an increase in the number and intensity of fluorescent DA cell bodies in the arcuate nucleus, and the rate of amine depletion in the nerve endings in the median eminence after α-MPT treatment was markedly accelerated. These changes were first observed soon after the start of pregnancy, and remained throughout the gestation period. Similar changes were observed during pseudopregnancy and during the first 15-20 days of lactation. Furthermore, the turnover of the DA in the nerve endings appears to be in

*Discussed in previous section.

phase with the estrous cycle; turnover is least during the proestrus period, i.e., the ovulation-related phase of the cycle. These cyclical changes are not observed after gonadectomy or in androgen-sterilized rats.

These observations suggested to Fuxe that the activity of the tuberoinfundibular neurons is increased in endocrine states associated with a blockade of ovulation. In recent studies (Fuxe and Hökfelt, 1970), Fuxe's group has obtained evidence that prolactin, and not FSH and LH, markedly activate tuberoinfundibular neurons; when prolactin secretion is decreased with 2-brom-α-ergokryptin the DA turnover is decreased in the median eminence. Therefore, Fuxe suggests a role for prolactin in the high activity of the tuberoinfundibular DA neurons in pregnancy, pseudopregnancy, and lactation.

As mentioned above, tuberoinfundibular DA turnover is least during the proestrus phase of the estrous cycle. In addition, Ahrén, Fuxe, et al. (1971) performed a correlative study, simultaneously measuring pituitary DA turnover and LH content, which led Fuxe to suggest that release of DA by tuberoinfundibular neurons acts to inhibit LH secretion. Also relevant to this are studies showing that the administration of estrogen or testosterone, but not progesterone, increases the DA turnover in the median eminence. Since estrogen and testosterone are known to suppress LH and FSH release by inhibitory feedback, Fuxe further suggests that tuberoinfundibular neurons mediate feedback suppression of LH release.

## Brain Lesion Effects on Monoamine
## Metabolism and on Pituitary Function

The placement of brain lesions has provided a very useful general method for analyzing the central control of anterior pituitary function; a variety of specific lesions has now been shown to interfere more or less specifically with the release of given pituitary hormones. Experimental brain lesions have also found widespread application in the study of brain monoamines; lesions have been used to localize the monoamines within brain regions, or within cell bodies or nerve terminals, and to illustrate their neurotransmitter activity at specific synapses. However, at the present time, little use has been made by endocrinologists of the information that can be obtained on the effects of their lesions on brain monoamine levels; similarly, neuroanatomists and neuropharmacologists, who have made lesions to examine the localization and neurophysiological significance of brain monoamines, have not determined the effects of these lesions on anterior pituitary function. It seems likely that one research strategy that would provide new insights into the roles of brain monoamines in endocrine function would be to determine in similarly lesioned animals the effects of given lesions both on anterior pituitary function and on monoamine synthesis, storage, and metabolism.

At the Work Session, Moore summarized information on effects of lesions on brain monoamine metabolism, especially lesions of the medial forebrain bundle. He stressed the difficulties in interpreting the changes in brain monoamine levels that follow a lesion, and the possibility that a regional decrease in monoamine content could reflect mechanisms other than the loss of monoaminergic nerve endings located in that brain region. Gorski then described a very useful new method for studying central influences on pituitary function by depriving the mediobasal hypothalamus (i.e., the portion of the hypothalamus that probably contains most of the neuroendocrine transducer cells that secrete hypophysiotropic hormones) of its connections with other parts of the brain. This technique uses a knife devised by Bela Halász.

### Interpretation of Effects of Lesions on Localization of Brain Monoamines: R. Y. Moore

Moore described anatomico-chemical studies relating lesions to brain monoamine levels. Heller, Harvey, and Moore (1962) first demonstrated an effect of central lesions on brain monoamine content. They found that destruction of the medial forebrain bundle, or ablation of areas known to contribute fibers to that tract in the lateral hypothalamus, produced significant decreases in whole-brain levels of 5-HT in the rat. Subsequent experiments (Heller and Moore, 1965) demonstrated that tegmental lesions could affect brain 5-HT, brain NE, or both monoamines, depending upon the exact placement of the particular lesion (Table 5). The initial interpretation of these findings was that the loss of amine occurred secondary to the destruction and subsequent degeneration of amine-containing neuronal elements. These findings and conclusions were similar to those of Andén, Dahlström, Fuxe, and their co-workers, using the techniques of histochemical fluorescence (reported in Section III by Dahlström).

TABLE 5

Selectivity of Central Nervous System
Lesion Effects on Brain Serotonin and Norepinephrine [Moore]

| Lesion Group | Difference from Sham Operated Group Amine Levels, percent | |
|---|---|---|
| | 5-HT | NE |
| Medial forebrain bundle | −33* | −26* |
| Dorsomedial midbrain tegmentum | −28* | −24* |
| Ventrolateral midbrain tegmentum | − 3 | −32* |
| Periaqueductal gray | −18* | − 7 |

*Differences statistically significant; p. <.01. All other differences, p.>.05.
Data taken from Heller and Moore, 1965.

The anatomists* then confirmed and extended the histo-chemical findings of Andén and co-workers on the dopaminergic nigrostriatal pathway in experiments on the cat. In this species, nigrostriatal axons of dopaminergic neurons in their pathway from the substantia nigra to the caudate nucleus appear to run in the medial internal capsule and in the adjacent medial forebrain bundle. Complete destruction of the nigrostriatal pathway leads to a complete, or nearly complete, loss of DA, tyrosine hydroxylase, and dopa decarboxylase from the caudate nucleus; and the loss of enzymes in this instance is proportional to the loss of DA. In these experiments they were also able to show anterograde degeneration of axon terminals in the caudate nucleus using the sensitive anatomic silver methods of Fink and Heimer (1967). This provided the first anatomic evidence for a pathway from the substantia nigra to the neostriatum. It also indicated that these monoaminergic neurons are not intractable to staining with conventional silver methods, a point that is significant for subsequent considerations. These studies, combined with the findings of the Swedish group, provided definite anatomical and biochemical evidence for a direct nigrostriatal projection.

In further studies, Moore and co-workers examined the regional distribution of decreases in NE, 5-HT, and 5-HTP-dopa decarboxylase in rat and cat brain following medial forebrain bundle lesions (Moore et al., 1965; Heller et al., 1966a,b; and unpublished observations). On the

*Moore and Heller, unpublished observations.

basis of similar denervation experiments in the peripheral nervous system, they expected that the loss of amines and enzyme following central lesions should occur only in areas that had been innervated by the axons transected by the lesion. The axons of the medial forebrain bundle, a phylogenetically old, complex pathway, interconnect the basal and medial midbrain with the lateral hypothalamus and limbic areas of the telencephalon according to anatomical studies of Moore and others (see Moore and Heller, 1967, for review). In contrast to this restricted distribution of axons, the distribution of amine and enzyme losses that followed destruction of the medial forebrain bundle was widespread (Moore et al., 1965; Heller et al., 1966a,b; Moore and Heller, 1967; and unpublished observations). Decreases in 5-HT, NE, and 5-HTP-dopa decarboxylase were observed throughout the telencephalon, including such a widespread area as all parts of the neocortex (Harvey et al., 1963; Heller et al., 1965; Moore et al., 1966).

The discrepancy between the anatomic distribution of axons destroyed by the lesion and the more widespread distribution of amine and enzyme loss produced by the lesions led them to question their original concept that the biochemical changes represented simply destruction and degeneration of monoamine-containing neurons. They introduced the concept that at least a part of the amine and enzyme loss might occur as a result of transsynaptic metabolic changes in morphologically intact neurons that are one or more synapses removed from the neurons destroyed by the lesion—at variance with the Stockholm group's view (Andén, Dahlström, Fuxe, et al., 1966). These workers view the loss of amine as resulting entirely from degeneration of monoamine-containing axons and terminals derived from cell bodies in the brainstem and traversing the medial forebrain bundle to innervate the entire telencephalon and diencephalon.

The points Moore and co-workers took to support their view are: First, there is the discrepancy, given above, between the distribution of the amine loss and the distribution of the degenerating axons following medial forebrain bundle lesions. Second, the time course of the loss of amines following lesions appears to be longer than can be accounted for by direct Wallerian axonal degeneration (Moore and Heller, 1967). Third, in contrast to the situation in the neostriatum, the loss of both amine and enzyme is never complete, and the degree of the two types of loss are not correlated, regardless of the size of tegmental or hypothalamic lesions (Heller and Moore, 1968; Heller et al., 1969). Regarding Moore's second point, Dahlström provided results obtained subsequent to the Work Session that suggest a mechanism for the slowness of the appearance of amine loss if the degeneration were presynaptic. The axoplasmic transport of amine granules appears to be 7-8 times slower in rat bulbospinal NE neurons than in the peripheral NE neurons of the rat (Häggendal and Dahlström, 1969). Axoplasmic transport is important in the maintenance of nerve endings, so that a system with slow flow would be likely to show effects of interruption of the flow later than a system with fast flow. The degeneration of central monoaminergic fibers, therefore, may be slower than the degeneration in the peripheral nervous system because of differences in the rates of axoplasmic transport.

Even though transsynaptic changes are difficult to conceptualize, Moore cited precedents for believing they exist, besides the well-known transneuronal degeneration noted principally in the primary sensory projections. Perhaps especially pertinent to this discussion of "Brain Monoamines and Endocrine Function" Moore and co-workers* have recently observed a clear example of a transsynaptic change in the content of the enzyme hydroxyindole-O-methyl transferase (HIOMT) in the rat pineal following preganglionic denervation of the gland. It seems reasonable to assume that a similar effect could be obtained from any central lesion interrupting the input to the cervical sympathetics, regardless of whether brain monoamines are involved or not.

More experimental support for the hypothesis that the flow of impulses across the synapse can influence the synthesis and storage of NE in the postsynaptic neuron has been obtained in studies on rats in which one olfactory bulb has been transected (Pohorecky, Zigmond, Heimer, and Wurtman, 1969). In the rat, essentially all of the primary projections from the olfactory bulb remain uncrossed, and terminate in the ipsilateral telencephalon (Heimer, 1968). Subsequent to olfactory bulb transection, the content of NE in the ipsilateral telencephalon declines by about 30 percent. That this decline reflects the loss of noradrenergic nerve endings is indicated by the fact that the ability of the ipsilateral telencephalon to take up $^3$H-NE from the cerebrospinal fluid is decreased by an equivalent fraction, compared with the unlesioned side of the brain. Of greater interest, however, is the observation that olfactory bulb transection causes a significant *increase* in the NE content of the ipsilateral brainstem. In animals subjected to bilateral olfactory bulb transection, brainstem NE levels are also greater

*Moore, Vick, and Nielson, unpublished observations.

than those present in unoperated animals. Since no nerve fibers are known to run from the olfactory bulb to the brainstem, and since, even if such fibers existed, their destruction could cause a decrease, not an increase, in brainstem NE levels, the changes observed in lesioned animals must result from a transsynaptic effect; i.e., the normal input of signals from the olfactory bulb must tend to depress brainstem NE levels (possibly by depressing catecholamine biosynthesis in this region, or by stimulating NE release or metabolism). Removal of this "inhibitory" input allows brainstem NE levels to rise. If one admits to the possibility that presynaptic inputs can modify neurotransmitter metabolism in postsynaptic neurons, it should not be surprising that these inputs can be either "excitatory" or "inhibitory"; i.e., they can either enhance catecholamine synthesis, release, or storage, or depress these processes. Dahlström, however, suggested that the increase in brainstem NE following transection of the olfactory bulb could be explained by an alternative mechanism not involving transsynaptic effects. If a noradrenergic neuron existed that provided axon collaterals to both the olfactory bulb and the brainstem, transection of the collateral to the bulb might eventually cause more NE to be sent to the brainstem because of increased axonal flow to the intact collateral (see Dahlström, 1969).

In the discussion of Moore's paper, Nauta and Dahlström both commented on the current controversy regarding the neural connections between the brainstem reticular formation and the cerebral cortex. It is difficult at present to reconcile data obtained by the histochemical fluorescence method with data obtained by "classical" techniques of axonal degeneration, including the very contemporary Fink-Heimer method for demonstrating degenerating terminal boutons. Both the biochemical and the histochemical methods demonstrate a profound loss of telencephalic NE and 5-HT following lesions of the medial forebrain bundle. However, the histochemical fluorescence methods have failed to provide evidence of monoaminergic cell bodies situated between the lesion and the telencephalon, even despite the use of pharmacological agents (dopa, MAO inhibitors) known to increase the fluorescence of NE and 5-HT neurons. This negative finding has naturally led to the notion that monoamine pathways to the cerebral mantle originate quantitatively from cell bodies located caudal to the hypothalamus, that is to say in either the midbrain or rhombencephalon, or both. (In addition, however, Dahlström reported that Ungerstedt, in studies to be published in 1971, has been able to trace uninterrupted fluorescent fibers to the brainstem in consecutive cross- or sagittal sections after stereotactic lesions.) In at least partial conflict with this conclusion, Nauta pointed out that studies by the use of fiber-degeneration methods in the rat have revealed only a relatively small number of degenerating axons ascending from lesions in the medial forebrain bundle to the cerebral cortex. The distribution of such fibers is limited to the cingulate region of the cortex,* and thus appears to indicate a considerably more restricted projection to the cortical mantle than would be suggested by the histochemical findings. The fiber-degeneration studies seem to lead to the conclusion that any substantial conduction system ascending to the cerebral cortex from the hypothalamus or more caudal brainstem levels must be interrupted very largely by synapses, and hence, that monoaminergic cell bodies must exist between the lesion and the telencephalon.

Nauta and Dahlström agreed that the task of reconciling these observations is rendered almost impossible by the fact that both derive interpretations from negative data. The failure of one method to demonstrate noradrenergic cell bodies rostral to the hypothalamic lesion, and of the other method to demonstrate widespread axon degeneration in the telencephalon could indicate that intercalated cell bodies or monosynaptic pathways, respectively, do not exist; however they could also simply reflect inadequacies in the state of the art. It is possible that an improved histochemical fluorescence method would demonstrate the putative cell bodies or that an improved method for displaying axonal degeneration would demonstrate the putative mono-synaptic pathway.

### Lesion Effects on Hypothalamohypophyseal Function: R. A. Gorski

Gorski described an experimental preparation, developed by Halász and Pupp (1965), for the study of the neurally isolated or deafferented medial basal hypothalamus (MBH). The MBH is particularly rich in catecholamines; moreover it is essentially equivalent to the "hypophysiotrophic area," a diffuse area of the hypothalamus that is capable of maintaining the appearance and function of pituitary tissue transplanted within it (Halász et al., 1965) and that is thought by most observers to contain the cells that synthesize hypophysiotropic hormones. Included within this hypothalamic region is part or all of the ventromedial nucleus, the arcuate nucleus, and the median eminence

*Hedreen, unpublished observations.

itself, which is in contact with the pituitary complex through the still intact hypophyseal portal system. It also may variably include adjacent regions of the hypothalamus. Such a neurally isolated preparation, despite its artificial nature, provides a useful method for determining the functions of the area, and for obtaining data on such questions as the site of hormonal feedback, the relative importance of hormonal feedback or neuronal control, and the general anatomical pathway of neuronal control of median eminence function.

Earlier attempts to elucidate these problems with this approach (Egdahl, 1960; Wise, Van Brunt, and Ganong, 1963; Ganong, 1963; Matsuda et al., 1964) generally involved a preparation from which all of the forebrain overlying a small "island" of hypothalamic tissue was physically removed. Study of the island preparation, though very fruitful, was severely limited, because the animal without its forebrain was often unresponsive, unable to eat or thermoregulate, and likely to die within a short time following surgery. The metabolic and endocrine status of such a preparation was clearly so disrupted that interpretation of the data became very difficult. Several alternate strategies have been used over the past several years, including placement of small lesions within various regions of the hypothalamus, destruction of specific afferent connections of the hypothalamus (e.g., medial forebrain bundle, stria terminalis, etc.), and intrahypothalamic implantation of hormones or other chemical substances. The effects of these manipulations on median eminence function and thus, pituitary secretion has been extensively studied (see Martini and Ganong, 1966, 1967; Szentágothai et al., 1968).

Sawyer briefly described such lesion and implantation studies relating to sexual physiology and behavior. First, reversal of the inhibition of prolactin release (presumably by preventing the release of PIF) was achieved both by lesions (Haun and Sawyer, 1960; Kanematsu, Hilliard, and Sawyer, 1963) and by implants of reserpine (Kanematsu and Sawyer, 1963) in the basal hypothalamus. Second, precocious puberty was induced in the female rat by lesions in the anterior hypothalamus (Donovan and van der Werff ten Bosch, 1956), by lesions in the amygdala and its projections to the hypothalamus (Elwers and Critchlow, 1961), and by implants of estrogen into the anterior hypothalamus (Smith and Davidson, 1968; see also Gorski's results to follow shortly). Finally, lesion and estrogen implantation studies showed that the hypothalamic sites that control some components of ovarian target-organ function are distinctly different from the sites that control estrous behavior. Although there are species differences in the exact location of the hypothalamic sites, the principle holds true for the rat (Lisk, 1962), the cat (Robison and Sawyer, 1957; Sawyer, 1960, 1963; Michael, 1966), and the rabbit (Sawyer, 1959; Palka and Sawyer, 1966a,b).

The method described by Gorski consists of inserting a curved wirelike knife into the midline of the hypothalamus under stereotaxic control (Figure 22D). By rotating this simple instrument 360°, all neural connections of the MBH can be severed without extensive damage to the surrounding brain. Since the "efferent" hypophyseal portal vessels are left intact, in terms of pituitary regulation this procedure is termed complete deafferentation (Figure 22A). The experimental animal is generally in good condition, with the exception of persistent diabetes insipidus. Partial transections (Figure 22B,C) may also be produced, thus isolating the MBH from neural input coming specifically from the anterior or posterior direction (Halász and Gorski, 1967). With this preparation it is now possible to study the autonomous capacity of the MBH to secrete the factors that regulate pituitary function, as well as to study the source of neural afferents

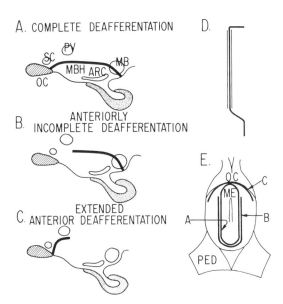

Figure 22. A, B, C: Schematic midsagittal representations of the three basic types of deafferentation of the MBH. D: Diagram of the knife held in its supporting tube. E: Schematic appearance of each deafferentation as seen from the base of the brain. Abbreviations: ARC, arcuate nucleus; MB, mamillary body; ME, median eminence; OC, optic chiasm; PED, cerebral peduncle; PV, paraventricular nucleus; SC, suprachiasmatic nucleus. [Gorski, 1970]

that influence the secretory activity of the hypothalamic neuroendocrine transducers.

Utilizing this deafferentation technique, Halász and Gorski
(1967) have been able to specify the afferent input necessary for
ovulation (Figure 23). Whereas the MBH appears to regulate the tonic
release of gonadotropin, a diffuse pathway from the septal region that
converges upon the median eminence controls the cyclic release of LH,
and thus, ovulation. This technique has also been applied to the study
of puberty in female rats (Ramaley and Gorski, 1967). The posterior
deafferentation produces no effect on puberty, but both the complete
and the anterior deafferentations produce precocious puberty. These
results suggest that the anterior afferent input provides an inhibitory

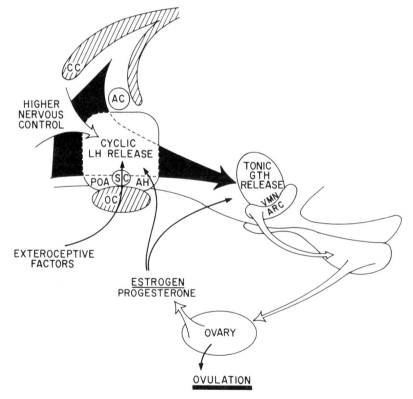

Figure 23. Schematic model of the neuroendocrine regulation of the estrous cycle in the
normal female rat. This concept of the localization of two distinct gonadotrophin
(GTH)-regulating mechanisms within the hypothalamus is supported by deafferentation studies.
Additional abbreviations: AC, anterior commissure; AH, anterior hypothalamus; CC, corpus
callosum; POA, preoptic area; VMN, ventromedial nucleus. [Gorski, 1970]

influence on the MBH, and its removal results in the precocious onset of adult ovarian activity. Such an input may arise in the amygdala (Critchlow and Bar-Sela, 1967).

This preparation has similarly been used to study the regulation of pituitary-adrenal function, including diurnal variations in ACTH (Halász, Vernikos-Danellis, and Gorski, 1967) and adrenocorticoid secretion (Halász, Slusher, and Gorski, 1967), the site of corticoid feedback, and the response to stress (Palka et al., 1968). Complete deafferentation of the MBH abolishes diurnal secretory patterns, and appears to remove an inhibitory restraint on basal ACTH secretion. The general effect of deafferentation of the MBH on the secretion of the several pituitary tropic hormones is summarized in Figure 24. In

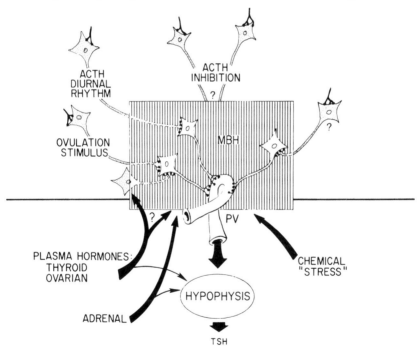

Figure 24. Highly schematic representation of the data obtained following deafferentation of the MBH (hatched area). Following transection of their afferent input, neurons of this area autoregulate the near-normal secretion of LH, FSH, and GH; but TSH secretion is reduced (Halász, Florsheim, Corcorran, and Gorski, 1967), and ACTH secretion is enhanced. Neurons (stippled) from outside of the MBH, perhaps from extrahypothalamic areas, provide afferent input necessary for normal thyroidal and gonadal hormone feedback, ovulation, diurnal ACTH rhythm, and the inhibition of ACTH secretion. The pathway for the last input is unknown; the other influences appear to reach the MBH over anterior pathways. [Gorski]

oversimplified terms, it would appear that the neuroendocrine trans-ducers within the median eminence region are capable of some degree of autoregulation, but are modulated by hypothalamic and extrahypo-thalamic neural inputs.

Nauta compared this preparation to Sherringtonian studies of the brainstem by means of progressive isolation of the spinal cord from the brain. In the spinal animal, the spinal cord has enormous residual capacity, and some functions even appear grossly exaggerated, e.g., the paraplegic flexor response. Nauta speculated that the flexor reflex, a highly organized withdrawal response, may represent a class of "basic survival mechanisms," which are relatively independent of "higher" central control. In the endocrine realm, the ACTH system might represent the mechanism most essential to an animal's survival. Nauta suggested that with progressive transections within the MBH, ACTH responses might prove to be the most resistant to obliteration. Gorski responded that it is likely that the technique of deafferentation is currently rather gross, and refinement in this and other techniques may yet reveal localization of function within the MBH.

### Joint Effects of Lesions on Monoamines and Endocrine Function

At the present time, no studies have been published on the effects of total or partial hypothalamic deafferentation on the metabolism or action of monoamines within the MBH or the rest of the brain. It was generally agreed that these would very likely be useful studies. As an example of monoamine metabolic effects, if the noradrenergic link in the central control of ovulation consisted of noradrenergic fibers originating outside the hypothalamus but termina-ting on the neuroendocrine transducer cells that release LRF, it would be anticipated that hypothalamic deafferentation would depress hypo-thalamic NE levels, and, probably, the ability of the hypothalamus to take up $^3$H-NE from the cerebrospinal fluid. As an example of monoamine action, if the hypothalamic cells that "sense" plasma cortisol levels release NE as their neurotransmitter, the administration of drugs that block the synthesis or release of NE (e.g., α-MPT or reserpine) in deafferented animals should open the feedback loop, while deafferentation alone should not.

However, Moore pointed out that the problems introduced by attempting to correlate lesion effects on brain monoamines with alterations in endocrine function are numerous. The usual problems in the interpretation of lesion effects are raised by the inevitable variability in the size and location of lesions and failings in providing accurate histological verification. On the other hand, lesions in a group of animals may be relatively homogeneous, but the effects obtained can be due to the incidental inclusion of some unexpected structure. With stereotaxic lesions, for example, this can occur with passage of the electrode. The manner in which a lesion is made may alter its effects. For example, the employment of different metals in the electrodes used for lesion-making may produce differing results.

Difficulties that can occur in correlating lesion-amine and lesion-endocrine effects can be exemplified by the following experiments: Destruction of the medial forebrain bundle bilaterally not only alters telencephalic monoamine metabolism but it abolishes the response of pineal HIOMT to environmental light (Axelrod, Snyder, Heller, and Moore, 1966). Such lesions also abolish the persistent vaginal estrous response in the female rat to continuous environmental light (Wurtman, Axelrod, Chu, Heller, and Moore, 1967), an effect that with the pineal HIOMT change might well be attributed to the changes in brain monoamine content. However, the endocrine effects are due not to the widespread neurochemical sequelae of the lesion but to the fact that a little-known critical component of the central retinal projection, the inferior accessory optic tract, runs within the medial forebrain bundle and would be transected by lesions destroying the bundle. This tract mediates the pineal response to environmental light (Moore, Heller, Wurtman, and Axelrod, 1967), but its function is not known to be influenced by changes in brain monoamines. Thus, Moore suggested that correlations between the effects of lesions on brain monoamine metabolism and on endocrine function must be approached with care.

## VI. CONCLUSIONS

The evidence described in this report seems to support the hypothesis that brain monoamines participate in controlling the secretion of hypophysiotropic hormones from neuroendocrine transducer cells ("releasing factor cells") in the median eminence of the hypothalamus. The most likely site of this participation is at synapses, either on the transducer cells themselves, or in multisynaptic neuronal pathways leading to these cells. Thus, intraventricularly administered DA or NE enhances GRF secretion, NE or epinephrine causes ovulation (perhaps by stimulating LRF release), DA stimulates LRF release, and NE inhibits ACTH secretion (probably by an action on the cells that make CRF). The pharmacological depletion of brain catecholamines blocks the secretion of LH from the pituitary, and accelerates the release of prolactin. Experimental manipulations that increase brain 5-HT levels, or decrease those of DA, block PMS-induced ovulation in immature rats; and brain 5-HT appears to exert a similar suppressive action on the lordosis behavioral response observed in estrous female rats. There is also abundant evidence that changes in the plasma levels of steroid hormones (particularly the gonadal hormones) cause characteristic changes in brain monoamine synthesis and turnover, and in the histochemical fluorescence of tuberoinfundibular neurons.

The growth of knowledge as to exactly how brain monoamines participate in the control of pituitary function is hampered by some of the same problems that plague neurobiology in general. Perhaps three of these problems should be singled out for being especially burdensome:

*1. The Anatomic Problem:* Too little information is available about every brain locus diagrammed in Figure 14. We do not yet know whether distinct populations of neuroendocrine transducer cells produce each of the hypophysiotropic hormones, much less the specific localization of such populations. We also have essentially no information about the number or location of the neurons and synapses in such multisynaptic pathways as those that are presumed to mediate the secretion of CRF in response to pain, of GRF in response to hypoglycemia, of LRF in response to the presumed cyclic stimulus from the preoptic area that causes ovulation, and so forth. Methodologically, as described earlier, in our attempts to map tracts by degeneration, histochemical, or chemical methods, we rapidly encounter problems that display the limitations of our techniques.

*2. The Neurotransmitter Problem:* If we postulate that a multi-synaptic pathway mediates one of the neuroendocrine reflexes mentioned above, we have no way of knowing whether or not the flow of information across all of those synapses is mediated by a single excitatory neurotransmitter or is suppressed by a single inhibitory neurotransmitter. It is possible to imagine that one such pathway might involve multiple synapses utilizing, for example, ACh, NE, and DA as excitatory transmitters and 5-HT as an inhibitory transmitter. In that event, what would it mean to state that the secretion of a particular releasing factor is under the control of a particular monoamine? It would be enormously helpful if all synapses ultimately enhancing the release of a given hypophysiotropic hormone utilized a single substance as their neurotransmitter. However, there is no reason to believe the brain is organized in this relatively simple way.

*3. The Neurophysiological Problem:* Underlying all research strategies in which brain monoamine levels or kinetics are measured in response to an experimental manipulation (e.g., castration) is the assumption that an increase in NE content or turnover is associated with a characteristic change (probably an increase) in the number of NE molecules released at the synapse. This hypothesis has never been adequately tested. It is at least possible that such changes could reflect increased, unchanged, or even decreased intrasynaptic NE (e.g., if the enhanced turnover results from accelerated intraneuronal catabolism, such as follows the administration of reserpine).

Because of these problems, and because of the unusual inaccessibility but great importance of hypothalamic neuroendocrine transducer cells, it seems reasonable to predict that the study of Brain Monoamines and Endocrine Function will continue to excite neurobiologists and endocrinologists for years to come.

## List of Abbreviations

| | |
|---|---|
| ACh | acetylcholine |
| ACTH | adrenocorticotropin |
| CNS | central nervous system |
| COMT | catechol-O-methyl transferase |
| CPZ | chlorpromazine |
| CRF | corticotropin releasing factor |
| cyclic AMP | cyclic 3′,5′-adenosine monophosphate |
| DA | dopamine |
| dopa | L-dihydroxyphenylalanine |
| dops | L-threodihydroxyphenylserine |
| EEG | electroencephalogram |
| FRF | follicle-stimulating hormone releasing factor |
| FSH | follicle-stimulating hormone |
| GH | growth hormone |
| GIF | growth hormone inhibitory factor |
| GRF | growth hormone releasing factor |
| HCG | human chorionic gonadotropin |
| HIOMT | hydroxyindole-O-methyl transferase |
| 5-HT | serotonin (5-hydroxytryptamine) |
| 5-HTP | 5-hydroxytryptophan |
| Hypox | hypophysectomized |
| ICSH | interstitial-cell-stimulating hormone |
| LGV | large granular vesicle |
| LH | luteinizing hormone |
| LOT | lateral olfactory tract |
| LRF | luteinizing hormone releasing factor |
| LtH | luteotropic hormone (prolactin) |
| MAO | monoamine oxidase |
| MBH | mediobasal hypothalamus |
| MIF | melanocyte-stimulating hormone inhibitory factor |
| α-MMT | α-methyl-*m*-tyrosine |
| α-MPT | α-methyl-*p*-tyrosine |
| MRF | melanocyte-stimulating hormone releasing factor |
| MSH | melanocyte-stimulating hormone |
| MUA | multiple unit activity |
| NE | norepinephrine |
| PCPA | *p*-chlorophenylalanine |
| PIF | prolactin inhibitory factor |
| PMS | pregnant mare's serum |
| PNMT | phenylethanolamine-N-methyl transferase |
| SGV | small granular vesicle |
| TRF | thyroid-stimulating hormone releasing factor |
| TSH | thyroid-stimulating hormone |

## BIBLIOGRAPHY

This bibliography contains two types of entries: (1) citations given or work alluded to in the report, and (2) additional references to pertinent literature by conference participants and others. Citations in group (1) may be found in the text on the pages listed in the right-hand column.

Page

Abrahams, V.C., Koelle, G.B., and Smart, P. (1957): Histochemical demonstration of cholinesterases in the hypothalamus of the dog. *J. Physiol.* 139:137-144. ... 216

Aghajanian, G.K. and Bloom, F.E. (1966): Electron-microscopic autoradiography of rat hypothalamus after H-3 norepinephrine. *Science* 153:308-310. ... 206

Aghajanian, G.K. and Bloom, F.E., (1967a): Electron-microscopic localization of tritiated norepinephrine in rat brain: Effect of drugs. *J. Pharmacol. Exp. Ther.* 156:407-416. ... 206

Aghajanian, G.K. and Bloom, F.E. (1967b): Localization of tritiated serotonin in rat brain by electron-microscopic autoradiography. *J. Pharmacol. Exp. Ther.* 156:23-30. ... 206

Aghajanian, G.K., Bloom, F.E., and Sheard, M.H. (1969): Electron microscopy of degeneration within the serotonin pathway of rat brain. *Brain Res.* 13:266-273. ... 206

Ahrén, K., Fuxe, K., Hamberger, L., and Hökfelt, T. (1971): Turnover changes in the tubero-infundibular DA neurons during the ovarian cycle. *Endocrinology.* (In press) ... 257

Andén, N.-E. (1967): Effects of reserpine and a tyrosine hydroxylase inhibitor on the monoamine levels in different regions of the rat central nervous system. *Europ. J. Pharmacol.* 1:1-5. ... 200

Andén, N.-E., Carlsson, A., Dahlström, A., Fuxe, K., Hillarp, N.-A., and Larsson, K. (1964): Demonstration and mapping out of nigro-neostriatal dopamine neurons. *Life Sci.* 3:523-530. ... 202

Andén, N.-E., Corrodi, H., Dahlström, A., Fuxe, K., and Hökfelt, T. (1966): Effects of tyrosine hydroxylase inhibition on the amine levels of central monoamine neurons. *Life Sci.* 5:561-568.

Andén, N.-E., Corrodi, H., and Fuxe, K. (1969): Turnover studies using synthesis inhibition. *In: Metabolism of Amines in the Brain.* Hooper, G., ed. London: Macmillan and Co., Ltd., pp. 38-47. ... 256

Andén, N.-E., Dahlström, A., Fuxe, K., and Larsson, K. (1965): Mapping out of catecholamine and 5-hydroxytryptamine neurons innervating the telencephalon and diencephalon. *Life Sci.* 4:1275-1279. ... 202

Page

Andén, N.-E., Dahlström, A., Fuxe, K., Larsson, K., Olson, L., and Ungerstedt, U.     199,200,
(1966): Ascending monoamine neurons to the telencephalon and diencephalon.     202,260
*Acta Physiol. Scand.* 67:313-326.

Andén, N.-E., Fuxe, K., and Ungerstedt, U. (1967): Monoamine pathways to the     200,202
cerebellum and cerebral cortex. *Experientia* 23:838-839.

Anton-Tay, F., Anton, S., and Wurtman, R.J. (1970): Mechanism of changes in
brain norepinephrine metabolism after ovariectomy. *Neuroendocrinology*
6:265-273.

Anton-Tay, F., Chou, C., Anton, S., and Wurtman, R.J. (1968): Brain serotonin     196
concentration: elevation following intraperitoneal administration of melatonin.
*Science* 162:277-278.

Anton-Tay, F., Pelham, R.W., and Wurtman, R.J. (1969): Increased turnover of     253
$^3$H-norepinephrine in rat brain following castration on treatment with ovine
follicle-stimulating hormone. *Endocrinology* 84:1489-1492.

Anton-Tay, F. and Wurtman, R.J. (1968): Norepinephrine: turnover in rat brains     250,253,
after gonadectomy. *Science* 159:1245.     254

Anton-Tay, F. and Wurtman, R.J. (1969): Regional uptake of $^3$H-melatonin from
blood or cerebrospinal fluid by rat brain. *Nature* 221:474-475.

Anton-Tay, F. and Wurtman, R.J. (1971): Brain monoamines and the control of
anterior pituitary function. *In: Frontiers in Neuroendocrinology.* Ganong, W.F.
and Martini, L., eds. New York: Oxford University Press. (In press)

Axelrod, J. (1965): The metabolism, storage, and release of catecholamines. *Recent
Progr. Hormone Res.* 21:597-622.

Axelrod, J. (1966): Control of catecholamine and indoleamine metabolism by
sympathetic nerves. *In: Mechanism of Release of Biogenic Amines.* Euler, U.S.
von, Rosell, S., and Uvnas, B., eds. New York: Pergamon Press, pp. 189-209.

Axelrod, J., Snyder, S.H., Heller, A., and Moore, R.Y. (1966): Light-induced     269
changes in pineal hydroxyindole-O-methyl-transferase: Abolition by lateral
hypothalamic lesions. *Science* 154:898-899.

Axelrod, J. and Weissbach, H. (1961): Purification and properties of hydroxy-     196
indole-O-methyl transferase. *J. Biol. Chem.* 236:211-213.

Baldessarini, R.J. and Kopin, I.J. (1967): The effect of drugs on the release of     212
norepinephrine-H$^3$ from central nervous system tissues by electrical stimulation
*in vitro. J. Pharmacol. Exp. Ther.* 156:31-38.

Barraclough, C.A and Cross, B.A. (1963): Unit activity in the hypothalamus of the     249
cyclic female rat: effect of genital stimuli and progesterone. *J. Endocr.*
26:339-359.

Bertler, A. and Rosengren, E. (1959): Occurrence and distribution of catechol     198
amines in brain. *Acta Physiol. Scand.* 47:350-361.

Page

Betz, D. and Ganong, W.F. (1963): Effect of chlorpromazine on pituitary-adrenal          236
function in the dog. *Acta Endocr.* 43:264-270.

Beyer, C., Ramirez, V.D., Whitmoyer, D.I., and Sawyer, C.H. (1967): Effects of          250
hormones on the electrical activity of the brain in the rat and rabbit. *Exp.
Neurol.* 18:313-326.

Bhattacharya, A.N. and Marks, B.H. (1969a): Effects of pargyline and ampheta-       241,242
mine upon acute stress responses in rats. *Proc. Soc. Exp. Biol. Med.*
130:1194-1198.

Bhattacharya, A.N. and Marks, B.H. (1969b): Reserpine- and chlorpromazine-          241
induced changes in hypothalamo-hypophyseal-adrenal system in rats in the
presence and absence of hypothermia. *J. Pharmacol. Exp. Ther.* 165:108-116.

Bhattacharya, A.N. and Marks, B.H. (1970): Effects of alpha methyl tyrosine and          239
p-chlorophenylalanine on the regulation of ACTH secretion. *Neuroendocrinology* 6:49-55.

Björklund, A., Ehinger, B., and Falck, B. (1968): A method for differentiating       198,202
dopamine from noradrenaline in tissue sections by microspectrofluorometry. *J.
Histochem. Cytochem.* 16:263-270.

Björklund, A., Falck, B., Hromek, F., Owman, C., and West, K.A. (1970):          203
Identification and terminal distribution of the tubero-hypophyseal monoamine
fibre systems in the rat by means of stereotaxic and microspectrofluorimetric
techniques. *Brain Res.* 17:1-23.

Bloom, F.E. (1968): Electrophysiological pharmacology of single nerve cells. *In:
Psychopharmacology. A Review of Progress 1957-1967.* Efron, D., editor-in-
chief. Washington, D.C.: Government Printing Office, pp. 355-373.

Bloom, F.E. (1970): The fine structural localization of biogenic monoamines in          207
nervous tissue. *Int. Rev. Neurobiol.* 13:27-66.

Bloom, F.E. and Aghajanian, G.K. (1968a): An electron microscopic analysis of       206,207
large granular synaptic vesicles of the brain in relation to monoamine content. *J.
Pharmacol. Exp. Ther.* 159:261-273.

Bloom, F.E. and Aghajanian, G.K. (1968b): An osmiophilic substance in brain       206,208
synaptic vesicles not associated with catecholamine content. *Experientia*
24:1225-1227.

Bloom, F.E. and Barrnett, R.J. (1966): Fine structural localization of norepineph-          205
rine in vesicles of autonomic nerve endings. *Nature* 210:599-601.

Bloom, F.E., Costa, E., and Salmoiraghi, G.C. (1964): Analysis of individual rabbit          213
olfactory neuron response to microelectrophoresis of acetylcholine, norepinephrine, and serotonin synergists and antagonists. *J. Pharmacol. Exp. Ther.* 146:16-23.

Bloom, F.E., Oliver, A.P., and Salmoiraghi, G.C. (1963): The responsiveness of          211
individual hypothalamic neurons to microelectrophoretically administered
endogenous amines. *Int. J. Neuropharmacol.* 2:181-193.

Page

Bondareff, W. and Gordon, B. (1966): Submicroscopic localization of norepineph-      205
rine in sympathetic nerves of rat pineal. *J. Pharmacol. Exp. Ther.* 153:42-47.

Bondy, P. (1969): *Duncan's Diseases of Metabolism.* Philadelphia: Saunders.

Bowers, C.Y., Schally, A.V., Enzmann, F., Böler, J., and Folkers, K. (1970):      221,222
Porcine thyrotropin releasing factor is (pyro)glu-his-pro(NH$_2$). *Endocrinology*
86:1143-1153.

Bunag, R.D., Page, I.H., and McCubbin, J.W. (1966): Neural stimulation of release      186
of renin. *Circ. Res.* 19:851-860.

Burgus, R., Dunn, T.F., Desiderio, D., Ward, D.N., Vale, W., and Guillemin, R.      221,222
(1970): Characterization of ovine hypothalamic hypophysiotropic TSH-releas-
ing factor. *Nature* 226:321-325.

Chase, T.N. and Kopin, I.J. (1968): Stimulus-induced release of substances from      212
olfactory bulb using the push-pull cannula. *Nature* 217:466-467.

Chirigos, M.A., Greengard, P., and Udenfriend, S. (1960): Uptake of tyrosine by rat      189
brain *in vivo. J. Biol. Chem.* 235:2075-2079.

Cooper, J.R., Bloom, F.E., and Roth, R.H. (1970): *The Biochemical Basis of
Neuropharmacology.* New York: Oxford University Press.

Coppola, J.A. (1968): The apparent involvement of the sympathetic nervous      252
system in the regulation of gonadotrophin secretion in rats. *J. Reprod. Fertil.*
16(Suppl. 4):35-45.

Coppola, J.A. (1969): Turnover of hypothalamic catecholamines during various      252
states of gonadotrophin secretion. *Neuroendocrinology* 5:75-80.

Coppola, J.A. (1971): Brain catecholamines and gonadotropin secretion. *In:
Frontiers in Neuroendocrinology.* Ganong, W.F. and Martini, L., eds. New York:
Oxford University Press, Inc. (In press)

Coppola, J.A., Leonardi, R.G., and Lippmann, W. (1966): Ovulatory failure in      245
rats after treatment with brain norepinephrine depletors. *Endocrinology*
78:225-228.

Coppola, J.A., Leonardi, R.G., Lippmann, W., Perrine, J.W., and Ringler, I. (1965):      245
Induction of pseudopregnancy in rats by depletors of endogenous catecholamines. *Endocrinology* 77:485-490.

Corrodi, H., Fuxe, K., Hamberger, B., and Ljungdahl, A., (1970): Studies on central      200
and peripheral noradrenaline neurons using a new dopamine-β-hydroxylase
inhibitor. *Europ. J. Pharmacol.* 12:145-155.

Costa, E. and Sandler, M., eds. (1968): *Advances in Pharmacology, Vols. 6A, 6B.*
New York: Academic Press.

Page

Coupland, R.E. and Hopwood, D. (1966): Mechanism of a histochemical reaction differentiating between adrenaline- and noradrenaline-storing cells in the electron microscope. *Nature* 209:590-591.    206

Crighton, D.B., Schneider, H.P.G., and McCann, S.M. (1970): Localization of LH-releasing factor in the hypothalamus and neurohypophysis as determined by an *in vitro* method. *Endocrinology* 87:323-329.    222

Critchlow, V. and Bar-Sela, M.E. (1967): Control of the onset of puberty. *In: Neuroendocrinology, Vol. II.* Martini, L. and Ganong, W.F., eds. New York: Academic Press, pp. 101-162.    267

Curtis, D.R. and Davis, R. (1962): Pharmacological studies upon neurons of the lateral geniculate nucleus of the cat. *Brit. J. Pharmacol.* 18:217-246.    212

Curzon, G. and Green, A.R. (1968): Effect of hydrocortisone on rat brain 5-hydroxytryptamine. *Life Sci.* 7:657-663.    188

Dahlström, A. (1969): Fluorescence histochemistry of monoamines in the CNS. *In: Basic Mechanisms of the Epilepsies.* Jasper, H.H., Ward, A.A., and Pope, A., eds. Boston: Little, Brown, pp. 212-227.    262

Dahlström, A. and Fuxe, K. (1964): Evidence for the existence of monoamine-containing neurons in the central nervous system. I. Demonstration of monoamines in the cell bodies of brain stem neurons. *Acta Physiol. Scand.* 62(Suppl. 232):1-55.    199,200

Dahlström, A., Fuxe, K., Olson, L., and Ungerstedt, U. (1964): Ascending systems of catecholamine neurons from the lower brain stem. *Acta Physiol. Scand.* 62:485-486.    202

Dahlström, A., Fuxe, K., Olson, L., and Ungerstedt, U. (1965): On the distribution and possible function of monoamine nerve terminals in the olfactory bulb of the rabbit. *Life Sci.* 4:2071-2074.    213

Dahlström, A. and Häggendal, J. (1966): Studies on the transport and life-span of amine storage granules in a peripheral adrenergic neuron system. *Acta Physiol. Scand.* 67:278-288.

Donoso, A.O. and Cukier, J.O. (1968): Oestrogen as depressor of noradrenaline concentration in the anterior hypothalamus. *Nature* 218:969-970.    252

Donoso, A.O. and Gutierrez Moyano, M.B. de (1970): Adrenergic activity in hypothalamus and ovulation. *Proc. Soc. Exp. Biol. Med.* 135:633-635.    251

Donoso, A.O., Gutierrez Moyano, M.B. de, and Santolaya, R.C. (1969): Metabolism of noradrenaline in the hypothalamus of castrated rats. *Neuroendocrinology* 4:12-19.    252

Donoso, A.O. and Santolaya, R.C. (1969): The effect of monoamine synthesis-inhibitors on the ovarian compensatory hypertrophy. *Experientia* 25:855-857.

Page

Donoso, A.O. and Stefano, F.J.E. (1967): Sex hormones and concentration of    251,252
noradrenalin and dopamine in the anterior hypothalamus of castrated rats.
*Experientia* 23:665-666.

Donoso, A.O., Stefano, F.J.E., Biscardi, A.M., and Cukier, J. (1967): Effects of    251,252
castration on hypothalamic catecholamines. *Amer. J. Physiol.* 212:737-739.

Donovan, B.T. and van der Werff ten Bosch, J.J. (1956): Precocious puberty in rats    264
with hypothalamic lesions. *Nature* 178:745.

Egdahl, R.H. (1960): The effect of brain removal, decortication and mid brain    264
transection on adrenal cortical function in dogs. *Acta Endocr.* 35(Suppl.
51):49-50.

Elwers, M. and Critchlow, V. (1961): Precocious ovarian stimulation following    264
interruption of stria terminalis. *Amer. J. Physiol.* 201:281-284.

Engberg, I. and Ryall, R.W. (1966): The inhibitory action of noradrenaline and    211
other monoamines on spinal neurons. *J. Physiol.* 185:298-322.

Falck, B., Hillarp, N.-A., Thieme, G., and Torp, A. (1962): Fluorescence of    197
catechol amines and related compounds condensed with formaldehyde. *J.
Histochem. Cytochem.* 10:348-354.

Falck, B. and Owman, C. (1965): A detailed methodologic description of the    198
fluorescence method for the cellular demonstration of biogenic monoamines.
*Acta Universitatis Lundensis* Section II, No. 7.

Fawcett, D.W., Long, J.A., and Jones, A.L. (1969): The ultrastructure of endocrine    184
glands. *Recent Progr. Hormone Res.* 25:315-380.

Fink, R.P. and Heimer, L. (1967): Two methods for selective silver impregnation of    259
degenerating axons and their synaptic endings in the central nervous system.
*Brain Res.* 4:369-374.

Fraschini, F., Mess, B., Piva, F., and Martini, L. (1968): Brain receptors sensitive to    239
indole compounds: function in control of luteinizing hormone secretion.
*Science* 159:1104-1105.

Fuxe, K. (1964): Cellular localization of monoamines in the median eminence and    200,202,
the infundibular stem of some mammals. *Z. Zellforsch.* 61:710-724.    203

Fuxe, K. (1965a): Evidence for the existence of monoamine neurons in the central    198,200
nervous system. III. The monoamine nerve terminal. *Z. Zellforsch.* 65:573-596.

Fuxe, K. (1965b): Evidence for the existence of monoamine neurons in the central
nervous system. IV. The distribution of monoamine nerve terminals in the
central nervous system. *Acta Physiol. Scand.* 64(Suppl. 247):37-85+.

Fuxe, K. and Hökfelt, T. (1966): Further evidence for the existence of tubero-    202
infundibular dopamine neurons. *Acta Physiol. Scand.* 66:245-246.

Page

Fuxe, K. and Hökfelt, T. (1967): The influence of central catecholamine neurons    256
on the hormone secretion from the anterior and posterior pituitary. *In:*
*Neurosecretion.* Stutinsky, F. New York: Springer-Verlag, pp. 165-177.

Fuxe, K. and Hökfelt, T. (1969): Catecholamines in the hypothalamus and the    250,254
pituitary gland. *In: Frontiers in Neuroendocrinology.* Ganong, W.F. and Martini,
L., eds. New York: Oxford University Press, pp. 47-96.

Fuxe, K. and Hökfelt, T. (1970): Participation of central monoamine neurons in    257
the regulation of anterior pituitary function with special regard to the neuroendocrine role of tubero-infundibular dopamine neurons. *In: Aspects of Neuro-*
*endocrinology.* Bargmann, W. and Scharrer, B., eds. Berlin: Springer-Verlag, pp.
192-205.

Fuxe, K., Hökfelt, T., Jonsson, G., and Ungerstedt, U. (1970): Fluorescence    202
microscopy in neuroanatomy. *In: Contemporary Research Methods in Neuro-*
*anatomy.* Nauta, W.J.H. and Ebbesson, S.O.E., eds. New York: Springer-Verlag,
pp. 275-314.

Fuxe, K., Hökfelt, T., and Nilsson, O. (1969a): Castration, sex hormones, and    256
tubero-infundibular dopamine neurons. *Neuroendocrinology* 5:107-120.

Fuxe, K., Hökfelt, T., and Nilsson, O. (1969b): Factors involved in the control of    256
the activity of the tubero-infundibular dopamine neurons during pregnancy and
lactation. *Neuroendocrinology* 5:257-270.

Fuxe, K., Hökfelt, T., and Nilsson, O. (1965): A fluorescence and electronmicro-    206
scopic study on certain brain regions rich in monoamine terminals. *Amer. J.*
*Anat.* 117:33-45.

Fuxe, K., Hökfelt, T., and Ungerstedt, U. (1968): Localization of indolealkyl-    200
amines in CNS. *Advances Pharmacol.* 6:235-251.

Fuxe, K. and Ungerstedt, U. (1968): Histochemical studies on the distribution of    204
catecholamines and 5-hydroxytryptamine after intraventricular injections.
*Histochemie* 13:16-28.

Gale, C.C. and Jobin, M. (1967): Further studies on CNS-endocrine responses to    233
hypothalamic cooling in unanesthetized baboons. *Fed. Proc.* 26:255. (Abstr.)

Gann, D.S., Ostrander, L.E., and Schoeffler, J.D. (1968): A finite state model for    231
the control of adrenal cortical secretion. *In: Systems Theory and Biology.*
Mesarović, M.D., ed. New York: Springer-Verlag, pp. 185-200.

Ganong, W.F. (1963): The central nervous system and the synthesis and release of    236,264
adrenocorticotropic hormone. *In: Advances in Neuroendocrinology.* Nalbandov,
A.V., ed. Urbana: University of Illinois Press, pp. 92-149.

Ganong, W.F. (1966): Neuroendocrine integrating mechanisms. *In: Neuroendocri-*
*nology, Vol. I.* Martini, L. and Ganong, W.F., eds. New York: Academic Press,
pp. 1-14.

Page

Ganong, W.F. (1967): Brain-endocrine relationships: some current concepts and their possible clinical implications. *Milit. Med.* 132:360-365.

Ganong, W.F. (1971): Control of ACTH and MSH secretion. *In: Integration of Endocrine and Nonendocrine Mechanisms in the Hypothalamus.* Martini, L., ed. New York: Academic Press. (In press)

Ganong, W.F., Biglieri, E.G., and Mulrow, P.J. (1966): Mechanisms regulating adrenocortical secretion of aldosterone and glucocorticoids. *Recent Progr. Hormone Res.* 22:381-430.

Ganong, W.F., Boryczka, A.T., Lorenzen, L.C., and Egge, A.S. (1967): Lack of effect of α-ethyltryptamine on ACTH secretion when blood pressure is held constant. *Proc. Soc. Exp. Biol. Med.* 124:558-559.                                   237

Ganong, W.F. and Lorenzen, L. (1967): Brain neurohumors and endocrine function. *In: Neuroendocrinology, Vol. II.* Martini, L. and Ganong, W.F., eds. New York: Academic Press, pp. 583-640.

Ganong, W.F. and Martini, L., eds. (1969): *Frontiers in Neuroendocrinology.* New York: Oxford University Press.

Ganong, W.F., Wise, B.L., Shackleford, R., Boryczka, A.T., and Zipf, B. (1965): Site at which α-ethyltryptamine acts to inhibit the secretion of ACTH. *Endocrinology* 76:526-530.                                   237

Gillis, C.N., Schneider, F.H., Van Orden, L.S., and Giarman, N.J. (1966): Biochemical and microfluorometric studies of norepinephrine redistribution accompanying sympathetic nerve stimulation. *J. Pharmacol. Exp. Ther.* 151:46-54.                                   205

Glowinski, J. and Axelrod, J. (1966): Effects of drugs on the disposition of $H^3$-norepinephrine in the rat brain. *Pharmacol. Rev.* 18:775-785.

Glowinski, J. and Iversen, L.L. (1966): Regional studies of catecholamines in the rat brain. I. The disposition of [3H]norepinephrine, [3H]dopamine and [3H]dopa in various regions of the brain. *J. Neurochem.* 13:655-669.                                   198

Gold, E.M. and Ganong, W.F. (1967): Effects of drugs on neuroendocrine processes. *In: Neuroendocrinology, Vol. II.* Martini, L. and Ganong, W.F., eds. New York: Academic Press, pp. 377-438.

Goodman, L. and Gilman, A. (1970): *The Pharmacological Basis of Therapeutics.* New York: Macmillan.

Gorski, R.A. (1970): Sexual differentiation of the hypothalamus. *In: The Neuroendocrinology of Human Reproduction.* Mack, H.C., ed. Springfield, Ill.: Charles C Thomas. (In press)                                   265,266

Greenspan, F.S., Li, C.H., Simpson, M.E., and Evans, H.M. (1949): Bioassay of hypophyseal growth hormone: the tibia test. *Endocrinology* 45:455-463.                                   233

Page

Grillo, M.A. (1966): Electron microscopy of sympathetic tissues. *Pharmacol. Rev.* 18:387-399.

205

Häggendal, C.J. and Dahlström, A. (1969): The transport and life-span of amine storage granules in bulbospinal noradrenaline neurons of the rat. *J. Pharm. Pharmacol.* 21:55-57.

261

Halász, B., Florsheim, W.H., Corcorran, N.L., and Gorski, R.A. (1967): Thyrotrophic hormone secretion in rats after partial or total interruption of neural afferents to the medial basal hypothalamus. *Endocrinology* 80:1075-1082.

267

Halász, B. and Gorski, R.A. (1967): Gonadotrophic hormone secretion in female rats after partial or total interruption of neural afferents to the medial basal hypothalamus. *Endocrinology* 80:608-622.

265,266

Halász, B. and Pupp, L. (1965): Hormone secretion of the anterior pituitary gland after physical interruption of all nervous pathways to the hypophysiotrophic area. *Endocrinology* 77:553-562.

263

Halász, B., Pupp, L., Uhlarik, S., and Tima, L. (1965): Further studies on the hormone secretion of the anterior pituitary transplanted into the hypophysiotrophic area of the rat hypothalamus. *Endocrinology* 77:343-355.

263

Halász, B., Slusher, M.A., and Gorski, R.A. (1967): Adrenocorticotrophic hormone secretion in rats after partial or total deafferentation of the medial basal hypothalamus. *Neuroendocrinology* 2:43-55.

267

Halász, B., Vernikos-Danellis, J., and Gorski, R.A. (1967): Pituitary ACTH content in rats after partial or total interruption of neural afferents to the medial basal hypothalamus. *Endocrinology* 81:921-924.

267

Harris, G.W. (1937): The induction of ovulation in the rabbit, by electrical stimulation of the hypothalamic-hypophysial mechanism. *Proc. Roy. Soc. B.* 122:374-394.

242

Harris, G.W. (1964): The central nervous system and the endocrine glands. *Triangle* 6:242-251.

179

Harris, G.W., Reed, M., and Fawcett, C.P. (1966): Hypothalamic releasing factors and the control of anterior pituitary function. *Brit. Med. Bull.* 22:266-272.

Harvey, J.A., Heller, A., and Moore, R.Y. (1963): The effect of unilateral and bilateral medial forebrain bundle lesions on brain serotonin. *J. Pharmacol. Exp. Ther.* 140:103-110.

260

Haterius, H.O. and Derbyshire, A.J., Jr. (1937): Ovulation in the rabbit following upon stimulation of the hypothalamus. *Amer. J. Physiol.* 119:329-330. (Abstr.)

242

Haun, C.K. and Sawyer, C.H. (1960): Initiation of lactation in rabbits following placement of hypothalamic lesions. *Endocrinology* 67:270-272.

264

Page

Heimer, L. (1968): Synaptic distribution of centripetal and centrifugal nerve fibers          261
in the olfactory system of the rat. An experimental anatomical study. *J. Anat.*
103:413-423.

Heller, A., Bhatnagar, R.K., and Moore, R.Y. (1969): Selective neural control of             260
telencephalic monoamines and enzymes involved in their biosynthesis. *In:*
*Progress in Neurogenetics.* Barbeau, A. and Brunette, J.R., eds. Amsterdam:
Excerpta Medica Foundation, pp. 283-288.

Heller, A., Harvey, J.A., and Moore, R.Y. (1962): A demonstration of a fall in brain         258
serotonin following central nervous system lesions in the rat. *Biochem. Pharmacol.* 11:859-866.

Heller, A. and Moore, R.Y. (1965): Effect of central nervous system lesions on          258,259
brain monoamines in the rat. *J. Pharmacol. Exp. Ther.* 150:1-9.

Heller, A. and Moore, R.Y. (1968): Control of brain serotonin and norepinephrine            260
by specific neural systems. *Advances Pharmacol.* 6:191-206.

Heller, A., Seiden, L.S., and Moore, R.Y. (1966a): Regional effects of lateral          259,260
hypothalamic lesions on brain norepinephrine in the cat. *Int. J. Neuropharmacol.* 5:91-101.

Heller, A., Seiden, L.S., Porcher, W., and Moore, R.Y. (1965): 5-Hydroxytrypto-
phan decarboxylase in the rat brain: effect of hypothalamic lesions. *Science*
147:887-888.

Heller, A., Seiden, L.S., Porcher, W., and Moore, R.Y. (1966b): Regional effects of     259,260
lateral hypothalamic lesions on 5-hydroxytryptophan decarboxylase in the cat
brain. *J. Neurochem.* 13:967-974.

Hellman, L., Bradlow, H.L., Zumoff, B., and Gallagher, T.F. (1961): The influence           229
of thyroid hormone on hydrocortisone production and metabolism. *J. Clin.
Endocr.* 21:1231-1247.

Hilton, J.G., Black, W.C., Athos, W., McHugh, B., and Westermann, C.D. (1962):              229
Increased ACTH-like activity in plasma of patients with thyrotoxicosis. *J. Clin.
Endocr.* 22:900-905.

Hökfelt, T. (1967): On the ultrastructural localization of noradrenaline in the             206
central nervous system of the rat. *Z. Zellforsch.* 79:110-117.

Iggo, A. and Vogt, M. (1960): Preganglionic sympathetic activity in normal and in           190
reserpine-treated cats. *J. Physiol.* 150:114-133.

Iversen, L.L. (1967): *The Uptake and Storage of Noradrenaline in Sympathetic*              198
*Nerves.* London: Cambridge University Press.

Iversen, L.L. and Glowinski, J. (1966): Regional studies of catecholamines in the           194
rat brain. II. Rate of turnover of catecholamines in various brain regions. *J.
Neurochem.* 13:671-682.

Page

Jacobowitz, D.M., Marks, B.H., and Vernikos-Danellis, J. (1963): Effect of acute   242
stress on the pituitary gland: uptake of serine-1-$C^{14}$ into ACTH. *Endocrinology*
72:592-597.

Javoy, F., Thierry, A.M., Kety, S.S., and Glowinski, J. (1968): The effect of
amphetamine on the turnover of brain norepinephrine in normal and stressed
rats. *Communications Behav. Biol.* 1:43-48.

Jequier, E., Robinson, D.S., Lovenberg, W., and Sjoerdsma, A. (1969): Further   196
studies on tryptophan hydroxylase in rat brainstem and beef pineal. *Biochem.
Pharmacol.* 18:1071-1081.

Kamberi, I.A., Mical, R.S., and Porter, J.C. (1969): Luteinizing hormone-releasing   225
activity in hypophysial stalk blood and elevation by dopamine. *Science*
166:388-390.

Kamberi, I., Schneider, H.P.G., and McCann, S.M. (1970): Action of dopamine to   225
induce release of FSH-releasing factor (FRF) from hypothalamic tissue *in vitro.*
*Endocrinology* 86:278-284.

Kanematsu, S., Hilliard, J., and Sawyer, C.H. (1963): Effect of hypothalamic   264
lesions on pituitary prolactin content in the rabbit. *Endocrinology* 73:345-348.

Kanematsu, S. and Sawyer, C.H. (1963): Effects of intrahypothalamic implants of   264
reserpine on lactation and pituitary prolactin content in the rabbit. *Proc. Soc.
Exp. Biol. Med.* 113:967-969.

Katz, S.H., Dhariwal, A.P.S., and McCann, S.M. (1967): Effect of hypoglycemia on   234
the content of pituitary growth hormone (GH) and hypothalamic growth
hormone-releasing factor (GHRF) in the rat. *Endocrinology* 81:333-339.

Kawakami, M. and Sawyer, C.H. (1959): Neuroendocrine correlates of changes in   249
brain activity thresholds by sex steroids and pituitary hormones. *Endocrinology*
65:652-668.

Kawakami, M. and Sawyer, C.H. (1967) Effects of sex hormones and antifertility
steroids in brain thresholds in the rabbit. *Endocrinology* 80:857-871.

Kety, S.S. (1966): Catecholamines in neuropsychiatric states. *Pharmacol. Rev.*
18:787-798.

Kety, S.S. (1967): The central physiological and pharmacological effects of the
biogenic amines and their correlations with behavior. *In: The Neurosciences: A
Study Program.* Quarton, G.C., Melnechuk, T., and Schmitt, F.O., eds. New
York: Rockefeller University Press, pp. 444-451.

Kety, S.S. (1970): The biogenic amines in the central nervous system: their possible
roles in arousal, emotion, and learning. *In: The Neurosciences: Second Study
Program.* Schmitt, F.O., editor-in-chief. New York: Rockefeller University Press,
pp. 324-336.

Page

Kety, S.S., Javoy, F., Thierry, A.M., Julou, L., and Glowinski, J. (1967): A sustained effect of electroconvulsive shock on the turnover of norepinephrine in the central nervous system of the rat. *Proc. Nat. Acad. Sci.* 58:1249-1254.

Kety, S.S. and Samson, F.E., Jr. (1967): Neural properties of the biogenic amines.                  180,188
*Neurosciences Res. Prog. Bull.* 5:1-119. Also *In: Neurosciences Research Symposium Summaries, Vol. 2.* Schmitt, F.O. et al., eds. Cambridge, Mass.: M.I.T. Press. 1967. pp. 399-514.

Kirshner, N. (1959): Biosynthesis of adrenaline and noradrenaline. *Pharmacol. Rev.*                  191
11:350-357.

Kitay, J.I. (1968): Effects of estrogen and androgen on the adrenal cortex of the                  229
rat. *In: Functions of the Adrenal Cortex, Vol. II.* McKerns, K.W., ed. New York: Appleton-Century-Crofts, pp. 775-811.

Kitay, J.I., Holub, D.A., and Jailer, M.W. (1959): "Inhibition" of pituitary ACTH                  236,240
release after administration of reserpine or epinephrine. *Endocrinology* 65:548-554.

Kobayashi, H. and Matsui, T. (1969): Fine structure of the median eminence and                  219
its functional significance. *In: Frontiers of Neuroendocrinology.* Ganong, W.F. and Martini, L., eds. New York: Oxford University Press, pp. 1-46.

Komisaruk, B.R., McDonald, P.G., Whitmoyer, D.I., and Sawyer, C.H. (1967):                  250
Effects of progesterone and sensory stimulation on EEG and neuronal activity in the rat. *Exp. Neurol.* 19:494-507.

Kopin, I.J., Hertting, G., and Gordon, E.K. (1962): Fate of norepinephrine-H$^3$ in                  193
the isolated perfused rat heart. *J. Pharmacol. Exp. Ther.* 138:34-40.

Kordon, C. (1969): Effects of selective experimental changes in regional hypotha-                  246
lamic monoamine levels on superovulation in the immature rat. *Neuroendocrinology* 4:129-138.

Kordon, C. and Glowinski, J. (1969): Selective inhibition of superovulation by                  246,247
blockade of dopamine synthesis during the "critical period" in the immature rat. *Endocrinology* 85:924-931.

Kordon, C., Javoy, F., Vassent, G., and Glowinski, J. (1968): Blockade of                  246,247
superovulation in the immature rat by increased brain serotonin. *European J. Pharmacol.* 4:169-174.

Krulich, L. and McCann, S.M. (1966): Influence of stress on the growth hormone (GH) content of the pituitary of the rat. *Proc. Soc. Exp. Biol. Med.* 122:612-616.

Kuhn, E., Krulich, L., Quijada, M., Illner, P., Kalra, P.S., and McCann, S.M. (1970):                  225
Effect of oxytocin and adrenergic agents on prolactin release *in vivo* and *in vitro*. Program, 52nd Meeting, The Endocrine Society, p. 180.

Page

Labrie, F. (1967): Interactions hormonale et role de la transcortive dans l'ajuste-    229
ment de l'activité hypophysosurrénaliene. Ph.D. Dissertation. Laboratoires
d'Endocrinologie, Département de Physiologie, Faculté de Médecine, Université
Laval, Quebec.

Lake, R.B. and Gann, D.S. (1970): Dynamic response of the intact, infused adrenal.    230
Fed. Proc. 29:947. (Abstr.)

Laverty, R. and Sharman, D.F. (1965): The estimation of small quantities of    198
3,4-dihydroxyphenylethylamine in tissues. Brit. J. Pharmacol. 24:538-548.

Lenn, N.J. (1967): Localization of uptake of tritiated norepinephrine by rat brain    206
in vivo and in vitro using electron microscopic autoradiography. Amer. J. Anat.
120:377-389.

Lichtensteiger, W. (1969): Cyclic variations of catecholamine content in hypothalamic nerve cells during the estrous cycle of the rat, with a concomitant study of
the substantia nigra. J. Pharmacol. Exp. Ther. 165:204-215.

Lindström, L.H. and Meyerson, B.J. (1967): The effect of pilocarpine, oxotremor-    248
ine and arecoline in combination with methyl-atropine or atropine on hormone
activated oestrous behavior in ovariectomized rats. Psychopharmacologia
11:405-413.

Lippmann, W., Leonardi, R., Ball, J., and Coppola, J.A. (1967): Relationship    245
between hypothalamic catecholamines and gonadotrophin secretion in rats. J.
Pharmacol. Exp. Ther. 156:258-266.

Lisk, R.D. (1962): Diencephalic placement of estradiol and sexual receptivity in the    265
female rat. Amer. J. Physiol. 203:493-496.

Loewi, O. (1921): Uber humorale Ubertragbarkeit der Herznervenwirkung. I.    209
Mitteilung. Pflugers Arch. ges. Physiol. 189:239-242.

Lorenzen, L.C. and Ganong, W.F. (1967): Effect of drugs related to α-ethyl-    237
tryptamine on stress-induced ACTH secretion in the dog. Endocrinology
80:889-892.

McCann, S.M., Antunes-Rodrigues, J., and Dhariwal, A.P.S. (1965): Physiology and
chemistry of hypothalamic factors which influence gonadotrophin secretion. In:
Excerpta Medica International Congress Series, No. 87. Amsterdam: Excerpta
Medica Foundation, pp. 292-299.

McCann, S.M. and Dhariwal, A.P.S. (1966): Hypothalamic releasing factors and the
neurovascular link between the brain and the anterior pituitary. In: Neuroendocrinology, Vol. I. Martini, L. and Ganong, W.F., eds. New York: Academic
Press, pp. 261-296.

McCann, S.M., Dhariwal, A.P.S., and Porter, J.C. (1968): Regulation of the
adenohypophysis. Ann. Rev. Physiol. 30:589-640.

Page

McCann, S.M. and Porter, J.C. (1969): Hypothalamic pituitary stimulating and        218,224
inhibiting hormones. *Physiol. Rev.* 49:240-284.

Machlin, L.J., Takahashi, Y., Horino, M., Hertelendey, F., Gordon, R.S., and          233
Kipnis, D. (1968): Regulation of growth hormone secretion in non-primate
species. *In: Growth Hormone.* Pecile, A. and Müller, E.E., eds. Amsterdam:
Excerpta Medica Foundation, pp. 292-305.

Maickel, R.P., Westermann, E.O., and Brodie, B.B. (1961): Effects of reserpine and    238,241
cold-exposure on pituitary-adrenocortical function in rats. *J. Pharmacol. Exp.
Ther.* 134:167-175.

Mangili, G., Motta, M., and Martini, L. (1966): Control of adrenocorticotropic        240
secretion. *In: Neuroendocrinology, Vol. I.* Martini, L. and Ganong, W.F., eds.
New York: Academic Press, pp. 297-370.

Manshardt, J. and Wurtman, R.J. (1968): Daily rhythm in the noradrenaline
content of rat hypothalamus. *Nature* 217:574-575.

Markee, J.E., Sawyer, C.H., and Hollinshead, W.H. (1946a): Activation of the          243
anterior hypophysis by electrical stimulation in the rabbit. *Endocrinology*
38:345-357.

Markee, J.E., Sawyer, C.H., and Hollinshead, W.H. (1946b): Adrenergic control of      243
the release of luteinizing hormone from the hypophysis of the rabbit. *Recent
Progr. Hormone Res.* 2:117-131.

Marks, B.H., Hall, M.M., and Bhattacharya, A.N. (1970): Psychopharmacological        241,242
effects and pituitary-adrenal activity. *Progr. Brain Res.* 32:58-70.

Marks, B.H. and Vernikos-Danellis, J. (1963): Effect of acute stress on the pituitary  242
gland: action of ethionine on stress-induced ACTH release. *Endocrinology*
72:582-587.

Martel, R.R., Westermann, E.O., and Maickel, R.P. (1962): Dissociation of             239
reserpine-induced sedation and ACTH hypersecretion. *Life Sci.* 4:151-155.

Martini, L. and Ganong, W.F., eds. (1966): *Neuroendocrinology, Vol. I.* New York:    264
Academic Press.

Martini, L. and Ganong, W.F., eds. (1967): *Neuroendocrinology, Vol. II.* New         264
York: Academic Press.

Matsuda, K., Duyck, C., Kendall, J.W., and Greer, M.A. (1964): Pathways by which      264
traumatic stress and ether induce increased ACTH released in the rat. *Endocrinology* 74:981-985.

Meites, J., ed. (1970): *Hypophysiotropic Hormones of the Hypothalamus.* Baltimore: Williams & Wilkins.

Page

Meyerson, B.J. (1964a): Central nervous monoamines and hormone induced estrus          247,248
behaviour in the spayed rat. *Acta Physiol. Scand.* 63(Suppl. 241):1-32.

Meyerson, B.J. (1964b): The effect of neuropharmacological agents on hormone-          247
activated estrus behaviour in ovariectomised rats. *Arch. Int. Pharmacodyn.*
150:4-33.

Meyerson, B.J. (1964c): Estrus behaviour in spayed rats after estrogen or progester-          247
one treatment in combination with reserpine or tetrabenazine. *Psychopharmacologia* 6:210-218.

Meyerson, B.J. (1966): The effect of imipramine and related antidepressive drugs          250
on estrus behaviour in ovariectomised rats activated by progesterone, reserpine
or tetrabenazine in combination with estrogen. *Acta Physiol. Scand.* 67:411-422.

Meyerson, B.J. (1967): Relationship between the anesthetic and gestanic action          249
and estrous behavior-inducing activity of different progestins. *Endocrinology*
81:369-374.

Meyerson, B.J. (1968a): Amphetamine and 5-hydroxytryptamine inhibition of
copulatory behaviour in the female rat. *Ann. Med. Exp. Biol. Fenn* 46:394-398.

Meyerson, B.J. (1968b): Female copulatory behaviour in male and androgenized          248
female rats after oestrogen-amine depletor treatment. *Nature* 217:683-684.

Meyerson, B.J. (1970): Monoamines and hormone activated oestrous behavior in
the ovariectomized hamster. *Psychopharmacologia* 18:50-57.

Meyerson, B.J. and Lewander, T. (1970): Serotonin synthesis inhibition and estrous          248,250
behavior in female rats. *Life Sci.* 9:661-671.

Meyerson, B.J. and Sawyer, C.H. (1968): Monoamines and ovulation in the rat.
*Endocrinology* 83:170-176.

Michael, R.P. (1966): Control of sexual behavior by gonadal steroids. I. Action of          265
hormones on the cat brain. *In: Brain and Behavior, Vol. III, The Brain and
Gonadal Function.* Gorski, R.A. and Whalen, R.E., eds. Los Angeles: University
of California Press, pp. 82-98.

Miller, F.P., Cox, R.H., Jr., and Maickel, R.P. (1968): Differential effects of          239
monoamine oxidase inhibitors on biogenic amine levels in discrete brain areas.
*Fed. Proc.* 27:273. (Abstr.)

Miller, R.E. and Yates, F.E. (1970): Adrenal and ACTH dynamics in the          230
unanesthetized dog. *Fed. Proc.* 29:948. (Abstr.)

Moir, A.T.B. and Eccleston, D. (1968): The effects of precursor loading in the          195
cerebral metabolism of 5-hydroxyindoles. *J. Neurochem.* 15:1093-1108.

Moore, R.Y., Bhatnagar, R.K., and Heller, A. (1966): Norepinephrine and DOPA          260
decarboxylase in rat brain following hypothalamic lesions. *Int. J. Neuropharmacol.* 5:287-291.

Page

Moore, R.Y. and Heller, A. (1967): Monoamine levels and neuronal degeneration in        260
rat brain following lateral hypothalamic lesions. *J. Pharmacol. Exp. Ther.*
156:12-22.

Moore, R.Y., Heller, A., Bhatnagar, R.K., Wurtman, R.J., and Axelrod, J. (1968):
Central control of the pineal gland: visual pathways. *Arch. Neurol.* 18:208-218.

Moore, R.Y., Heller, A., Wurtman, R.J., and Axelrod, J. (1967): Visual pathway          269
mediating pineal response to environmental light. *Science* 155:220-223.

Moore, R.Y., Wong, S.L.R., and Heller, A. (1965): Regional effects of hypotha-      259,260
lamic lesions on brain serotonin. *Arch. Neurol.* 13:346-354.

Mueller, R.A., Thoenen, H., and Axelrod, J. (1969a): Increase in tyrosine               190
hydroxylase activity after reserpine administration. *J. Pharmacol. Exp. Ther.*
169:74-79.

Mueller, R.A., Thoenen, H., and Axelrod, J. (1969b): Inhibition of trans-synaptical-    190
ly increased tyrosine hydroxylase activity by cycloheximide and actinomycin D.
*Molec. Pharmacol.* 5:463-469.

Müller, E.E. (1970): Brain catecholamines and growth hormone release. *In: Aspects*     234
*of Neuroendocrinology.* Bargmann, W. and Scharrer, B., eds. Berlin: Springer-
Verlag, pp. 206-219.

Müller, E.E., Arimura, A., Sawano, S., Saito, T., and Schally, A.V. (1967a): Growth     234
hormone-releasing activity in the hypothalamus and plasma of rats subjected to
stress. *Proc. Soc. Exp. Biol. Med.* 125:874-878.

Müller, E.E., Dal Pra, P., and Pecile, A. (1968): Influence of brain neurohumors        233
injected into the lateral ventricle of the rat on growth hormone release.
*Endocrinology* 83:893-896.

Müller, E.E., Pecile, A., Felici, M., and Cocchi, D., (1970): Norepinephrine and        234
dopamine injection into lateral brain ventricle of the rat and growth hormone-
releasing activity in the hypothalamus and plasma. *Endocrinology*
86:1376-1382.

Müller, E.E., Pecile, A., Naimzada, M.K., and Ferrario, G. (1969): The involvement      235
of cyclic 3´,5´-adenosine monophosphate in the growth hormone release
mechanism (s). *Experientia* 25:750-751.

Müller, E.E., Pecile, A., and Smirne, S. (1965): Substances present at the              233
hypothalamic level and growth hormone releasing activity. *Endocrinology*
77:390-392.

Müller, E.E., Saito, T., Arimura, A., and Schally, A.V. (1967b): Hypoglycemia,
stress and growth hormone release: blockade of growth hormone release by
drugs acting on the central nervous system. *Endocrinology* 80:109-117.

Page

Müller, E.E., Sawano, S., Arimura, A., and Schally, A.V. (1967c): Blockade of    233,234
release of growth hormone by brain norepinephrine depletors. *Endocrinology*
80:471-476.

Musacchio, J.M., Julou, L., Kety, S.S., and Glowinski, J. (1969): Increase in rat
brain tyrosine hydroxylase activity produced by electroconvulsive shock. *Proc.
Nat. Acad. Sci.* 63:1117-1119.

Nagatsu, T., Levitt, M., and Udenfriend, S. (1964): Tyrosine hydroxylase. The    189
initial step in norepinephrine biosynthesis. *J. Biol. Chem.* 239:2910-2917.

Nauta, W.J.H. (1963): Central nervous organization and the endocrine motor
system. *In: Advances in Neuroendocrinology.* Nalbandov, A.V., ed. Urbana:
University of Illinois Press, pp. 5-27.

Nauta, W.J.H. and Haymaker, W. (1970): Hypothalamic nuclei and fiber connections. *In: The Hypothalamus.* Haymaker, W., Anderson, E., and Nauta, W.J.H.,
eds. Springfield, Ill.: Charles C Thomas, pp. 136-209.

Nauta, W.J.H., Koella, W.P., and Quarton, G.C. (1966): Sleep, wakefulness, dreams    180
and memory. *Neurosciences Res. Prog. Bull.* 4:1-103. Also *In: Neurosciences
Research Symposium Summaries, Vol. 2.* Schmitt, F.O. et al., eds. Cambridge,
Mass.: M.I.T. Press. 1967. pp. 1-90.

Nugent, C.A., Warner, H.R., Estergreen, V.L., and Eik-Nes, K.B. (1964): The    229,230
distribution and disposal of cortisol in humans. *In: Excerpta Medica International Congress Series, No. 83.* Amsterdam: Excerpta Medica Foundation, pp.
257-261.

Palka, Y.S., Coyer, D.D., and Critchlow, V. (1968): Hypothalamic deafferentation    267
and adrenal function. *Fed. Proc.* 27:217.

Palka, Y.S. and Sawyer, C.H. (1966a): The effects of hypothalamic implants of    265
ovarian steroids on oestrous behaviour in rabbits. *J. Physiol.* 185:251-269.

Palka, Y.S. and Sawyer, C.H. (1966b): Induction of estrous behavior in rabbits by    265
hypothalamic implants of testosterone. *Amer. J. Physiol.* 211:225-228.

Pellegrino de Iraldi, A., Zieher, L.M., and De Robertis, E. (1965): Ultrastructure    205
and pharmacological studies of nerve endings in the pineal organ. *Progr. Brain
Res.* 10:389-421.

Peterson, R.E. (1958): The influence of the thyroid on adrenal cortical function. *J.    229
Clin. Invest.* 37:736-743.

Piezzi, R.S., Larin, F., and Wurtman, R.J. (1970): Serotonin, 5-hydroxyindole-    239
acetic acid (5-HIAA), and monoamine oxidase in the bovine median eminence
and pituitary gland. *Endocrinology* 86:1460-1462.

Piezzi, R.S. and Wurtman, R.J. (1970): Pituitary serotonin content: Effects of
melatonin or deprivation of water. *Science* 169:285-286.

Page

Pohorecky, L.A., Larin, F., and Wurtman, R.J. (1969): Mechanism of changes in          212
brain norepinephrine levels following olfactory bulb lesions. *Life Sci.*
8:1309-1317.

Pohorecky, L.A., Zigmond, M.J., Heimer, L., and Wurtman, R.J. (1969): Olfactory        261
bulb removal: effects on brain norepinephrine. *Proc. Nat. Acad. Sci.*
62:1052-1055.

Pohorecky, L.A., Zigmond, M., Karten, H., and Wurtman, R.J. (1969): Enzymatic
conversion of norepinephrine to epinephrine by the brain. *J. Pharmacol. Exp.*
*Ther.* 165:190-195.

Potter, L.T. and Axelrod, J. (1963): Properties of norepinephrine storage particles    191
of the rat heart. *J. Pharmacol. Exp. Ther.* 142:299-305.

Pronczuk, A.W., Baliga, B.S., Triant, J.W. and Munro, H.N. (1968): Comparison of       195
the effect of amino acid supply on hepatic polysome profiles in vivo and in
vitro. *Biochim. Biophys. Acta* 157:204-206.

Rall, T.W. and Gilman, A.G. (1970): The role of cyclic AMP in the nervous system.   212,235
*Neurosciences Res. Prog. Bull.* 8:221-323.

Rall, W. (1970): Dendritic neuron theory and dendrodendritic synapses in a simple      213
cortical system. *In: The Neurosciences: Second Study Program.* Schmitt, F.O.,
editor-in-chief. New York: Rockefeller University Press, pp. 552-565.

Rall, W. and Shepherd, G.M. (1968): Theoretical reconstruction of field potentials     213
and dendrodendritic synaptic interactions in olfactory bulb. *J. Neurophysiol.*
31:884-915.

Rall, W., Shepherd, G.M., Reese, T.S., and Brightman, M.W. (1966): Dendroden-          213
dritic synaptic pathway for inhibition in the olfactory bulb. *Exp. Neurol.*
14:44-56.

Ramaley, J.A. and Gorski, R.A. (1967): The effect of hypothalamic deafferenta-         266
tion upon puberty in the female rat. *Acta Endocr.* 56:661-674.

Ramirez, V.D., Komisaruk, B.R., Whitmoyer, D.I., and Sawyer, C.H. (1967):              250
Effects of hormones and vaginal stimulation on the EEG and hypothalamic
units in rats. *Amer. J. Physiol.* 212:1376-1384.

Ratner, A. and Meites, J. (1964): Depletion of prolactin-inhibiting activity of rat
hypothalamus by estradiol or suckling stimulus. *Endocrinology* 75:377-382.

Redmond, W.C. (1968): Ovulatory response to brain stimulation or exogenous            250
luteinizing hormone in progesterone-treated rats. *Endocrinology* 83:1013-1022.

Reis, D.J. and Wurtman, R.J. (1968): Diurnal changes in brain noradrenalin. *Life*     252
*Sci.* 7:91-98.

Reivich, M. and Glowinski, J. (1967): An autoradiographic study of the distribu-       204
tion of $C^{14}$-norepinephrine in the brain of the rat. *Brain* 90:633-646.

Page

Richardson, K.C. (1966): Electron microscopic identification of autonomic nerve    205,206
endings. *Nature* 210:756.

Rinne, U.K. and Sonninen, V. (1968): The occurrence of dopamine and noradren-    204
aline in the tubero-hypophyseal system. *Experientia* 24:177-178.

Robison, B.L. and Sawyer, C.H. (1957): Loci of sex behavioral and gonadotrophic    265
centers in the female cat hypothalamus. *Physiologist* 1:72. (Abstr.)

Roth, J., Glick, S.M., Yalow, R.S., and Berson, S.A. (1963): Hypoglycemia: a    232,233
potent stimulus to secretion of growth hormone. *Science* 140:987-988.

Rubinstein, L. and Sawyer, C.H. (1970): Role of catecholamines in stimulating the    245
release of pituitary ovulating hormone(s) in rats. *Endocrinology* 86:988-995.

Salmoiraghi, G.C. and Bloom, F.E. (1964): Pharmacology of individual neurons.    211
*Science* 144:493-499.

Salmoiraghi, G.C., Bloom, F.E., and Costa, E. (1964): Adrenergic mechanisms in    212
rabbit olfactory bulb. *Amer. J. Physiol.* 207:1417-1424.

Samorajski, T. and Marks, B.H. (1962): Localization of tritiated norepinephrine in    240
mouse brain. *J. Histochem. Cytochem.* 10:392-399.

Saul, G.R. and Sawyer, C.H. (1961): EEG-monitored activation of the hypo-    243
thalamo-hypophysial system by amygdala stimulation and its pharmacological
blockade. *Electroenceph. Clin. Neurophysiol.* 13:307.

Sawyer, C.H. (1952): Stimulation of ovulation in the rabbit by the intraventricular    243
injection of epinephrine or norepinephrine. *Anat. Rec.* 112:385. (Abstr.)

Sawyer, C.H. (1955): Rhinencephalic involvement in pituitary activation by    244
intraventricular histamine in the rabbit under Nembutal anesthesia. *Amer. J.
Physiol.* 180:37-46.

Sawyer, C.H. (1959): Effects of brain lesions on estrous behavior and reflexogenous    265
ovulation in the rabbit. *J. Exp. Zool.* 142:227-246.

Sawyer, C.H. (1960): Reproductive behavior. *In: Handbook of Physiology: Neuro-*    265
*physiology, Vol. II.* Field, J., Magoun, H.W., and Hall, V.E., eds. Washington,
D.C.: American Physiological Society, pp. 1225-1240.

Sawyer, C.H. (1963): Induction of estrus in the ovariectomized cat by local    265
hypothalamic treatment with estrogen. *Anat. Rec.* 145:280. (Abstr.)

Sawyer, C.H. (1964): Control of secretion of gonadotropins. *In: Gonadotropins.*    243
Cole, H.H., ed. San Francisco: W.H. Freeman and Company, pp. 113-159.

Sawyer, C.H. and Kawakami, M. (1961): Interactions between the central nervous    249
system and hormones influencing ovulation. *In: Control of Ovulation.* Villee,
C.A., ed. New York: Pergamon Press, pp. 79-97.

Page

Sawyer, C.H., Kawakami, M., and Kanematsu, S. (1966): Neuroendocrine aspects
of reproduction. *Res. Publ. Ass. Res. Nerv. Ment. Dis.* 43:59-85.

Sawyer, C.H., Markee, J.E., and Townsend, B.F. (1949): Cholinergic and adrenergic        243
components in the neurohumoral control of the release of LH in the rabbit.
*Endocrinology* 44:18-37.

Schally, A.V., Arimura, A., Bowers, C.Y., Kastin, A.J., Sawano, S., and Redding,
T.W. (1968): Hypothalamic neurohormones regulating anterior pituitary function. *Recent Progr. Hormone Res.* 24:497-588.

Scharrer, B. (1965): Recent progress in the study of neuroendocrine mechanisms in
insects. *Arch. Anat. Micr. Morph. Exp.* 54:331-342.

Scharrer, B. (1966): Ultrastructural study of the regressing prothoracic glands of
blattarian insects. *Z. Zellforsch.* 69:1-21.

Scharrer, B. (1967): The neurosecretory neuron in neuroendocrine regulatory        186
mechanisms. *Amer. Zool.* 7:161-169.

Scharrer, B. (1968a): Neuroendocrine factors in the control of reproduction. *In:*
*Perspectives in Reproduction and Sexual Behavior.* Diamond, M., ed. Bloomington: Indiana University Press, pp. 145-149.

Scharrer, B. (1968b): Neurosecretion. XIV. Ultrastructural study of sites of release
of neurosecretory material in blattarian insects. *Z. Zellforsch.* 89:1-16.

Scharrer, B. (1969a): Comparative aspects of neurosecretory phenomena. *In:*        186
*Progress in Endocrinology. Excerpta Medica International Congress Series, No.*
*184.* Amsterdam: Excerpta Medica Foundation, pp. 365-367.

Scharrer, B. (1969b): Current concepts in the field of neurochemical mediation.        186
*Med. College Virginia Quarterly* 5:27-31.

Scharrer, B. (1969c): Neurohumors and neurohormones: Definitions and terminol-        186
ogy. *J. Neurovisc. Relat.* Suppl. IX:1-20.

Scharrer, B. (1970): General principles of neuroendocrine communication. *In: The*        186
*Neurosciences: Second Study Program.* Schmitt, F.O., editor-in-chief. New
York: Rockefeller University Press, pp. 519-529.

Scharrer, B. and Kater, S.B. (1969): Neurosecretion. XV. An electron microscopic
study of the corpora cardiaca of *Periplaneta americana* after experimentally
induced hormone release. *Z. Zellforsch.* 95:177-186.

Scharrer, B. and Weitzman, M. (1970): Current problems in invertebrate neuro-        186
secretion. *In: Aspects of Neuroendocrinology.* Bargmann, W. and Scharrer, B.,
eds. Berlin: Springer-Verlag, pp. 1-23.

Scharrer, E. (1965): The final common path in neuroendocrine integration. *Arch.*        186
*Anat. Micr. Morph. Exp.* 54:359-370.

Page

Scharrer, E. (1966): Principles of neuroendocrine integration. *Res. Publ. Ass. Res. Nerv. Ment. Dis.* 43:1-35.     186

Scharrer, E. and Scharrer, B. (1963): *Neuroendocrinology*. New York: Columbia University Press.     186

Schildkraut, J.J. and Kety, S.S. (1967): Biogenic amines and emotion. *Science* 156:21-37.

Schneider, H.P.G., Crighton, D.B., and McCann, S.M. (1969): Suprachiasmatic LH-releasing factor. *Neuroendocrinology* 5:271-280.     223

Schneider, H.P.G. and McCann, S.M. (1969): Possible role of dopamine as transmitter to promote discharge of LH-releasing factor. *Endocrinology* 85:121-132.     225

Schneider, H.P.G. and McCann, S.M. (1970a): Mono- and indolamines and control of LH secretion. *Endocrinology* 86:1127-1133.     225

Schneider, H.P.G. and McCann, S.M. (1970b): Release of LH-releasing factor (LRF) into the peripheral circulation of hypophysectomized rats by dopamine and its blockage by estradiol. *Endocrinology* 87:249-253.     225

Shepherd, G.M. (1970): The olfactory bulb as a simple cortical system: Experimental analysis and functional implications. *In: The Neurosciences: Second Study Program.* Schmitt, F.O., editor-in-chief. New York: Rockefeller University Press, pp. 539-551.     213

Shoemaker, W.J. and Wurtman, R.J. (1970): Effect of perinatal undernutrition on development of brain catecholamines in the rat. *Fed. Proc.* 29:496. (Abstr.)     189

Shoemaker, W.J. and Wurtman, R.J. (1971): Perinatal undernutrition: Accumulation of catecholamines in rat brain. *Science.* (In press)     189

Smith, E.R. and Davidson, J.M. (1968): Role of estrogen in the cerebral control of puberty in female rats. *Endocrinology* 82:100-108.     264

Stefano, F.J.E. and Donoso, A.O. (1967): Norepinephrine levels in the rat hypothalamus during the estrous cycle. *Endocrinology* 81:1405-1406.     251,252

Steiner, A.L., Peake, G.T., Utiger, R.D., Karl, I.E., and Kipnis, D.M. (1970): Hypothalamic stimulation of growth hormone and thyrotropin release *in vitro* and pituitary 3′,5′-adenosine cyclic monophosphate. *Endocrinology* 86:1354-1360.     235

Steiner, F.A., Ruf, K., and Akert, K. (1969): Steroid-sensitive neurones in rat brain: anatomical localization and responses to neurohumours and ACTH. *Brain Res.* 12:74-85.     211,216, 242

Sutherland, E.W., Rall, T.W., and Menon, T. (1962): Adenyl cyclase. I. Distribution, preparation, and properties. *J. Biol. Chem.* 237:1220-1227.     235

Page

Szentágothai, J., Flirkó, B., Mess, B., and Halász, B. (1962): *Hypothalamic Control*          202
*of the Anterior Pituitary*. Budapest: Akadémiai Kiadó.

Szentágothai, J., Flirkó, B., Mess, B., and Halász, B. (1968): *Hypothalamic Control*          264
*of the Anterior Pituitary, 3rd Ed.* Budapest: Akadémiai Kiadó.

Talwalker, P.K., Ratner, A., and Meites, J. (1963): *In vitro* inhibition of pituitary
prolactin synthesis and release by hypothalamic extract. *Amer. J. Physiol.*
205:213-218.

Taxi, J. and Droz, B. (1966): Etude de l'incorporation de noradrénaline-3H              205
(NA-3H) et de 5-hydroxytryptophane-3H (5-HTP-3H) dans les fibres nerveuses
du canal déférent et de l'intestin. *C.R. Acad. Sci. [D]* 263:1237-1240.

Terasawa, E. and Sawyer, C.H. (1970): Diurnal variation in the effects of              250
progesterone on multiple unit activity in the rat hypothalamus. *Exp. Neurol.*
27:359-374.

Thierry, A.M., Javoy, F., Glowinski, J., and Kety, S.S. (1968): Effects of stress on
the metabolism of norepinephrine, dopamine and serotonin in the central
nervous system of the rat. I. Modifications of norepinephrine turnover. *J.*
*Pharmacol. Exp. Ther.* 163:163-171.

Thoenen, H., Mueller, R.A., and Axelrod, J. (1969a): Increased tyrosine hydroxyl-      190
ase activity after drug-induced alteration of sympathetic transmission. *Nature*
221:1264.

Thoenen, H., Mueller, R.A., and Axelrod, J. (1969b): Trans-synaptic induction of
adrenal tyrosine hydroxylase. *J. Pharmacol. Exp. Ther.* 169:249-254.

Tranzer, J.P. and Thoenen, H. (1967): Significance of 'empty vesicles' in post-        205
ganglionic sympathetic nerve terminals. *Experientia* 23:123-124.

Tullner, W.W. and Hertz, R. (1964): Suppression of corticosteroid production in        236
the dog by monase. *Proc. Soc. Exp. Biol. Med.* 116:837-840.

Turner, C.D. and Bagnara, J. (1970): *General Endocrinology.* Philadelphia:
Saunders.

Urquhart, J. (1970a): Blood-borne signals. The measuring and modelling of              236
humoral communication and control. *The Physiologist* 13:7-41.

Urquhart, J. (1970b): Endocrinology and the systems paradigm. *Behavioral Sci.*        230
15:57-71.

Urquhart, J., Krall, R.L., and Li, C.C. (1968): Analysis of the Koritz-Hall            230
hypothesis for the regulation of steroidogenesis by ACTH. *Endocrinology*
83:390-394.

Urquhart, J. and Li, C.C. (1968): The dynamics of adrenocortical secretion. *Amer.*    229,230
*J. Physiol.* 214:73-85.

Page

Van Orden, L.S., Bensch, K.G., and Giarman, N.J. (1967): Histochemical and     205
functional relationships of catecholamines in adrenergic nerve endings. II.
Extravesicular norepinephrine. *J. Pharmacol. Exp. Ther.* 155:428-439.

Van Orden, L.S., Bloom, F.E., Barrnett, R.J., and Giarman, N.J. (1966): Histo-     205
chemical and functional relationships of catecholamines in adrenergic nerve
endings. I. Participation of granular vesicles. *J. Pharmacol. Exp. Ther.*
154:185-199.

Vernikos-Danellis, J. (1965): The regulation of the synthesis and release of ACTH.     242
*Vitamins Hormones* 23:97-152.

Vernikos-Danellis, J. (1968): The pharmacological approach to the study of the     240,241
mechanisms regulating ACTH secretion. *In: Pharmacology of Hormonal Polypeptides and Proteins.* Back, N., Martini, L., and Paoletti, R., eds. New York:
Plenum Press, pp. 175-189.

Vernikos-Danellis, J. and Harris, C.G., III (1968): The effect of *in vitro* and *in vivo*     240
caffeine, theophylline and hydrocortisone on the phosphodiesterase activity of
the pituitary median eminence, heart, and cerebral cortex of the rat. *Proc. Soc.
Exp. Biol. Med.* 128:1016-1021.

Vernikos-Danellis, J. and Marks, B.H. (1962): Epinephrine-induced release of     217,240
ACTH in normal human subjects: a test of pituitary function. *Endocrinology*
70:525-531.

Vogt, M. (1954): The concentration of sympathin in different parts of the central     197,198
nervous system under normal conditions and after the administration of drugs.
*J. Physiol.* 123:451-481.

Watanabe, S. and McCann, S.M. (1968): Localization of FSH-releasing factor in the     222
hypothalamus and neurohypophysis as determined by *in vitro* assay. *Endocrinology* 82:664-673.

Weil-Malherbe, H., Whitby, L.G., and Axelrod, J. (1961): The uptake of circulating     233
[$^3$H] norepinephrine by the pituitary gland and various areas of the brain. *J.
Neurochem.* 8:55-64.

Weiner, N. (1970): Regulation of norepinephrine biosynthesis. *Ann. Rev. Pharma-*     190
*col.* 10:273-290.

Weiner, N. and Rabadjija, M. (1968): The effect of nerve stimulation on the     190,194
synthesis and metabolism of norepinephrine in the isolated guinea-pig hypo-
gastric nerve-vas deferens preparation. *J. Pharmacol. Exp. Ther.* 160:61-71.

Weiner, R., Rubinstein, L., and Sawyer, C.H. (1970): Intraventricular epinephrine     245
induces changes in electrical activity of the rat median eminence. Program, 52nd
Meeting, The Endocrine Society, p. 62.

Westermann, E.O., Maickel, R.P., and Brodie, B.B. (1962): On the mechanism     238,239
of pituitary-adrenal stimulation by reserpine. *J. Pharmacol. Exp. Ther.*
138:208-217.

Page

Wise, B.L., Van Brunt, E.E., and Ganong, W.F. (1963): Effect of removal of various    264
parts of the brain on ACTH secretion in dogs. *Proc. Soc. Exp. Biol. Med.*
112:792-795.

Wolfe, D.E., Potter, L.T., Richardson, K.C., and Axelrod, J. (1962): Localizing    205
tritiated norepinephrine in sympathetic axons by electron microscopic autoradiography. *Science* 138:440-442.

Wurtman, R.J. (1966): *Catecholamines.* Boston: Little, Brown.                   186,189,
                                                                                      193

Wurtman, R.J. (1970a): Brain catecholamines and the control of secretion from the
anterior pituitary gland. *In: Hypophysiotropic Hormones of the Hypothalamus:
Assay and Chemistry.* Meites, J., ed. Baltimore: Williams & Wilkins, pp.
184-194.

Wurtman, R.J. (1970b): The effects of endocrine, synaptic, and nutritional inputs
on catecholamine-containing neurons. *In: University of Wisconsin-Parkside
Symposium on Biochemistry of Brain and Behavior.* Bowman, R. and Datta,
S.P., eds. New York: Plenum Press, pp. 91-96.

Wurtman, R.J. (1970c): Neuroendocrine transducer cells in mammals. *In: The*    182
*Neurosciences: Second Study Program.* Schmitt, F.O., editor-in-chief. New
York: Rockefeller University Press, pp. 530-538.

Wurtman, R.J. (1971): The role of brain and pineal indoles in neuroendocrine
mechanisms. *In: Integration of Endocrine and Nonendocrine Mechanisms in the
Hypothalamus.* Martini, L., ed. New York: Academic Press. (In press)

Wurtman, R.J., Anton-Tay, F., and Anton, S. (1969): On the use of synthesis    194,254
inhibitors to estimate brain norepinephrine synthesis in gonadectomized rats.
*Life Sci.* 8:1015-1022.

Wurtman, R.J. and Axelrod, J. (1965): The pineal gland. *Sci. Amer.* 213(1):50-60.    183

Wurtman, R.J., Axelrod, J., Chu, E.W., Heller, A., and Moore, R.Y. (1967): Medial    269
forebrain bundle lesions: blockade of effects of light on rat gonads and pineal.
*Endocrinology* 81:509-514.

Wurtman, R.J., Axelrod, J., and Kelly, D.E. (1968a): *The Pineal.* New York:    186,196
Academic Press.

Wurtman, R.J., Axelrod, J., and Reis, D.J. (1968b): Metabolic cycles of monoamines
and their modification by drugs. *In: Cycles Biologiques et Psychiatrie.*
Ajuriaguerra, J. de, ed. Paris: Masson & Cie.

Wurtman, R.J., Chou, C., and Rose, C.M. (1970): The fate of $C^{14}$-dihydroxy-    191
phenylalanine ($C^{14}$-dopa) in the whole mouse. *J. Pharmacol. Exp. Ther.*
174:351-356.

Wurtman, R.J. and Fernstrom, J.D. (1971): L-Tryptophan, L-tyrosine, and the    195
control of brain monoamine synthesis. *In: Perspectives of Neuropharmacology.*
Snyder, S.H., ed. New York: Oxford University Press. (In press)

Page

Wurtman, R.J., Rose, C.M., Chou, C., and Larin, F. (1968c): Daily rhythms in the    189,195
concentrations of various amino acids in human plasma. *New Eng. J. Med.*
279:171-175.

Wurtman, R.J., Rose, C.M., Matthysse, S., Stephenson, J., and Baldessarini, R.
(1970): L-Dihydroxyphenylalanine: effect on S-adenosylmethionine in brain.
*Science* 169:395-397.

Wurtman, R.J., Shoemaker, W.J., and Larin, F. (1968): Mechanism of the daily
rhythm in hepatic tyrosine transaminase activity: role of dietary tryptophan.
*Proc. Nat. Acad. Sci.* 59:800-807.

Yamamoto, W.S. and Brobeck, J.R., eds. (1965): *Physiological Controls and
Regulations.* Philadelphia: Saunders Co.

Yates, F.E. (1967): Physiological control of adrenal cortical hormone secretion. *In:*    236
*The Adrenal Cortex.* Eisenstein, A.B., ed. Boston: Little, Brown, pp. 133-183.

Yates, F.E. and Brennan, R.D. (1969): Study of the mammalian adrenal gluco-    231
corticoid system by computer simulation. *In: Hormonal Control Systems.* Stear,
E.G. and Kadish, A.H., eds. New York: American Elsevier, pp. 20-87.

Yates, F.E., Brennan, R.D., and Urquhart, J. (1969): Application of control    236
systems theory to physiology. Adrenal glucocorticoid control system. *Fed. Proc.*
28:71-83.

Yates, F.E., Brennan, R.D., Urquhart, J., Dallman, M.F., Li, C.C., and Halpern, W.    230,231
(1968): A continuous system model of adrenocortical function. *In: Systems
Theory and Biology.* Mesarović, M.D., ed. New York: Springer-Verlag, pp.
141-184.

Yokoyama, A., Halász, B., and Sawyer, C.H. (1967): Effect of hypothalamic
deafferentation on lactation in rats. *Proc. Soc. Exp. Biol. Med.* 125:623-626.

Zarrow, M.X., Yochum, J.M., McCarthy, J.L., and Sanborn, R.C. (1964): *Experimental Endocrinology.* New York: Academic Press.

Zigmond, M.J. and Wurtman, R.J. (1970): Daily rhythm in the accumulation of    193
brain catecholamines synthesized from circulating $H^3$-tyrosine. *J. Pharmacol.
Exp. Ther.* 172:416-422.

# INDEX

# Carriers and Specificity in Membranes

A report based on an NRP Work Session
held February 9-11, 1969

by

**Manfred Eigen**
Max Planck Institute for Biophysical Chemistry
Göttingen-Nikolausberg, West Germany

and

**Leo De Maeyer**
Max Planck Institute for Biophysical Chemistry
Göttingen-Nikolausberg, West Germany

Dorothy W. Bishop
NRP Writer-Editor

CONTENTS

LIST OF PARTICIPANTS

Dr. Leo De Maeyer
Max Planck Institute for
 Biophysical Chemistry
D-3400 Göttingen-Nikolausberg
West Germany

Dr. Manfred Eigen
Max Planck Institute for
 Biophysical Chemistry
D-3400 Göttingen-Nikolausberg
West Germany

Dr. George Eisenman
Department of Physiology
The University of California
Los Angeles, California 90024

Dr. Humberto Fernandez-Moran
Department of Biophysics
The University of Chicago
5640 South Ellis Avenue
Chicago, Illinois 60637

Dr. Ernst Grell
Max Planck Institute for
 Biophysical Chemistry
D-3400 Göttingen-Nikolausberg
West Germany

Dr. Ching-hsien Huang
Department of Biochemistry
University of Virginia School of Medicine
Charlottesville, Virginia 22901

Dr. Wayne L. Hubbell
Department of Chemistry
University of California
Berkeley, California 94720

Dr. Georg Ilgenfritz
Max Planck Institute for
 Biophysical Chemistry
D-3400 Göttingen-Nikolausberg
West Germany

Dr. Aharon Katchalsky
Polymer Department
Weizmann Institute of Science
Rehovoth, Israel

Dr. Heinrich Müldner
Max Planck Institute for
 Biophysical Chemistry
D-3400 Göttingen-Nikolausberg
West Germany

Dr. Lars Onsager
Sterling Chemical Laboratory
Yale University
New Haven, Connecticut 06520

Dr. Berton C. Pressman
Department of Pharmacology
University of Miami Medical Center
Miami, Florida 33152

Dr. Francis O. Schmitt
Neurosciences Research Program
280 Newton Street
Brookline, Massachusetts 02146

Dr. Wilhelm Simon
Laboratory for Organic Chemistry
Swiss Federal Institute of Technology
Universitätstrasse 6/8
8006 Zurich, Switzerland

Dr. Daniel C. Tosteson
Department of Physiology
  and Pharmacology
Duke University Medical Center
Durham, North Carolina 27706

Dr. Hermann Träuble
Max Planck Institute for
  Biophysical Chemistry
D-3400 Göttingen-Nikolausberg
West Germany

Dr. Victor P. Whittaker
Department of Biochemistry
Cambridge University
Tennis Court Road
Cambridge, England

Mrs. Ruthild Winkler
Max Planck Institute for
  Biophysical Chemistry
D-3400 Göttingen-Nikolausberg
West Germany

Note: NRP Work Session summaries are reviewed and revised by participants prior to publication.

## I. INTRODUCTION: Manfred Eigen

Two recent developments in the field of molecular biology of membranes deserve special attention with respect to the neurosciences.

1. *Artificial and Natural Membranes and Membrane Fragments.* Comparative studies of both natural and artificial membranes and of membrane fragments have provided a deep insight into an important building principle of nature: the self-assembly of certain types of molecules into two-dimensional structures. Here the isolation of natural membranes, their characterization by various physical techniques (e.g., electron microscopy, X-ray diffraction, or spin labeling), is paralleled by the investigation of artificially constructed membranes such as lipid bilayers (using the Mueller-Rudin technique) or vesicles of various forms and sizes.

2. *Carriers.* The dynamic aspect of membranes may be studied with the help of identified, specific carriers. For many biologists, the carrier has been a most useful concept in explaining specificity of transport across membranes, although for a long time no one could actually demonstrate that it existed. Perhaps, for the neurobiologist, the most striking example of specificity of transport is the fast exchange of sodium and potassium during the excitation of the nerve membrane. When this problem was presented to the physical chemists, no simple explanation could be given, except that apparently a membrane constituent with a highly sophisticated structure must be present, which manages to distinguish between two ions that are chemically so similar.

Only recently, several research groups have found out that membrane constituents of such subtle structure really do exist (see Pressman in Section II). These compounds are antibiotics. They induce and facilitate quite specifically the transport of a given ion, e.g., $K^+$ or $Rb^+$, across natural and artificial membranes, fulfilling all the specifications once assigned to the concept, "carrier." The only question still unanswered is whether these or similar substances, which were usually isolated from microorganisms, actually carry out this function in membranes of higher organisms. Nevertheless, these substances are available for investigation, and a large part of this report deals with such studies.

The basic feature of transport across membranes is the coupling of a diffusion or flow process with a single chemical reaction, or a sequence of chemical reactions. Our report includes a theoretical approach to this very basic problem by Katchalsky. He reviews recent work from his own laboratory, and from Prigogine's group, that is based on the thermodynamics of irreversible processes, and he shows that coupling between chemical reactions and transport processes can lead to peculiar structures that are maintained by the dissipation of energy. Any quantitative experiments on the dynamic behavior of membranes must start from those considerations. (See Section VI.)

Several contributions deal more specifically with alkali ion carriers. It is obvious that the unique properties of these carriers show up in the spatial structure of these compounds. Simon describes the different types of antibiotics, their chemical and spatial structure. Dynamics and mechanisms of action as well as the principle on which distinction of various ions is based are discussed by Eigen and Winkler. Tosteson, Eisenman, and Pressman show how the carriers behave in natural and artificial membranes.

The role of proteins in ion transport across membranes deserves much attention. Little work so far has been done on isolated and well-characterized membrane proteins. The work reported by Müldner on a bacterial adenosine triphosphatase (ATPase) is one of the few examples discussed at the Work Session.

The other papers deal more specifically with properties of natural and artificial membranes. Whittaker describes his work with synaptic vesicles and demonstrates that well-characterized membrane fragments can be isolated and studied under defined conditions.

The preparation and characterization of artificial membranes is elegantly demonstrated in three short contributions. Huang describes the preparation of membrane vesicles, which he then uses for various dynamic studies that complement the classical studies of Mueller and Rudin and those now being done in Thompson's laboratory. An even more sophisticated technique of "molecular engineering" of specific membrane vesicles is presented by Träuble and Grell. The study of the behavior of lipid molecules in membranes by spin-label probes—a powerful technique with many potentialities—is demonstrated by Hubbell.

Certainly, most of our knowledge about membrane structure comes from work with the electron microscope. This powerful tool has been refined to the extent that molecular structures now are becoming visible. So it seems only appropriate to include a review of electron microscopy techniques in a report on membrane structure with an outlook on further work in this field; Fernandez-Moran's review tells much that is in progress now and anticipates exciting new developments.

The general discussion of the Work Session is summarized in a concluding paper by De Maeyer. It is quite obvious that the time is ripe for a thorough study of membrane phenomena at the molecular level. For some time, the isolation and characterization of membrane constituents have lagged behind the analogous phase in protein and nucleic acid research. Much of the delay is due to the lack of suitable separation techniques for nonpolar media, and these techniques are especially relevant if we consider the specific constituents of nerve cell membranes, especially those of the synaptic regions. However, as this *Bulletin* shows, the tools are ready, and experience with well-defined model systems is available for application to comparative studies of well-defined natural systems. By combining all these efforts, much more can be learned about the molecular biology of the central nervous system.

## II. CHARACTERIZATION OF CARRIERS

### PROPERTIES OF ION-SPECIFIC CARRIERS:
### W. Simon

In 1964 Moore and Pressman discovered that valinomycin shows very great cation specificity in the transport of ions into mitochondria. Since that time, many other antibiotics have been studied. In general, they are metabolites isolated from cultures of microorganisms. At present, two categories of such cation-specific antibiotics are known, the valinomycin and the nigericin group.

The first, or valinomycin group, comprises valinomycin, the enniatins (A, B), gramicidins (A, B, C, S), and macrotetrolides. (See Lardy et al., 1967.) These compounds may be characterized as follows:

1. They are electrically neutral.

2. They form charged complexes with alkali metal cations and may therefore act as "carriers."

3. They induce transport of alkali metal ions into intact mitochondria.

4. They increase the cation permeability of artificial lipid membranes (bilayers) by several orders of magnitude.

5. They show a pronounced selectivity for $K^+$ ($Rb^+$) over $Na^+$.

Nigericin and monensin are the best-known members of the second category, the nigericin group (Lardy et al., 1967), and they have the following characteristics:

1. They are negatively charged at physiological pH due to a dissociated carboxylic group.

2. They form electrically neutral complexes with alkali metal cations.

3. They interfere with the uptake of alkali metal cations into mitochondria when uptake is induced by antibiotics of the valinomycin group.

4. The selectivity order for monensin is $Na^+ > K^+$, for nigericin $K^+ > Na^+$ (Lutz et al., 1970).

Chemical Structure of Some Carrier Substances

*Compounds of the Valinomycin Group*

Nonactin and its homologs monactin, dinactin, and trinactin are known as macrotetrolides (Keller-Schierlein and Gerlach, 1968). All the macrotetrolides can be isolated from *Actinomycete* cultures, and they differ only by their substituents $R^2$ to $R^4$, i.e., by one to three additional methylene groups as shown in Figure 1. They are tetralactones, forming 32-membered polycyclic ring systems.

The other compounds of the valinomycin group are chemically quite different from the macrotetrolides. The gramicidins A, B, and C (Hotchkiss and Dubos, 1941; Sarges and Witkop, 1965a,b,c) are open chain compounds (acyclic formyl pentadecapeptide ethanolamides), in contrast to the cyclic decapeptide structure of gramicidin S, which forms a 30-membered macrocyclic ring (Gause and Brazhnikova, 1944; Schwyzer and Sieber, 1957). (See Figure 2.)

Enniatin A or B is a cyclohexadepsipeptide made up from three α-amino acid units and three α-hydroxy acid units, linked by alternating ester and amide bonds (Plattner et al., 1948, 1963). (See Figure 3.) The enniatins are 18-membered ring compounds.

Valinomycin similarly is a cyclic depsipeptide but has twice the ring size of the enniatins (Brockmann and Schmidt-Kastner, 1955; Brockmann et al., 1963). The alternating sequence of 6 α-amino and 6 α-hydroxy acid units leads to a 36-membered ring. (See Figure 4.) The valinomycin group of carriers also includes some synthetic substances such as certain cyclic polyethers, frequently referred to as "crown compounds" (Pedersen, 1967). A representative type is shown in Figure 5.

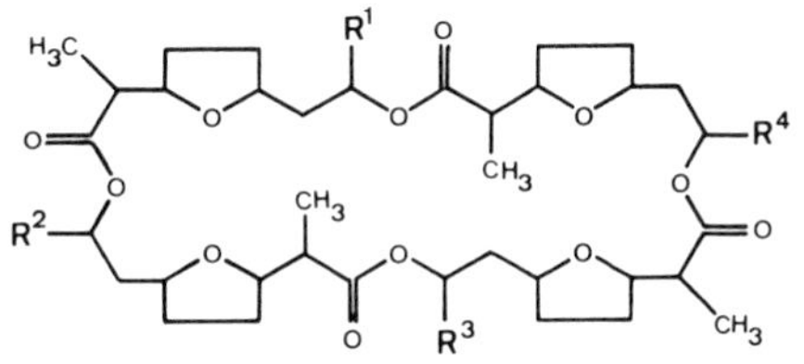

| | | |
|---|---|---|
| $R^1 = R^2 = R^3 = R^4 = CH_3$ | | Nonactin |
| $R^1 = R^2 = R^3 = CH_3$ | $R^4 = C_2H_5$ | Monactin |
| $R^1 = R^3 = CH_3$ | $R^2 = R^4 = C_2H_5$ | Dinactin |
| $R^1 = CH_3$ | $R^2 = R^3 = R^4 = C_2H_5$ | Trinactin |

Figure 1. Structure of the macrotetrolides. [Adapted from Keller-Schierlein and Gerlach, 1968]

```
                    L·Pro ——— L·Val
                   /                 \
              D·Phe                    L·Orn
              /                           \
          L·Leu                             L·Leu
              \                             /
          L·Orn                         D·Phe
               \                         /
                L·Val ——— L·Pro
```

Gramicidin  S

HC=O
|
[ 1 ]— Gly —L·Ala —D·Leu—L·Ala — D·Val— L·Val — D·Val

NH —L·Try —D·Leu —L·Try—D·Leu—[ 11 ]— D·Leu—L·Try
|
(CH₂)₂
|
OH

|                              | 1:     | 11:   |
|-------------------------------|--------|-------|
| Valine - Gramicidin A         | L-Val  | L-Try |
| Isoleucine - Gramicidin A     | L-Ileu | L-Try |
| Valine - Gramicidin B         | L-Val  | L-Phe |
| Isoleucine - Gramicidin B     | L-Ileu | L-Phe |
| Valine - Gramicidin C         | L-Val  | L-Tyr |
| Isoleucine - Gramicidin C     | L-Ileu | L-Tyr |

Figure 2. Structure of the gramicidins. [Simon]

ENNIATIN A    ┌A — A — A┐

ENNIATIN A'   ┌A — A — B┐

ENNIATIN B'   ┌A — B — B┐

ENNIATIN B    ┌B — B — B┐

A:
```
        CH₃CH₂   CH₃     CH₃   CH₃
             \CH           \CH
              |              |
       -N—CH-C —O—CH—C-
        |    ||        ||
        CH₃  O         O
```

B:
```
         CH₃  CH₃    CH₃  CH₃
           \CH         \CH
            |            |
       -N—CH-C —O—CH—C-
        |    ||       ||
        CH₃  O        O
```

Figure 3. Structure of the enniatins. [Simon]

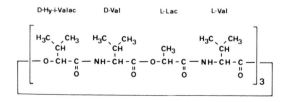

D-Hy-i-Valac      D-Val          L-Lac        L-Val

Figure 4. Structure of valinomycin. [Simon]

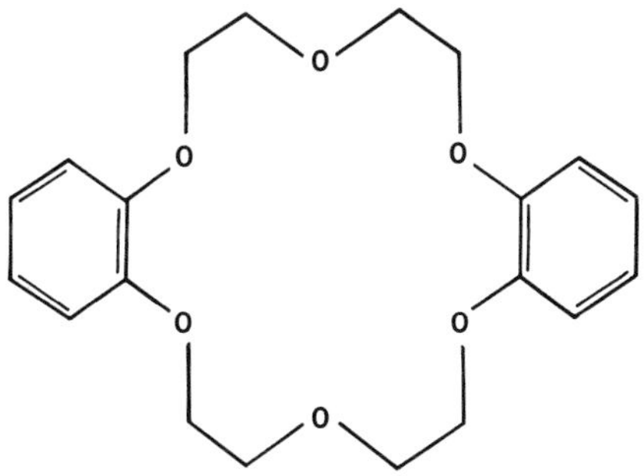

Figure 5. Dibenzo-18-crown-6 as representative of the cyclic polyethers. [Adapted from Pedersen, 1967]

## Compounds of the Nigericin Group

The structure of nigericin or monensin is very different from any compound of the valinomycin group (Agtarap et al., 1967). (See Figure 6.) The carboxyl group that is typical for this class of compounds is dissociated around ph 7. These substances will therefore be negatively charged at physiological pH and can form electrically neutral complexes with alkali metal ions. It is believed that the free anions as well as the complex molecules in solution exist in a cyclic conformation stabilized by two hydrogen bonds that lie between the two terminal hydroxyl groups and the carboxylate group at the other end of the molecule (Lutz et al., 1970).

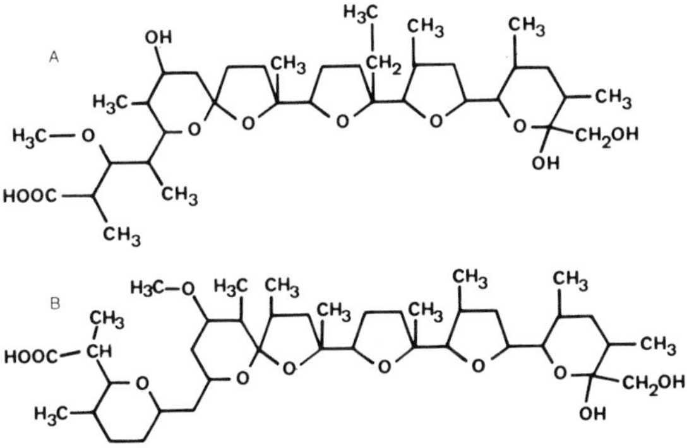

Figure 6. Structure of A, monensin and B, nigericin. [Adapted from Agtarap et al., 1967]

## Stability Constants

The complex formations of nonactin and monactin consistently show that both compounds bind $K^+$ more strongly than they bind $Na^+$ (Pioda et al., 1967; Stefanac and Simon, 1966). (See Table 1.) The complex formation constants in the solvent methanol are smaller than

TABLE 1

Complex Formation Constants of Nonactin and Monactin with $Na^+$ and $K^+$ [Simon]

| Antibiotic | Method | Solvent | Temperature (°C) | Complex formation constants $[M^{-1}]$ with NaSCN | with KSCN | References |
|---|---|---|---|---|---|---|
| Nonactin | VPO* | MeOH | 30 | $1.6 \pm 0.3 \times 10^2$ | $6.3 \pm 0.9 \times 10^3$ | Pioda et al., 1967 |
| | Glass electrode | MeOH | 25 | — | $3.9 \pm 0.7 \times 10^3$ | Wipf et al., 1968 |
| | VPO | EtOH | 30 | $2.9 \pm 0.3 \times 10^3$ | $5.5 \pm 1.0 \times 10^4$ | Wipf, 1970 |
| | VPO | EtOH | 60 | — | $5.5 \pm 1.3 \times 10^3$ | Wipf, 1970 |
| Monactin | VPO | MeOH | 30 | $1.4 \pm 0.1 \times 10^3$ | $3.2 \pm 1.3 \times 10^5$ | Pioda et al., 1967 |
| | VPO | MeOH | 30 | — | $2.8 \pm 0.3 \times 10^3$ | Wipf, 1970 |
| | Indicator | MeOH | 30 | $5.0 \quad \times 10^2$ | — | Winkler, 1969 |
| | VPO | EtOH | 30 | $3.9 \pm 0.1 \times 10^3$ | $2.6 \pm 0.1 \times 10^4$ | Wipf, 1970 |
| | VPO | EtOH | 60 | — | $8.3 \pm 0.3 \times 10^3$ | Wipf, 1970 |

*VPO = vapor pressure osmometry.

TABLE 2

Complex Formation Constants of Antibiotics of the Valinomycin Group with Na$^+$ and K$^+$ [Simon]

| Antibiotic | Method | Solvent | Anion | Temperature (°C) | Complex formation constants [M$^{-1}$] with NaX | with KX | References |
|---|---|---|---|---|---|---|---|
| Nonactin | VPO* | EtOH | SCN$^-$ | 30 | $2.9 \pm 0.3 \times 10^3$ | $5.5 \pm 1.0 \times 10^4$ | Wipf, 1970 |
| Monactin | VPO | EtOH | SCN$^-$ | 30 | $3.9 \pm 0.1 \times 10^3$ | $2.6 \pm 0.1 \times 10^4$ | Wipf, 1970 |
| Dinactin | VPO | EtOH | SCN$^-$ | 30 | $9.9 \pm 1.8 \times 10^3$ | — | Wipf, 1970 |
| Trinactin | VPO | EtOH | SCN$^-$ | 30 | $4.4 \pm 0.3 \times 10^3$ | — | Wipf, 1970 |
| Enniatin A | Glass electrode | MeOH | I$^-$ | 25 | — | $1.2 \pm 0.1 \times 10^3$ | Wipf et al., 1968 |
| Enniatin B | Glass electrode | MeOH | I$^-$ | 25 | $2.4 \pm 0.5 \times 10^2$ | $8.4 \pm 1.0 \times 10^2$ | Wipf et al., 1968 |
| | Conductivity | EtOH | Cl$^-$ | 25 | $1.3 \pm 0.3 \times 10^3$ | $3.7 \pm 0.8 \times 10^3$ | Shemyakin et al., 1967 |
| | ORD[†] | EtOH | SCN$^-$ | 25 | $2.6 \times 10^3$ | $6.5 \times 10^3$ | Shemyakin et al., 1967 |
| Valinomycin | Glass electrode | MeOH | I$^-$ | 25 | $1.2 \pm 1.7 \times 10^1$ | $>8 \times 10^3$ | Wipf et al., 1968 |
| | Conductivity | MeOH | Cl$^-$ | 25 | 0 | $2.7 \pm 0.5 \times 10^4$ | ‡ |
| | Conductivity | EtOH | Cl$^-$ | 25 | 0 | $2.0 \pm 0.8 \times 10^6$ | ‡ |

*VPO = vapor pressure osmometry.

†ORD = optical rotatory dispersion.

‡Yu. A. Ovchinnikov, personal communication. These values for the valinomycin-potassium complex formation constants replace the previously published results of this author (Ovchinnikov et al., 1969).

those in the solvent ethanol by about one order of magnitude.* Almost all of these constants have been determined by vapor pressure osmometry (VPO) (Pioda et al., 1967; Wipf et al., 1968; Wipf, 1970). Divalent ions such as $Ba^{2+}$ with an ionic radius comparable to that of $K^+$ also seem to show a tendency to complex with the macrotetrolides, but no such binding was found with $Sr^{2+}$, $Ca^{2+}$, or $Mg^{2+}$.

Table 2 compares formation constants of all compounds of the valinomycin group investigated so far. In comparing the values for various carriers, the enniatins clearly show the smallest selectivity of $K^+$ over $Na^+$. Nonactin and its higher homologues are more selective; they bind $K^+$ about 20 times more strongly than $Na^+$. Valinomycin is clearly the most selective carrier. The constant for the valinomycin-$K^+$ complex is by at least three orders of magnitude larger than for the valinomycin-$Na^+$ complex (Shemyakin et al., 1967; Wipf et al., 1968).

### Structure and Conformation of the Carrier Complexes

Crystalline 1:1 complexes of the macrotetrolides with various cations can be isolated (Wipf, 1970). Figure 7 shows the crystal structure of the nonactin-$K^+$ complex obtained by Dunitz and his colleagues from X-ray diffraction analysis of the crystals (Kilbourn et al., 1967). All polar groups point towards the cation so that the $K^+$ is surrounded by 8 oxygen atoms. The nonpolar groups are turned

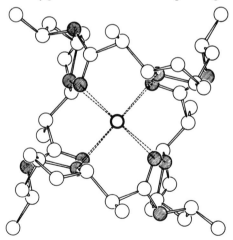

Figure 7. Crystal structure of the nonactin-$K^+$ complex. [Adapted from Kilbourn et al., 1967]

*A range of magnitude extending from some value to ten times that value.

towards the outside, providing the molecule with a hydrophobic exterior. The $K^+$....O distance is about 2.7 A, corresponding very accurately to the sum of the radii of oxygen and $K^+$. This close contact of the potassium ion with the ligand atoms indicates how perfectly the ion fits into the cavity of the cyclic carrier molecule. Figure 8 shows how the nonactin molecule is wrapped around the ion like the seam of a tennis ball. In this way, the ligand atoms, 4 tetrahydrofuran and 4 carbonyl oxygen atoms, surround the metal ion in quasi-cubic symmetry. Figure 9 is a model of this complex. The tennis ball seam structure is quite flexible and may easily open up with a change in conformation. In fact, the uncomplexed nonactin seems to have a flat ring conformation.

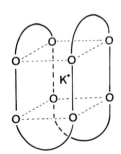

Figure 8. Schematic representation of the nonactin-$K^+$ complex. [Simon]

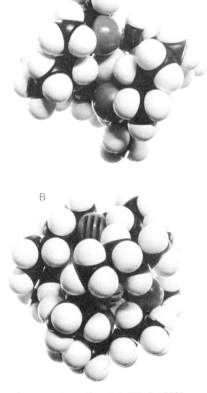

Figure 9. Nonactin-$K^+$ complex, two views of model. [Wipf, 1970]

In the enniatin-$K^+$ complex the cation is bound to 6 ligand atoms (carbonyl oxygen) in almost octahedral symmetry. Again, all the nonpolar groups are on the outside of the molecule. The enniatin molecule is wrapped around the cation (Figure 10) and forms a doughnut-shaped structure, in contrast to the almost spherical macrotetrolide complex (Dobler et al., 1969).

Recent X-ray studies by Pinkerton, Steinrauf, and Dawkins (1969) show that valinomycin also forms a complex in octahedral coordination with the potassium ion (Figure 11A). Again, the polar groups are oriented towards the central cation, while the nonpolar substituents form the hydrophobic exterior of the complex ion. The macrocyclic structure of the complexing antibiotic molecule is folded into three loops enclosing the potassium ion in approximately threefold symmetry. A system of 6 hydrogen bonds helps to stabilize this conformation (Figure 11B).

In the alkali metal complexes of the nigericin group, the cation is also coordinated by oxygen atoms of the antibiotic molecule. Recent X-ray diffraction studies show that the deprotonated molecule is wrapped around the cation in a manner similar to the complexes of the

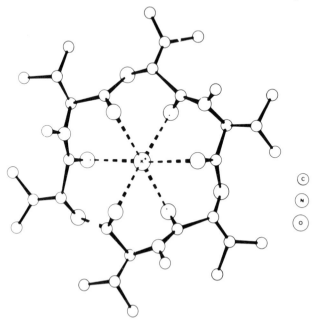

Figure 10. Crystal structure of the enniatin B-$K^+$ complex. [Adapted from Dobler et al., 1969]

valinomycin group. Two hydrogen bonds hold the complexing anion in the proper conformation. In Figure 12, a projection of the monensin-Ag$^+$ complex is presented.

The X-ray work confirms and illustrates the correlation between structure and function of these carriers. Selective transport is facilitated by the binding of the metal ion. Here the polar groups point to the cation, while the lipophilic groups remain at the periphery of the spherical complex. This means that a carrier complex can easily enter and penetrate a lipophilic medium such as a cell membrane. A highly selective transport of potassium ions is obtained by applying a potential difference across a synthetic membrane impregnated with macrotetrolide antibiotics or valinomycin (Wipf and Simon, 1970).

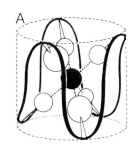

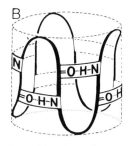

Figure 11. Schematic representation of the valinomycin-K$^+$ complex: A, coordination of the cation; B, system of hydrogen bonds. [Simon]

**Biological Activity**

Table 3 shows the wide variation in biological activity of the macrotetrolides. Monactin, for example, has only one methylene group more than nonactin, but it is about 10 times more active. Lardy has observed similar facts for the influence of different macrotetrolides on the hydrolysis of ATP in mitochondria (Figure 13) (Graven et al., 1966); again, monactin is much more active than nonactin. Dinactin and trinactin show activities similar to monactin. The presence of K$^+$ produces much larger effects with monactin than the presence of Na$^+$. These results correlate very well with the values found for the stability constants given in Table 1.

Electromotive force (EMF) measurements on bulk membranes in the presence of nonactin and its homologues show a pronounced specificity for the different alkali ions (Figure 14); these are in perfect agreement with the data on complex formation reported here, as well as with the biological activity of these antibiotics in mitochondria (Stefanac and Simon, 1966, 1967; Wipf and Simon, 1970).

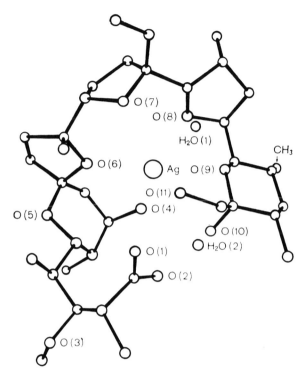

Figure 12. Crystal structure of the monensin-Ag$^+$ complex salt. [Adapted from Agtarap et al., 1967]

TABLE 3

Biological Activity of Nonactin and Its Homologues
[Adapted from Keller-Schierlein and Gerlach, 1968]

| Macrotetrolide | Biological activity (M.I.C.* in $\mu g/ml$) | |
|---|---|---|
| | *Staphylococcus aureus* 209P | *Mycobacterium bovis* BCG strain |
| Nonactin | 0.95 | 0.96 |
| Monactin | 0.08 | 0.12 |
| Dinactin | 0.05 | 0.04 |
| Trinactin | 0.04 | 0.04 |

*M.I.C. = Minimal Inhibitory Concentration

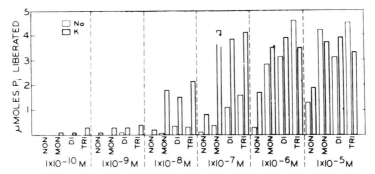

Figure 13. Influence of different macrotetrolides on hydrolysis of ATP in mitochondria. [Graven et al., 1966]

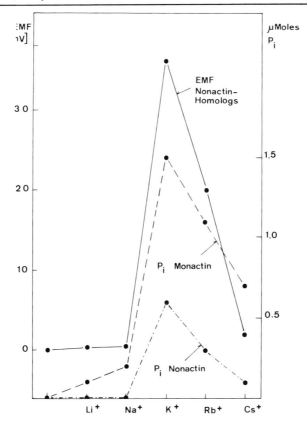

Figure 14. Comparison of results obtained by electromotive force (EMF) measurements with data on ATPase activity in mitochondria. [Adapted from Stefanac and Simon, 1966]

## THE ACTION OF IONOPHOROUS AGENTS:
### B. C. Pressman

Pressman's observations of the phenomenon of antibiotic-induced mitochondrial ion transport are related to the formation of an alkali ion complex by such antibiotic compounds. To describe the transport properties of these antibiotics generally, he introduced the term "ionophorous" agents. Detailed accounts of Pressman's studies of complex formation using bulk phase ion migration and nuclear magnetic resonance (NMR) spectroscopy can be found in Moore and Pressman, 1964; Pressman, 1965; Harris et al., 1967a,b; Pressman et al., 1967; Pressman, 1968a,b; Pressman and Haynes, 1969; Haynes et al., 1969. A brief description of his recent work follows.

Rat liver mitochondria are normally relatively impermeable to potassium ions. After the addition of valinomycin or related ionophorous agents, a decrease of the potassium ion concentration in the medium can be observed by means of ion-selective electrodes, which indicates movement of these ions into mitochondria (Eisenman, 1962). Simultaneously, a countermovement of hydrogen ions can be measured, apparently arising from the necessity for preserving gross electroneutrality across the mitochondrial membrane. A concomitant decrease in light scattering indicates an increase in mitochondrial volume, i.e., water uptake, and an accelerated drop in oxygen level is indicative of an increase in energy expenditure consistent with the considerable concentration gradient against which potassium ions move when entering the mitochondria. A second group of antibiotics, represented by nigericin and monensin, also affects mitochondrial ion transport. However, these compounds appear to reverse most of the processes initiated by the valinomycin-type agents (Lardy et al., 1967).

Any biological vesicle that has a potassium ion gradient across its membrane will establish a transmembrane equilibration between potassium and hydrogen ions when nigericin is applied. After the addition of nigericin to human erythrocytes, potassium ions leave the system, and hydrogen ions enter simultaneously in an approximately equivalent amount (Harris and Pressman, 1967), establishing that the effects of nigericin are not entirely unique to mitochondria.

The question then arises as to what molecular structural features can explain the apparent antagonistic behavior of the two types of antibiotic compounds represented by valinomycin and nigericin. A diagrammatic reduction of the structures, useful for explaining some of their main properties, is shown in Figure 15. The well-known

valinomycin types of antibiotics exhibit a cyclic structure, but show no
charged groups on the molecule. Compounds of the monensin type, including nigericin, consist of linear
chains of oxygen that have hetero-
cyclic rings. In the case of the monensin silver ion complex, the cyclic conformation of the chain is stabilized by
hydrogen bonding between the depro-

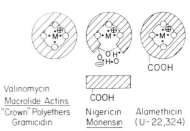

Valinomycin
Macrolide Actins
"Crown" Polyethers
Gramicidin

COOH

Nigericin    Alamethicin
Monensin    (U-22,324)

Figure 15. Diagrammatic representation
of subclasses of ionophore complexes.
[Pressman and Haynes, 1969]

tonated carboxyl group and hydroxyl groups on the opposite end of
the molecule (Pinkerton and Steinrauf, 1970). The expectation in this
case would be that the complex formation would be pH dependent and
in addition that the monensin-type compounds would carry alkali ions
as electrically neutral zwitterions and protons in their undissociated
carboxylic acid form.

The antibiotic alamethicin also induces energy-linked potassium
ion accumulation in mitochondria and exhibits some interesting additional properties. Although it possesses an ionizable carboxyl group,
this covalently cyclic peptide forms an alkali-ion complex over a wide
pH range, indicating that the state of ionization of the carboxyl group
is not critical for complex formation. The fact that alamethicin
complexes can exist in two forms, the zwitterionic at high pH, and the
positively charged at low pH, led Pressman to suggest that this double
property might have something to do with the unusual conductivity
behavior of alamethicin in bilayers as reported by Mueller and Rudin
(1968a,b).

To examine the alkali ion complex formation and the pH
dependence and anionic requirements of antibiotics, Pressman has
carried out the following two-phase distribution studies: A bulk system
with a radioisotope of the test cation, usually $^{86}Rb^+$, was added to the
water phase, while the test antibiotic was added to the organic phase
consisting of 30% butanol and 70% toluene. Migration of radioactivity
into the organic phase was taken as a quantitative measure of complex
formation. As expected, a sharp pH-dependent complexing of nigericin
with rubidium ions could be demonstrated. The aqueous pH, supporting half-maximal complex formation, was 8.5. Chloride ions, generally
regarded as impermeant in mitochondrial transport, exhibited weak
permeant properties in this bulk model system, and a systematic survey
of common inorganic anions was undertaken. The test procedure was to
measure the ability of various anions to support the migration of
labeled rubidium ions from water into the butanol-toluene phase. The

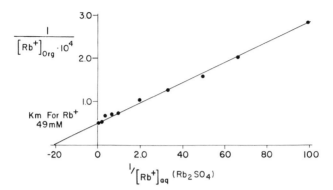

Figure 16. Double reciprocal plot of the saturation of valinomycin by $Rb^+$ in a two-phase system. The organic phase contained 70% toluene, 30% n-butanol, and $2.0 \times 10^{-4}$ M valinomycin. The aqueous phase contained glycine-tricine buffer (pH 7.0) and 50 mM $Mg(SCN)_2$. As the concentration of aqueous $^{86}Rb_2SO_4$ was varied, the concentration of complex formed was measured by the migration of $^{86}Rb^+$ into the organic phase. [Pressman and Haynes, 1969]

anions phosphate and sulfate are without significant permeant properties, while nitrate is weakly active. The permeant properties of the halides increase markedly with increasing atomic weight and are greatest in the case of the quasi-halide, thiocyanate.

Dissociation constants for the complexes were measured in the same bulk system. To test if the complexing of the antibiotic by the cation follows an ideal Langmuir saturation isotherm,* the concentration of cation in the aqueous phase can be varied and its reciprocal plotted against the reciprocal of complex formed, with the latter measured as the quantity of cation migrating into the organic phase. The resulting plot should be a straight line whose slope is a measure of the affinity for the test cation. The intercept of the ordinate is a measure of the stoichiometry between the cation and the ionophorous ligand at saturation. Indeed, this ideal relationship was realized for the saturation of valinomycin by rubidium ions (Figure 16). It was necessary to add the salt magnesium thiocyanate to the aqueous phase to supply a lipid-compatible anion to accompany the complex cation. Otherwise, cation migration would set up strong interphase potentials, distorting the ideal saturation behavior. A similar ideal saturation isotherm could be obtained with nigericin, which is facilitated by conducting the experiment at a constant high pH (e.g., 10), while pH

* $$\frac{1}{[M \cdot I^+]_{org}} = \frac{1}{[M^+]_{H_2O}} \cdot \frac{K_{D_2}}{[I]_{org}} + \frac{1}{[I]_{org}}$$

where $[M \cdot I^+]_{org}$ is measured as the concentration of aqueous cation which migrates to the organic phase, $[M^+]_{H_2O}$ is the concentration of complexing cation present in the aqueous phase, $[I]_{org}$ is the concentration of ionophore originally added to the organic phase, and $K_{D_2}$ is the apparent two-phase dissociation constant of the ionophore complex.

does not markedly affect valinomycin complexation. In contrast to the case of valinomycin-type ionophores where the anion concentration must be held constant, the anionic composition of the aqueous phase for the experiments with nigericin is not critical, which is consistent with the zwitterionic properties of the complex. From the slope of this type of Langmuir plot, a two-phase complex dissociation constant can be computed, and the intercept indicates a 1:1 complex between the alkali ion and the antibiotic compound in all the examples presented. Complex formation is increased at increasing concentrations of thiocyanate added as magnesium thiocyanate; but the common intercept of the ordinate indicates that, at saturation, a 1:1 complex is formed in each case. To obtain affinities for alkali ions other than rubidium, it was in certain cases advantageous to measure their abilities to displace $^{86}Rb^+$ by competition from a given complex.

The two-phase dissociation constants that are pertinent to the equilibrium conditions governing the movement of a cation across an interphase can be converted into a one-phase dissociation constant for comparison with single-phase systems. The two-phase dissociation constant $K_{D_2}$ can be considered the product of the one-phase dissociation constant $K_{D_1}$ and the distribution coefficient of free alkali ion in

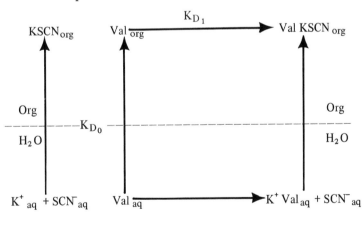

$$K_{D_1} = \frac{[Val]_{org} [KSCN]_{org}}{[Val\ KSCN]_{org}} \qquad K_{D_2} = \frac{[K^+]_{aq} [Val]_{org}}{[Val\ KSCN]_{org}}$$

$$K_{D_2} = K_{D_1} K_{D_0}$$

Figure 17. Resolution of the process involved in complexation of alkali ions by valinomycin in a two-phase system. The two-phase dissociation constant $K_{D_2}$ can be used to determine the single-phase constant in the organic phase $K_{D_1}$ by means of the distribution coefficient of the cation in the absence of ionophore $K_{D_0}$. [Pressman and Haynes, 1969]

the absence of ionophore $K_{D_0}$. This treatment is shown for the formation of the valinomycin-potassium, thiocyanate complex in Figure 17.

A wide comparison of cation affinities, bilayer activities, and
effects on biological systems with regard to the corresponding selectivities of the antibiotic compounds is possible. In Figure 18, the
logarithms of the relative two-phase dissociation constants for various
cations, using several ionophorous agents, have been plotted, the ion of
greatest affinity arbitrarily being assigned a value of 2. The selectivities
range from the compound X-537, which has a slight preference for
cesium ions, to those of monensin, which exhibits the greatest relative
affinity for sodium ions. In the first numerical column, the absolute
values of the two-phase dissociation constants for the potassium ion

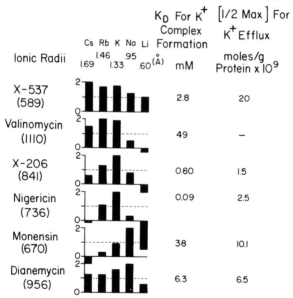

Figure 18. Ion selectivities of various ionophores. Valinomycin data were obtained under
conditions described in Figure 16, using various concentrations and species of alkali sulfates.
For the carboxylate ionophores, the pH was adjusted to 10.0 with tetramethylammonium
hydroxide, and the $Mg(SCN)_2$ was omitted. Selectivity coefficients for $^{134}Cs^+$ and $^{86}Rb^+$ by
direct titration; the coefficients for $K^+$, $Na^+$, and $Li^+$ were calculated from their ability to
displace $^{86}Rb^+$ or $^{134}Cs^+$ from the ionophore complexes in the organic phase. In each case, the
affinity for the ion most favored for complexation was assigned a value of 100 (i.e., log = 2)
and the bars were drawn proportional to the logarithms of the relative affinities for each
particular cation. The absolute affinities are indicated by the two-phase dissociation constants
$K_D$ for $K^+$. The concentration of $K^+$ necessary to half-saturate the organically dissolved
ionophore has also been compared to the concentration of ionophore necessary to achieve a
half-maximal rate of release of engodenous $K^+$ from mitochondria (last column). [Pressman and
Haynes, 1969]

complexes are given, and the wide range of their values is obvious. For the sake of comparison, the intrinsic equilibrium affinities in the model system have been compared with the relative abilities of these same antibiotics to release endogenous potassium ions from mitochondria deenergized with a respiratory inhibitor (to prevent energy-linked potassium uptake). The range of antibiotic concentrations required to cause a half-maximum rate of potassium ions in the third column shows considerably less variation, the extreme range from X-206 to monensin being only fourteenfold. This indicates that antibiotics with intrinsically low cation affinities must have compensatory advantages in dynamic biological membrane test systems, presumably arising from favorable kinetic parameters of complex formation and dissociation.

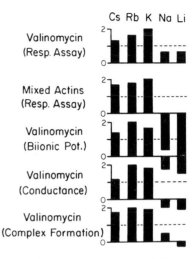

Figure 19. Comparison between different assays for measurement of ion selectivity of valinomycin. The respiratory assay depends on the ion-dependent stimulation of mitochondrial respiration; complex formation is measured as in Figure 18. Data for the biionic potentials and conductance were taken from Mueller and Rudin (1967). Data expressed in the same logarithmic format as Figure 18. [Pressman and Haynes, 1969]

Valinomycin-induced ionic selectivities were tested in the laboratory by determining equilibrium-complex formation in organic solvents, electrometric properties of the artificial lipid bilayers, and mitochondrial respiration. Data have been expressed logarithmically in Figure 19. Although the same relative cation selectivity patterns were obtained with all of the assay methods used, the absolute selectivity values were somewhat different for the various test systems.

Direct measurement was made of two-phase cation migration with the available alkali isotopes ($^{134}Cs^+$, $^{86}Rb^+$, $^{42}K^+$, $^{22}Na^+$); this did not permit the possibility of obtaining information about $Li^+$. The results, however, are a more accurate reflection of the true ionic affinities. These data are shown in Figure 20 for three ionophorous agents, dicyclohexyl-18-crown-6, alamethicin, and valinomycin. Note that the extreme $K^+$:$Na^+$ selectivity (about 10,000:1) corresponds to that obtained with the dynamic mitochondrial respiratory system (see Figure 19).

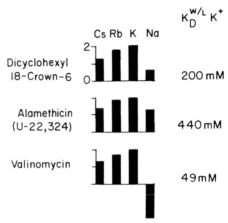

Figure 20. Ion specificities based on noncompetitive titration. The data (expressed in same logarithmic format as Figures 18 and 19) were obtained by direct saturation of the ionophore with $^{134}$Cs, $^{86}$Rb, $^{42}$K, or $^{22}$Na; no convenient isotope of Li is available. It should be noted that the K:Na selectivity ratio for valinomycin is considerably greater than that shown in Figure 18 and is based on the indirect displacement of labeled cation from ionophore complex. [Pressman and Haynes, 1969]

Finally, Pressman and his collaborators were able to demonstrate that antibiotics such as nigericin truly carry ions across a lipophilic phase, as shown in the following experiment:

A glass vessel with a septum across the top is filled with carbon tetrachloride so as to insulate two upper compartments; these are then filled with buffer solution (Figure 21). No transport of labeled rubidium ions from one of the upper compartments to the other can be detected until nigericin is added to the organic phase. After this addition, a constant rate of transfer of rubidium ions could be obtained. The nigericin-mediated bulk-phase transport is not influenced by the application of an electrical potential, nor could this antibiotic induce electrical conductivity across the organic phase. These experiments are consistent with nigericin's carrying ions across the organic phase as a complexed zwitterion devoid of any net charge.

Pressman concluded with an illustration of field-dependent ion transport in a bulk system (Figure 22). The organic phase for this experiment was carbon tetrachloride containing 10% nitrobenzene. The rubidium ions moved across the organic phase only during the interval when a potential of 45 V was applied. This is consistent with the fact that bi-ionic potentials can be obtained across bulk phases in a manner similar to that customarily obtained with black lipid membranes. The experiments also confirm that valinomycin can carry cations across a bulk phase as a charged complex, and that the term "ionophorous," or

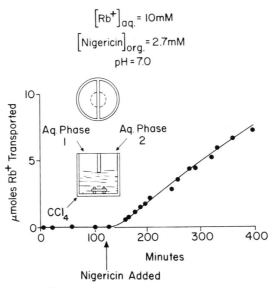

Figure 21. Transport of $^{86}Rb^+$ across a bulk phase of $CCl_4$ by nigericin. The glass reaction vessel was 3.5 cm tall x 2.4 cm i.d., with a sealed-in septum extending 1.5 cm below the top. The $^{86}Rb^+$ was placed in one aqueous compartment and its rate of passage to the opposing compartment through the $CCl_4$ measured before and after the addition of nigericin. [Pressman]

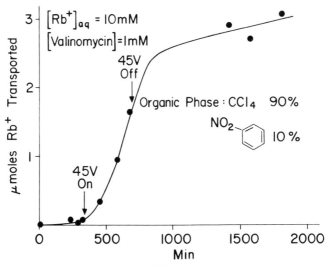

Figure 22. Voltage-dependent transport of $^{86}Rb^+$ by valinomycin. The system resembled that of Figure 21 except for the inclusion of 10% nitrobenzene in the organic phase, which was unstirred but subjected to a 45 V field between the aqueous compartments. The resistance in ohms was also determined for calculation of the transference number during the electrophoretic transport of $^{86}Rb^+$. [Pressman]

ion-bearing, is appropriate to both valinomycin-like and nigericin-like agents.

In order to obtain information on conformational changes of the antibiotic during complexation, proton NMR spectroscopy was applied. The spectra of valinomycin and nonactin in deuterated chloroform and their corresponding potassium thiocyanate complexes can be compared in Figure 23. The observed differences in chemical shifts for the different types of protons and the changes of coupling constants indicate radial conformational changes in the antibiotics after complexes were formed.

The initial broadening of certain resonances of the antibiotic upon titration with KCNS was used to obtain the rate constants for both complexation and decomplexation reactions. The upper limit for the exchange rate constant for the reaction

Ionophore + Ionophore* KCNS ⟷ Ionophore KCNS + Ionophore*

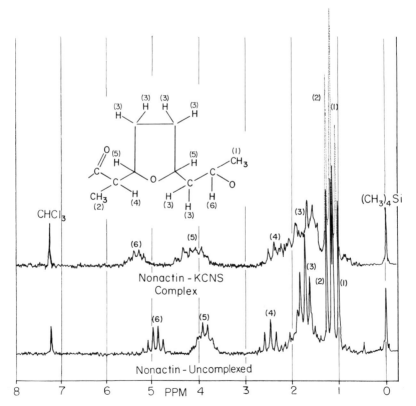

Figure 23. Alteration of conformation of ionophores in solution by complexation as detected by proton nuclear magnetic resonance (NMR) spectroscopy. A = nonactin; B = valinomycin. [A. Haynes, 1970; B. Pressman and Haynes, 1969]

is less than 25 $sec^{-1}M^{-1}$ for valinomycin in the nonpolar solvent deuterochloroform. A much faster exchange rate can be observed in the more strongly polar solvent composed of 80% methanol, 20% chloroform. The potassium ion was exchanged by the mechanism

$$\text{Ionophore} + K^{+} \underset{k_{off}}{\overset{k_{on}}{\longrightarrow}} \text{Ionophore} \cdot K^{+}$$

and the $k_{on}$ value was $\geqslant 1 \times 10^{6}$ $sec^{-1}M^{-1}$; the $k_{off}$ was $21 \pm 5$ $sec^{-1}$

The low ionophore-cation exchange rates in a nonpolar solvent are incompatible with a mechanism in which the cation migrates through the interior of a lipid membrane via a channel composed of several ionophore molecules. The picture completed by noting the high

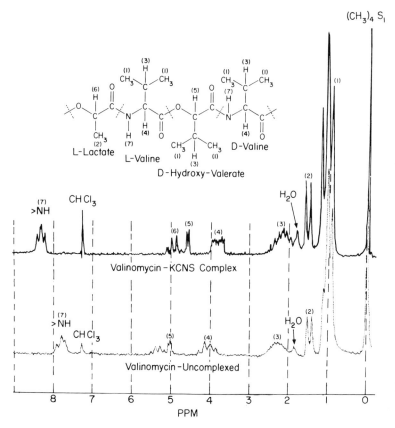

Figure 23B. Valinomycin.

exchange rates in a polar solvent is consistent with dynamic complexation-decomplexation at the membrane interfaces, while the low exchange rates in solvents of low polarity suggest that the cation traverses the membrane interior as a stable complex. Thus the NMR data obtained for bulk phases supports the mobile carrier mechanism for ionophore-mediated membrane transport.*

### ALKALI ION CARRIERS: SPECIFICITY, ARCHITECTURE, AND MECHANISMS:
#### An Essay by M. Eigen and R. Winkler

The "carrier" is an entity that facilitates a "dynamic" process, namely, the transport of some molecule or ion through a membrane. An understanding of its functional mechanism requires studies of its dynamic behavior. Where Simon (see Section II, Properties of Ion-Specific Carriers) dealt mainly with structural properties of carriers, the present contribution is concerned with that dynamic behavior from which the mechanism of uptake and release of ions by carrier molecules can be derived.

It may be helpful in understanding the behavior of alkali ions to give a short summary of what is known about the dynamics of ions in solution. Figure 24 contains a summary of characteristic rates of substitution of water molecules from the inner coordination shell of various metal ions (Eigen, 1963a; Eigen and Wilkins, 1965). This step turns out to be rate limiting for metal complex formation reactions. The presentation of such a periodic table of rate constants is meaningful only if the rates can be correlated specifically to the properties (i.e., the electronic structure) of the metal ion, independent of the nature of the incoming ligand.† This has indeed turned out to be so for most metal ions.

There are two facts to be deduced from Figure 24 that are of importance in the further discussion of alkali ion carriers, and, as will be seen, they hold true for the formation of alkali ion complexes where ligands have been substituted for all the water molecules of the hydration sphere.

*The NMR experiments were carried out in collaboration with D. H. Haynes and A. Kowalsky (Haynes et al., 1969).

†Ligand: a group, ion, or molecule coordinated to the central atom in a coordination complex.

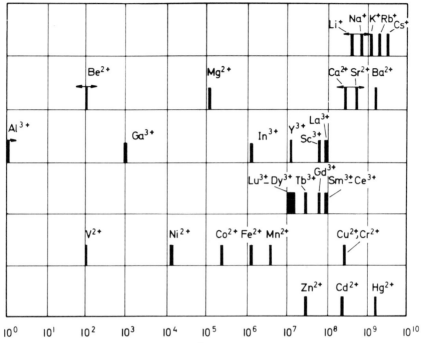

Figure 24. Characteristic rate constant ($sec^{-1}$) for substitution of inner sphere $H_2O$ of various aquo metal ions. [Eigen and Winkler]

1. Any specificity in rates not following a simple metal ion radius dependence can be found only in the transition metal ions, where, typically, chemical effects govern the substitution behavior (for example, in the nonmonotonic radius dependence of substitution rates of transition metal ions as shown in Figure 25) (Eigen, 1963b). In particular, the well-studied substitution processes for alkaline earth ions, irrespective of the ligand, always show high rate constants for $Ca^{2+}$ and more than three orders of magnitude lower values for $Mg^{2+}$ (Diebler et al., 1969).

2. All alkali ions are very fast in substituting single solvent molecules. The time constants are all in the neighborhood of $10^{-9}$ sec, with a slight radius dependence, i.e., rates decreasing from $Cs^+$ to $Li^+$ (Diebler et al., 1969).

The second finding especially may be surprising if it is correlated with the well-known solvation behavior of alkali ions. Figure 26 shows the free energy of solvation for the different alkali ions as a

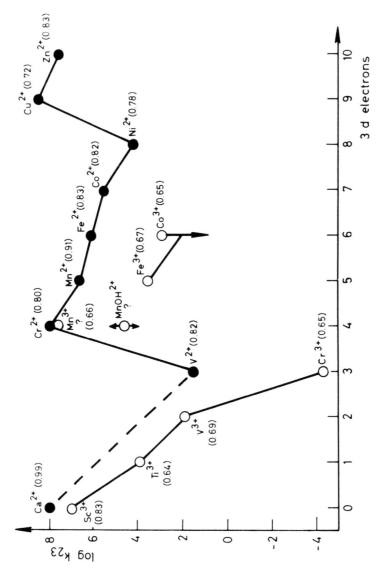

Figure 25. Radius dependence of substitution rate constant ($sec^{-1}$) of transition metal ions. [Diebler et al., 1969]

function of the ionic radius. The increase of the free energy of solvation with decreasing radius is monotonic, as can be expected from any simple electrostatic picture. What is surprising is the relatively small correlation between solvation energy and rate. Despite the high solvation energy values, substitution is an extremely rapid process involving only a few kcal/mole of free energy of activation. The two dotted curves in Figure 26 indicate standardized free energies to be expected for complexes with ligands that are either more or less tightly bound than solvent molecules. No specific carrier behavior can be deduced from such curves. The free energy of complex formation in the given solvent (here water) would be proportional to the differences between the broken and the solid line. For simple ligands, only monotonic behavior, i.e., no maximum at any intermediate metal ion radius could be expected. Thus the metal ion does not appear to have any specific property that would explain the specific behavior of a carrier. The metal ion specificity must therefore be the consequence of the peculiar property of the carrier molecule utilizing the differences in solvation energy for the different alkali ions.

Figure 27 shows how this type of behavior can be envisaged. The two upper curves represent the free energy of binding for two different chelating agents. They consist of multidentate ligands, which for complexation have to substitute for the entire solvation sphere of the metal ion in order to enclose that metal ion in a cavity. At large metal ion radius, the free energy of interaction will increase monotonically with decreasing metal ion radius (even relative to the solvent, provided that the ligand is favored with respect to the solvent molecule as expressed by the higher interaction energy). A decrease of the metal ion radius is accompanied by a shrinkage of the size of the cavity. Due to steric hindrance and repulsion between the binding groups of the carrier, the cavity soon will approach the minimum size that will optimally fit a given metal ion radius. A further decrease of the metal ion size will, then, not result in any appreciable increase of binding energy because the binding groups are "frozen" into fixed positions. Thus, the difference between ligand binding and solvation energy will pass through a maximum at a given size of the metal ion.

We must therefore conclude that binding specificity is a consequence of the specific architecture of the carrier, which utilizes the difference between free energy of solvation and ligand binding. This difference involves appreciable entropy increments in favor of the chelating ligand. The question remains: How is the process facilitated

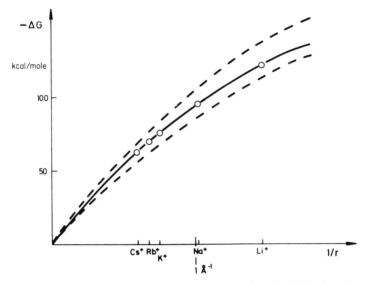

Figure 26. Free energy of hypothetical ligand binding (– – –) and solvation (–o–) experimental value) as function of the reciprocal radius of alkali ions. (The ligand binding curve is related to a fixed ligand concentration.) [Eigen and Winkler]

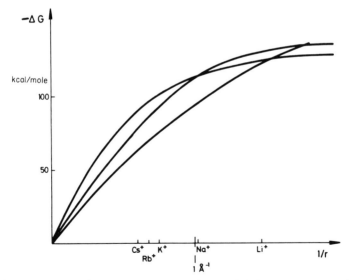

Figure 27. Free energy of solvation (lower curve) and chelation (upper curves for two hypothetical cases) as function of the reciprocal radius of alkali metal ions. [Eigen and Winkler]

quickly? Even though one single solvent molecule is to be substituted rapidly, *total* desolvation always requires quite high free energies of activation. On the other hand, as will be seen, successful carrier action depends on quick uptake and release of the metal ion.

Equilibrium and rate studies have been carried out with sodium and monactin as a model system. (Macrotetrolide monactin has already been described by Simon in the preceding pages.) Several problems had to be solved to make such measurements possible:

1. A specific indicator for alkali ions in methanol had to be found. (Because lipophilic carrier molecules are not sufficiently soluble in water, measurements had to be carried out with methanol as solvent.)

2. Relaxation methods had to be adjusted to the particular reaction system in methanol. Amplitudes of relaxation provide the information about equilibrium parameters such as stability constant, reaction enthalpy, etc., whereas relaxation times yield the rate constants for uptake and release of the metal ion.

Murexide, the ammonium salt of purpuric acid (Figure 28), turned out to be an ideal indicator for alkali ions in methanol (see Schwarzenbach and Gysling, 1949, and Winkler, 1969). A titration curve is shown in Figure 29, demonstrating the characteristic color change upon addition of sodium ions. Figure 30 shows that the addition of monactin results in a detectable absorption change. Murexide can therefore be used as an indicator for the reactions of alkali ions with carriers. It is ideal for the present system, because (a) the stability constant of the $Na^+$-murexide is in a range most suitable for competition with the carrier (with "half-binding" at concentrations between $10^{-3}$ and $10^{-4}$ M), and (b) indication occurs very rapidly. The reaction of murexide with $Na^+$ is almost diffusion controlled (Winkler, 1969), as found from relaxation studies using an electrical traveling wave technique developed by Ilgenfritz (1966). The relaxation time is around 100 nsec. Table 4 summarizes the properties of murexide. A detailed description of the measurements is found in Winkler, 1969.*

A special technique utilizing differences in temperature jump amplitude has been worked out. This procedure allowed a simultaneous determination of stability constants and reaction enthalpies for the metal ion carrier complex. The technique is quite precise and of special

*See also Winkler and Eigen, 1970, unpublished observations, and Diebler et al., 1969.

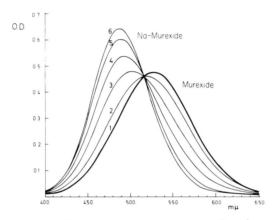

Figure 28. Chemical structure of the murexide anion. [Schwarzenbach and Gysling, 1949]

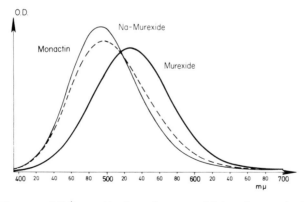

Figure 29. Spectrophotometric titration of murexide with $Na^+$ ($25^\circ C$; $c_{Mu} = 4 \times 10^{-5}$ M). [Eigen and Winkler]

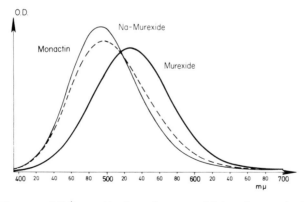

Figure 30. Decrease of $Na^+$-murexide absorption upon addition of monactin ($---$). [Eigen and Winkler]

TABLE 4

Stability and Rate Constants of Alkali-Murexide Complexes in MeOH [Simon]

| 25°C | $k_R$ [$M^{-1}$ $sec^{-1}$] | $k_D$ [$sec^{-1}$] | $K$ [$M^{-1}$] |
|---|---|---|---|
| $Li^+$ | $5.5 \times 10^9$ | $7.7 \times 10^6$ | $7.1 \times 10^2$ |
| $Na^+$ | $1.5 \times 10^{10}$ | $5.9 \times 10^6$ | $2.55 \times 10^3$ |
| $K^+$ | $\sim 2 \times 10^{10}*$ | $\geq 10^7$ | $1.1 \times 10^3$ |

*Diffusion controlled.

advantage in the study of biochemical reactions involving substances that tend to denature. It is described in more detail elsewhere (Winkler, 1969).

Using this technique in combination with other relaxation techniques, such as E-field pulse (Ilgenfritz, 1966) and sound absorption (Eggers, 1968), for the determination of relaxation times, the following results were obtained for the sodium-monactin system:

Dissociation constant
of the $Na^+$ monactin complex:     $K = 2 \cdot 10^{-3}$ [M]
Reaction enthalpy:                  $\Delta H \approx -6$ [kcal/mole]
Complex formation rate constant:    $k_R = 3 \cdot 10^8$ [$M^{-1} sec^{-1}$]
Lifetime of complex:                $1/k_D = 1.5$ [$\mu sec$]

The most surprising result is the high value for the rate constant of complex formation. A detailed analysis (Winkler, 1969) shows that each solvent molecule of the $Na^+$ solvation shell is substituted within less than $10^{-9}$ sec, in agreement with the data shown in Figure 24 for single ligand substitution of alkali ions. The mechanism must be a stepwise substitution process in which each solvent molecule in the coordination shell of the alkali ion is replaced stepwise by the polar groups of the carrier, keeping the total coordination number essentially constant. Such a mechanism requires quite an open form of the carrier molecule in which all the polar groups are easily accessible to the solvated alkali ion. Thus the uptake of the alkali ion is accompanied by a conformational change of the carrier from an open ring to a closed sphere with the cavity containing the metal ion inside, well shielded by hydrophobic groups. The twisted tennis ball seam structure allows the existence of the two alternate conformations, depending on the compensation of the negative charge of the polar groups by the metal ion.

Four rules for the design of a carrier follow from these mechanistic studies:

1. The carrier molecule should possess electrophilic groups that are able to compete with the solvent molecules for metal ion binding. These groups should be located inside an otherwise lipophilic structure that dissolves easily in membranes.

2. As many solvent molecules of the inner coordination sphere as possible should be replaced by the coordinating sites of the carrier molecule. For two ions of different sizes, the reference state may then involve as much as the total difference of free energy of solvation.

3. The ligand should form a cavity adapted to the size of the metal ion. Optimal fit is seen as an arrangement where there is maximal difference between the free energies of ligand binding and of solvation. The optimal fit frequently almost coincides with "fittest" geometrical arrangement. Cavity formation involves ligand-ligand repulsion as well as steric fixation of the chelate.

4. The carrier molecule should possess sufficient flexibility in order to allow for stepwise substitution of the solvent molecules. Otherwise—that is, if complete or substantial desolvation were required for the ion to slip into the cavity—the activation barrier would be rather high, and the reaction rate quite low.

Rule 1 fulfills the biological requirement, i.e., to gate the ion through a (lipid) membrane. Rules 2 and 3 take care of a high selectivity, and rule 4 allows for fast loading and unloading of the carrier. Not too many low-molecular-weight structures are known that would simultaneously conform to all four rules. Almost all of the classical complexing agents are poor in one respect or another.

The high rate of complex formation is of great significance for such carriers in biological membranes. The carrier can be selective only if the overall transport rate is not limited by the rate of metal ion release. If the latter process were rate limiting, the high selectivity reflected by a high binding constant would be compensated by the slower rate of release. (The ratio of rate constants for uptake and release yield the binding constant.) For $Na^+$-monactin the time constant for release lies in the $\mu$sec range; for the more selectively bound $K^+$ ion it would reach almost the msec range. The release times would be appreciably longer if the recombination process did not have a rate constant as high as $\sim 10^8 (M^{-1} sec^{-1})$. It turns out that the carrier can play its selective role in alkali ion transport across molecular bilayer membranes only due to its high—almost diffusion-controlled—recombination rate.

## III. CARRIER-FACILITATED TRANSPORT

### MACROCYCLIC COMPOUNDS AND IONIC MOVEMENT
### THROUGH LIPID MEMBRANES: D. C. Tosteson

Tosteson discussed three main aspects of ion transport across thin lipid bilayers (formed from lipids extracted from high-potassium sheep red-cell lipids dissolved in decane) separating two aqueous phases and in the presence of macrocyclic compounds (a combination of 70% monactin, 30% dinactin).

The first aspect of these experiments bears on the question of ion movement across bilayers—whether the ions form lipid-soluble complexes with macrocyclic compounds which then diffuse independently through the membrane, or whether they diffuse in single file through the central hole of a stack of several macrocyclic compounds.

The second aspect is connected with his investigations on the selectivity of macrocyclic compounds in the transport of ions through the bilayer when there is a net movement of ions through the membrane. These tests were made with isotopes to measure ionic current fluxes under different experimental conditions. When the net ionic flux of $K^+$ was measured in the presence of an electrical potential difference (but with no difference in $K^+$ concentration between the bathing solutions), it was found that only about half of the current was carried by potassium ions.

Third, Tosteson described experiments directed towards defining the rate-limiting step for ion transport through membranes in the presence of macrocyclic compounds. The considerations here were three main processes: (1) the phase transition at one surface, (2) the diffusion of ions generally in the form of complexes through the aliphatic chains in the interior of the membrane, and (3) the movement across the opposite phase boundary. The experimental evidence is based upon measurements of (1) the electrical properties of thick lipid membranes and (2) diffusion of ions in solutions of red-cell lipids in decane (Tosteson, 1968; Tosteson et al., 1968b).

An analysis of the mechanism (independent versus single file) of ionic movement across the thin lipid membrane was made by measuring the monactin-dinactin-induced potassium fluxes in the presence of a tenfold ratio of KCl concentrations and at zero electrical potential difference. (The membrane voltage of this system at zero current is

−58 mV.) (See Table 5a.) The bilayer system under these conditions behaves as if it were permeable only to potassium ions. The total membrane current actually measured was $1.3 \times 10^{-5}$ amp/cm$^2$, while that calculated from the net potassium flux was $1.6 \times 10^{-5}$ amp/cm$^2$. Under these conditions, the transference number for potassium is unity within the limits of the experimental error. The flux ratio is 10, which is the value expected of K$^+$ complexes moving across thin lipid membranes independently.

For single-file diffusion, Tosteson pointed out that the flux ratio should be 10 raised to some power greater than 1, depending upon the number of ions that might be lined up in a stack of carrier molecules (Hodgkin and Keynes, 1955; Tosteson et al., 1968a). The flux ratio for K$^+$ in sheep red cells exposed to valinomycin was also found to equal the electrochemical activity for K$^+$ and thus be consistent with the hypothesis that K$^+$-valinomycin complexes also move independently through biological membranes (Tosteson et al., 1968b).

If one imposes the potential in the direction opposite to the KCl concentration gradient (passing the current from the compartment with the $10^{-2}$ M KCl to the adjacent one with $10^{-1}$ M KCl), an interesting effect is observed. The transference number of potassium falls to a value of 0.4 (Table 5a), and more than half the current is carried by some ion other than potassium. Even smaller values for the transference number of potassium resulted in those experiments where the concentration of KCl was equal on both sides of the bilayer and ionic movement was driven exclusively by a voltage gradient (Table 5b).

The most obvious possibility for the nature of the other ion would be that anions move across the membrane. But this could also be ruled out, because the replacement of KCl by KBr or K$_2$SO$_4$ leads to the same main observation (Table 5c). The transference numbers of these anions are negligible compared to the value for potassium; under these conditions, no net movement of anions was observed in the presence of a difference in electric potential.

The question then to consider is whether hydrogen or hydroxyl ions could account for the difference between total current and the potassium current; Tosteson therefore presented some experiments with the KCl concentration equal in both compartments, but with the usual pH 5.7 in one instance and, in the other instance, pH 8 on both sides of the membrane (Table 5d). This increase of pH from 5.7 to 8 brought the transference number for potassium almost up to unity. Tosteson therefore concluded tentatively that in the presence of

TABLE 5a

(a) Monactin-Dinactin-Induced $K^+$ Fluxes in the Presence of a KCl Concentration Gradient and an Electrical Potential Difference

| KCl Concentration M | | $V_m$ mV | $K^+$ Fluxes pmoles/cm$^2$ sec | | $I_m^*$ $10^{-5}$ amp/cm$^2$ | $I_k^†$ | $tK^‡$ | Flux Ratio | Theoretical Flux Ratio |
|---|---|---|---|---|---|---|---|---|---|
| Back | Front | | $bf_M$ | $fb_M$ | | | | | |
| $10^{-1}$ | $10^{-2}$ | 0 | 198 ±33 | 22 ±6 | 1.3 ±0.5 | 1.6 ±0.3 | 1.2 ±0.3 | 9.0 ±2 | 10 |
| $10^{-1}$ | $10^{-2}$ | +30 | 246 ±22 | 7.5 ±0.8 | 2.4 ±0.2 | 2.4 ±0.2 | 1.0 ±0.1 | 33 +5 | 33 |
| $10^{-1}$ | $10^{-2}$ | −150 | 11.1 +2 | 98 ±12 | 1.9 ±0.1 | 0.8 ±0.1 | 0.4 ±0.1 | 0.11 ±0.02 | 0.03 |

(b) MD-Induced $K^+$ Fluxes at Equal KCl Concentrations in the Presence of an Electrical Potential Difference

| Back | Front | | $bf_M$ | $fb_M$ | | | | | |
|---|---|---|---|---|---|---|---|---|---|
| $10^{-3}$ | $10^{-3}$ | +60 | 33 ±1.2 | 0.9 ±0.5 | 0.19 ±0.1 | 0.03 ±0.01 | 0.14 ±0.1 | 3.8 ±1.9 | 10 |
| $10^{-2}$ | $10^{-2}$ | +60 | 82 ±12 | 12 ±2 | 1.6 ±0.2 | 0.04 ±0.1 | 0.4 ±0.1 | 6.8 ±1.3 | 10 |
| $10^{-1}$ | $10^{-1}$ | +60 | 244 ±29 | 66 ±9 | 6.1 ±0.2 | 1.80 ±0.03 | 0.3 ±0.02 | 4.0 ±0.7 | 10 |

*Direct Electrical Measurement
†Calculation from Net $K^+$ Flux
‡$K^+$ Transference Number = $I_k/I_m$

Note: Tables 5a-5e [Tosteson]

TABLE 5

(c) MD-Induced $K^+$ and $Br^-$ Fluxes at Equal Concentrations in the Presence of an Electrical Potential Difference

| Ion Concentration M | $V_m$ mV | Fluxes pmoles/cm² sec | | $I_m$* | $I_k$† | $tK$‡ | Flux Ratio |
|---|---|---|---|---|---|---|---|
| | | $bf_M$ | $fb_M$ | | | | |
| $K^+ = 10^{-1}$ | +60 | 125 ±16 | 56 ±14 | 1.7 ±0.3 | 0.70 ±0.2 | 0.4 ±0.1 | 2.0 ±0.6 |
| $Br^- = 10^{-1}$ | +60 | 61 ±24 | 67 ±21 | 2.4 ±1.0 | 0.06 ±0.2 | 0.02 ±0.07 | 1.1 ±0.5 |
| $K^+ = 10^{-1}$ | +60 | 22 ±6 | 9 ±2 | 0.45 ±0.2 | 0.14 ±0.04 | 0.3 ±0.1 | 2.5 ±0.7 |
| $SO_4^= = 5 \times 10^{-2}$ | +60 | 1.23 ±0.3 | 1.25 ±0.3 | 1.6 ±0.2 | $0.2 \times 10^{-3}$ ±$4 \times 10^{-3}$ | $10^{-4}$ ±$10^{-2}$ | 1.0 ±0.3 |

*Direct Electrical Measurement
†Calculated from $K^+$, $Br^-$, or $SO_4$ Fluxes
‡Transference Number for $K^+$, $Br^-$, or $SO_4$

Note: Tables 5a-5e [Tosteson]

## TABLE 5

**(d) The Effect of pH on MD-Induced K⁺ Fluxes at Equal KCl Concentrations in the Presence of an Electrical Potential Difference**

| | KCl Concentration M Back | Front | $V_m$ mV | K⁺ Fluxes pmoles/cm² sec $bf_M$ | $fb_M$ | $I_m$* | $I_k$† | $tK$‡ | Flux Ratio |
|---|---|---|---|---|---|---|---|---|---|
| pH | $10^{-1}$ 5.7 | $10^{-1}$ 5.7 | +60 | 244 ±29 | 66 ±9 | 6.1 ±0.2 | 1.80 ±0.03 | 0.3 ±0.02 | 4.0 ±0.7 |
| pH | $10^{-1}$ 8.2 | $10^{-1}$ 5.0 | +60 | 242 ±50 | 42 ±5 | 2.2 ±0.5 | 1.9 ±0.2 | 0.8 ±0.2 | 6.0 ±2 |
| pH | $10^{-1}$ 8.0 | $10^{-1}$ 8.0 | +60 | 536 ±74 | 140 ±25 | 4.4 ±0.4 | 4.0 ±0.9 | 0.9 ±0.2 | 4.0 ±0.9 |

**(e) The Effect of pH on MD-Induced K⁺ Fluxes in the Presence of a KCl Concentration Gradient**

| | Back | Front | $V_m$ | $bf_M$ | $fb_M$ | $I_m$ | $I_k$ | $tK$ | Flux Ratio |
|---|---|---|---|---|---|---|---|---|---|
| pH | $10^{-1}$ 5.7 | $10^{-2}$ 5.7 | 0 | 198 ±33 | 22 ±6 | 1.3 ±0.5 | 1.6 ±0.3 | 1.2 ±0.3 | 9.0 ±2 |
| pH | $10^{-1}$ 5.0 | $10^{-2}$ 8.2 | 0 | 386 ±18 | 46 ±6 | 5.4 ±0.3 | 3.3 ±0.2 | 0.6 ±0.06 | 8.0 ±1 |

*Direct Electrical Measurement
†Calculated from K⁺ Fluxes
‡K⁺ Transference Number = $I_k/I_m$

Note: Tables 5a-5e [Tosteson]

monactin-dinactin, when no current is passing, the bilayer behaves as a
perfect potassium electrode; but when a potential is imposed across the
membrane, a very substantial fraction of the current is carried by a
pH-dependent component.

In order to explain these very interesting effects, Tosteson
suggested the following possibilities. In the presence of the macrocyclic
compounds, one must expect a certain concentration of potassium
complexes within the membrane. In the absence of a potential difference across the membrane, the distribution of these complexes will be
symmetrical. When a potential is imposed across the bilayer, which
separates two aqueous solutions of equal salt concentration, the
complexes can move to the negative side of the membrane by taking up
a position at the interface and producing an asymmetrical distribution
across the bilayer. Thus the charged complexes would reduce the
effective negative charge referable to the phosphate groups of the
structurally fixed phospholipids on the negative side of the bilayer.
Because of this asymmetrical distribution of the macrocyclic-
compound-ion complexes induced by the applied potential, one must
expect a perturbation of the equilibrium partition of the mobile,
positive counterions of the negatively charged lipid phosphate groups,
for example, the potassium and hydrogen ions in the aqueous solutions
at both interfaces. This effect could produce a change of the surface pH
in the direction of increasing the hydrogen ion concentration on the
positive side of the membrane.

Some effort has been made to test these arguments by experiments in which the pH was changed on only one side of the membrane.
An increase of pH on the positive side of the membrane that separates
KCl solutions of equal concentration is sufficient to bring the transference number of potassium close to unity (Table 5d). On the other
hand, with a tenfold KCl concentration gradient, an increase of pH on
the dilute side of the bilayer reduces the transference number from 1.2
(as it had been with equal pH on both sides) to a value of 0.6 (Table
5e). Tosteson suggests that these effects may be due to the mobile
charges in the bilayer that are directly related to the macrocyclic
compound-ion complexes. But the question of whether or not the
proton transport actually is responsible for the discrepancy between
total current and potassium current still remains unanswered.

Katchalsky, in the discussion, pointed out that under these
circumstances, if one compares the very low hydrogen ion concentration ($\sim 10^{-6}$ M) with the high potassium ion concentration ($10^{-1}$ M),

the ion conductance within the membrane would have to be approximately $10^5$ times higher than the corresponding conductance of potassium ions to produce any real competition between $H^+$ and $K^+$. Eigen also recalled that the potassium ion bound to big carrier molecules could be quite immobile in the membrane; on the other hand, high mobility of the protons within the membrane cannot be excluded as a consideration.

Apropos of these remarks, Tosteson recalled an interesting and still unsolved effect produced by valinomycin: if a bilayer separates two NaCl solutions of equal concentration, a stable potential difference can be observed after the addition of valinomycin on only one side of the membrane (Andreoli et al., 1967).

He next turned to the question of location of the rate-limiting step for the diffusion of cations across bilayers. Does this step take place at the interface, or in the core of the membrane? Experiments have been conducted with thick, complex, artificial membrane systems formed from sheep red-cell lipids dissolved in decane and placed between two cellophane membranes separated by a polyethylene spacer (Tosteson, 1968; Tosteson et al., 1968a; Andreoli and Tosteson, 1971). The results of a series of measurements of membrane resistance showed that in the absence of red-cell lipids, pure decane has a resistance in this system of approximately $10^{15}$ $\Omega$ cm$^2$. The addition of only a small amount of lipid (less than 0.5 mg/ml of decane) produces a great fall in membrane resistance to less than $10^{10}$ $\Omega$ cm$^2$. Further addition of red-cell lipid produces relatively little further reduction in membrane resistance. If phospholipids function in decane as an electrolyte does in a solvent, the conductance of the decane phase should be proportional to the concentration of phospholipid, and the resistance should be proportional to the reciprocal of the phospholipid concentration (Figure 31). However, the extrapolated value of the resistance of this thick membrane in the presence of an infinite concentration of lipids is not zero, as would be expected for an ideal salt solution, but rather about $10^8$ $\Omega$ cm$^2$ when expressed as a surface resistivity. If the hypothesis is correct that the deviation of the extrapolated resistance from zero is referable to the special properties of the decane-water interfaces, the resistance of these thick membranes should be proportional to the thickness of the system. Figure 32 shows that this result is observed, and the slope of the line relating thickness to resistance depends on the lipid concentration. However, the extrapolated value of membrane resistance at zero membrane thickness is independent of

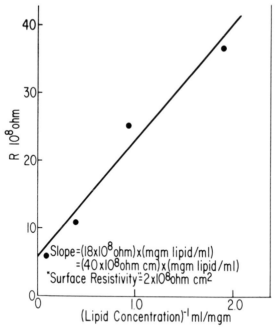

Figure 31. Effect of lipid concentration on electrical resistance of thick lipid membranes. [Tosteson]

lipid concentration, and again it is equal to about $10^8 \, \Omega \, cm^2$ when expressed in terms of surface resistivity. These results are consistent with the view that the electrical resistance of thick lipid membranes consists of two resistances arranged in series. At the interfaces bordering the two sides of the thick membrane, there are surface resistivities that are presumably dependent on the special features of these border interfaces.

All of the surface resistances are separated by the bulk resistance of the decane-phospholipid phase and are approximately $10^8 \, \Omega \, cm^2$, remarkably similar to the resistance of thin lipid bilayer membranes. It would appear that bilayer membranes have electrical resistance properties characteristic of two monolayers arranged in series and relatively independent of the thickness of the bulk lipid phase separating them.

Having defined the electrical properties of thick lipid membranes, it is now possible to ask whether the macrocyclic compounds affect the surface resistivity primarily, or the resistivity of the bulk phase separating the monolayers. The results of such an investigation

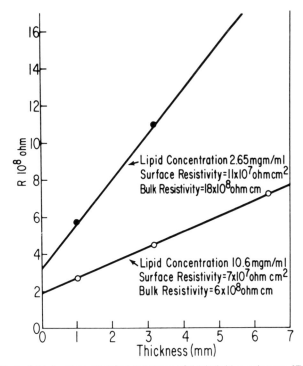

Figure 32. Effect of thickness on electrical resistance of thick lipid membranes. [Tosteson]

for monactin-dinactin are shown in Figure 33. The intercepts of the plot of the membrane resistance as a function of its thickness is markedly reduced by monactin, whereas the slope of the line is not appreciably affected by its presence (Tosteson et al., 1968a; Andreoli and Tosteson, 1971). Similar results have been obtained with valinomycin, and the selectivity for sodium and potassium ions of these thick artificial membranes in the presence of the macrocyclic compounds was also confirmed. From these effects, Tosteson concluded that, in this system, these compounds act primarily to reduce surface resistivity. This effect could be due to catalysis of the transition of ions between aqueous and nonaqueous phases or to the increase in concentration of ions (presumably as complexes with macrocyclic compounds) in the region of the hydrocarbon chains of phospholipids oriented at, and normal to, the interface.

In order to decide between these two possible (not mutually exclusive) explanations for the effect of macrocyclic compounds on alkali metal ion transport across thin lipid bilayer membranes or at the

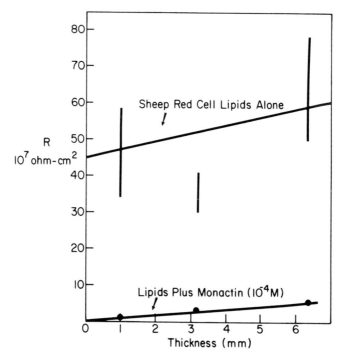

Figure 33. Effect of monactin on electrical resistance of thick lipid membranes. [Tosteson]

surface of thick lipid membranes, it is important to know the concentration and mobility both of free ions and of complexes in the aliphatic chain regions of the membranes.

Returning to his first approach to this problem, namely, attempts to estimate mobilities, Tosteson described some experiments for the determination of the self-diffusion coefficient of aqueous and nonaqueous potassium decane-phospholipid phases: self-diffusion coefficients were measured by the capillary method of Wang (1951). A fluid (either water or lipid dissolved in decane) containing $^{42}K$ was placed in a capillary open at one end. The capillary was suspended vertically (open end at the top) in a large volume of well-stirred fluid identical to that in the capillary except that it contained $^{39}K$. For example, aliquots of lipid-decane solutions were equilibrated by prolonged shaking with 10 mM $^{42}KCl$ and 10 mM $^{39}KCl$. The radioactive lipid-decane mixture was then placed in the capillary, while the nonradioactive lipid-decane solution was placed outside the capillary; the self-diffusion coefficient was calculated from measurements of the rate of appearance of $^{42}$K in the fluid outside the capillary. The results given in Table 6 show that the self-diffusion coefficient for K$^+$ in 1.0 mg/ml lipid in decane was about 1/6 that for K$^+$ in water in the presence or absence of valinomycin. Similar results were obtained when the valinomycin was replaced with monactin-dinactin.

From the data in Table 6, it is possible to compute the K$^+$ flux across a bilayer in the absence and presence of valinomycin, assuming that the concentration and the self-diffusion coefficient for K$^+$ in the membrane are the same as those observed in lipid-decane solutions that have been equilibrated with aqueous solutions of KCl. Thus, for a membrane 100 A thick, the unidirectional flux of K$^+$ when the bathing solutions are 10 mM KCl should be about $7 \times 10^{-8}$ moles/cm$^2$ x sec in the absence and $3 \times 10^{-7}$ moles/cm$^2$ x sec in the presence of valinomycin. The data in Table 6 show that the observed fluxes across bilayers separating identical aqueous phases containing 10 mM KCl and $10^{-6}$ M valinomycin are at least 2 orders of magnitude less. The discrepancy is several times greater in the absence of a macrocyclic compound. Thus, if the rate-limiting step in K$^+$ movement across the bilayer is diffusion in the interior of the membrane, either concentration or mobility (or both) are much reduced compared to the values observed in dilute lipid-decane solutions. (The bilayers are usually formed from solutions containing 10 to 30 mg/ml lipid in decane.) Furthermore, because macrocyclic compounds produce a change of several orders of magnitude in the K$^+$ permeability of the bilayer, it is necessary to assume that they have a different effect on the concentration of K$^+$ in that system as compared with bulk systems of the type

TABLE 6

Self-Diffusion Coefficients of K$^+$ in H$_2$O and Decane at 23°C [Tosteson]

| Solvent | Lipid (mg/ml) | Valinomycin (M) | K$^+$ (M) | D ($10^{-6}$ cm$^2$/sec) |
|---|---|---|---|---|
| H$_2$O | 0 | 0 | $10^{-1}$ | 35 |
| H$_2$O | 0 | $10^{-6}$ | $10^{-1}$ | 41 |
| Decane | 1.0 | 0 | $1.4 \times 10^{-5}$ | 6.5 |
| Decane | 1.0 | $10^{-5}$ | $4.5 \times 10^{-5}$ | 9.6 |
| Decane | 0.1 | 0 | $0.7 \times 10^{-5}$ | 0.87 |
| Decane | 0.1 | $10^{-5}$ | $1.4 \times 10^{-5}$ | 0.88 |

shown in Table 6; in the latter case, valinomycin increases the product of mobility and concentration by no more than a factor of 10. This may be due to the fact that the lipid molecules in the bilayer have restricted mobility and thus produce a membrane interior that is made up exclusively of aliphatic chains and is free of phospholipid polar groups. Alternatively, it is possible that the rate-limiting step for ion penetration through bilayers is not by diffusion through the hydro-carbon interior but rather by the transition of ions through the interface between water and nonpolar "phases." If the latter hypothesis is correct, the macrocyclic compounds could function to catalyze the phase transition. Tosteson feels that further work is needed to resolve this point.

## ANTIBIOTIC EFFECTS ON ION DISTRIBUTION:
### G. Eisenman

Eisenman summarized work that has been carried out in collaboration with Ciani and Szabo (Eisenman et al., 1968,1969; Ciani et al., 1969; Szabo et al., 1969,1970), as follows:

To develop a quantitative theoretical treatment for the effects of neutral macrocyclic antibiotics on the electrical properties of phospholipid bilayer membranes, Eisenman and his colleagues started from the known ability of such molecules to form stoichiometric lipid-soluble complexes with cations and deduced the electrical proper-ties that a simple organic solvent phase would have to have if it were made into a membrane of the same thinness as the phospholipid bilayer. In essence, they postulated that the primary barrier to ion movement across a bilayer membrane is its quasi-liquid hydrocarbon interior and that the neutral macrocyclic antibiotics bind monovalent cations and solubilize them in the membrane as mobile, positively charged complexes. Using the Poisson-Boltzmann equation to describe the equilibrium profile of the electrical potential, they have shown (Ciani et al., 1969; Appendix to Eisenman et al., 1968) that an excess of the positive complexes over all the other ions is expected (1) as a net space charge for appropriate conditions of membrane thickness and (2) as values of the partition coefficients of the various ionic species. This concept facilitates certain integrations of the flux equations. Describing the fluxes of these complexes by the Nernst-Planck equation, and neglecting the contribution of uncomplexed ions to the electric current, theoretical expressions are derived for the membrane potential in ionic mixtures and for the limiting value of the membrane conductance at zero current when the membrane is interposed between identical solutions. The expressions are given in terms of the ionic activities and antibiotic concentrations in the aqueous solutions so as to make the data accessible to direct experimental test (Ciani et al., 1969; Eisenman et al., 1969; Szabo et al., 1969).

This approach, which may at first glance seem oversimplified, is reasonable in view of the following facts:

1. The hydrocarbon tails in the interior of the membrane are quasi-liquid, and the interior of artificial bilayers contains significant amounts of solvents such as decane.

2. The presence of the charged polar-head groups of the lipid can be shown theoretically to be unimportant over a significant range of experimental conditions (Szabo et al., 1970).*

3. The rate of formation and dissociation of the complexes in aqueous solutions is so rapid that these processes are unlikely to be rate limiting in the aqueous phase (Diebler et al., 1969).

The possibility that such complexes, once formed in the membrane interior, may not dissociate within the lifetime of diffusion across the membrane, does not alter the expectations deduced in the present section.

Using no arbitrary assumptions as to electroneutrality or as to profiles of concentration or electric potential within the membrane, the work of Eisenman, Ciani, and Szabo (1968) first presented a theoretical analysis of the effects of neutral macrocyclic molecules on the electrical properties of a simple model in which the phospholipid bilayer membrane is represented as a thin liquid hydrocarbon phase, some 60 A thick, interposed between two aqueous solutions. Expressions were then deduced for the membrane potential and membrane resistance at zero current as a function of the concentrations of antibiotic and ions in the aqueous solutions. In addition, quantitative interrelationships between such properties as membrane potential and electric resistance were predicted.

An expression for the membrane potential in ionic mixtures can be developed in terms of the aqueous concentrations of ions and of

*Note: Subsequent to the Work Session, Eisenman and his colleagues have successfully extended their analysis to take the effects of the charged polar-head groups explicitly into account (McLaughlin et al., 1970).

antibiotic. Under usual conditions, this equation is identical in form to the Goldman-Hodgkin-Katz equation (see Hodgkin and Katz, 1949), with the permeability ratio representing combinations of such membrane parameters as the mobilities of the complexed cations, their partition coefficients, and the formation constants of the complexes in aqueous solution.

An equation for membrane conductance in the limit of zero current has also been derived for a membrane exposed on both sides to the same solution; this indicates that the membrane conductance is expected to be directly proportional to the total concentration of cations for dilute solutions. Membrane conductance is found to depend on the same parameters as did membrane potential. Indeed, it is shown that the ratio of conductance measured in single NaCl solutions for two different cations should be identical to their permeability ratio, thereby providing an immediate test application of the theoretical treatment. (Incidentally, this identity between permeability and conductance ratios is also expected for systems obeying the "Independence Principle" of Hodgkin and Huxley.)

If the overall size of the complex is approximately the same regardless of the particular cation bound to it (as is likely for the class of macrocyclic antibiotics), the mobilities of the complexes will be the same for all cations. In this event, the permeability and conductance ratios are expected to depend only on equilibrium selectivity parameters, which are shown to be measurable independently by the equilibrium extraction of appropriate salts into an appropriate bulk organic solvent phase (Eisenman et al., 1969). The comparison of membrane electrical properties with appropriate salt extraction equilibria provides a means for distinguishing neutral carriers from neutral "tunnels."

Eisenman, Ciani, and Szabo (1969) examined the equilibrium chemistry of macrotetrolide actin molecules and showed how the salt-extraction properties conferred by these molecules on bulk organic solvent phases are related to electrical properties that are measurable for phospholipid bilayer membranes. The effects of such molecules on the equilibria of ionic distribution between aqueous solutions and organic solvents were first deduced theoretically, then measured experimentally, and an appropriate set of equilibrium constants was characterized for these antibiotics. A variety of effects on bilayer membranes could be predicted within this framework. Measurements were made in spectrophotometric studies of the equilibrium extraction of picrates and dinitrophenolates of Li, Na, K, Rb, Cs, and $NH_4$ in hexane, in dichloromethane, and in hexane-dichloromethane mixtures; a wide range of ionic and antibiotic concentrations was used. The results indicate that the simple chemistry postulated for the model is indeed characteristic of nonactin, monactin, dinactin, and trinactin. The equilibrium constant for the extraction of each cation by a given macrotetrolide actin antibiotic was measurable with sufficient precision to yield meaningful differences due to the varied number of methyl groups in this series. The ratios among the various cations observed to be virtually independent of the solvent and chromophore anion were found to be selectively characteristic of given antibiotics.

In 1969, Szabo, Eisenman, and Ciani also characterized the observed effects of the macrotetrolide actin antibiotics on the electrical properties of phospholipid bilayer membranes and compared these

TABLE 7

The Correspondence Between Bilayer Membrane and Salt Extraction Parameters*
[Szabo et al., 1969]

| Macro-tetrolide | Ion | $K_i$ | $G_0(I)$ (x 2.08 x $10^6$) | $K_i/K_{Rb}$ | $P_i/P_{Rb}$ | $G_0(I)/G_0(Rb)$ |
|---|---|---|---|---|---|---|
| Nonactin | Li | 0.05 | 0.077 | 0.00056 | 0.0021 | 0.00088 |
| | Na | 3.2 | 1.2 | 0.036 | 0.015 | 0.014 |
| | K | 190 | 190 | 2.1 | 2.1 | 2.1 |
| | Rb | 90 | 88 | 1.0 | 1.0 | 1.0 |
| | Cs | 11.5 | 7.1 | 0.13 | 0.077 | 0.082 |
| | $NH_4$ | 9,000 | 580 | 100 | – | 6.7 |
| Monactin | Li | 0.10 | 0.23 | 0.00034 | 0.001 | 0.00073 |
| | Na | 8.0 | 4.4 | 0.028 | 0.015 | 0.014 |
| | K | 850 | 920 | 2.93 | 2.0 | 2.9 |
| | Rb | 290 | 310 | 1.0 | 1.0 | 1.0 |
| | Cs | 25 | 13 | 0.086 | 0.047 | 0.042 |
| | $NH_4$ | 16,000 | – | 55.2 | – | – |
| Dinactin | Li | 0.15 | 0.48 | 0.00019 | 0.0014 | 0.00042 |
| | Na | 25 | 19 | 0.031 | 0.016 | 0.017 |
| | K | 2,000 | 2,300 | 2.5 | 2.4 | 2.1 |
| | Rb | 800 | 1,200 | 1.0 | 1.0 | 1.0 |
| | Cs | 46 | 31 | 0.058 | 0.033 | 0.027 |
| | $NH_4$ | 24,000 | – | 30 | – | – |
| Trinactin | Li | 0.23 | (0.011) | 0.0002 | (0.0018) | (0.000087) |
| | Na | 42 | (14) | 0.036 | (0.028) | (0.011) |
| | K | 4,000 | 3,100 | 3.4 | 3.1 | 2.6 |
| | Rb | 1,170 | 1,200 | 1.0 | 1.0 | 1.0 |
| | Cs | 75 | 42 | 0.064 | 0.047 | 0.034 |
| | $NH_4$ | 46,000 | – | 39 | – | – |

*Parenthesized values were obtained on imperfectly thinned membranes.

effects with the quantitative expectations of the theory, as here discussed. Remarkably good agreement was found for bilayers, not only between theory and experiment but also between the observed bilayer electrical properties and those "predicted" from the equilibrium measurements described briefly in the preceding paragraph (Szabo et al., 1969). In particular, it was possible to correlate the effects of the molecular structure of the antibiotic molecules in bulk phases with their effects on the electric properties of the bilayer membrane. Not only was the expected proportionality between membrane conductance and antibiotic concentration observed experimentally, but so was the proportionality between permeant ion concentration and conductance when ionic strength was held constant with a relatively impermeant salt. Holding ionic strength constant was necessitated by the discovery that an apparent "saturation" at high salt concentration, not expected theoretically, was due to a balance between a pure ionic strength effect* that decreases membrane conductance, the opposite of the effect expected from an increase in concentration of the permeant cation.

Most notably, as required by the theory, the relative effects of Li, Na, K, Rb, and Cs were found to be quantitatively identical for each antibiotic (but different from antibiotic to antibiotic) in three different types of measurements: bulk-phase equilibrium salt extractions, bilayer membrane conductance studies, and bilayer membrane potentials. The identity predicted among the conductance ratios, permeability ratios, and solvent extraction selectivity ratios for these ions was indeed observed (Table 7). These results strongly support the initial postulate that neutral antibiotics such as the macrotetrolide actins produce their effects on lipid bilayer membranes by acting as molecular cation carriers. The extent to which the parameters governing the equilibrium selectivity of the macrotetrolides suffices to account for their observed effects on membrane conductance is illustrated in Figures 34 and 35.†

---

*These observed effects of ionic strength have subsequently been shown to be expected from the negative surface charge of the lipid Asolectin used in these experiments (McLaughlin et al., 1970).

†In view of the importance of equilibrium interactions demonstrated here, it should be pointed out that subsequent to the Work Session Eisenman has extended his considerations of the equilibrium selectivity of monopolar exchange sites (Eisenman, 1961) to dipolar models appropriate to the ligand groups of the present neutral sequestering molecules (Eisenman, 1969). From these considerations he has concluded "that the energies of interaction of cations with the dipolar ligand groups, in competition with the hydration energies of the ions, can account for the salient features of the selectivity among the alkali metal cations characteristic of neutral sequestering molecules."

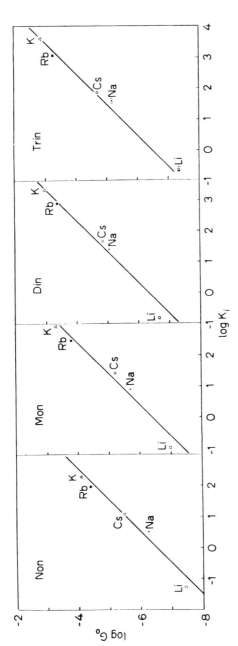

Figure 34. The proportionality between the conductances of bilayer membranes and the equilibrium constants for salt extraction. Abscissa: Logarithm of the salt extraction equilibrium constants for the indicated macrotetrolide antibiotics from paper II. Ordinate: Logarithm of membrane conductance measured at $10^{-2}$ M salt and the presence of $10^{-7}$ M macrotetrolide concentration. (Data from Table 7.) The lines of unit slope are drawn to indicate the proportionality expected. [See also Szabo et al., 1969, Table 2 and Equation 12.]

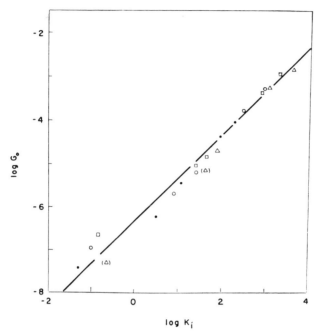

Figure 35. The single proportionality between $G_0(l)$ and $K_i$. The data of Figure 34 are condensed on a single log-log plot to show that a single proportionality constant relates the bilayer conductances $G_0(l)$ to the corresponding salt extraction equilibrium constants, $K_i$, for all of the macrotetrolide actins and all cations. • = nonactin; ○ = monactin; □ = dinactin; and △ = trinactin. The solid line of unit slope has an intercept of $0.48 \times 10^{-6}$, which is the value of the proportionality constant relating the conductances to the salt extraction equilibrium constants. [Szabo et al., 1969]

By varying the cholesterol content of lecithin membranes, Szabo, Eisenman, and Ciani (1969, 1970) have been able to demonstrate that the effectiveness of a given antibiotic depends markedly on the lipid composition of the membranes. Moreover, the striking ability of cholesterol to decrease the effectiveness of monactin on membrane conductance is exactly what would be expected if the lipid tails were immobilized by cholesterol when monactin is a cation carrier. Despite this effect, the permeability ratios and conductance ratios of the cations appear to be wholly independent of the lipid composition, as would be expected from the theory proposed here.

Last, by carrying out a series of membrane conductance and membrane potential measurements in the presence and absence of a gradient in monactin concentration, but with the same concentrations of KCl on either side of the membrane, it has also been possible to show that the diffusion of the antibiotic molecule within the membrane is rapid as compared to its rate of diffusion across membrane-solution interfaces (Szabo et al., 1970).

## MOLECULAR MOTION IN PHOSPHOLIPID BILAYERS AND BIOMEMBRANES: W. L. Hubbell

One particularly challenging problem in structural chemistry is the molecular organization of the biological membrane. Straightforward application of spectroscopic techniques is hampered by the fact that one is obliged to obtain spectroscopic data on all of the complex membrane components simultaneously. Such data contain a wealth of information, but the difficulty of analysis often prohibits meaningful interpretation. This difficulty may be overcome by the use of probe molecules that can be introduced into selected regions in complex systems to provide spectroscopic signals with little or no interference, yielding information on the local environment of the probe.

The most recent of the probe techniques, the spin-labeling method of McConnell (Hamilton and McConnell, 1968; Griffith and Waggoner, 1969; McConnell and McFarland, 1970), makes use of a stable, organic free radical having a molecular structure and/or chemical reactivity that directs it to a particular site in the

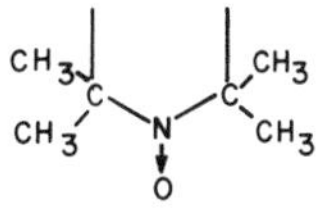

Figure 36. Protected nitroxide group. [Hubbell]

system of interest. Spin labels are commonly designed around the protected nitroxide group, which is the paramagnetic center (Figure 36).

The paramagnetic resonance spectrum of a nitroxide depends on the rate and anisotropy of rotational motion, the polarity of the environment, and, if the radical has a fixed orientation in space, on orientation. It is this spectral sensitivity that allows one to infer something about the local environment of the radical in the system.

Spin labels have recently been used to study structural features of a number of biological membranes (Hubbell and McConnell, 1968, 1969a,b; Keith et al., 1968, 1970). Of particular interest with regard to these studies are the axonal, sarcoplasmic reticular, and retinal rod outer disk membranes. The paramagnetic resonance spectrum of any of these membranes suspended in the appropriate saline containing small amounts of the nitroxide I shows that the radical is in dynamic

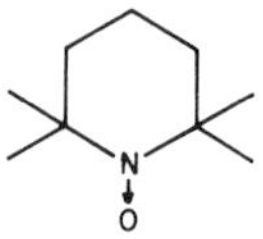

Figure 37. Nitroxide I. [Hubbell]

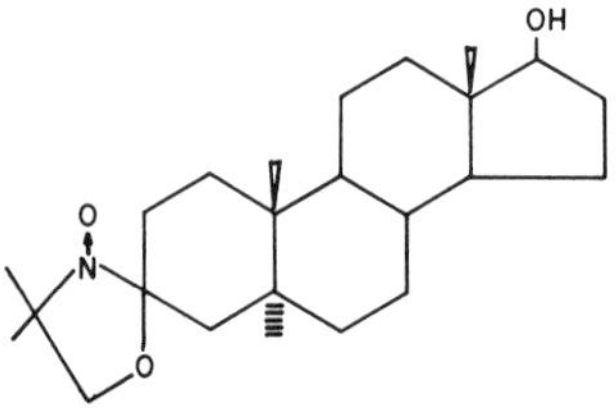

Figure 38. Steroid spin label II. [Hubbell]

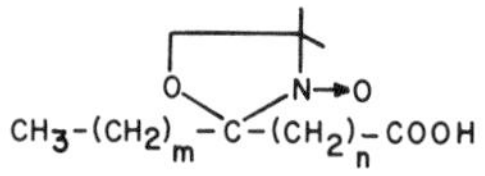

Figure 39. Amphiphilic spin labels III(m,n). [Hubbell]

equilibrium between the aqueous solution and a fluid, hydrophobic phase in the membranes (Figure 37). The label I is also distributed between the aqueous phase and a fluid hydrophobic phase in an aqueous dispersion of phospholipid lamellar structures (liposomes), suggesting that the fluid hydrophobic regions in the membrane may be formed by an association of the hydrocarbon chains of the phospholipids.

The steroid spin label II (Figure 38) is readily taken up by the membrane preparations. The paramagnetic resonance spectrum of II in any of the membranes mentioned previously can be interpreted in terms of a rapid rotational diffusion of II about its long axis with a correlation time in the order-of-magnitude range $10^{-7}$ to $10^{-10}$ sec, but with slow or no motion of the long axis itself. Another steroid with the same structure as II but with no −OH group at the C-17 position shows a paramagnetic resonance spectrum in the membrane that indicates rapid motion, but little or no anisotropy. These data suggest that the steroid nucleus of II is intercalated between the hydrocarbon chains of an associated lipid structure, with the −OH group "anchoring" the molecule at the polar interface. The paramagnetic resonance spectrum of II in liposomes shows the same rapid motion as in the membranes, but the motion is nearly isotropic. This lack of anisotropy may be due to additional degrees of freedom as a result of thermal motions in the polar head group region of the bilayer. These motions could well be damped out in the membrane due to a surface coating of protein as in the Danielli-Davson model (1935).

The high motional freedom of II and other steroid spin labels in membranes is considered to lend some plausibility to models of transport that involve rotation and/or translation of carriers across the membrane.

Additional information on the nature of the membrane hydrophobic regions has come from a study of the amphiphilic spin labels III(m, n) (Figure 39). The resonance spectrum of III(12, 3) incorporated in membranes indicates a rapid anisotropic motion of the nitroxide about the long molecular˙axis, which is parallel to the methylene chain axis in its extended configuration. The same motion is obtained when III(12, 3) is incorporated in liposomes. A striking feature of the motion of the III(m, n) in liposomes is that the motional restrictions on the nitroxide are dramatically reduced as it is moved away from the polar end of the chain. Thus, III(12, 3) shows highly restricted anisotropic motion, while III(5, 10) shows nearly isotropic motion. This indicates that the motions of the hydrocarbon chains in a bilayer become increasingly more chaotic as the distance from the polar interface is increased. In contrast to this marked dependence on n, the motional characteristics are only slightly affected by changes in m. This motional dependence on n can be accounted for in terms of a realistic model of the polymethylene chain in a bilayer in which motions other than axial rotation are the result of rapid configurational isomerizations about carbon-carbon single bonds and "rigid stick" motions of the (average) chain axis. The analysis of the resonance data in terms of this model makes it possible to estimate the probability of *trans* ($P_t$) and the probability of *gauche* ($P_g$) configurations about the carbon-carbon single bonds (Hubbell and McConnell, 1971). For III(m, n) in liposomes of egg lecithin:cholesterol (2:1 mole ratio), the experimental data are accounted for by taking all carbon-carbon bonds to be equivalent in the sense that $P_t$ and $P_g$ are the same for each single bond, with $P_t = 0.976$ and $P_g = 1 - P_t = 0.024$. (The probability of the sterically hindered *cis* configuration is assumed to be so small as to be negligible.) This means that for a chain of 15 carbon-carbon bonds the probability of the *all trans* configuration is $(0.976)^{15} = 0.7$. Thus one may think of the chains in this system as relatively rigid rods. In egg lecithin liposomes without cholesterol, the probability of the *all trans* configuration is considerably lower, clearly showing the so-called condensing effect of cholesterol observed in monolayer studies.

This series of III(m, n) labels has been studied in the membranes of the unmyelinated walking leg axons of the Maine lobster, *Homarus americanus*. Here the III(m, n) labels show the same striking spectral dependence on chain position as was found in the egg lecithin:cholesterol liposomes. The experimental data are accounted for using the same model for chain motion and a value of $P_t = 0.96$. Thus in this membrane the hydrocarbon chains may be thought of as nearly rigid rods.

Another important class of lipid spin labels is IV(m, n) (Figure 40). The IV(m, n) are simply lecithin molecules with a nitroxide group

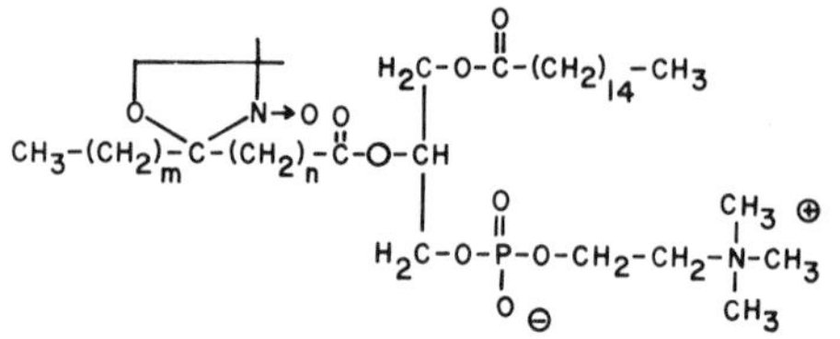

Figure 40. Lecithin spin labels IV(m,n). [Hubbell]

at various positions on the fatty acid chain. The paramagnetic resonance line shape of IV(m, n) incorporated in egg lecithin:cholesterol (2:1 mole ratio) liposomes indicates rapid axial motion, just as for the III(m, n) in this system. However, to account in a simple way for the resonance data of the IV(m, n), it must be assumed that the value of $P_t$ depends on the position of the bond in the polymethylene chain. It appears that for $n > 8$, $P_t$ decreases more rapidly than expected on the basis of the III(m, n) results. It must be realized that the motions derived from the resonance data on III(m, n) are directly applicable only to the motions of the III(m, n) chain itself, and not to the motions of the fatty acid chains on the phospholipids. In IV(m, n), direct information is obtained on the motion of the fatty acid chains on the phospholipids. Experiments with the IV(m, n) in membranes have not yet been carried out, but considering the results from the liposomes, it can be concluded that the motions of the III(m, n) chain provide at least a crude indication of the motion of the fatty acid chains on the membrane phospholipids. Certainly the conformational limits of the III(m, n) chain are dictated by its local environment, and the values of $P_t$ determined for the III(m, n) chain provide a feeling for the fluidity of that environment.

The geometry of the phospholipid bilayer makes it possible to obtain highly oriented arrays. It has been shown that spin labels of the type III(m, n) in these oriented lamellae are themselves highly oriented, with the long molecular axis perpendicular to the plane of the bilayer, as might be expected (Libertini et al., 1969; Hsia et al., 1970). In a few cases, it is possible to produce anisotropic arrays of biological membranes. For example, erythrocytes, being biconcave disks, can be partially oriented by hydrodynamic shear, and the walking leg axons of *Homarus americanus* have the membrane distributed in a cylindrical array. In a spin-label study of these two oriented membrane systems, it has been possible to show that the preferred orientation of spin labels of the type II and III(m, n) is one in which the long molecular axis is perpendicular to the local membrane surface. The orientations are far from perfect, but evidence suggests that the dispersion of orientations is dominated by the dispersion of membrane orientations, rather than by dispersions of label orientations relative to the membrane (Hubbell and McConnell, 1969b).

From these detailed, quantitative, comparative studies of the phospholipid bilayer system and the membranes, and of the anisotropic label distribution in the membrane, it is clear that the fluid hydrophobic regions in the membranes in which the labels locate themselves are formed by the association of the hydrocarbon chains of phospholipids in a bilayer configuration. Further investigation will be required to determine the extent of these structures in the membrane.

### KINKS AS CARRIERS IN MEMBRANES: H. Träuble

Träuble pointed out that in addition to the highly selective carrier-induced permeability already discussed at the Work Session, all biological membranes show an intrinsic nonspecific permeability both to "lipophilic" molecules and to water, and that the process of nonspecific permeation depends on (1) the partition coefficient of the permeating molecules between the aqueous phase and the hydrocarbon region of the membrane, and (2) the molecular weight of the permeating molecules. A satisfactory molecular theory for this general type of nonselective permeation of molecules through membranes is not as yet available. In earlier work, the nonspecific permeation has been explained in terms of channels or pores within the membranes (Solomon, 1952, 1960; Stein, 1967). More recently, the diffusion of molecules within membranes has been discussed in connection with thermal fluctuations of the hydrocarbon chains of the membrane-forming lipids. Such fluctuations might result in the formation of transient pockets of free volume through which molecules might permeate (Lieb and Stein, 1969).

Zwolinski et al. (1949) applied the absolute rate theory to the problem of diffusion across membranes. Their theory regards the flow of molecules as a series of successive jumps from one equilibrium position to the next. Figure 41 shows a possible energy profile for a molecule diffusing through a membrane. However, if applied to biological systems, the difficulty lies in interpreting the elementary processes.

Träuble proposed a detailed mechanism for the molecular diffusion within lipid membranes. This model is based on recent

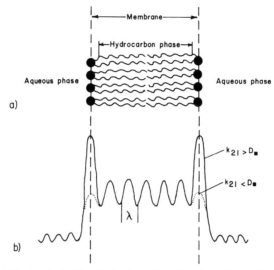

Figure 41. (a) Bimolecular leaflet. The hydrocarbon phase is made up of fatty acid chains (CH₂ chains). (b) Possible potential energy profile for a molecule diffusing through a membrane. [Träuble]

experimental and theoretical investigations of the mechanical relaxation of paraffins, polyethylene, and other polymer materials (Pechhold et al., 1963; Pechhold et al., 1966; Blasenbrey and Pechhold, 1967; Pechhold and Blasenbrey, 1967). These studies strongly support the view that polymer materials both in the crystalline and in the liquid-crystalline state contain certain types of mobile structural defects, so-called kinks, which result from conformational changes in the hydrocarbon chains. Kinks, if present in the hydrocarbon region of a membrane, produce small, mobile pockets of free volume in different sizes depending upon the type and the arrangement of the kinks. A molecule present in the aqueous phase adjacent to the membrane may jump into the free volume of a kink at the membrane surface; it may then diffuse across the membrane together with the mobile kink—in a sort of "hitch-hiking" process.

A permeation theory based on this picture requires knowledge of the molecular structure and the process of formation and diffusion of kinks in a hydrocarbon phase. The molecular structure of the simplest, so-called 2g1 kink (see Figure 43) in a $CH_2$ chain is shown in Figure 42. This kink can be formed from a straight hydrocarbon chain by rotating about a particular C–C bond by an angle of $+120°$ and rotating either of the two next nearest neighboring C–C bonds by

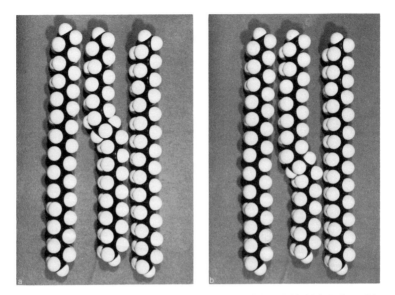

Figure 42. A "2gl" kink in a $CH_2$ chain shown in two positions. This kink is formed by two rotations about C–C bonds which are separated by one chain unit. Rotation angles $\rho_1 = +120°$, $\rho_2 = -120°$. By formation of one kink the chain is shortened by one $CH_2$ unit length. Therefore kinks cannot be generated and cannot disappear inside of the hydrocarbon bulk phase. They are more easily generated at the surface, whence they migrate into the bulk phase. [Träuble]

$-120°$. By this procedure, two *trans* configurations in the $CH_2$ chain are transformed into *gauche* configurations. The dependence of the potential energy on the angle of rotation about a C–C bond in such a chain has been calculated by Harris and Harris (1959) and by Volkenstein (1963) for a butane molecule, as shown in Figure 43. The energy difference $\Delta E$ between *trans* ($\rho = 0°$) and *gauche* configuration ($\rho = 120°$) is 0.8 kcal/mole. Thus the formation of one kink increases the self-energy of an isolated chain by 2 x 0.8 = 1.6 kcal/mole. However, the activation energy for the transition from *trans* to *gauche* configuration is about 2.4 kcal/mole; twice this energy, i.e., 4.8 kcal/ mole, must be supplied by thermal energy to move a kink along the chain. Within a polymer material there is a further contribution to the kink's self-energy due to the distortion of the polymer chain in the neighborhood of the kink. This additional energy has been estimated by Pechhold (1968) to be about 2 kcal/mole. The equilibrium concentration of kinks is determined by the change in free enthalpy $\Delta G = \Delta H - T\Delta S$ of the polymer. The formation of kinks within a polymer is accompanied by a considerable increase in entropy or

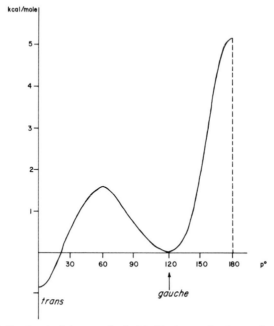

Figure 43. Rotational potential energy for isolated butane molecule as a function of rotation angle $\rho$ about a C–C bond in the middle of the molecule. Sum of exchange and van der Waals energy. The maximum at $\rho = 180°$ is due to strong van der Waals interaction between approaching H atoms. The normal *trans* position ($\rho = 0°$) and the *gauche* position at $\rho = 120°$ are separated by an activation barrier of about 2.4 kcal/mole. In a 2gl kink two *gauche* positions are separated by one *trans* configuration. [Träuble]

disorder (positive $\Delta S$); thus kinks are favored thermodynamically. Further, the free energy of formation and migration of kinks is comparable to the thermal energy at room temperature (RT $\approx 0.6$ kcal/mole); the hydrocarbon region of the membranes can therefore be expected to contain many kinks that move continuously up and down the hydrocarbon chains.

Pechhold and Blasenbrey (1967; Blasenbrey and Pechhold, 1967; Pechhold, 1968) applied statistical methods to calculate absolute values of the kink concentration, expressed as the fraction $\xi$ of $CH_2$ units of the polymer in *gauche* positions. The $\xi$ in general increases with increasing temperature. In paraffins and linear polymers, as well as in lipids, the kink concentration undergoes two sharp increases with increasing temperature. One occurs at a temperature $T_t$ below the melting point of the material. This temperature $T_t$ is associated with the so-called "rotational phase transition." The second sharp increase in kink concentration with increasing temperature occurs at the melting point $T_m$. For $T \ll T_t$, the $\xi$ values are in the range from 0.01 to 0.05, whereas for $T > T_t$, $\xi$ ranges from 0.1 to 0.5. At room temperature, most membrane-forming lipids are above the transition temperature $T_t$. Thus the kink concentration of lipid membranes is probably between 0.1 and 0.5.

Figure 44 shows the possible effect of kinks on the structure of the hydrocarbon phase within a lipid bilayer; with the appropriate combination of kinks, pockets are formed that are large enough to accommodate molecules of the size of benzene or propane.

The model is here developed for the nonspecific permeation through membranes in which kinks serve as carriers for the permeating molecules. It is assumed that at the polar surface of the membrane, complexes are formed between the permeants and the free kinks, which

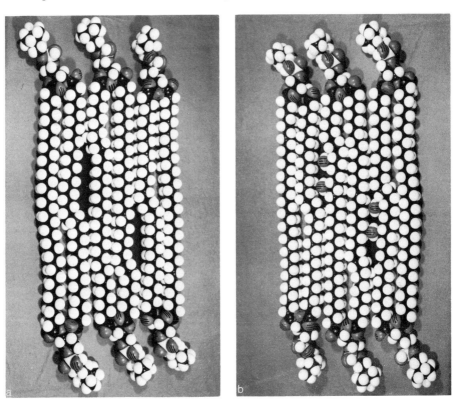

Figure 44. Phosphatidylcholine molecules in planar bilayer arrangement. A: By combination of two 2gl kinks fairly large pockets of free volume can be formed. B: $H_2O$ molecules or larger molecules fit into these pockets. [Träuble]

then move across the membrane. A net flow of occupied kinks across the membrane is established if there is a difference in concentration of the permeant across the membrane. The net flow is calculated for a single neutral permeant, assuming that only one type of kink, representing a fixed free volume, is involved in the act of permeation. For this purpose, the coupling between (a) the diffusional flow $\Phi_\blacksquare$ of occupied kinks and (b) the formation and dissociation of kink-permeant complexes at the membrane surface will be considered. For this case (see Figure 45) the two reactions at the boundaries are

$$\bullet + \square \underset{k_{21}}{\overset{k_{12}}{\rightleftharpoons}} \blacksquare + \Phi_\blacksquare \qquad \text{at side (') of membrane} \tag{1}$$

$$\bullet + \square \underset{k_{21}}{\overset{k_{12}}{\rightleftharpoons}} \blacksquare - \Phi_\blacksquare \qquad \text{at side ('') of membrane} \tag{2}$$

where $\bullet$ denotes the permeating molecule, $\square$ the free kink, and $\blacksquare$ the occupied kink; $k_{12}$ and $k_{21}$ are the overall rate constants for the formation and dissociation of kink-permeant complexes. These constants are determined by the partition ratio of the permeating molecules between the aqueous phase and the hydrocarbon phase and the potential barrier at the polar interphase between membrane and solution. The dissociation equilibrium constant $K$ is given by the ratio $k_{21}{:}k_{12}$ for this reaction; $K$ has the dimension of a concentration. A similar permeation problem has been formulated by Britton (1964).

Defining:
$c_\bullet{}', c_\bullet{}''$ = concentrations of permeant                    on side (') and
$c_\square{}', c_\square{}''$ = concentrations of free kinks                 side ('') of the
$c_\blacksquare{}', c_\blacksquare{}''$ = concentrations of occupied kinks          membrane
$\delta$ = membrane thickness
$D_\blacksquare$ = diffusion coefficient of occupied kinks

$$\Phi_\blacksquare = \frac{D_\blacksquare}{\delta}(c_\blacksquare{}' - c_\blacksquare{}'') = \text{net flow of occupied kinks across the membrane for the steady state.}$$

The rate expressions for reactions 1 and 2 are given by

$$c_\bullet{}'\, c_\square{}'\, \epsilon k_{12} = c_\blacksquare{}'\, \epsilon k_{21} + D_\blacksquare \frac{c_\blacksquare{}' - c_\blacksquare{}''}{\delta} \tag{3}$$

$$c_\bullet{}''\, c_\square{}''\, \epsilon k_{12} = c_\blacksquare{}''\, \epsilon k_{21} - D_\blacksquare \frac{c_\blacksquare{}' - c_\blacksquare{}''}{\delta} \tag{4}$$

where $\epsilon$ denotes the thickness of the surface sheet of the membrane within which complex formation and dissociation takes place (see Figure 45). The value of $\epsilon$ is given by the dimensions of the free volume of a kink.

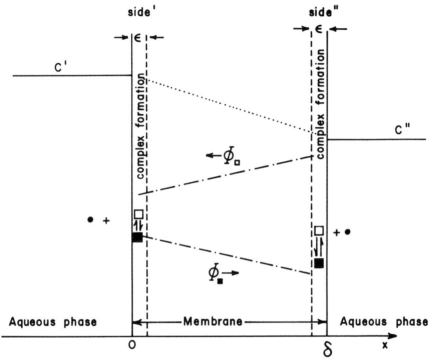

Figure 45. Model used for calculation of kink-mediated transport of molecules across membrane. Symbols: ● permeant, □ free kink, ■ occupied kink, $\Phi$ flows. Molecules (■) jump on left side into free volumes of kinks and are carried across membrane (flow $\Phi_\blacksquare$) together with the kinks. The formation and dissociation of permeant-kink complexes takes place within a surface sheet of thickness $\epsilon$. [Träuble]

By subtraction, Equation 3 minus Equation 4 yields

$$\Phi_\blacksquare = D_\blacksquare \frac{c_\blacksquare' - c_\blacksquare''}{\delta} = \frac{D_\blacksquare}{\delta} \frac{k_{12}}{k_{21} + \dfrac{2D_\blacksquare}{\epsilon\delta}} (c_\square' c_\bullet' - c_\square'' c_\bullet'') \quad (5)$$

The net flow $\Phi_\blacksquare$ can be expressed as a function of the permeant concentrations $c_\bullet'$, $c_\bullet''$ and of the total kink concentration $c_k$ by eliminating $c_\blacksquare'$, $c_\blacksquare''$, $c_\square'$, and $c_\square''$ from Equation 3. For this purpose, it is assumed that

$\Phi_\square = \Phi_\blacksquare$, i.e.,   the back flow of free kinks is taken equal   *Condition* I
to the net flow of occupied kinks, and
that

$D_\square = D_\blacksquare$, i.e.,   the diffusion coefficient of free and   *Condition* II
occupied kinks are taken to be equal.

Further, the kink concentration $c_k$, given by

$$c_k = \frac{1}{\delta} \int_0^\delta [c_\blacksquare(x) + c_\square(x)] \; dx \text{ is taken as constant.} \qquad \textit{Condition} \text{ III}$$

Combining these three conditions with Equation 5 and setting the flow of permeant $\Phi_\bullet$ equal to $\Phi_\blacksquare$ $n$, where $n$ denotes the number of molecules per kink, yields the following expression describing the flow of permeant:

$$\Phi_\bullet = \frac{D_\blacksquare}{\delta} \Delta c_\bullet c_k \frac{nK}{(K + c_\bullet'' + \frac{2D_\blacksquare K}{\epsilon \delta k_{21}}) \, (K + c_\bullet'') + (K + c_\bullet'' + \frac{D_\blacksquare K}{\epsilon \delta k_{21}}) \, \Delta c_\bullet} \qquad (6)$$

where $\Delta c_\bullet = c_\bullet' - c_\bullet''$ represents the difference in permeant concentrations across the membrane.

Equation 6 states that the kink-mediated flow of molecules through the membrane is the product of three terms: (1) the classical Ficks expression $\Phi = D/\delta \; \Delta c$, in this case for the diffusion of kinks, (2) the total kink concentration $c_k$, and (3) a factor that includes the characteristic features of the kink-mediated transport. According to Equation 6, at low values of $\Delta c_\bullet$ the flow increases with increasing $\Delta c_\bullet$. The flow $\Phi_\bullet$, however, reaches a maximum value $\Phi_\bullet \text{max}$ at high values of $\Delta c_\bullet$, the difference between the permeant concentrations on either side of the membrane. The maximum value $\Phi_\bullet \text{max}$ is given by

$$\Phi_\bullet \text{max} = \frac{D_\blacksquare \, K \, n \, c_k}{\delta} \; \frac{1}{K + c_\bullet'' + \frac{D_\blacksquare \, K}{\epsilon \delta k_{21}}} \qquad (7)$$

It depends upon the absolute value of the concentration of the permeant in the aqueous phase.

Equation 6 is here considered for two limiting cases:

(a)   $\dfrac{2D_\blacksquare}{\epsilon \delta} \gg k_{21}$,

and (b)   $\dfrac{2D_\blacksquare}{\epsilon \delta} \ll k_{21}$.

In case (a), the diffusion is fast compared to the dissociation of the kink-substrate complex. Kinks and permeant on the membrane surface are not in equilibrium. The dissociation rate-constant $k_{21}$ is the rate-limiting factor. In case (b), diffusion is slow by comparison with the kink-permeant dissociation; thus, equilibrium between permeant and carrier can be established ($\bullet + \square \rightleftharpoons \blacksquare$). The diffusion coefficient $D_\blacksquare$ of the kinks is the rate-limiting factor. Condition (a), together with $\Delta c_\bullet \ll c_\bullet$, leads to

$$\Phi_\bullet = \frac{D_\blacksquare}{\delta} \left(\frac{nc_k}{K + c_\bullet{}''}\right) \frac{K}{\dfrac{D_\blacksquare}{\epsilon \delta k_{12}} + c_\bullet{}''} \Delta c_\bullet \tag{8.1}$$

which for $D_\blacksquare / \delta \epsilon k_{12} \gg c_\bullet{}''$ simplifies to

$$\Phi_\bullet = k_{21} \, \epsilon \left(\frac{nc_k}{K + c_\bullet{}''}\right) \Delta c_\bullet \tag{8.2}$$

Condition (b) leads to

$$\Phi_\bullet = \frac{D_\blacksquare}{\delta} \left(\frac{nc_k}{K + c_\bullet{}''}\right) \frac{K}{K + c_\bullet{}'} \Delta c_\bullet, \tag{9.1}$$

which for $c_\bullet{}' \ll K$ reduces to

$$\Phi_\bullet = \frac{D_\blacksquare}{\delta} \left(\frac{nc_k}{K + c_\bullet{}''}\right) \Delta c_\bullet \tag{9.2}$$

The term $nc_k/(K + c_\bullet{}'')$ in these equations can be identified with the equilibrium partition coefficient $K_p$ of the permeant between the aqueous phase and the membrane.

For the case of equilibrium, Equation 1 reads: $\bullet + \square \rightleftharpoons \blacksquare$. The corresponding rate equation is $c_\bullet c_\blacksquare k_{12} = c_\blacksquare k_{21}$. With $c_\square = c_k - c_\blacksquare$ and $K = k_{21}/k_{12}$ we obtain

$$\frac{nc_k}{K + c_\bullet} = \frac{nc_\blacksquare}{c_\bullet} = K_p. \tag{10}$$

The term $nc_\blacksquare/c_\bullet$ is the concentration ratio of the permeant between the membrane phase and the aqueous phase. This value thus represents the partition coefficient $K_p$ of the system.

Equations 8.2 and 9.2 are formally identical. However, in case (a) the rate-limiting factor is $k_{21}\epsilon$, whereas in case (b) the ratio $D_\blacksquare:\delta$ between the diffusion coefficient of the kinks and the membrane thickness is rate limiting. The diffusion coefficient $D_\blacksquare$ of the kinks can be regarded as an intrinsic property of the hydrocarbon phase of the membrane, whereas $k_{21}$ depends both on the polar head groups of the membrane and on the affinity of the permeating molecule to the hydrocarbon phase of the membrane. Also, the apparent activation energies are different for the two cases; they are determined by the partition coefficient $K_p$ and by the factors $k_{21}$ and $D_\blacksquare$, respectively. A plot of $\ln\,\Phi_\bullet/T$ against $1/T$ will give a straight line only if $K \gg c_\bullet$. In the general case, described by Equation 6, a plot of $\ln\,\Phi_\bullet/T$ against $1/T$ will be curved, because the expression for $\Phi_\bullet$ contains sums of exponential functions.

In order to test the model against permeability data from lipid bilayers, the absolute values for the diffusion coefficient $D_\blacksquare$ and the concentration $c_k$ of kinks must be known. The diffusion coefficient $D_\blacksquare$ for kinks can be expressed as

$$D_\blacksquare = p\lambda^2 = v \cdot e^{\Delta S/R} \cdot e^{-Q/RT}\lambda^2 \tag{11}$$

where $\lambda$ refers to the distance between two successive energy minima for a kink moving along a $CH_2$ chain (see Figure 41); thus $\lambda = 1.27$ A. The term $p = v \cdot e^{\Delta S/R} \cdot e^{Q/RT}$ denotes the probability for a jump per second; $v$ is the frequency of the thermal oscillations and can be taken equal to the Debye frequency $v = 10^{13}/\text{sec}$. The activation energy $Q$ for the movement of a kink is given by 4.8 kcal/mole, or two times the value of the energy barrier for the *trans → gauche* conformational change (see Figure 43). The entropy factor $e^{\Delta S/R}$ has a value of about 10 (Pechhold, 1968). This leads to

$$D_\blacksquare = 0.5\ 10^{-5}\ \text{cm}^2/\text{sec}. \tag{12}$$

We note that $D_\blacksquare$ has the same order of magnitude as the diffusion coefficient of molecules like ethanol or phenol in water. Thus the diffusion of kinks within a hydrocarbon phase is a very fast process.

The kink concentration $c_k$, which is defined by the fraction $\xi$ of $CH_2$ groups in *gauche* position, can be taken equal to the corresponding value calculated for paraffins in the liquid crystalline state (Pechhold, 1968); thus $\xi \approx 0.1$. Taking for $\delta$, the membrane thickness, $\delta = 50$ A, and assuming that the hydrocarbon chains of the lipids are 20 $CH_2$ groups in length and have a cross-sectional area of 20 [A]$^2$ each (Van

Deenen et al., 1962), yields for the concentration $c_k$ of kinks (expressed in moles per cm$^3$ of the hydrocarbon phase):

$$c_k = \xi\ 8.5 \times 10^{-2}\ \frac{\text{kinks in mole}}{\text{cm}^3 \text{ hydrocarbon phase}}. \tag{13}$$

Except for values with respect to water, the data for the permeation of molecules through lipid bilayer membranes are rather meager. The values for the water permeability coefficient $P$ [cm/sec] obtained by several investigators using various kinds of phospholipids range from $10^{-2}$ to $5 \times 10^{-4}$ cm/sec; however, most values are concentrated at $2 \times 10^{-3}$ cm/sec.

For a quantitative comparison with the theory, the permeability coefficient $P_{\text{theor}}$ is calculated from Equation 9.1, assuming equilibrium at the membrane-water interphase to be established. In order to estimate the constant $K$ in Equation 9.1, we identify the partition coefficient $nc_k/(K + c_{\bullet}'') = K_p$, with the partition coefficient $K_p^+ = 0.64 \times 10^{-4}$ between water and hexadecane, as measured by Schatzberg (1963, 1965). Using $c_k$ according to Equation 13 with $\xi = 0.1$, the value of $K$ can be calculated; together with $D_{\blacksquare}$ as given by expression Equation 12, we obtain $P_{\text{theor}} = 0.65 \times 10^{-3}$ cm/sec, which is compatible with the measured values $P_{\text{exp}} \approx 2 \times 10^{-3}$ cm/sec.

So far we have considered only fully saturated hydrocarbon chains. However, most naturally occurring lipids are partially unsaturated, the double bonds being of the *cis* type. Thus these molecules will fit into a regular arrangement only if kinks are formed near the double bond. From this it is expected that the kink concentration, and consequently the permeability, increases with increasing degrees of unsaturation. In fact, Finkelstein and Cass (1968) reported that membranes formed from egg lecithin completely saturated by hydrogenation have a $P$ value of $1.7 \times 10^{-3}$ cm/sec, compared with $P = 4.2 \times 10^{-3}$ cm/sec for the partially unsaturated egg lecithin.

Another phenomenon of interest is the decrease of water permeability with increasing cholesterol concentration. Finkelstein and Cass (1968) observed a decrease in water permeability from $4.2 \times 10^{-3}$ to $0.75 \times 10^{-3}$ cm/sec as the cholesterol-phospholipid molar ratio increases from 0:1 to 8:1. This effect can be attributed to the decrease in kink concentration accompanying the decrease in hydrocarbon contents.

To prove the validity of the proposed kink-mediated permeation mechanism requires, of course, a systematic investigation of the characteristic functional dependencies expressed by Equation 6; for example, the dependence of the permeant flow $\Phi_\bullet$ on concentration, or the predicted saturation behavior of $\Phi_\bullet$ with increasing concentration difference $\Delta c_\bullet$. A further test could be provided by measuring the temperature dependence of the flow $\Phi_\bullet$ in membranes that undergo a phase transition with increasing temperature. At the phase transition, the kink concentration is expected to increase sharply, which should result in a comparable increase of permeability. For a more detailed discussion of these points, the reader is referred to the recent articles by Träuble (1971a,b) and Träuble and Haynes (1971).

Note: Acknowledgment is made to D. H. Haynes for his critical reading of this material.

## IV. MODEL VESICLES AND MEMBRANES

### THE FORMATION OF ASYMMETRICAL SPHERICAL LECITHIN VESICLES: H. Träuble and E. Grell

Because biological membranes are immensely complex, one practical approach is to study simpler, relevant, model membranes. Much effort has therefore been devoted to the investigation of planar bilayer membranes that are comprised mainly of the phospholipids of natural membranes (Mueller et al., 1962a,b; Huang and Thompson, 1965, 1966; Thompson and Henn, 1970). These membranes are formed by the thinning-down of a droplet of a membrane-forming solution,* which is placed over an aperture in a septum separating two aqueous phases. There is evidence that the molecules in these membranes are organized as bimolecular lamellae according to the Danielli-Davson hypothesis (1935). Studies of the electrical and permeability properties of such membranes may be of considerable value in understanding some of the properties of biological membranes. However, caution should be exercised. Planar membranes are surrounded by a torus consisting of a membrane-forming solution, which may influence the properties of the system. The stability and, presumably, some of the properties of planar bilayer membranes depend on the presence of the previously mentioned lipid-like molecules (stabilizers) in the system. Other limitations are given by the smallness of the membrane surface area, which makes it difficult to perform binding or spectroscopic studies.

Apparently some of these drawbacks could be overcome if very small, closed membrane spheres were available. (Concentrated vesicle dispersions may provide very large membrane surface areas in a small volume.)

#### Spherical Bilayer Vesicles

In fact, Huang (1969) succeeded in forming small spherical bilayer vesicles (with an average diameter of about 250 A) by prolonged ultrasonic irradiation of pure egg-yolk phosphatidylcholine in aqueous

*Most experimenters use solutions containing cholesterol, n-tetradecane or DL-tocopherol in addition to phospholipids. The solvent is usually a chloroform-methanol mixture or n-decane.

solution. (See Huang, later in this section.) This new model system makes it possible to apply chemical relaxation techniques to the study of very fast processes associated with membranes.

Because there was much evidence that biological membranes are structurally asymmetric, it seemed desirable to develop an additional technique for the formation of vesicles with asymmetric lipid bilayer membranes, i.e., membranes with chemically different inner and outer layers. It turned out that this formation can be achieved in a two-step procedure. In the first step, prevesicles of the type shown in Figure 46A are formed; an aqueous inner phase and an organic outer phase are separated by a closed spherical monomolecular lipid layer. In the second step, another monomolecular layer, which may be chemically different from the first one, is wrapped around the prevesicles, forming an asymmetric lipid bilayer membrane as shown in Figure 46B. Simultaneously, the vesicles are converted into an aqueous phase.

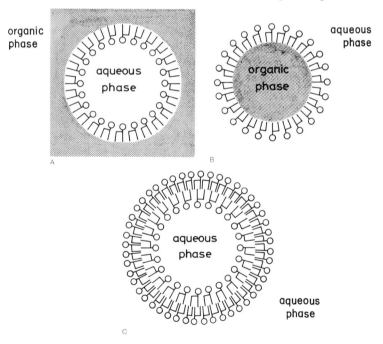

Figure 46A. Prevesicle in organic solvent (e.g., benzene). A phospholipid monolayer separates the inner aqueous phase from the outer organic phase. The polar groups of the lipid molecules are directed towards the aqueous phase, the hydrocarbon chains towards the organic solvent. B. Bilayer vesicle in aqueous solution. C. Monolayer vesicle in aqueous solvent containing a microdroplet of organic solvent. [Träuble and Grell]

*Procedures*

*Step 1.* Purified egg-yolk lecithin is dissolved in an apolar (organic) solvent such as benzene or di-n-butylether (quantities: 10 mg lecithin per 20 ml of organic solvent). A defined, small amount of aqueous solution (for example 0.05 ml of 1 M CsCl) is added. This solution is treated by ultrasonic irradiation under nitrogen atmosphere for about 15 min at 35°C. During this treatment, prevesicles of the type shown in Figure 46A are formed in a process of self-organization. Figure 47 is an electron micrograph of these prevesicles. The prevesicles are very stable; they can be destroyed in the electron microscope only by the application of a very intense electron beam, and this process is demonstrated in Figure 48. Figure 49 gives a typical size distribution curve of the prevesicles. Stability and size distribution depend on the initial amount of lipid and aqueous phase, the character of the organic solvent, and the type and concentration of the added salt.

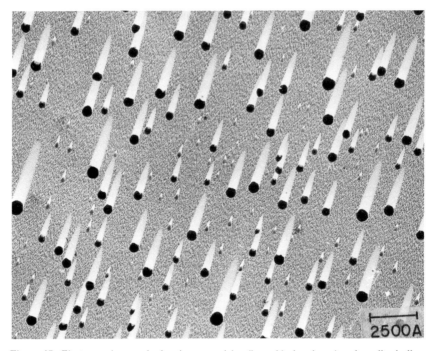

Figure 47. Electron micrograph showing prevesicles (large black spheres) and small micelles. Shadow casting with germanium. The prevesicles contain 1 M CsCl solution. The much greater contrast of the prevesicles compared to the micelles is due to the scattering of the electrons on the cesium and chloride ions. [Träuble and Grell]

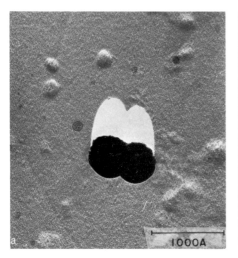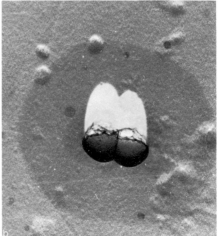

Figure 48. Destruction of prevesicles in the electron microscope. [Träuble and Grell]

A. Pair of intact, closed prevesicles containing 1 M CsCl solution. Germanium shadowing, Formvar-Carbon grid.

B. Same prevesicles after destruction by an intense electron bombardment. Note the loss in contrast. Destruction led to pouring out and evaporation of aqueous phase, leaving behind a CsCl halo around the vesicle skeleton.

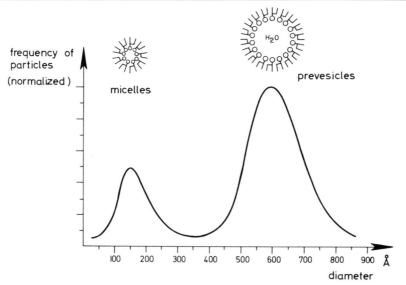

Figure 49. Size distribution of prevesicles and micelles in benzene solution after Step 1 of the preparation (see text). Evaluation of electron micrographs. [Träuble and Grell]

Preliminary experiments with pure synthetic phosphatidylcholines showed that a self-organization into prevesicles takes place only if the ultrasonic irradiation is carried out at a temperature above the first polymorphic transition point of the lipid (Chapman, 1968).

The question arises whether the prevesicles are in fact enclosed by a monolayer as indicated, or, eventually, by a multilayer membrane. This was checked in the following way. The average prevesicle diameter $d$ can be calculated from the number $n$ of lipid molecules in the organic solvent and from the aqueous volume $V$, under the assumption that monolayer membranes are formed around the prevesicles. This leads to an average prevesicle diameter $d = 6\ V/O$, where $O$ denotes the total available membrane surface*; $O$ is related to the number $n$ of lipid molecules by $O = nF$, where $F$ denotes the average surface per molecule as derived from monolayer surface-pressure diagrams. The average diameter $d$ as measured on electron micrographs differed from the calculated value by not more than 15%. This indicates that the prevesicles in fact do have monolayer membranes.

*Step 2* can be accomplished by various procedures. The simplest consists of centrifuging the prevesicles from the organic phase through a monolayer interface into an aqueous medium. This procedure is shown schematically in Figure 50. This step requires an adaptation of the density and the osmotic properties of the aqueous subphase to the inner phase of the prevesicles. Centrifugation was carried out for about 3 hours at $10^5\ g$. The density of the subphase corresponded to 0.7 M CsCl. Prior to centrifugation, a calculated amount of lipid, necessary for the continuous formation of a monolayer at the interface, was added to the organic phase. (Experimental conditions: aqueous subphase, 3 ml of 0.7 M CsCl; upper organic phase, 2 ml of prevesicle solution + 0.5 ml organic solvent containing 0.5 to 1 mg lecithin.)

It was shown by electron microscopy that the aqueous phase after the centrifugation contained spherical vesicles with an average diameter that was about 50 A larger than that of the prevesicles. This result is in agreement with the increase in diameter expected if a second monolayer is wrapped around the prevesicles. The size distribution of the prevesicles and of the bilayer vesicles are given in Figure 51. Compared to the prevesicles, the bilayer vesicles are relatively unstable; they are especially sensitive to changes in osmotic pressure. For example, they are destroyed easily if the changes in osmotic conditions

---

*In most of the experiments, the total surface area $O$ in a volume of 10 ml of organic solvent was roughly $3m^2$.

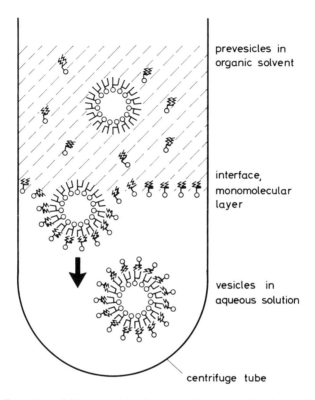

Figure 50. Formation of bilayer vesicles from monolayer prevesicles (schematic). The prevesicles with an aqueous phase inside are centrifuged through an interfacial phospholipid monolayer into an aqueous medium. As the prevesicles penetrate the interface they are coated by a second lipid layer. [Träuble and Grell]

occurring during preparation for electron microscopy take place too rapidly.

As will become clear from examination of Figure 50, the procedure just outlined leads to asymmetric bilayer membrane vesicles if the lipid molecules forming the interface monolayer are chemically different from those forming the prevesicle membranes.

Compared to planar membranes, the following points make these vesicles promising as membrane model systems:

1. It is possible to have a relatively large membrane surface in a small volume (some $m^2$ in 1 ml solution).
2. Vesicles are comprised of pure lipid molecules, i.e., they contain no additional stabilizers.

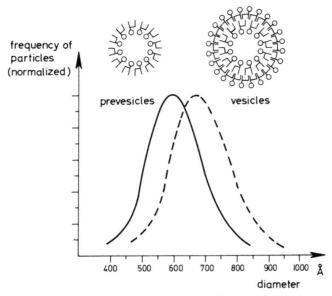

Figure 51. Size distribution of prevesicles and vesicles. Evaluation of electron micrographs. [Träuble and Grell]

3. The membranes of asymmetrical vesicles are supposedly more similar to natural biological membranes than are symmetrical bilayers.

Bilayer vesicles can be used for the study of permeation processes across lipid bilayer leaflets. In such experiments, appropriate indicators for the permeating molecules or ions must be present inside the vesicles. Then the permeation process can be studied with optical methods using perturbation techniques.

### Monolayer Vesicles with an Inner Organic Phase

This type of monolayer vesicle is shown in Figure 46C. A microdroplet of an organic solvent is enclosed by a lipid monolayer with the lipid polar head groups facing the outer aqueous phase. The structure of these vesicles is inverse to the one shown in Figure 46A.

The vesicles are prepared by the following procedure: The lipid is dissolved in an appropriate organic solvent of density 1, for example, in a mixture of chloroform - di-n-butylether (lipid concentration: 100 mg per ml of organic solvent). A small amount (0.1 ml) of this solution is added to 20 ml of 2 mM CsCl and dispersed ultrasonically for 10 to 15 min under nitrogen atmosphere. Nearly clear solutions are obtained when the irradiation is carried out at a temperature above the first polymorphic transition point of the lipid employed. The presence of spherical vesicles with an average diameter of 500 A in the irradiated solution has been demonstrated by electron microscopy. The solutions are usually centrifuged for about 10 min at 25 x $10^3$ $g$ prior to further use. Vesicles of this type may be used for the following experiments:

1. Studies of the distribution of molecules or ions between an aqueous and an organic phase separated by a lipid monolayer barrier. Provided that the permeants have different optical properties in the aqueous and in the organic phase, the kinetics of these processes can be studied with perturbation methods. Such experiments provide informa-tion about the process of entry of molecules or ions into the hydrophobic part of a membrane.

2. Similar to bilayer vesicles, the monolayer vesicles provide a large interface between the lipid polar head groups and the aqueous phase. They can therefore be used to study the binding of ions or molecules to the lipid membrane surface.


## PHOSPHATIDYLCHOLINE VESICLES: FORMATION, PHYSICAL CHARACTERISTICS, AND DYE-LIPID INTERACTION: C. Huang

Huang described a method for the separation of phosphatidyl-choline vesicles formed by prolonged ultrasonic irradiation under nitrogen in 0.1 M buffered NaCl solution. This method involves molecu-lar sieve chromatography on large-pore agarose gels. One fraction of the separated vesicles was found to be homogeneous by the criteria of gel filtration, of sedimentation velocity ultracentrifugation, of sulfhydryl titration, and of electron microscopy. He first presented data that sup-port the thesis that it is a homogeneous vesicle, shell-like in structure, about 250 A in diameter, with a continuous phosphatidylcholine bilayer surrounding a volume of solvent (Huang, 1969; Huang et al., 1970). The following physical parameters were established for the homo-geneous vesicles:

$s_{20,w}, 2.6 \ S; D°_{20,w}, 1.87 \times 10^{-7} \ cm^2 \ sec^{-1}$;
$\phi', 0.9885 \ mlg^{-1}; [\eta], 0.041 \ dlg^{-1}$
vesicle weight, $2.1 \times 10^6$ daltons

These phosphatidylcholine vesicles were also selected as model membranes for the interaction studies with dyes. A red shift of the dye (rhodamin 6G) was observed in absorption and circular dichroism (CD) spectra upon the addition of phosphatidylcholine vesicles, indicating the binding of the dye to the model membrane. The gel-fraction method of Hummel and Dreyer (1962) and spectrophotometric titration method have also been used for the determination of the association constants for the binding reaction (Huang et al., 1969). In the dye concentration range of $5 \times 10^{-7}$ to $2 \times 10^{-5}$ M, the association constant obtained from both methods was $3.3 - 4.5 \times 10^4$ $M^{-1}$.

## INTERACTION OF A BACTERIAL ADENOSINE TRIPHOSPHATASE WITH PHOSPHOLIPID BILAYERS: H. Müldner

Although considerable information is available about the importance of membranes in the regulation of ion transport, very little is known about the molecular mechanisms. It is still extremely difficult to isolate in a pure and active form those membrane constituents which participate in transport through biological membranes.

Following the discovery of a large group of toxic antibiotics that can induce alkali-cation movement through biological as well as protein-phospholipid bilayers, much attention has been turned to this class of compounds. Although it has been suggested that biological membranes may contain similar substances, it is improbable that they will be found, because most of these antibiotics, which are circular peptide molecules, contain d-amino acids, and this configuration has never been found in the proteins of animal membranes.

So the main question remains: What are the carriers in real membranes? Of the many substances that have been isolated from membranes, most are complexed with lipids, and of those that have been tested on bilayers, only a highly purified adenosine triphosphatase (ATPase) from *Streptococcus faecalis* spheroblasts interacts and increases the electrical conductance of the bilayer $10^2$- to $10^4$-fold; we also know that the magnitude of the increase is dependent on the presence of $Mg^{2+}$ and $K^+$.

According to Abrams and his co-workers, who discovered this enzyme and have studied it over many years, this ATPase is apparently associated with the active transport of $K^+$. (See Abrams, 1965; Abrams and Baron, 1967, 1968.) The similarity between this interaction and the dependence of ATPase activity on $Mg^{2+}$, $K^+$, and adenosine triphosphate (ATP) (which contributes an additional tenfold increase in conductance) in phospholipid bilayers suggests that the bilayer-ATPase interactant complex may be similar in structure and properties to the membrane-ATPase complex in the intact organism.

The bimolecular lipid film used in the experiments described here is very similar to the structural core of the biological membrane, a core that confers on the membrane a limited set of physical properties. All other properties of the biological membrane are gained by local modifications of the lipid film through interactions with specific proteins. This concept, which is the essence of the Davson-Danielli membrane model (1935), has been the basis for a considerable body of experimental work on phospholipid model systems, including bilayer and liquid crystal (or vesicle) dispersions. The evidence brought together in these investigations reveals a marked similarity between the physical properties of the model systems and the established physical properties of a wide variety of biological membranes. Additional support is provided by the analogous effects on the system properties both of the biological membranes and of the model systems mentioned above caused by a number of substances such as the macrocyclic antibiotics. If this concept of biological membrane structure is correct, then an obvious extension of the work on model phospholipid films is an investigation of the interaction of these models with membrane-bound proteins, coupled with an examination of the biologically relevant system properties of the interactant complexes.

**Methods**

Of course, the isolated systems that interact with the bilayer must be nonparticulate. A big advantage of the ATPase now under study is the simple method of releasing it from bacterial ghosts; several washings of the spheroblasts with an $Mg^{2+}$-free low-ionic-strength buffer releases most of the enzyme. After precipitation with 80% saturated ammonium sulphate at pH 7.5, it is redissolved in 0.1 M tris buffer, following extensive dialysis. The enzyme can be purified by repeated chromatography on agarose and diethylaminoethyl (DEAE) cellulose.

**Measurements**

The enzyme has not been characterized completely because of the minute amounts of proteins that have been isolated so far.

Preliminary measurements indicate a molecular weight of 300,000 and the existence of two subunits after treatment with 7 M urea. The protein binds strongly to liquid crystal dispersions composed of 50% egg lecithin and 50% bacterial cardiolipin in the presence of $Mg^{2+}$. Because lecithin is not a component of bacterial membranes, it is of considerable interest to examine the interaction of the ATPase with bacterial lipids. Vorbeck and Marinetti (1965) have shown that the lipid components of this organism are glycosyl diglycerides containing glucose and galactose, phosphatidylglycerol, and amino acid esters; phosphatidic acid and cardiolipin are present as minor components. Liquid crystal dispersions from total lipid extracts prepared by ultrasonic treatment again have a strong affinity to the ATPase in the presence of $Mg^{2+}$.

### Binding Constants

The determination of the binding constants is now under way. Preliminary experiments with liquid crystal dispersions of total bacterial lipid extracts indicate that after binding of the ATPase in the presence of $Mg^{2+}$ and ATP, $K^{2+}$ specifically is pumped through the membrane into the holes of the liquid crystals. Compared to the results of Harold et al. (1969), it seems evident that we can prepare a simple bacterial spheroblast with at least one "pump" driven by ATP.

Mainly electrical studies have been carried out. These were done in a simple two-compartment cell similar to the one described by Hanai et al. (1964). The ac potential across the black films was held below 50 mV. In this range, both capacitance and conductance were independent of the applied voltage. Some dc measurements were made by the method described by Miyamoto and Thompson (1967).

The NMR-relaxation measurements indicate binding of ATPase with $Mg^{2+}$ in a binary complex. Further, enhancement of the spin lattice relaxation in the presence of ATP indicates the formation of a tertiary complex. Changes in the $T_2$-relaxation point to changes in the coordination sphere of the $Mg^{2+}$ ion. More detailed experiments are under way to determine binding constants and ion-proton distances in the various complexes as well as NMR-relaxation effects in the presence of ATPase-depleted bacterial membranes and liquid crystal dispersions.

### Preparation of the Membranes

The bilayer membranes were formed by the brush technique after the method of Mueller et al. (1962b) from an n-decane solution of any of the following lipids: (1) 5 mg pure egg lecithin and 1 mg pure cholesterol; (2) 6 mg pure 1,2-diphytanoyl-3-sn-diphytanoylcholine; or (3) 6 mg of the total lipid extract of bacterial spheroblasts per ml n-decane. The aqueous phase consisted of 0.001 to 0.1 M NaCl with KCl buffered at pH 7.5 with 0.05 M to 0.1 M tris HCl. Additions to the aqueous phase were made symmetrically in both compartments after the membranes had thinned to bilayer configuration. Without the addition of ATPase and $Mg^{2+}$, bilayer resistance was in the range reported for this system. Addition of the enzyme plus $Mg^{2+}$ caused a marked decrease in the bilayer resistance. Although this decrease in resistance was dependent on the addition of $Mg^{2+}$, it is apparent that the addition of $Mg^{2+}$ alone was without effect on the system. It is pertinent that the ATPase activity, as well as the binding of the enzyme to depleted membrane fragments, has been shown to be dependent upon $Mg^{2+}$. Final resistance was usually in the range of 0 x $10^5$ to 5 x $10^6$ $\Omega$ $cm^2$. The resistance change occurred over a time interval of 5 to 30 min. Formation of a bilayer with all components present in the aqueous phase produced in most cases a membrane of initially high resistance in the range of 6 x $10^8$, which decreased to a stable low value after 30 min. When the ATPase was added to the aqueous phase before membrane formation, the rate of spontaneous thinning of the bilayer formation was markedly decreased. At protein concentrations higher than 100 $\mu$g/ml, it was difficult to form bilayers.

The bilayer membranes formed from pure diphytanoyl-lecithin were considered more satisfactory for experimentation than the egg lecithin system, because spontaneous autoxidation could not occur with the fully saturated synthetic phospholipid. Furthermore, it was not necessary to add cholesterol to improve the stability of the system.

The dc resistances of these bilayers were greater than $10^8$ $\Omega$ $cm^2$ and did not decrease over a period of 90 min. Concentrations of ATPase of 1 $\mu$g/ml in the aqueous phase lowered the bilayer resistance to 5 x $10^4$ $\Omega$ $cm^2$. A slight decrease in membrane capacitance was also observed.

All the lipids mentioned above are nonbacterial, but bacterial lipids were also used for bilayer formation. Bilayers formed from bacterial ghosts were stable for 1 to 2 hours, but drainage behavior was quite different. Drainage to the bilayer configuration in the bacterial lipid system required about 20 min in comparison with about 5 min for egg lecithin in n-decane. Figure 52 shows a representative dc current-voltage curve obtained for a bacterial lipid bilayer. However, the curve

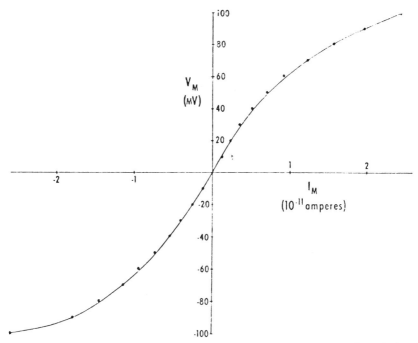

Figure 52. Current-voltage characteristics formed from the total lipid extract from the lysed bacterial ghosts in n-decane. Aqueous phase: 0.05 M tris HCl, pH 7.5, 0.1 M NaCl, 0.1 M KCl, 0.02 M MgCl$_2$ at 21°C. [Müldner]

has the characteristic sigmoidal shape reported for bilayers of a wide variety of compositions. The specific resistance of the linear portion of the current voltage curve (−50 to +50 mV) was 2 x 10$^7$ to 7 x 10$^7$ Ω cm$^2$. This range is about ten times lower than the values usually obtained for egg lecithin in decane. Dielectric breakdown occurred when the voltage was raised between 150 and 200 mV.

Figure 53 is a typical dc current-voltage curve obtained in the presence of a 1 μg/ml concentration of ATPase. The sigmoidal curve is preserved. It is apparent that the resistance in the linear range has been lowered to about 7 x 10$^3$ Ω cm$^2$. The interaction of the ATPase with the bilayer formed from the total lipid extract produced a decrease in resistance by a factor of about 10$^3$ to 10$^4$.

The data clearly demonstrate a resistance decrease attributable to the molecular species in solution that displays the ability to catalyze hydrolysis of ATP. With molecular weight at about 3 x 10$^5$ and a concentration of enzyme in the system at 1 μg/ml, true concentration is less than 3 x 10$^{-9}$ M/l. The marked decrease in membrane resistance

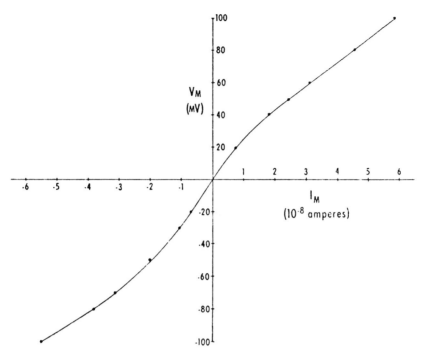

Figure 53. Same conditions, but bilayer with interacting ATPase; 1 μg per ml ATPase per ml. [Müldner]

suggests that the result of the interaction of the ATPase with the bilayer membrane is an alteration in ionic conductivity.

# V.  NEURON  MEMBRANES

### MEMBRANE  PHENOMENA  AT  THE  SYNAPSE:  V. P. Whittaker

Whittaker stated that the transfer of information, which is the main business of the neuron, can be divided into two main processes, both of which rely on membrane phenomena of considerable complexity and sophistication.

First, there is the flow of information from one part of the nerve cell to another. This involves the well-known propagated axon potential accompanied by transient changes in permeability to $Na^+$ and $K^+$ and is related to and maintained by the ability of the neuronal plasma membrane to bring about differences in the ionic composition between cytoplasm and extracellular fluid. Although this flow of information is usually unidirectional from dendrites and soma to axon and axon terminals, we must realize that this unidirectionality is not a built-in property. In principle, this flow of information can take place in both directions, and it is the unidirectional properties of the synapse and the fact that most synapses are in the dendritic field or applied to the soma that ensures unidirectionality of the action potential propagation. Second, there is the flow of information from one cell to another across the specialized region of contact, the synapse. With a few interesting but relatively unimportant exceptions, this type of transfer is believed to be mediated by the release from the presynaptic terminal of specific chemical transmitter substances. We can show in favorable cases both by the application of histochemical techniques and by the actual isolation of the presynaptic terminals that the transmitter exists as a local store in the presynaptic terminal. Here, then, is an explanation for the duality of transfer processes in the CNS; antidromic impulses may invade the soma but cannot travel further because there is no postsynaptic store of transmitter.

There is a good deal of evidence that there are in fact two stores of transmitter in the terminal—an easily accessible store and a less easily accessible store. This has been shown by exchange experiments with radioactively labeled acetylcholine (ACh). In the case of ACh, and possibly of NE and other transmitter substances too, Whittaker believes that the easily labeled store in this type of experiment is cytoplasmic transmitter, whereas the less easily labeled store is transmitter stored within the 500 A diameter synaptic vesicles that are a prominent feature of the cytoplasm of most presynaptic nerve terminals. These vesicles vary in morphology in different types of synapse, and partially successful attempts have been made to correlate the morphology with the type of transmitter present. The best correlation is between monoamine transmitter NE and the presence within the terminal of dense-cored particles. Acetylcholine seems to be associated with endings containing only small vesicles with clear centers, and certain inhibitory transmitters may be associated with endings containing vesicles that appear elongated or polymorphic in certain fixation procedures.

Another constant morphologic feature of chemical synapses is the 200 A gap or "cleft" between presynaptic and postsynaptic membrane. The function of this gap may be to provide a low-resistance pathway for action currents, thus attenuating electrotonic spread across the gap.

Chemical transmission involves membrane function in various ways:

1. The uptake of transmitters and/or their precursors by the external presynaptic nerve terminal and by the vesicle. Some pharmacological agents may exert specific blocking effects here; others, called "false transmitters," act by displacing the transmitter from its presynaptic store.

2. When the action potential arrives, the transmitter is released. The presence of $Ca^{2+}$ seems to be necessary for this process, but details of its function are hardly understood. There is also the well-known phenomenon of "quantized release." Has this to do with the vesicles? Are they the morphological counterparts of the quantized release, or is this a property of carriers or patches in the excited membrane?

3. Finally, there is the interaction of the transmitter with the postsynaptic membrane. The postsynaptic membrane responds by an increase in permeability to univalent cations, accompanied by depolarization, or by increase in permeability to chloride ions (or perhaps a decrease in permeability to potassium ions), leading to hyperpolarization, which forms the basis for central inhibition.

Interactions with the postsynaptic membrane are highly specific and stereospecific, both for transmitter substances and their antagonists.

Whittaker felt that the possibilities for more detailed studies of the membranes and other components of the synapse have been greatly enhanced by the preparation of isolated synaptic structures. A subcellular fraction, obtained by differential and density-gradient centrifugation of brain tissue (cortical grey matter) homogenized under carefully controlled conditions, consists mainly of presynaptic nerve terminals that have been torn away from their attachments. These retain all the structural features of the nerve endings, together with their content of transmitter substances, and have been called "synaptosomes." They are sealed structures; obviously the membrane recloses after rupture of the presynaptic structure from the nerve ending. They carry with them the portion of the postsynaptic membrane that faces the "gap," which must therefore consist of a fairly rigid, connective material that fixes together the pre- and postsynaptic membranes. After disruption of the synaptosomes in water, synaptic vesicles can be isolated by sucrose gradient centrifugation, together with other synaptosome subfractions (Figure 54). Eserine sulphate may be added to the water used to disrupt the synaptosomes, and the effect of the eserine is then to stabilize that part of the ACh which is not protected by its occlusion in the vesicle.

On transfer to mildly hypoosmotic solutions, the synaptosomes become transiently permeable to small molecules and then reseal; this process may be hastened by the addition of lecithin. This may be followed by measurements of lactic dehydrogenase (LDH) occlusion. The synaptosomes are metabolically active, as is shown by increases in

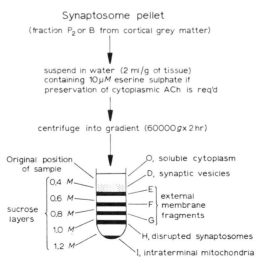

Figure 54. Scheme summarizing technique for subfractionating osmotically ruptured synaptosomes by density-gradient centrifuging. [Whittaker, 1969]

ATP or phosphocreatinine content on incubation with glucose and succinate. Fast uptake of choline (Ch) will take place during incubation. The Ch influx shows saturation behavior at high external Ch concentration. The influx is also strongly inhibited by hemicholinium (a structural analog of Ch). Both observations seem to indicate a carrier-facilitated mechanism. The uptake mechanism for Ch will also take up ACh, but the latter probably has a lower affinity for the hypothetical carrier. The two substrates compete. The uptake of ACh can only be demonstrated in the presence of an acetylcholinesterase (AChE) blocking agent. Even when this enzyme is almost completely inhibited ($>94\%$), there is the possibility that any apparent ACh uptake might be due to the uptake of Ch liberated by the nonenzymic hydrolysis of ACh, followed by intrasynaptosomal synthesis of ACh. It was found, however, that labeled acetate was not converted to ACh under the conditions of the experiments.

Whittaker reported that Colburn and his colleagues (1968) have demonstrated the uptake of NE in synaptosomes. This is $Na^+$ and $K^+$ dependent, the optimum being 140 mM $Na^+$ and 2 to 3 mM $K^+$. The uptake is inhibited by reserpine $10^{-8}$ M and has a maximum velocity of 0.1 $\mu g/min/g$ and a $K_m$ of 0.56 x $10^{-6}$ M. The uptake is also inhibited by ouabain and desmethylimiprimine. (A 200:1 concentration ratio can be obtained after a 10 min incubation at $37°C$ in the presence of substrates.)

Grahame-Smith and Parfitt (1970), using a millipore filtration technique, find that respiring synaptosomes will take up tryptophan. The uptake of radiotryptophan is greatly enhanced by preloading with tryptophan but is blocked by externally applied phenylalanine. Preloading with radiotryptophan and addition of cold tryptophan or any one of a number of related amino acids cause a high increase in the efflux of radiotryptophan.

Thus, there may be at least two transport systems or carriers for amino acids in the synaptosomal membrane, with overlapping specificities, only one of which is linked to $Na^+$. The degree of coupling between the two carriers may well depend on the presence of substrates interacting with both carriers and therefore be dependent on the metabolic state of the synaptosomes.*

---

*However, recent work suggests that the protein synthesis in the synaptosome fraction may be due mainly to contaminating particles consisting of endoplasmic reticulum enclosed within an external membrane (Morgan, 1970).

Synaptosomes not only possess mechanisms for uptake but also for incorporation of amino acids in proteins. The incorporation of radioleucine in proteins is dependent on an intact external membrane. Transfer of the leucine across the synaptosome membrane is a necessary preliminary to incorporation.

Kuriyama, Roberts, and Kakefuda (1968) have demonstrated an $Na^+$-dependent binding of $\gamma$-aminobutyric acid (GABA) to brain microsomes and synaptic vesicles. They propose that GABA could be transferred across neuronal membranes in regions of high to low $Na^+$-content.

Ling and Abdel-Latif (1968) have observed an $Mg^{2+}$ and energy-dependent transport system for $Na^+$ in synaptosomes. The efflux, but not the influx, of $Na^+$ is sensitive both to ouabain and to an iodoacetic acid plus cyanide ($IAA + CN^-$) combination. Hypoosmotic shock destroys the uptake. Uptake and efflux is temperature dependent. On adding $K^+$ to synaptosomes loaded with radioactive $Na^+$, there is an immediate release of $Na^+$, suggesting an exchange of $K^+$ and $Na^+$.

These examples cited by Whittaker demonstrate that synaptosomes have many of the carrier-mediated membrane-transport systems possessed by whole-cell preparations. Because the external membranes can be readily separated and isolated, synaptosomes would appear to be an excellent starting material for the separation of the membrane carriers, provided suitable test systems can be obtained for detecting them. Suitable methods must be worked out for transferring the carrier from the synaptosome membrane to artificial membranes that may eventually be developed for use in such a test system.

He also pointed out that an attractive feature of the synaptosome is the possibility for manipulation in vivo, with comparison of its properties to controls.


## HIGH-RESOLUTION ELECTRON MICROSCOPY APPLIED TO THE STUDY OF NERVE MEMBRANES: H. Fernandez-Moran

Elucidation of the molecular organization of cell membranes is a fundamental problem of biomedical research and a major challenge to further progress in molecular biology.

General characteristics of membrane organization have been formulated from correlated ultrastructural and biochemical studies. These features include coherent paucimolecular layers, which extend laterally and indefinitely and which appear to consist of a periodic hydrated lipoprotein substrate, integrated with specific macromolecular repeating subunits. These subunits are organized within the plane of the layers in asymmetric "paracrystalline" arrays (Fernandez-Moran, 1967). The elementary particle of the mitochondrion is considered to be a prototype of these repeating subunits or macromolecular assemblies found in association with membranes of all types (Fernandez-Moran et al., 1964; Schmitt, 1963).*

Based on these ultrastructural and biochemical studies (Fernandez-Moran, 1962; Fernandez-Moran et al., 1964; Green and Perdue, 1966), Changeux and his colleagues (1967) have identified and characterized the following major aspects of membrane organization:

1. Membranes are made up by the association of repeating macromolecular lipoprotein units.

2. The conformation of these units differs when (a) they are organized into a membrane structure or (b) dispersed in solution.

3. Many biological or artificial lipoprotein membranes respond in vivo, as well as in vitro, to the binding of specific ligands by some modification of their properties that reflects rearrangement of the membrane organization and presumably of the repeating units' conformation.

Specific enzymes or enzyme complexes are also associated with all membrane systems and are significant in determining their structural and functional organization (Fernandez-Moran, 1967).

The detection of DNA in mitochondria and in chloroplasts has provided important leads to the problem of membrane biosynthesis (André, 1965; André and Marinozzi, 1965; Nass, 1966; Swift et al., 1964; van Bruggen et al., 1966; Woodcock and Fernandez-Moran, 1968).

The research program at Fernandez-Moran's laboratory includes the following related problems of nerve membrane ultrastructure that are particularly suitable for correlated electron microscopic investigations:

1. Characterization of multienzyme and other macromolecular components closely associated with cell membranes for carrying out energy and information transduction functions.

---

*Since the time of the Work Session, relevant new findings have been published in Fernandez-Moran (1970a,b) and Fernandez-Moran et al. (1970).

2. Study of the association of nucleic acids and the protein synthetic machinery with cell membranes to gain a better understanding of membrane biosynthesis, including (a) study of deoxyribonucleic acid (DNA) and ribonucleic acid (RNA) conformations associated with membranes in chloroplasts, mitochondria, and in nerve cells, and (b) study of RNA polymerase and its participation in the differential RNA transcription upon DNA templates.

High-resolution electron microscopy and electron optics have now progressed to a stage where they can contribute significantly to these studies. Several new approaches in instrumentation and preparation techniques are responsible for these advances, which have resulted in the attainment of point-to-point resolutions of 2 to 3 A and which bring direct readout of molecular structures closer to reality. These technical advances and their applications have been described recently by Fernandez-Moran (1969a).

High-voltage electron microscopy is particularly promising for examining biological specimens because of the increased penetration power, reduction in radiation damage, and improved resolution. Fernandez-Moran has obtained resolutions of about 2.8 A to 6.0 A in crystalline lattices and 4 A point resolutions in 250 to 350 A-thick biological specimens. High-resolution electron diffraction carried out with the 200 kV microscope results in 50 to 100 diffractions for biological specimens as compared with the typical 5 to 10 diffractions obtained from lower voltage microscopes.

The development of cryoelectron microscopes operating with high-field superconducting solenoid lenses at liquid helium temperatures represents one of the most significant advances (Fernandez-Moran, 1965; 1966a,b). The instrument provides superstable lenses, ultrahigh vacuum, minimized specimen damage, contamination, and thermal noise, and enhanced image contrast. In earlier experiments, biological specimens were recorded at 4.2°K and revealed new electron optical phenomena (Figure 55).

A unique opportunity now opens up to apply simultaneously the advantages of superconducting microscopy and high-voltage electron microscopy to biological investigations. The new installations have provided us with the only existing high-voltage electron microscope with superconducting lenses operating routinely with a Collins closed-cycle superfluid helium refrigeration system. During preliminary experiments with this facility, an anomalous transparency effect for

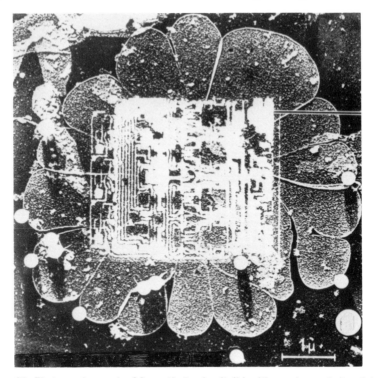

Figure 55. Electron micrograph of frog retinal rod unit disk with superimposed ultraminiaturized integrated circuit mask and text, demagnified $10^{-4}$ X electron-optically on organometallic thin film substrate. Prepared by Fernandez-Moran in collaboration with C. Hough.

200,000 V electrons in thick lead and niobium films (1000 to 2000 A) was discovered at liquid helium temperatures (i.e., $1.8°$ to $4.2°K$) (Figure 56). This remarkable effect extends previous observations by Boersch (1964) and Fernandez-Moran's group at 30 to 200 kV. It poses interesting theoretical and experimental questions, particularly because it is coupled with a marked decrease in radiation damage in both organic and inorganic specimens.

    With superconducting electron microscopy carried out on a regular practical basis, Fernandez-Moran felt it should be possible to pursue Gabor's wavefront reconstruction microscopy and high-resolution holography (Gabor, 1949), enhanced by the high stability of superconducting lenses and coherent microbeam illumination.

    In collaboration with Träuble, Fernandez-Moran's group plans to conduct correlated high-voltage electron microscopic studies of ion permeation in cell membranes and model systems with cyclic antibiotics (Pressman, 1965). Special vacuum-tight microchambers for examining thin membranes in their natural liquid state will also be used.

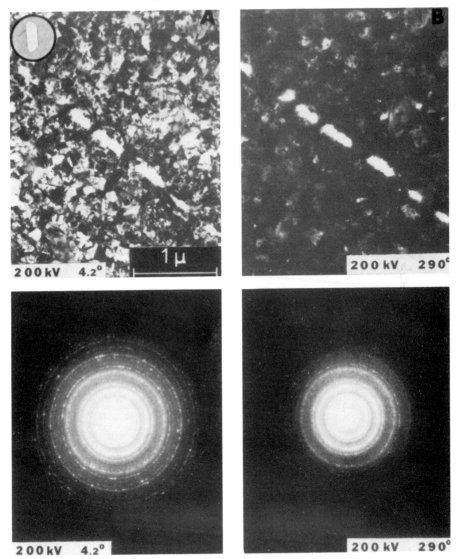

Figure 56. High-voltage electron micrographs and selected-area electron diffraction patterns of thick (about 1000 A) Pb film showing anomalous transparency at 4.2°K. [Fernandez-Moran]

In the optimal projection of these techniques, one can foresee integrated ultramicroelectric circuits, near the size of macromolecular assemblies, incorporated into key junctional sites of living nerve membranes without causing serious perturbation. These "submicroscopic prosthetic sensors," whose envelopes would be composed of

biosynthetically produced protein coats so as to form integral components of the nervous system, could be of unique service. Because they
could be produced in large number and inserted throughout the central
nervous system, the sensors could effect a direct operational link at the
macromolecular level between the central nervous system and man-
made information-processing systems, such as computers of commensu-
rate complexity (Fernandez-Moran, 1967).

The present state of knowledge about nerve membranes is but a
harbinger of many as yet unknown and unexpected properties of these
basic cell constituents. Electron microscopy in its broadest sense will
undoubtedly play an important role in this fascinating quest.

## VI. BIOLOGICAL FLOW STRUCTURES AND THEIR RELATION TO CHEMICODIFFUSIONAL COUPLING

### An Essay by A. Katchalsky

**Introduction**

Living beings are dynamic systems that undergo constant chemical and physical change even if their overall geometric pattern retains a stationary structure. According to the "central dogma" of molecular biophysics, all information about biological structure and function is coded in DNA, and the phenomenon of life is only the unfolding of the genetic script. Granted that the dogma may be accepted literally, the dynamics of biological structures still remains an open problem that deserves further consideration.

Recent work on the self-organization of suitable biopolymers into two-dimensional units, such as microtubules and membrane-like structures, lends support to the recognition that short-range forces play an important role in cellular organization. Biopolymer structures retain their functional capacity even when they are cooled in liquid air and then lyophilized; however, it is the dynamic aspects apparent only under the ordinary conditions of life that concern us in this paper. Examples of dynamic processes are the mobility of cell membranes seen in tissue culture, the contractility of the microtubules of the mitotic spindle, and the morphogenetic redistribution of cellular material in the fertilized egg. Another group of phenomena includes the periodic metabolic processes that impose a clocklike behavior on cells, and the pumping capability of biomembranes that underlies active transport. All these examples appear to be governed by flows and forces that may be treated macroscopically without explicitly considering short-range atomic or molecular forces.

The structure and maintenance of flow patterns by the coupling of macroscopic flows is well known in classical hydrodynamics and will be mentioned briefly.

The first to recognize that the coupling of chemical reaction and diffusional flow may also lead to instabilities and the formation of dynamic patterns was A. M. Turing. In his fundamental paper published in 1952 and provocatively entitled "The Chemical Basis of Morphogenesis," he presented the first mathematical model for biological dynamic structures. For many years Turing's paper failed to attract the interest of either theoreticians or experimentalists, but recently both chemical engineers and thermodynamicists engaged in biophysical research have found that the pioneering ideas of Turing lead to new insights into the nature of the dynamic structures that can be observed both in living systems and in industrial reactors. Turing's model considered stationary states far from equilibrium that may undergo transition from one structure to another. Fuller treatment of such systems requires mathematical methods more powerful than those used conventionally in the linear description of quasi-equilibrium processes. The difficulties of the nonlinear approach are amply rewarded by the interesting results, which provide novel intellectual approaches to the interpretation of living structures.

### The Bénard Phenomenon and Related Phenomena

The classic example of the maintenance of dynamic structures in hydrodynamics is the phenomenon of Bénard, which he discovered in 1900. In this experiment, the coupling of heat flow with convectional flow creates an instability that leads to formation of well-defined flow structures. To observe the Bénard phenomenon, one slowly and uniformly heats the bottom of a round vessel containing a dense liquid such as spermaceti. At a critical temperature, instability develops in the liquid, and circular regions having a lower refractive index become visible near the wall of the container. Subsequently, these regions break up into smaller cylindrical patches that wander toward the center of the vessel and gradually fill the whole liquid, forming a hexagonal array that strikingly resembles a honeycomb structure. It can be shown that each cell of the honeycomb is formed by the circulation of a single convectional flow, moving between the heated bottom and the cooler surface of the liquid. Evidently, the heating of the bottom gives rise not only to a heat flow upwards but also establishes a density gradient that causes a downward convection current. While the heat flow destabilizes the homogeneous structure of the liquid, the convection current moderates the heat effect and maintains stability. However, moderation is possible only as long as the heat input at the bottom is compensated by the output through the liquid surface. When the critical temperature is reached, the moderation fails, and the coupled heat and convection flows break the symmetry of the liquid and transform it into a new honeycomb pattern that can absorb further excesses of heat energy.

The flow pattern in the Bénard phenomenon exists only as long as energy is supplied for the maintenance of the dissipative flows participating in its formation. From this point of view, these "dissipative structures" differ markedly from "equilibrium structures," require no energy, and have greater structural stability at low temperatures due to decreased thermal motion. The notion of dissipative structures, introduced recently by Prigogine (1969), not only requires a source of dissipative energy but also indicates a narrow range of stability. While equilibrium structures are usually maintained for long periods and have a wide range of external parameters, the correct matching of coupled flows to give stationary dissipative structures occurs only within a narrow range of driving forces with well-defined upper and lower bounds (Rayleigh, 1916). Because the dynamic patterns of life are also based on a heat-energy input and can exist only within narrow limits, it is attractive to conjecture that these patterns, too, are dissipative structures.

The quantitative aspect of the Bénard phenomenon need not concern us here; it will suffice to mention that Lord Rayleigh studied it carefully in 1916 and that it was discussed in great detail by Chandrasekhar (1961) in his well-known treatise on hydrodynamic and hydromagnetic stability.* In this treatise the reader will find experimental details and some fine photographs of the phenomenon. Pictures of the development of the Bénard pattern are also shown in a recent paper by Gmitro and Scriven (1966). (See Figures 57 and 58.)

More recently it has been observed that *Euglena viridis* grown on shallow dishes exhibits hexagonal distributions of microorganisms that bear striking resemblance to the honeycomb structures of Bénard; another group of findings, related to nucleation and crystallization processes, was discussed and evaluated quantitatively in the colloid chemical literature by Cahn (1965). Although in each case the flows and forces are different, there is always a common denominator: formally, the causative factors are coupled, and the existence (or development) of the pattern is based on a supply of energy dissipated for the maintenance of the structure. Because the most important flows within the cell are those of chemical reaction and diffusional transport, our next step will be to consider the possibility of structuring through the coupling of these flows.

*Note: The Bénard instability occurs when the dimensionless Rayleigh number $gh^3/D\eta \; \Delta\rho/\rho$ reaches a critical value of about 650. Here $g$ is the gravitational constant, $h$ the depth of the liquid, $D$ the self-diffusion constant, $\eta$ the viscosity, $\rho$ the density, and $\Delta\rho$ the difference between the top and bottom densities.

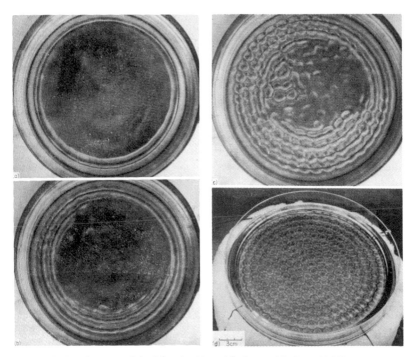

Figure 57. Development of the Bénard pattern. [Gmitro and Scriven, 1966]

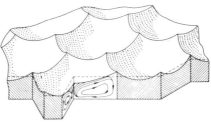

Figure 58. Schematic diagram of Bénard cells, showing streamlines of flow within a single "honeycomb" cell. [Gmitro and Scriven, 1966]

## Turing's Model and the Stability of Stationary States

The system considered by Turing was a series of identical cells created in a homogeneous medium by diffusional flows. For the sake of simplicity, we shall consider only two cells, each containing two reacting species of concentrations $x$ and $y$. If the permeability coefficients across the intercellular membranes are $P_x$ and $P_y$, then the rate of loss of $x$ and $y$ from cell 1 to 2 are given in the simplest case by $-P_x(x_1 - x_2)$ and $-P_y(y_1 - y_2)$. Turing used for his model an imaginary reaction, which has no real counterpart but which demonstrates clearly the basic ideas. The rate of reaction of $x$ and $y$ in cell 1 is given by

$$\left(\frac{dx_1}{dt}\right)^{\text{reaction}} = 5x_1 - 6y_1 + 1$$

$$\left(\frac{dy_1}{dt}\right)^{\text{reaction}} = 6x_1 - 7y_1 + 1$$

The reader will realize that the rate constants are normalized; the number 1 is introduced for mathematical convenience. The normalized values of 0.5 and 4.5 were chosen for $P_x$ and $P_y$, respectively; therefore the total change of the number of moles $x_1$ and $y_1$ per unit time is

$$\frac{dx_1}{dt} = 5x_1 - 6y_1 + 1 - 0.5(x_1 - x_2)$$

$$\frac{dy_1}{dt} = 6x_1 - 7y_1 + 1 - 4.5(y_1 - y_2)$$

(1)

and a similar pair of equations can be written for cell 2.

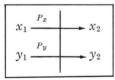

It will now be observed that the pruperties of Equation 1 have a stationary solution when $x_1 = x_2 = 1$ and $y_1 = y_2 = 1$. In this case, $dx_1/dt = 0$ and $dy_1/dt = 0$ so that the concentrations in both cells are equal and do not change with time. The interesting problem is whether this stationary and homogeneous solution is stable or whether a slight shift in the concentrations, due, say, to a thermal fluctuation, will cause a drift toward another, inhomogeneous solution. An adequate test for the stability of the steady state is the sign of $dx_1/dt$ and $dy_1/dt$ for a positive fluctuation in $x_1$ and $y_1$. If a small increase in $x_1$ and $y_1$ makes $dx_1/dt$ and $dy_1/dt$ negative, then the system will revert on its own to the initial value, and the steady state is stable. If, on the other hand, $dx_1/dt$ and $dy_1/dt$ become positive, the fluctuation will be amplified, and the system will not return to its initial state but will proceed toward a new state, which may be inhomogeneous. It may readily be seen that for extremely small fluctuations $dx_1/dt$ and $dy_1/dt < 0$; a fluctuation of a few percent, however, makes $dx_1/dt$ and $dy_1/dt > 0$. Thus, for $x_1 = 1.06$ and $x_2 = 0.94$ as well as for $y_1 = 1.02$ and $y_2 = 0.98$, $dx_1/dt = 0.12$ and $dy_1/dt = 0.04 > 0$, which indicates that matter will accumulate in cell 1 and be depleted in cell 2, thus breaking the symmetry of the system and establishing an inhomogeneous distribution of the components.

Prigogine and his co-workers Nicolis and Lefever studied a more realistic process than that of Turing, based on the following scheme (Prigogine, 1969).

$$A \xrightarrow{K_1} x \tag{2a}$$

$$2x + y \xrightarrow{K_2} 3x \tag{2b}$$

$$B + x \xrightarrow{K_3} y + D \tag{2c}$$

$$x \xrightarrow{K_4} E \tag{2d}$$

$$A + B \longrightarrow D + E$$

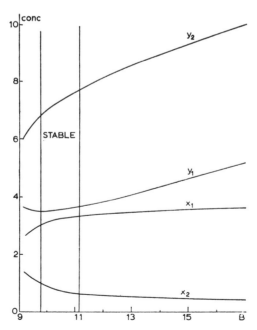

Figure 59. Steady states for systems beyond the critical point. See text and Equation 2. [Prigogine, 1969]

In this process, the reactants $A$ and $B$ undergo a transformation to the reaction products $D$ and $E$ via the intermittent forms $X$ and $Y$. It will be observed that the scheme comprises an autocatalytic step (Equation 2b), which may be moderated by a diffusional process across the intercellular membrane. The equations were solved by an appropriate choice of rate and permeability coefficients. It was shown that at a given critical concentration of the reactant $B$, holding $A$ constant, an instability ensues, and the concentrations of $X$ and $Y$ in compartments 1 and 2 begin to differ. However, there is a whole range of $B$ concentrations in which the inhomogeneous solution is stationary (see Figure 59).

Generalizing this two-component analysis, Lefever (1968) calculated the stable distribution of matter in a 50-cell system for the reaction given by Equation 2. Figure 60 illustrates his results and convincingly demonstrates the "morphogenetic" possibilities of chemicodiffusional coupling.

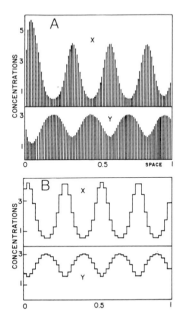

Figure 60A: Space-dependent steady state distribution for a system with fixed $X$ and $Y$ concentrations at the boundaries (points 1 and 101, where the values of $X$ and $Y$ are maintained equal to the homogeneous steady-state value; $X = 2$ and $Y = 2.62$). B: Steady-state distribution for a system of 50 boxes with periodic boundary conditions. Numerical values in both cases: $D_X = 0.0016, D_Y = 0.008, A = 2, B = 5.24$. [Lefever, 1968]

Our next problem is to determine which of the stationary solutions are stable, for it is only the stable stationary states that are biophysically significant. Further, we may ask whether there is only one inhomogeneous stationary stable state, or whether there is a whole spectrum of steady states with characteristic bandwidths and specific transition probabilities from one stable state to another. To date, only a partial answer has been given, and in the next paragraph a thermodynamic approach is outlined.

### A Thermodynamic Criterion for the Stability of Stationary States

For many purposes, the mathematical criterion of stability used by Turing (1952) is a satisfactory tool. It is, however, natural to look for a physical principle that may determine whether a stationary state is stable or not. Although no general statement can be made at present, an interesting advance was made by Prigogine (1969) and his colleagues, which is briefly summarized below. This treatment by the Brussels school is based on the extension of nonequilibrium thermodynamics concepts into the nonlinear range.

As is well known, the basic function of the thermodynamics or irreversible processes is the entropy product $d_iS/dt$ or the dissipation function $\phi = T(d_iS/dt)$, the latter representing the degradation of free energy by dissipative flows. In the range of validity of Gibbs's equation (which is wider than the linear range of the Onsager relations, but not wide enough to describe very rapid processes), the dissipation function equals the sum of the products of irreversible flows $J_i$ and their conjugate forces $X_i$; i.e.,

$$\phi = \sum_i J_i X_i \qquad (3)$$

The flow $J_i$ may be a diffusional flow, a flow of electric current, or the advancement of a chemical reaction, and the conjugate forces $X_i$ will be the negative gradient of the corresponding chemical potential, the electrical field, and the affinity of the reaction.

The total change of $\phi$ is given by

$$d\phi = \sum J_i dX_i + \sum X_i dJ_i = d_X\phi + d_J\phi \qquad (4)$$

If the boundaries of the system under consideration are fixed, and the potentials on the boundary are held constant, the system will of itself tend toward a steady state. It can be shown that if the flows depend linearly on the forces, then the development of the system toward stationarity is governed by the requirement that

$$d\phi \leqslant 0 \qquad (5)$$

This means that the dissipation function decreases with time and reaches a minimal value at the steady state. This is the well-known principle of minimum dissipation. However, Equation 5 does not hold for fast or for nonlinear processes, and it is doubtful whether it may be applied to living systems. In an attempt to "save the phenomena," Glansdorff and Prigogine (1954) proved that although in the general case the total dissipation function $\phi$ does not reach a minimum at steady state, and Equation 5 does not hold, part of Equation 4 still develops regularly, and one may write

$$d_X\phi = \Sigma J_i dX_i \leqslant 0 \qquad (6)$$

Thus it is the term $d_X\phi$ that is regarded by the authors as a general criterion for the evaluation of natural systems. At steady state, $d_X\phi = 0$ and $\phi$ is minimal, so that if the system is stable, any variation will increase $\phi$, or the criterion of stability is

$$(\delta_X\phi) \geqslant 0 \qquad (7)$$

at steady state.

Equation 7 may be written explicitly as

$$\delta_X\phi = \sum_i J_i \delta X_i \geqslant 0$$

Expanding $J_i$ in a Taylor series around its steady-state value $J^{SS}$,

$$J_i = J_i{}^{SS} + \sum_j \left(\frac{\delta J_i}{\delta X_j}\right)^{SS} \delta X_j$$

and

$$\delta_X\phi = \sum_i (J_i)^{SS} \delta X_i + \sum_{ij} \left(\frac{\delta J_i}{\delta X_j}\right)^{SS} \delta X_j \delta X_i \geqslant 0$$

However, at steady state, $\sum J_i{}^{SS} \delta X_i = (\delta_X\phi)^{SS} = 0$

and $\sum_j \left(\frac{\delta J_i}{\delta X_j}\right)^{SS} \delta X_j = \delta J_i$ therefore $\delta_X\phi = \Sigma \delta J_i \delta X_i \geqslant 0 \qquad (8)$

This last expression (Equation 8) was used by Prigogine, Nicolis, and Lefever in selecting a stable range of stationary states from the continuum of states shown in Figure 59.

To make Equation 8 more explicit, we shall illustrate its use in the following examples:

Let us consider first a bimolecular reaction

$$x + B \xrightarrow{K} C$$

in which $B$ and $C$ are constant concentrations. Assuming that the reaction proceeds from left to right, the flow of reaction will be given by

$$J_r = K \cdot x \cdot B \text{ and } \delta J_r = K \cdot B \cdot \delta_x$$

For the affinity of the reaction we can write

$$A = \mu_x + \mu_B - \mu_C = \mu_x{}^{\circ} + \mu_B{}^{\circ} - \mu_C{}^{\circ} + RT \ln \frac{x \cdot B}{C}$$

Hence $\delta A = \delta x (RT/x)$, because all other terms are constant. Inserting $J_r$ and $\delta A$ into Equation 8, we find that

$$\delta_x \phi = \delta J_r \delta A = \frac{KBRT}{x} (\delta x)^2 \geqslant 0$$

Thus $\delta_x \phi$ is positive, whatever the sign of $\delta x$ may be. The simple bimolecular reaction is therefore stable in the entire range. On the other hand, an autocatalytic reaction is found to be destabilizing; for example,

$$x + B \xrightarrow{K} 2x$$

Here, too,

$$J_r = KxB \qquad \delta J_r = KB \delta x$$

but the affinity is different

$$A = \mu_x + \mu_B - 2\mu_x = \mu_B - \mu_x = \mu_B{}^{\circ} - \mu_x{}^{\circ} + RT \ln \frac{B}{x}$$

$$\delta A = - \frac{RT}{x} \delta x$$

so that

$$\delta_x \phi = \delta J_r \delta A = - \frac{KBRT}{x} (\delta x)^2 \leqslant 0$$

Thus certain reaction steps may be stabilizing, while others are destabilizing, and the sum total of the terms in Equation 8 may be positive or negative as a function of concentration.

The general procedure for the prediction, a priori, of the number of possible stationary states, the stability of these states, and the regime of stable stationary states, is still under investigation.

**Continuous Systems**

*Conservation of Mass*

The treatment in the preceding paragraphs is readily applicable to continuous systems. Such systems are adequate models for the reaction column of the chemical engineers and also for the behavior of a continuous medium such as a biological membrane.

The starting point for the latter calculation is the equation of continuity that is another expression for the local conservation of mass. Denoting the local concentration of the $i$th component by $c_i$, the concentration change with time $\partial c_i / \partial t$ is given by the negative divergence of the diffusional flow $-\text{div } J_i^d$ and the change of the $i$th component through chemical reaction $\nu_i J_r$ (where $\nu_i$ is the stoichiometric coefficient and $J_r$ the unit chemical change); i.e.,

$$\frac{\partial c_i}{\partial t} = -\text{div } J_i^d + \nu_i J_r \tag{9}$$

If the $i$th component participates in several reactions, with a stoichiometric coefficient $\nu_{ik}$ included for the $k$th chemical process, Equation 9 should be written as

$$\frac{\partial c_i}{\partial t} = -\text{div } J_i^d + \sum_k \nu_{ik} J_r^k \tag{10}$$

Without going into detailed analysis of the treatment of the $n$ equations (Glansdorff and Prigogine, 1954) that correspond to the $n$ components participating in the processes, we propose to outline the general procedure.

In the first step one assigns an explicit form to the diffusional and reaction flows. A convenient form for $J_i^d$ is

$$J_i^d = \sum_j D_{ij} \nabla(-c_j) \tag{11}$$

where $D_{ii}$ is a straight diffusion coefficient, relating the diffusional flow of the $i$th component to the gradient of its own concentration $(-c_i)$, and $D_{ij}$ is a coupling coefficient relating $J_i^d$ to $\nabla(-c_j)$. It is arbitrarily assumed that all the $D$s are constant, so that

$$-\text{div } J_i^d = \sum_j D_{ij} \nabla^2(c_j) \tag{12}$$

For the rate of the $k$th chemical process,

$$J_r{}^k = \left(K_f{}^k \prod_{j=1}^{q_k} c_j{}^{-\nu_{jk}}\right) - \left(K_r{}^k \prod_{j=q_k+1}^{n} c_j{}^{\nu_{jk}}\right) \tag{13}$$

where $K_f{}^k$ and $K_r{}^k$ are the rate constants for the forward and backward reactions, respectively, and $q_k$ denotes the number of reactants.

For the stationary state during which $\partial c_i/\partial t = 0$, we next solve Equation 10. There are several stationary solutions for $c_i$; it is then necessary to find out which are physically valid and which are stable. Physical validity is determined by the requirement that $\bar{c}_i$ cannot be negative, and for stability one must consider small deviations $\alpha_i$ from the stationary state; namely,

$$c_i = \bar{c}_i + \alpha_i \tag{14}$$

where

$$\alpha_i \ll \bar{c}_i \tag{15}$$

Correspondingly, the reaction rate may be expanded about its steady state, using Taylor's theorem. Thus, for the $k$th reaction,

$$J_r{}^k = J_r{}^k + \sum_j \left(\frac{\partial J_r{}^k}{\partial c_j}\right)^{SS} \cdot \alpha_j \tag{16}$$

By inserting Equations 14 and 16 into 10 and noting that the equations are linear and homogeneous, we obtain with the aid of Equation 15 the following deviation equation:

$$\frac{\partial \alpha_i}{\partial t} = \sum_j D_{ij} \nabla^2 \alpha_j - \sum_j K_{ij} \alpha_j \tag{17}$$

where the generalized rate constants

$$K_{ij} = \sum_k \nu_{ik} \left(\frac{\partial J_r{}^k}{\partial c_j}\right) \alpha_j \tag{18}$$

Equation 17 may be expressed in a more elegant form by introducing the vector $\vec{\alpha}$ of the deviation from the steady state

$$\vec{\alpha} = \begin{pmatrix} \alpha_1 \\ \alpha_2 \\ \vdots \\ \alpha_r \end{pmatrix}$$

the matrices for the diffusion coefficients

$$(D) = \begin{pmatrix} D_{11} & \ldots & D_{1n} \\ & & \\ D_{n1} & \ldots & D_{nn} \end{pmatrix}$$

and for the generalized rate constants

$$(K) = \begin{pmatrix} K_{11} & \ldots & K_{1n} \\ & & \\ K_{n1} & \ldots & K_{nn} \end{pmatrix}$$

Thus

$$\frac{\partial \vec{\alpha}}{\partial t} = (D)\nabla^2 \vec{\alpha} - (K)\vec{\alpha} \tag{19}$$

### Organization in Space and Time

The solution of Equation 19 is separable into space-dependent and time-dependent parts whose exact forms need not concern us here. It is important to note, however, that if more than one chemical reaction is involved in the process, the time-independent part of the solution predicts a spatial distribution of matter, with formation of dissipative structures. (See Figure 60.) In addition, there may be periodic, time-dependent solutions that predict chemical oscillations, the foundation for biological clocks.

Some of the space-dependent expressions are related to the phenomena of spherical harmonics that are well known to the physical chemist familiar with quantum mechanics. Gmitro and Scriven (1966) have presented a simple case, which is shown in Figure 61. As these authors point out, the spatial distribution of reactants on a contractile biopolymer system may lead to the development of mechanical stresses; moreover, the periodic spatial solutions of Equation 19 may represent both standing and propagating waves. The propagation of mechanical waves in cells or across biomembranes can play a role in intracellular communication of chemical information, or function as a pumping device for active transport.

The existence of time-dependent oscillations, implied in Equation 19, has been a long-standing problem in theoretical biology. Although they had been predicted many decades ago by Lotka (1957), only recently have chemical oscillations been demonstrated in the test tube. This experimental breakthrough was achieved by Chance and his

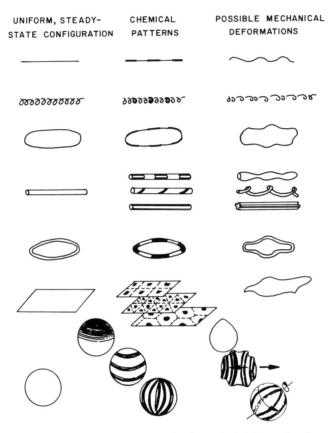

Figure 61. Some space-dependent diagrams related to spherical harmonics; the propagation of mechanical waves. [Gmitro and Scriven, 1966]

co-workers (1965) who succeeded in isolating an oscillating enzymatic system. Several years ago a synthetic "chemical clock" was prepared by Zhabotinsky (1967) in the Soviet Union; because most of his work has been published in Russian, it is summarized here.

Zhabotinsky's experiment is based on the catalytic oxidation of malonic, bromomalonic, or citric acid by $KBrO_3$ in a strongly acidic solution; low concentrations of $Ce_2(SO_4)_3$ are used as a catalyst. We repeated his observation successfully, using 1 M malonic acid; 0.27 M $KBrO_3$, and $10^{-3}$ M $Ce_2(SO_4)_3$ dissolved in 3 M $H_2SO_4$. After an induction period of 20 min, the adsorption band of tetravalent cerium at 3650 A showed a pronounced oscillation, with a period of about 1 min when the experiment was carried out at room temperature.

At higher temperatures, the frequency increased exponentially. Although the overall reaction scheme may be written as

$$Ce^{3+} \xrightarrow{KBrO_3} Ce^{4+}, \quad Ce^{4+} \xrightarrow{Malonate} Ce^{3+}$$

intermediate compounds whose concentrations reach a steady state during the induction period seem to be formed. The oscillations continue with a relatively constant amplitude for over an hour and may be used to demonstrate a "man-made biological clock" in university classes. Zhabotinsky studied the appearance of oscillations as a function of the concentrations of the components and found that a well-defined concentration range exists for the formation of time-dependent dissipative structures. (See Figure 62.)

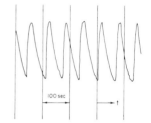

Figure 62. The oscillations of the light absorption at 3650 A (filter UFS6) produced by the periodic change in the ions $Ce^{4+}$. [Zhabotinsky, 1967]

While the Zhabotinsky experiment is performed with efficient stirring, a space-dependent structure may be obtained if the reaction mixture is kept undisturbed. This was recently shown by Prigogine and his co-workers* who observed a beautiful band structure in which the cerium catalyst accumulates parallel to the bottom of the test tube.

## Dissipative Peptide Synthesis

As shown recently in the laboratory, the consideration of dissipative structures may help in the understanding of the prebiotic synthesis of polypeptides. In 1943, Bernal pointed out that the formation of monomers under the conditions of the primeval ocean could not account fully for prebiological synthesis. The formation of primitive biopolymers in a homogeneous aqueous medium such as the ocean is highly improbable, and, according to Bernal, it would require the presence of a heterogeneous catalytic surface that could both accumulate the monomers and protect them from hydrolysis. Although the interesting experiments of Fox (1965; Krampitz and Fox, 1969) and of Steinman (1967; Steinman and Cole, 1967) make plausible some polycondensation mechanisms, further work is required for the development of a realistic mechanism for prebiotic polymer development.

*Private communication to A. Katchalsky.

In my laboratory at the Weizmann Institute we started the polycondensation of active amino acid derivatives in aqueous media at room temperature and at pH's close to 7. The main derivatives are the adenylates of amino acids that are known to be intermediates in protein synthesis in the living cell. Because both the synthesis of amino acids and of ATP under primeval conditions have been demonstrated by other workers (Ponnamperuma et al., 1963), the spontaneous formation of amino acid adenylates is plausible. Paecht-Horowitz, Berger, and Katchalsky (1970) have shown that in homogeneous aqueous solution, the adenylates polymerize to low polypeptides whose molecular weights are, however, too low to be considered as serious candidates for the prebiological evolution of biopolymers; but when montmorillonite is added to the reaction mixture, relatively high polypeptides are formed rapidly and in good yield. It is known that montmorillonite is an infinitely swelling clay that in aqueous media gives a suspension of platelets of very high specific surface tension. Preliminary experiments have demonstrated that while neither the amino acid nor adenylic acid adsorb onto the montmorillonite surface, the amino acid adenylate adsorbs strongly and is thus protected from the hydrolytic reaction that disturbs the polycondensation process.

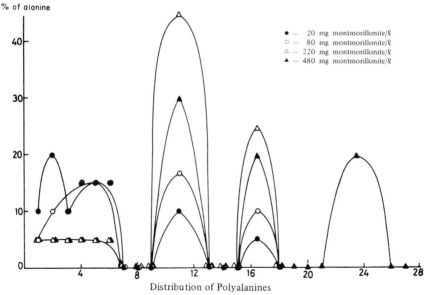

Figure 63. Distribution of the molecular weights of polyalanines, obtained when alanine-adenylate is introduced into an aqueous solution at pH 8.5, in the presence of various amounts of montmorillonite. [Katchalsky]

The most remarkable finding in this study was that the distribution of molecular weights in the synthesized polypeptides is not continuous, as is the case in aqueous polymerization, but forms a discontinuous band structure, as shown in Figure 63.

Having shown earlier that a low pH will diminish the rate of polycondensation, it is reasonable to assume that the adenylic acid liberated in the process inhibits peptide formation. However, when adenylic acid diffuses out of the material surrounding the montmorillonite particles, the polycondensation process is renewed until enough adenylic acid is formed to stop the process again. Polymerization is therefore controlled by the coupling of a chemical and a diffusional process, and the polypeptide formed reflects the existence of a dissipative structure that develops on the surface of the heterogeneous catalyst. It is therefore clear that dissipative structures may not only play a major role in the maintenance of certain cellular patterns, but they may also have participated in the development of those early structures from which life arose.

## VII. FINAL REMARKS: by L. De Maeyer

### Concepts and Definitions

The definitions and classifications of selective transport processes seem to be quite clear with regard to energies; the important distinction made in biological transport processes is between "active" and "passive" transport. Active transport is characterized by a coupling between transport and an overall chemical change, which supplies (or absorbs) the energy required for (or produced by) the movement of the transported species. This does not necessarily mean that the chemical reaction takes place directly at the site of transport, but the coupling of reaction and transport is important. If the coupling is not to vanish in reversible equilibrium, an anisotropic system is required, whereas in isotropic systems such couplings can exist in stationary states only.

The distinction between "active" and "passive" therefore applies to the energy source (or driving force), rather than to specific mechanisms of transport processes. The "active" transport of a chemical species must be driven directly by a chemical affinity. "Passive" transport may involve a variety of driving forces, especially those arising from gradients in electrochemical potentials and from couplings between fluxes of different species. (Eventually it also involves those that are, or have been, "actively" transported.) One of the ultimate aims of the study of biological transport phenomena should be to determine which species are involved in the primary active transport processes. These will be the species that derive their energy from metabolic chemical reactions and thereby become the "prime movers" for the directed exchange of matter between cells, or membrane-enclosed cell components, and their environment. This classification problem clearly has not yet been resolved, but it is related to the subject of this *Bulletin* from two aspects:

1. Do carrier molecules of the type discussed, or of a more complex type (that eventually may undergo chemical transformation during transport), provide possible mechanisms for active transport processes?

2. Is it possible to use these simple carrier molecules, by interfering with natural transport phenomena, either in real or in model systems, to elucidate these phenomena in more detail?

It seems that the answer to the first question cannot yet be affirmative. Carrier molecules of relatively small molecular weight, such as those that have been studied in detail so far, do provide mechanisms for facilitated and selective transport. There is no overall chemical reaction involved, and they act merely as transport-catalysts. However, more complex carrier molecules of this kind have not yet been characterized. Having analyzed the main structural features and architecture of simple carriers may of course help us in a search for other membrane constituents that may be related to more complex transport phenomena. But as long as such substances have not been identified, it seems relatively useless to anticipate mechanisms for complex phenomena.

Our interest should be focused on the second question. Investigations of this kind are reported in this *Bulletin*. At this time they aim, more or less directly, toward the properties of selectivity with which natural membranes are endowed; i.e., they examine the differential behavior of transported species of different kinds.

These properties of selectivity are hard to define in an appropriate way. Although quantitatable definitions could be stated in different forms (e.g., selectivity ratio = ratio of species fluxes per unit driving force), the quantities thus defined are dependent on an unknown number of factors and parameters and do not necessarily lead to unequivocal characterization of different situations. Thus, the selectivity ratio for the facilitated transport of $K^+$ versus $Na^+$ by a given carrier molecule may still depend upon the composition of the membrane. A simultaneous measurement of the net fluxes of the different species under specified conditions of driving force may be rather difficult, especially with nonstationary states that might be involved in kinetic studies.

It is therefore necessary that any measure of selectivity given to characterize a set of experimental observations should be interpreted carefully with respect to the methods and conditions used in the experiment, including the nature of the driving forces, the composition of the membrane, the bulk solutions, and so on.

Besides comparative studies of carrier-facilitated transport of ions of different kinds, detailed studies of the mechanism of single-ion transport by carrier substances are needed. The artificial bilayer technique has been used for this purpose, but it seems that experimental approaches using bilayer vesicles, obtained by the different techniques that have been discussed, can provide additional information, especially about kinetic parameters.

The complexity of the mechanism of carrier-facilitated transport even in simple systems is borne out by the experiments, reported here by Tosteson, with the bilayer technique in the presence of monactin and $K^+$ ions. At zero current the potential difference between both sides of the membrane is entirely determined by the potassium concentration, and one might conceive that the flow of potassium ions should be a simple function of the driving force. It was shown, however, that under certain circumstances more than half of the current is carried by hydrogen ions.

Clarification of the detailed mechanism of carrier-facilitated transport therefore requires elaborate studies not only of steady state or of integral fluxes, but separate measurements of individual contributions, especially during transient perturbations. For this purpose, new experimental techniques may have to be elaborated.

The validity of statistical-mechanical approaches to the description of organized systems comprising only a small number of molecules has been dealt with briefly, but most experts among the participants agreed that this does not pose special problems, as long as the predictions are not concerned with the behavior of the system in extremely short time periods (shorter than required for the repartition of energy). This also pertains with regard to the concept of diffusion, which remains valid even if the diffusional processes involve movements over separations of a few molecular distances, as in phase boundaries or in lipid bilayers.

The ultimate description of the metabolism-coupled transport phenomena in, and into, a living cell, enabling the cell to maintain steady-state dynamic equilibria with respect to its environment, will require an approach more fundamental than the rather simplified models and artificial systems from which we have gained our present understanding of the structure-function-mechanism relationships of the biologically important classes of molecules. The dynamic theory of dissipative nonequilibrium processes is needed here. Its current development and its importance with respect to the origin of organizing forces responsible for accumulation and dispersion of the molecular components of a living organism is reviewed by Katchalsky in a contribution that has been included in extenso in this Work Session report.

The predictions of this theory about the behavior of macroscopic observables in states far removed from equilibrium provide us with reliable criteria to judge the models and hypotheses about the molecular processes involved in the microscopic machinery of living systems in much the same way that thermodynamic equilibrium theory has supplied the main guidelines for the investigation of the static inanimate physical world.

## Prospects for Biological Transport Processes

The concept of carrier-facilitated transport may explain, at least on an experimental basis (and supported by our understanding of the mechanism of specific ionophore-ion binding), some of the features of selectivity and specificity observed in biological transport processes. Without specific mechanisms in mind, the established terminology, especially in nerve literature (e.g., the Hodgkin-Huxley model) deals with these selective transport processes in terms of channels (sodium channels, potassium channels, etc.). It is of course tempting to look for any hints or evidence for the existence of carrier molecules in biological membranes, especially in nerve membranes.

Recently, McIlwain demonstrated that the reentry of K after nerve excitation is inhibited by basic proteins (histones, protamin, poly-L-lysine). But excitability can be restored by adding gangliosides that have been extracted by chloroform-methanol from nervous tissues. Also, hyperpolarization induced by clupein is restored by gangliosides (McIlwain, 1960). Although such findings are not inconsistent with the idea that some lipophilic material is extracted that is connected with the function, this is certainly not sufficient proof of the existence of extractable molecules with a carrier function. If there were small molecules with carrier functions similar to those of the antibiotics that have been discussed, in all probability they would have been found.

An argument against the hypothesis that the carrier functions would be the business of small molecules, of molecular weight of a few thousand or less, is that they could hardly be confined to certain membranes only, because they would be released too easily and interfere with the assigned functions of other membranes. A better strategy seems to be to look for larger substances, proteins or lipo-proteins, perhaps that show the specific structure of carbonyl groups that for example would enable monactin or valinomycin to interact specifically with alkali ions. It would also be interesting to study from this viewpoint the alkali-ion activated enzymes, where the ions may be required to shape the polypeptide structure in a specific conformation by multiple interactions with oxygen atoms.

There was consensus among participants of the Work Session that the higher molecular weight substances are the more probable candidates for specific transport properties, especially for those properties that require the subtler control. Eisenman stressed the fact, however, that it would be premature to rule out a priori the participation of light molecules; their function and synthesis in the fungi that are now known to produce them could be studied more carefully. The argument that small molecules would be ubiquitous and tend to make uniform the transport properties of all membranes does not necessarily hold if their function is controlled by membrane composition; this may also be the reason for the observed species-sensitivity effect of valinomycin on bacteria.

A hierarchy of carriers with increasing complexity will be required. Not only simple ionic species such as alkali ions are to be transported selectively but also a large variety of organic molecules, including the transmitter substances at the postsynaptic membrane. These would probably require a transport mechanism with very sensitive means of control. Another class comprises the actively transported substances that require coupling to ATPase.

The techniques for extraction of proteins, for example from erythrocyte ghosts, have recently been much improved, and special chemical methods (such as poly-d, l, alanylization) have been developed for protein solubilization. It may therefore be expected that substances will become available in the near future for characterization.

An interesting approach, especially for the ATP-linked transport system in bacteria, is the study of mutants with deficient transport systems.

The problem of reincorporation may possibly be related to the problem of assaying techniques for identifying isolated carrier and membrane components, defining their role in transport chains or their receptor properties.

For small molecular weight substances, the bilayer technique as developed by Mueller and Rudin (1962a,b) seems to be an excellent and sensitive assay (e.g., able to detect $10^{-11}$ M monactin). For larger substances, refinements may be needed, eventually including immunological characterization.

It may be concluded that if our basic understanding of the mechanism of selective transport phenomena in natural systems is still very vague, our arsenal of techniques, model systems, and concepts has greatly improved in recent years. The availability of macroscopic artificial bilayers and of small bilayer vesicles allows a variety of experiments aiming not only at the specificity but also at the controllability of the transport phenomena by external parameters. These may be the direct effects of large electric fields, or of charge accumulation, or of indirect field-induced conformational changes, or of field- or ligand-dependent cooperative phenomena. Small selective carrier molecules have been isolated, the principle of their specificity has been defined, and they can even be purposely synthesized with specific properties in mind. Fractionation of natural systems has been perfected, although the problem of isolation and meaningful assay techniques is still critical. Even if the transport phenomena in the membrane are among the simplest aspects of the specific functioning of the neuron, they provide a challenging task to the biochemist and physical chemist who would explore the molecular mechanisms that make "intelligence" possible.

## List of Abbreviations

| | |
|---|---|
| ACh | acetylcholine |
| AChE | acetylcholinesterase |
| ATP | adenosine triphosphate |
| ATPase | adenosine triphosphatase |
| CD | circular dichroism |
| Ch | choline |
| CNS | central nervous system |
| DEAE | diethylaminoethyl |
| DNA | deoxyribonucleic acid |
| GABA | $\gamma$-aminobutyric acid |
| LDH | lactic dehydrogenase |
| MD | monactin-dinactin |
| M.I.C. | minimal inhibitory concentration |
| NE | norepinephrine |
| NMR | nuclear magnetic resonance |
| ORD | optical rotatory dispersion |
| RNA | ribonucleic acid |
| VPO | vapor pressure osmometry |

## BIBLIOGRAPHY

This bibliography contains two types of entries: (1) citations given or work alluded to in the report, and (2) additional references to pertinent literature by conference participants and others. Citations in group (1) may be found in the text on the pages listed in the right-hand column.

Page

Abrams, A. (1965): The release of bound adenosine triphosphatase from isolated bacterial membranes and the properties of the solubilized enzyme. *J. Biol. Chem.* 240:3675-3681.    381

Abrams, A. and Baron, C. (1967): The isolation and subunit structure of streptococcal membrane adenosine triphosphatase. *Biochemistry* 6:225-229.    381

Abrams, A. and Baron, C. (1968): Reversible attachment of adenosine triphosphatase to streptococcal membranes and the effect of magnesium ions. *Biochemistry* 7:501-507.    381

Agtarap, A., Chamberlin, J.W., Pinkerton, M., and Steinrauf, L. (1967): The structure of monensic acid, a new biologically active compound. *J. Amer. Chem. Soc.* 89:5737.    311,318

André, J. (1965): Quelques données récentes sur la structure et la physiologie des mitochondries: glycogène, particules élémentaires, acides nucléiques. *Arch. Biol. (Liège)* 76:277-304.    392

André, J. and Marinozzi, V. (1965): Présènce, dans les mitochondries, de particules ressemblant aux ribosomes. *J. Microscopie* 4:615-626.    392

Andreoli, T.E., Tieffenberg, M., and Tosteson, D.C. (1967): The effect of valinomycin on the ionic permeability of thin lipid membranes. *J. Gen. Physiol.* 50:2527-2545.    345

Andreoli, T.E. and Tosteson, D.C. (1971): The effect of valinomycin on the electrical properties of solutions of red cell lipids in n-decane. *J. Gen. Physiol.* (In press)    345,347

Bernal, J.D. (1967): *Origin of Life.* Cleveland: World Publishing Co.

Bénard, H. (1901): Les tourbillons cellulaires dans une nappe liquide transportant de la chaleur par convection en régime permanent. *Ann. Chim. Phys.* 23:62-144.    398

Blasenbrey, S. and Pechhold, W. (1967): Eine neue Theorie zur Rotationsumwandlung der n-Paraffine (Kinkentheorie). *Rheologica Acta* 6:174-185.    362,364

Blomstrom, D.C., Knight, E., Jr., Phillips, W.D., and Weiher, J.F. (1964): The nature of iron in ferredoxin. *Proc. Nat. Acad. Sci.* 51:1085-1092.

Page

Boersch, M. (1964): Temperaturabhängigkeit der Transparenz dünner Schichten für        394
schnelle Electronen. *Z. Physik.* 180:407-414.

Bogdanski, D.F., Tissari, A., and Brodie, B.B. (1968): Role of sodium, potassium,
ouabain and reserpine in uptake, storage and metabolism of biogenic amines in
synaptosomes. *Life Sci.* 7:419-428.

Britton, H.G. (1964): Permeability of the human red cell to labelled glucose. *J.*        366
*Physiol.* 170:1-20.

Brockmann, H. and Schmidt-Kastner, G. (1955): Valinomycin I. XXVII Mitteil.              309
Uber Antibiotica aus Actinomyceten. *Chem. Ber.* 88:57-61

Brockmann, H., Springorum, M., Träxler, G., and Höfer, I. (1963): Molekularge-          309
wicht des Valinomycins. *Naturwissenschaften* 50:689

Cahn, J.W. (1965): Phase separation by spinodal decomposition in isotropic              399
systems. *J. Chem. Phys.* 42:93-99.

Caswell, A.H. and Pressman, B.C. (1968): Transient permeability changes of
mitochondria induced by uncoupling agents. *Biochem. Biophys. Res. Commun.*
30:637-642.

Chance, B., Estabrook, R.W., and Williamson, J.R., eds. (1965): *Control of Energy*     409-410
*Metabolism.* New York: Academic Press.

Chandrasekhar, S. (1961): *Hydrodynamic and Hydromagnetic Stability.* New York:        399
Oxford University Press.

Changeux, J.P., Thiéry, J., Tung, Y., and Kittel, C. (1967): On the cooperativity of    392
biological membranes. *Proc. Nat. Acad. Sci.* 57:335-341.

Chapman, D. (1968): Physical studies of biological membranes and their constitu-        377
ents. *In: Membrane Models and the Formation of Biological Membranes.* Bolis,
L. and Pethica, B.A., eds. Amsterdam: North-Holland Publishing Co., pp. 6-18.

Ciani, S., Eisenman, G., and Szabo, G. (1969): A theory for the effects of neutral      350,351
carriers such as the macrotetrolide actin antibiotics on the electric properties of
bilayer membranes. *J. Membrane Biol.* 1:1-36.

Clementi, F., Whittaker, V.P., and Sheridan, M.N. (1966): The yield of synaptosomes from the cerebral cortex of guinea pigs estimated by a polystyrene bead
"tagging" procedure. *Z. Zellforsch.* 72:126-138.

Cockrell, R.S., Harris, E.J., and Pressman, B.C. (1966): Energetics of potassium
transport in mitochondria induced by valinomycin. *Biochemistry* 5:2326-2335.

Cockrell, R.S., Harris, E.J., and Pressman, B.C. (1967): Synthesis of ATP driven by
a potassium gradient in mitochondria. *Nature* 215:1487-1488.

Colburn, R.W., Goodwin, F.K., Murphy, D.L., Bunney, W.E., and Davis, J.M.              390
(1968): Quantitative studies of norepinephrine uptake by synaptosomes.
*Biochem. Pharmacol.* 17:957-964.

Page

Danielli, J.F. and Davson, H. (1935): A contribution to the theory of permeability    358,373,
of thin films. *J. Cell Comp. Physiol.* 5:495-508.          382

Delbrück, M. (1956): Cellular mechanisms in differentiation and growth. *In: Society for the Study of Development and Growth, Symposium 14.* Princeton, N.J.: Princeton University Press, pp. 3-44.

Delbrück, M. (1965): Primary transduction mechanisms in sensory physiology and the search for suitable experimental systems. *Israel J. Med. Sci.* 1:1363-1365.

Delbrück, M. (1970): Lipid bilayers as models of biological membranes. *In: The Neurosciences: Second Study Program.* Schmitt, F.O., editor-in-chief. New York: Rockefeller University Press, pp. 677-684.

Delbrück, M. and Varju, D. (1961): Photoreactions in *Phycomyces.* Responses to the stimulation of narrow test areas with ultraviolet light. *J. Gen. Physiol.* 44:1177-1188.

Diebler, H., Eigen, M., Ilgenfritz, G., Maass, G., and Winkler, R. (1969): Kinetics    331,332,
and mechanism of reactions of main group metal ions with biological carriers.    335,351
*Pure Appl. Chem.* 20:93-115.

Dobler, M., Dunitz, J.D., and Krajewski, J. (1969): Structure of the $K^+$ complex    316
with enniatin B., a macrocyclic antibiotic with $K^+$ transport properties. *J. Molec. Biol.* 42:603.

Eggers, F. (1968): Eine Resonatormethode zur Bestimmung von Schall-    337
Geschwindigkeit und Dämpfung an geringen Flüssigkeitsmengen. *Acustica* 19:323-329.

Eigen, M. (1963a): Fast elementary steps in chemical reaction mechanisms. *Pure*    330
*Appl. Chem.* 6:97-115.

Eigen, M. (1963b): Ionen- und Ladungsübertragungsreaktionen in Lösungen. *Ber.*    330
*Bunsenges. Physik. Chem.* 67:753-762.

Eigen, M. (1964): Information, ihre Speicherung und Verarbeitung in bimoleku-laren Systemen. *Ber. Bunsenges.* 68:889-894.

Eigen, M. (1967): Kinetics of reaction control and information transfer in enzymes and nucleic acids. *In: Fast Reactions and Primary Processes in Chemical Kinetics, Fifth Nobel Symposium.* Claesson, S., ed. New York: Interscience, pp. 333-369.

Eigen, M. (1968): Die "unmessbar" schnellen Reaktionen. *In: Les Prix Nobel, 1967.* Stockholm: Norstedt & Söner, pp. 151-180.

Eigen, M. (1968): New looks and outlooks on physical enzymology. *Quart. Rev. Biophys.* 1:3-33.

Eigen, M. (1971): Selforganization of matter and the evolution of biological macro-molecules. *Naturwissenschaften.* (In press)

  Page

Eigen, M. and De Maeyer, L.C.M. (1966): Chemical means of information storage
    and readout in biological systems. *Naturwissenschaften* 53:50-57.

Eigen, M. and Wilkins, R.G. (1965): The kinetics and mechanism of formation of          330
    metal complexes. *Adv. Chem. Ser.* 49:55.

Eigen, M. and Winkler, R. (1970): Alkali-ion carriers: dynamics and selectivity. *In:*
    *The Neurosciences: Second Study Program.* Schmitt, F.O., editor-in-chief. New
    York: Rockefeller University Press, pp. 685-696.

Eisenman, G. (1961): On the elementary atomic origin of equilibrium ionic              354
    specificity. *In: Symposium on Membrane Transport and Metabolism.* Klein-
    zeller, A. and Kotyk, A., eds. Prague: Czechoslovak Academy of Sciences, pp.
    163-179.

Eisenman, G. (1962): Cation selective glass electrodes and their mode of operation.     320
    *Biophys. J.* 2:259-323.

Eisenman, G. (1965): Some elementary factors involved in specific ion permeation.
    *In: Proceedings of the XXIIIrd International Congress of Physiological Sciences.*
    Excerpta Medica International Congress Series No. 87:489-506.

Eisenman, G. (1968): Ion permeation of cell membranes and its models. *Fed. Proc.*
    27:1249-1251.

Eisenman, G. (1968): Similarities and differences between liquid and solid ion
    exchangers and their usefulness as ion specific electrodes. *Anal. Chem.*
    40:310-320.

Eisenman, G. (1969): Theory of membrane electrode potentials: An examination           354
    of the parameters determining the selectivity of solid and liquid ion exchangers
    and of neutral ion-sequestering molecules. *In: Ion-Selective Electrodes.* Durst,
    R.A., ed. Washington, D.C.: National Bureau of Standards Special Publication
    314, pp. 1-56.

Eisenman, G., Ciani, S.M., and Szabo, G. (1968): Some theoretically expected and       350,351
    experimentally observed properties of lipid bilayer membranes containing
    neutral molecular carriers of ions. *Fed. Proc.* 27:1289-1304.

Eisenman, G., Ciani, S., and Szabo, G. (1969): The effects of the macrotetrolide       350,351,
    actin antibodies on the equilibrium extraction of alkali metal salts into organic      352
    solvents. *J. Membrane Biol.* 1:294-345.

Eisenman, G., Sandblom, J.P., and Walker, J.L., Jr. (1967): Membrane structure
    and ion permeation. *Science* 155:965-974.

Ferguson, R.C. and Phillips, W.D. (1967): High-resolution nuclear magnetic reso-
    nance spectrosscopy. Advances in instrumentation in this field are leading to
    new applications in chemistry and biology. *Science* 157:257-267.

Fernandez-Moran, H. (1962): Cell-membrane ultrastructure. Low-temperature elec-         392
    tron microscopy and X-ray diffraction studies of lipoprotein components in
    lamellar systems. *Circulation* 26:1039-1065.

Page

Fernandez-Moran, H. (1965): Electron microscopy with high-field superconducting solenoid lenses. *Proc. Nat. Acad. Sci.* 53:445-451.    393

Fernandez-Moran, H. (1966a): Applications of improved point cathode sources to high resolution electron microscopy. *In: Electron Microscopy 1966, Vol. I.* (Sixth International Congress for Electron Microscopy) Uyeda, R., ed. Tokyo: Maruzen Co., Ltd., pp. 27-28.    393

Fernandez-Moran, H. (1966b): High-resolution electron microscopy with superconducting lenses at liquid helium temperatures. *Proc. Nat. Acad. Sci.* 56:801-808.    393

Fernandez-Moran, H. (1967): Membrane ultrastructure in nerve cells. *In: The Neurosciences: A Study Program.* Quarton, G.C., Melnechuk, T., and Schmitt, F.O., eds. New York: Rockefeller University Press, pp. 281-304.    392,396

Fernandez-Moran, H. (1969a): Data "reduction" for information retrieval. *In: Applying Emerging Technologies.* (Proceedings of the Fifth Annual Conference on Industrial Research, Chicago, Ill., September 18-19, 1969). Chicago: University of Chicago Press.    393

Fernandez-Moran, H. (1969b): The world of inner space. A science year report. *In: The World Book Encyclopedia.* Chicago: Field Enterprises Education Corporation.

Fernandez-Moran, H. (1970a): Cell fine structure and function—past and present. *Exp. Cell Res.* 62:90-101.    392

Fernandez-Moran, H. (1970b): High voltage electron microscopy at liquid helium temperatures. *In: Proceedings of the Seventh International Congress on Electron Microscopy.* Grenoble, France, 2:91-92.    392

Fernandez-Moran, H., Marchalonis, J.J., and Edelman, G.M. (1968): Electron microscopy of a hemagglutinin from *Limulus polyphemus. J. Molec. Biol.* 32:467-469.

Fernandez-Moran, H., Oda, T., Blair, P.V., and Green, D.E. (1964): A macromolecular repeating unit of mitochondrial structure and function. Correlated electron microscopic and biochemical studies of isolated mitochondria and submitochondrial particles of beef heart muscle. *J. Cell Biol.* 22:63-100.    392

Fernandez-Moran, H., Ohtsuki, M., and Hough, C. (1970): High resolution electron microscopy of cell membranes and derivatives. *In: Proceedings of the Seventh International Congress on Electron Microscopy.* Grenoble, France, 3:9-10.    392

Finkelstein, A. and Cass, A. (1968): Permeability and electrical properties of thin lipid membranes. *J. Gen. Physics* 52:145-172.    371

Fonnum, F. (1967): The "compartmentation" of choline acetyltransferase within the synaptosome. *Biochem. J.* 103:262-270.

Fonnum, F. (1968): Choline acetyltransferase binding to and release from membranes. *Biochem. J.* 109:389-398.

  Page

Fox, S. (1965): A theory of macromolecular and cellular origins. *Nature*        411
205:328-340.

Gabor, D. (1949): Microscopy by reconstructive wave-fronts. *Proc. Roy. Soc. A*  394
197:454-487.

Gambetti, P., Autilio-Gambetti, L., and Gonatas, N.K. (1970): Ultrastructural
radioautographic study of protein synthesis in synaptosomal fractions. *J. Cell
Biol.* 47:68a.

Gause, G.F. and Brazhnikova, M.G. (1944): Gramicidin S and its use in the        309
treatment of infected wounds. *Nature* 154:703.

Glansdorff, P. and Prigogine, I. (1954): Sur les propriétés différentielles de la   405,407
production d'entropie. *Physica* 20:773-780.

Gmitro, J.L. and Scriven, L.E. (1966): A physicochemical basis for pattern and   399,400,
rhythm. *In: Intracellular Transport.* Warren, K.B., ed. New York: Academic      409,410
Press, pp. 221-255.

Grahame-Smith, D.G. and Parfitt, A.G. (1970): Tryptophan transport across the    390
synaptosomal membrane. *J. Neurochem.* 17:1339-1353.

Graven, S.N., Lardy, H.A., Johnson, D., and Ruter, A. (1966): Antibiotics as tools  317,319
for metabolic studies. V. Effect of nonactin, monactin, dinactin, and trinactin
on oxidative phosphorylation and adenosine triphosphatase induction. *Biochemistry* 5:1729-1735.

Green, D.E. and Perdue, J.F. (1966): Membranes as expressions of repeating units.  392
*Proc. Nat. Acad. Sci.* 55:1295-1302.

Griffith, O.H. and Waggoner, A.S. (1969): Nitroxide free radicals: spin labels for  357
probing biomolecular structure. *Acct. Chem. Res.* 2:17.

Hamilton, C.L. and McConnell, H.M. (1968): Spin labels. *In: Structural Chemistry*  357
*and Molecular Biology.* Rich, A. and Davidson, N., eds. San Francisco: W. H.
Freeman and Co., pp. 115-149.

Hanai, T. and Haydon, D.A. (1966): The permeability to water of biomolecular
lipid membranes. *J. Theor. Biol.* 11:370-382.

Hanai, T., Haydon, D.A., and Taylor, J. (1964): An investigation by electrical    383
methods of lecithin in hydrocarbon films in aqueous solutions. *Proc. Roy. Soc.
A* 281:377-391.

Harold, F.M., Baarda, J.R., Baron, C., and Abrams, A. (1969): Inhibition of       383
membrane-bound adenosine triphosphatase and of cation transport in *Streptococcus faecalis* by N,N'-dicyclohexylcarbodiimide. *J. Biol. Chem.*
244:2261-2268.

  Page

Harris, E.J., Dam, K. van, and Pressman, B.C. (1967a): Dependence of uptake of          320
succinate by mitochondria on energy and its relation to potassium retention.
*Nature* 213:1126-1127.

Harris, E.J., Höfer, M.P., and Pressman, B.C. (1967b): Stimulation of mitochondrial     320
respiration and phosphorylation by transport-inducing antibiotics. *Biochemistry*
6:1348-1360.

Harris, E.J. and Pressman, B.C. (1967): Obligate cation exhanges in red cells.          320
*Nature* 216:918-920.

Harris, G.M. and Harris, F.E. (1959): Valence bond calculation of the barrier to        363
internal rotation in molecules. *J. Chem. Phys.* 31:1450-1453.

Haynes, D.H., Kowalsky, A., and Pressman, B.C. (1969): Application of nuclear       320,330
magnetic resonance to the conformational changes in valinomycin during
complexation. *J. Biol. Chem.* 244:502-505.

Helene, C., Haug, A., Delbrück, M., and Douzou, P. (1964): Mise en évidence de
formes tautomères de la cytosine. *C.R. Acad. Sci.* 259:3385-3388.

Henn, F.A. and Thompson, T.E. (1969): Synthetic lipid bilayer membranes. *Ann.
Rev. Biochem.* 38:241-262.

Hodgkin, A.L. and Katz, B. (1949): The effect of temperature on the electrical          352
activity of the giant axon of the squid. *J. Physiol.* 109:240-249.

Hodgkin, A.L. and Keynes, R.D. (1955): The potassium permeability of a giant            340
nerve fibre. *J. Physiol.* 128:61-88.

Hopfer, U., Lehninger, A.L., and Thompson, T.E. (1968): Protonic conductance
across phospholipid bilayer membranes induced by uncoupling agents for
oxidative phosphorylation. *Proc. Nat. Acad. Sci.* 59:484-490.

Hopkin, J.M., Horton, E.W., and Whittaker, V.P. (1968): Prostaglandin content of
particulate and supernatant fractions of rabbit brain homogenates. *Nature*
217:71-72.

Hotchkiss, R.D. and Dubos, R.J. (1941): The isolation of bacterial substances from      309
cultures of *Bacillus brevis. J. Biol. Chem.* 141:155-162.

Hsia, J.C., Schneider, H., and Smith, I.C.P. (1970): Spin label studies of oriented     360
phospholipids: Egg lecithin. *Biochim. Biophys. Acta* 202:399-401.

Huang, C. (1969): Studies on phosphatidylcholine vesicles. Formation and physical   373,380
characteristics. *Biochemistry* 8:344-352.

Huang, C., Charlton, J.P., and Litman, B.L. (1969): Studies on phosphatidyl            381
choline vesicles: dye-lipid interaction. *Biophys. J.* 9:A-37.

Page

Huang, C., Charlton, J.P., Shyr, C.I., and Thompson, T.E. (1970): Studies on            380
phosphatidylcholine vesicles with thiocholesterol and a thiocholesterol-linked
spin label incorporated in the vesicle wall. *Biochemistry* 9:3422-3426.

Huang, C. and Thompson, T.E. (1965): Properties of lipid bilayer membranes             373
separating two aqueous phases: determination of membrane thickness. *J. Molec.
Biol.* 13:183-193.

Huang, C. and Thompson, T.E. (1966): Properties of lipid bilayer membranes             373
separating two aqueous phases: water permeability. *J. Molec. Biol.* 15:539-554.

Huang, C., Wheeldon, L., and Thompson, T.E. (1964): The properties of lipid
bilayer membranes separating two aqueous phases: formation of a membrane of
simple composition. *J. Molec. Biol.* 8:148-160.

Hubbell, W.L. and McConnell, H.M. (1968): Spin-label studies of the excitable          357
membranes of nerve and muscle. *Proc. Nat. Acad. Sci.* 61:12-16.

Hubbell, W.L. and McConnell, H.M. (1969a): Motion of steroid spin labels in            357
membranes. *Proc. Nat. Acad. Sci.* 63:16-22.

Hubbell, W.L. and McConnell, H.M. (1969b): Orientation and motion of amphi-        357,361
philic spin labels in membranes. *Proc. Nat. Acad. Sci.* 64:20-27.

Hubbell, W.L. and McConnell, H.M. (1971): Molecular motion in spin-labeled             359
phospholipids and membranes. *J. Amer. Chem. Soc.* 93:314-326.

Hummel, J.P. and Dreyer, W.J. (1962): Measurement of protein-binding phenom-          381
ena by gel filtration. *Biochim. Biophys. Acta* 63:530-532.

Ibbott, F.A. and Abrams, A. (1964): The phospholipids in membrane ghosts from
*Streptococcus faecalis* protoplasts. *Biochemistry* 3:2008-2012.

Ilgenfritz, G. (1966): Chemische Relaxation in starken elektrischen Feldern.        335,337
Doctoral Dissertation, University of Göttingen.

Keith, A., Bulfield, G., and Snipes, W. (1970): Spin-labeled *Neurospora* mitochon-    357
dria. *Biophys. J.* 10:618-629.

Keith, A., Waggoner, A.S., and Griffith, O.H. (1968): Spin-labeled mitochondrial       357
lipids in *Neurospora crassa. Proc. Nat. Acad. Sci.* 61:819-826.

Keller-Schierlein, W. and Gerlach, H. (1968): Makrotetrolide. *Fortschr. Chem.*     309,318
*Organ. Naturst.* 26:161-189.

Kilbourn, B.T., Dunitz, J.D., Pioda, L.A.R., and Simon, W.(1967): Structure of the     314
$K^+$ complex with nonactin, a macrotetrolide antibiotic possessing highly specific
$K^+$ transport properties. *J. Molec. Biol.* 30:559-563.

Krampitz, G. and Fox, S.W. (1969): The condensation of the adenylates of the           411
amino acids common to protein. *Proc. Nat. Acad. Sci.* 62:399-406.

Page

Kuriyama, K., Roberts, E., and Kakefuda, T. (1968): Association of the $\gamma$-amino-butyric acid system with a synaptic vesicle fraction from mouse brain. *Brain Res.* 8:132-152.     391

Ladbrooke, B.D. and Chapman, D. (1969): Thermal analysis of lipids, proteins and biological membranes: A review and summary of some recent studies. *Chem. Phys. Lipids* 3:304-356.

Lardy, H.A., Graven, S.N., and Estrada, S. (1967): Specific induction and inhibition of cation and anion transport in mitochondria. *Fed. Proc.* 26:1355-1360.     308,320

Lefever, R. (1968): Dissipative structures in chemical systems. *J. Chem. Phys.* 49:4977-4978.     403

Lehninger, A.L. (1970): Mitochondria and their neurofunction. *In: The Neurosciences: Second Study Program.* Schmitt, F.O., editor-in-chief. New York: Rockefeller University Press, pp. 827-839.

Libertini, L.J., Waggoner, A.S., Jost, P.C., and Griffith, O.H. (1969): Orientation of lipid spin labels in lecithin multilayers. *Proc. Nat. Acad. Sci.* 64:13-19.     360

Lieb, W.R. and Stein, W.D. (1969): Biological membranes behave as non-porous polymeric sheets with respect to the diffusion of non-electrolytes. *Nature* 224:240-243.     361

Ling, C.-M. and Abdel-Latif, A.A. (1968): Studies on sodium transport in rat brain nerve-ending particles. *J. Neurochem.* 15:721-729.     391

Lotka, A.J. (1957): *Elements of Mathematical Biology.* 2nd Ed. New York: Dover Publications, Inc.     409

Lutz, W.K., Wipf, H.-K., and Simon, W. (1970): Alkalikationen-Spezifität und Träger-Eigenschaften der Antibiotica Nigericin und Monensin. *Helv. Chim. Acta* 53:1741-1746.     308,311

Maddy, A.H., Huang, C., and Thompson, T.E. (1966): Studies on lipid bilayer membranes: a model for the plasma membrane. *Fed. Proc.* 25:933-936.

Marchbanks, R.M. (1967): The osmotically sensitive potassium and sodium compartments of synaptosomes. *Biochem. J.* 104:148-157.

Marchbanks, R.M. (1968): Exchangeability of radioactive acetylcholine with the bound acetylcholine of synaptosomes and synaptic vesicles. *Biochem. J.* 106:87-95.

Marchbanks, R.M. (1968): The uptake of [$^{14}$C] choline into synaptosomes *in vitro. Biochem. J.* 110:533-541.

Marchbanks, R.M. and Whittaker, V.P. (1967): Some properties of the limiting membranes of synaptosomes and synaptic vesicles. *Abstracts of the First International Meeting for Neurochemistry, Strasbourg,* p. 147.

Page

Marchbanks, R.M. and Whittaker, V.P. (1969): The biochemistry of synaptosomes. *In: The Biological Basis of Medicine, Vol. 5.* Bittar, E.E. and Bittar, N., eds. New York: Academic Press, pp. 39-76.

McConnell, H.M. and McFarland, B.G. (1970): Physics and chemistry of spin labels. *Quart. Rev. Biophys.* 3:91-136.                                                                  357

McDonald, C.C. and Phillips, W.D. (1967): Manifestations of the tertiary structures of proteins in high-frequency nuclear magnetic resonance. *J. Amer. Chem. Soc.* 89:6332-6341.

McDonald, C.C., Phillips, W.D., and Lazar, J. (1967): Nuclear magnetic resonance determination of thymine nearest neighbor base frequency ratios in deoxyribonucleic acid. *J. Amer. Chem. Soc.* 89:4166-4170.

McIlwain, H. (1960): Characterization of constituents of blood plasma and of the    417
brain which restore excitability to isolated cerebral tissues. *Biochem. J.* 76:16P.

McLaughlin, S.G.A., Szabo, G., Eisenman, G., and Ciani, S. (1970): Surface charge   351,354
and the conductance of phospholipid membranes. *Proc. Nat. Acad. Sci.* 67:1268-1275.

Mellanby, J. and Whittaker, V.P. (1968): The fixation of tetanus toxin by synaptic membranes. *J. Neurochem.* 15:205-208.

Miyamoto, V.K. and Thompson, T.E. (1967): Some electrical properties of lipid       383
bilayer membranes. *J. Colloid Interface Sci.* 25:16-25.

Moore, C. and Pressman, B.C. (1964): Mechanism of action of valinomycin on          308,320
mitochondria. *Biochem. Biophys. Res. Commun.* 15:562-567.

Morgan, I.G. (1970): Protein synthesis in brain mitochondrial and synaptosome preparations. *FEBS Letters* 10:273-275.

Mueller, P., Rudin, D.O., Tien, H.T., and Wescott, W.C. (1962a): Reconstitution of   373,418
excitable cell membrane structure in vitro. *Circulation* 26:1167-1171.

Mueller, P., Rudin, D.O., Tien, H.T., and Wescott, W.C. (1962b): Reconstitution of   373,383,
cell membrane structure in vitro and its transformation into an excitable system.    418
*Nature* 194:979-980.

Mueller, P. and Rudin, D.O. (1968a): Action potentials induced in biomolecular       321
lipid membranes. *Nature* 217:713-719.

Mueller, P. and Rudin, D.O. (1968b): Resting and action potentials in experimental   321
biomolecular lipid membranes. *J. Theor. Biol.* 18:222-258.

Nass, M.M.K. (1966): The circularity of mitochondrial DNA. *Proc. Nat. Acad. Sci.*   392
56:1215-1222.

Ovchinnikov, Yu. A., Ivanov, V.T., Evstratov, A.V., and Laine, L.A. (1969): *In: Proceedings of the Tenth European Peptide Symposium, Abano Therme, September 1969.* (In press)

Page

Paecht-Horowitz, M., Berger, J., and Katchalsky, A. (1970): Prebiotic synthesis of          412
polypeptides by heterogeneous condensation of amino-acid adenylates. *Nature*
228:636-639.

Pagano, R. and Thompson, T.E. (1968): Spherical lipid bilayer membranes:
electrical and isotopic studies of ion permeability. *J. Molec. Biol.* 38:41-57.

Panar, M. and Phillips, W.D. (1968): Magnetic ordering of polygamma-benzyl-
L-glutamate solutions. *J. Amer. Chem. Soc.* 90:3880-3882.

Pechhold, W. (1968): Molekülbewegung in Polymeren. I Teil: Konzept einer          363,364,
Festokörperphysik makromolekularer Stoffe. *Kolloid Z.* 228:1-38.          370

Pechhold, W. and Blasenbrey, S. (1967): Kooperative Rotationsisomerie in Poly-          362,364
meren. *Z. Polymere* 216:235-244.

Pechhold, W., Blasenbrey, S., and Woerner, S. (1963): Eine niedermolekulare          362
Modellsubstanz für lineares Polyäthylen. Voschlag des "kinkenmodells" zur
Deutung des γ und α Relaxations-prozesses. *Kolloid Z.* 189:14-22.

Pechhold, W., Dollhopf, W., and Engel, A. (1966): Untersuchung der Rotationsum-          362
wandlung reiner Paraffine und Paraffinmischungen mit Hilfe des komplexen
Schubmoduls. *Acustica* 17:61-72.

Pedersen, C.J. (1967): Cyclic polyethers and their complexes with metal salts. *J.*          309,311
*Amer. Chem. Soc.* 89:7017-7036.

Pinkerton, M., Steinrauf, L.K., and Dawkins, P. (1969): The molecular structure          316
and some transport properties of valinomycin. *Biochem. Biophys. Res.*
*Commun.* 35:512-518.

Pinkerton, M. and Steinrauf, L.K. (1970): Molecular structure of metal cation          321
complexes of monensin. *J. Molec. Biol.* 49:533-546.

Pioda, L.A.R., Wachter, H.A., Dohner, R.E., and Simon, W. (1967): Komplexe von          311,314
Nonactin und Monactin mit Natrium-, Kalium- und Ammonium-Ionen. *Helv.*
*Chim. Acta* 50:1373-1376.

Plattner, P.A., Nager, U., and Boller, A. (1948): Uber die Isolierung neuartiger          309
Antibiotika aus Fusarien. *Helv. Chim. Acta* 31:584-602.

Plattner, P.A., Vogler, K., Studer, R.O., Quitt, P., and Keller-Schierlein, W. (1963):          309
Synthesen in der Depsipeptid-Reihe: 1. Synthese von Enniatin B. *Helv. Chim.*
*Acta* 46:927-935.

Ponnamperuma, C., Sagan, C., and Mariner, R. (1963): Synthesis of adenosine          412
triphosphate under possible primitive earth conditions. *Nature* 199:222-226.

Pressman, B.C. (1965): Induced active transport of ions in mitochondria. *Proc. Nat.*          320,394
*Acad. Sci.* 53:1076-1083.

Page

Pressman, B.C. (1967): Biological applications of ion-specific glass electrodes. *In: Methods in Enzymology, Vol. 10.* Estabrook, W. and Pullam, M.E., eds. New York: Academic Press, pp. 714-726.

Pressman, B.C. (1968a): An apparatus for observing multiparameter changes in cation transport systems. *Ann. N.Y. Acad. Sci.* 148:285-287.                               320

Pressman, B.C. (1968b): Inophorous antibiotics as models for biological transport. *Fed. Proc.* 27:1283-1288.                                                           320

Pressman, B.C. (1969): Control of mitochondrial substrate metabolism by regulation of cation transport. *In: Mitochondria–Structure and Function.* (FEBS Symposia, Vol. 17). Ernster, L. and Drahota, Z., eds. London: Academic Press, pp. 315-333.

Pressman, B.C. (1969): Coupling of cation and anion transport in mitochondria. *In: The Energy Level and Metabolic Control in Mitochondria.* Papa, S., Tager, J.M., Quagliariello, E., and Slater, E.C., eds. Bari, Italy: Adriatica Editrice, pp. 87-96.

Pressman, B.C., Harris, E.J., Jagger, W.S., and Johnson, J.H. (1967): Antibiotic-    320
mediated transport of alkali ions across lipid barriers. *Proc. Nat. Acad. Sci.* 58:1949-1956.

Pressman, B.C. and Haynes, D.H. (1969): Ionophorous agents as mobile ion     320,321,
carriers. *In: The Molecular Basis of Membrane Function.* Tosteson, D.C., ed.   322,323,
Englewood Cliffs, N.J.: Prentice-Hall, pp. 221-246.                              324,325,
                                                                                 326,329
Price, H.D. and Thompson, T.E. (1969): Properties of lipid bilayer membranes separating two aqueous phases: temperature dependence of water permeability. *J. Molec. Biol.* 41:443-457.

Prigogine, I. (1969): Structure, dissipation and life. *In: Theoretical Physics and*   399,402,
*Biology.* Marois, M., ed. Amsterdam: North-Holland, pp. 23-52.                        404

Quitt, P., Studer, R.O., and Vogler, K. (1963): Synthesen in der Depsipeptid-Reihe: 2. Synthese von Enniatin A. *Helv. Chim. Acta.* 46:1715-1729.

Rayleigh, Lord (1916): On convective currents in a horizontal layer of fluid, when    399
the higher temperature is on the under side. *Phil. Mag.* 32:529-546.

Redwood, W.R., Müldner, H., and Thompson, T.E. (1969): Interaction of a bacterial adenosine triphosphatase with phospholipid bilayers. *Proc. Nat. Acad. Sci.* 64:989-996.

Rothfield, L. and Finkelstein, A. (1968): Membrane biochemistry. *Ann. Rev. Biochem.* 37:463-496.

Sarges, R. and Witkop, B. (1965a): Gramicidin A. V. The structure of valine- and    309
isoleucine-gramicidin A. *J. Amer. Chem. Soc.* 87:2011-2020.

Page

Sarges, R. and Witkop, B. (1965b): Gramicidin. VII. The structure of valine- and       309
isoleucine-gramicidin B. *J. Amer. Chem. Soc.* 87:2027-2030.

Sarges, R. and Witkop, B. (1965c): Gramicidin S. The structure of valine- and       309
isoleucine-gramicidin C. *Biochemistry* 4:2491-2494.

Schatzberg, P. (1963): Solubilities of water in several normal alkanes from $C_7$ to
$C_{16}$. *J. Phys. Chem.* 67:776-779.

Schatzberg, P. (1965): Diffusion of water through hydrocarbon liquids. *J. Polymer Sci.* 10:92.

Schmitt, F.O. (1963): The macromolecular assembly–a hierarchical entity in       392
cellular organization. *Develop. Biol.* 7:546-559.

Schmitt, F.O. (1971): Molecular membranology. (Gesellschaft für biologische
Chemie, 21st Colloquium, Dynamic Structure of Cell Membranes, Mosbach,
April 15-17, 1971.) *Hoppe-Seylers Z. Physiol. Chem.* (In press)

Schwarzenbach, G. and Gysling, H. (1949): Murexid als Indikator auf Calcium- und       335,336
andere Metall-Ionen. Komplexbildung und Lichtabsorption. *Helv. Chim. Acta*
32:1314-1325.

Schwyzer, R. and Sieber, P. (1957): Die Synthese von Gramicidin S. *Helv. Chim.*       309
*Acta* 40:624-639.

Shemyakin, M.M., Aldanova, N.A., Vinogradova, E.I., and Feigina, M.Yu. (1963):
The structure and total synthesis of valinomycin. *Tetrahedron Lett.* 28:1921.

Shemyakin, M.M., Ovchinnikov, Yu.A., Ivanov, V.T., Antonov, V.K., Shkrob, A.M.,       314
Mikhaleva, I.I., Evstratov, A.V., and Malenkov, G.G. (1967): The physicochemical basis of the functioning of biological membranes: conformational specificity
of the interaction of cyclodepsipeptides with membranes and of their complexation with alkali metal ions. *Biochem. Biophys. Res. Commun.* 29:834.

Simon, W. (1969): Molecular structure on monovalent cation complexes of
antibiotics and their ion specific electrochemical behaviour. Presented at the
Symposium on Specific Ion Electrodes. (Theoretical Division of the Electrochemical Society.) May 4-9, 1969, New York.

Simon, W., Pioda, L.A.R., and Wipf, H.-K. (1969): Cation specificity of inhibitors.
Presented at the 20th Colloquium der Gesellschaft für Biologische Chemie. April
14-16, 1969, Mosbach, West Germany.

Solomon, A.K. (1952): Permeability of human erythrocytes to sodium and       361
potassium. *J. Gen. Physiol.* 36:57-110.

Solomon, A.K. (1960): Pores in the living cell. *Sci. Amer.* 203:146-156.       361

Stefanac, Z. and Simon, W. (1966): In-vitro-Verhalten von Makrotetroliden in       311,317,
Membranen als Grundlage für hochselektive kationenspezifische Elektroden-       321
systeme. *Chimia* 20:436.

Page

Stefanac, Z. and Simon, W. (1967): Ion specific electrochemical behavior of macrotetrolides in membranes. *Microchem. J.* 12:125-132.   317,319

Stein, W.D. (1967): *The Movement of Molecules Across Cell Membranes.* New York: Academic Press.   361

Steinman, G. (1967): Sequence generation in prebiological peptide synthesis. *Arch. Biochem. Biophys.* 119:76-82 and 121:533-539.   411

Steinman, G. and Cole, M. (1967): Synthesis of biologically pertinent peptides under possible primordial conditions. *Proc. Nat. Acad. Sci.* 58:735-742.   411

Steinrauf, L.K., Pinkerton, M. and Chamberlin, J.W. (1968): The structure of nigericin. *Biochem. Biophys. Res. Commun.* 33:29-31.

Swift, H., Kislev, N., and Bogorad, L. (1964): Evidence for DNA and RNA in mitochondria and chloroplasts. *J. Cell Biol.* 23:91A.   392

Szabo, G., Eisenman, G., and Ciani, S. (1969): The effects of the macrotetrolide actin antibiotics on the electrical properties of phospholipid bilayer membranes. *J. Membrane Biol.* 1:346-382.   350,353, 354,355, 356

Szabo, G., Eisenman, G., and Ciani, S.(1970): Ion distribution equilibria in bulk phases and the ion transport properties of bilayer membranes produced by neutral macrocyclic antibiotics. *In: Physical Principles of Biological Membranes.* Snell, F., Wolen, J., Iverson, G., and Lam, J., eds. New York: Gordon and Breach, pp. 79-133.   350,351, 356,357

Thompson, T.E. and Henn, F.A. (1970): Experimental phospholipid model membranes. *In: Membranes of Mitochondria and Chloroplasts.* Racker, E., ed. New York: Reinhold, pp. 1-52.   373

Thompson, T.E. and Huang, C. (1966); The water permeability of lipid bilayer membranes. *Ann. N.Y. Acad. Sci.* 137:740-744.

Tosteson, D.C. (1968): Effect of macrocyclic compounds on the ionic permeability of artificial and natural membranes. *Fed. Proc.* 27:1269.   339,345

Tosteson, D.C., Andreoli, T.E., and Tieffenberg, M. (1968a): The effects of macrocyclic compounds on cation transport in sheep red cells and thin and thick lipid membranes. *J. Gen. Physiol.* 51(Suppl):373-384.   340,345, 347

Tosteson, D.C., Cook, P., and Andreoli, T.E. (1967): The effect of valinomycin on potassium and sodium permeability of HK and LK sheep red cells. *J. Gen. Physiol.* 50:2513-2525.

Tosteson, D.C., Cook, P., and Blount, R. (1965): Separation of adenosine triphosphatase of HK and LK sheep red cell membranes by density gradient centrifugation. *J. Gen. Physiol.* 48:1125-1143.

Page

Tosteson, D.C., Tieffenberg, M., and Cook, P. (1968b): The effect of macrocyclic compounds on ionic permeability of HK and LK sheep red cell membranes and on artificial thin lipid membranes. *In: Metabolism and Membrane Permeability of Erythrocytes and Thrombocytes.* Deutsch, E., Gerlach, E., and Moser, K., eds. Stuttgart: G. Thieme, pp. 424-428.     339,340

Träuble, H. (1971a): The movement of molecules across lipid membranes: a molecular theory. *J. Membrane Biol.* 4:193-208.     372

Träuble, H. (1971b): Phasenumwandlungen in Lipiden, Mögliche Schaltprozesse in biologischen Membranen. *Naturwissenschaften.* (In press)     372

Träuble, H. and Haynes, D.H. (1971): The volume change in lipid bilayer lamellae at the crystalline liquid crystalline phase transition. *Chem. Phys. Lipids.* (In press)     372

Turing, A.M. (1952): The chemical basis of morphogenesis. *Phil. Trans. Roy. Soc. B.* 237:37-72.     397,404

Van Bruggen, E.F.J., Borst, P., Ruttenberg, G.J.C.M., Gruber, M., and Kroon, A.M. (1966): Mitochondrial DNA. *Biochim. Biophys. Acta* 119:437-439.     392

Van Deenen, L.L.M., Houtsmuller, U.M.T., de Haas, G.H., and Muldner, E. (1962): Monomolecular layers of synthetic phosphatides. *J. Pharm. Pharmacol.* 14:429-444.

Varjú, D., Edgar, L., and Delbrück, M. (1961): Interplay between the reactions to light and to gravity in *Phycomyces. J. Gen. Physiol.* 45:47-58.

Varon, S., Weinstein, H., and Roberts, E. (1967): Sodium-dependent metabolism and transport of aminobutyric acid in subcellular particles from brain. *Protoplasma* 63:318-321.

Volkenstein, M.V. (1963): *Configurational Statistics of Polymeric Chains.* New York: Interscience (Wiley) Publishers.     363

Vorbeck, M.L. and Marinetti, G.V. (1965): Intracellular distribution and characterization of the lipids of *Streptococcus faecalis* (ATCC 9790). *Biochemistry* 4:296-305.     383

Wang, J.H. (1951): Self-diffusion and structure of liquid water. I. Measurement of self-diffusion of liquid water with deuterium as tracer. *J. Amer. Chem. Soc.* 73:510.

Wenner, C.E., Harris, E.J., and Pressman, B.C. (1967): Relationship of the light scattering properties of mitochondria to the metabolic state in intact ascites cells. *J. Biol. Chem.* 242:3454-3459.

Whittaker, V.P. (1965): The application of subcellular fractionation techniques to the study of brain function. *Progr. Biophys.* 15:39-96.

Page

Whittaker, V.P. (1966): Some properties of synaptic membranes isolated from the
central nervous system. *Ann. N.Y. Acad. Sci.* 137:982-998.

Whittaker, V.P. (1968): The morphology of fractions of rat forebrain synaptosomes
separated on continuous sucrose density gradients. *Biochem. J.* 106:412-417.

Whittaker, V.P. (1968): Structure and function of animal-cell membranes. *Brit.
Med. Bull.* 24:101-106.

Whittaker, V.P. (1968): Synaptic transmission. *Proc. Nat. Acad. Sci.* 60:1081-1091.

Whittaker, V.P. (1969): The nature of acetylcholine pools in brain tissue. *Progr.*          389
*Brain Res.* 31:211-222.

Winkler, R. (1969): Kinetik und Mechanismus der Alkali- und Erdalkalimetall-          335,337
komplexbildung in Methanol. Doctoral Dissertation, Max-Planck-Institut, Göttingen, and Technical University, Vienna.

Wipf, H.-K. (1970): Komplexbildung von Antibiotika der Valinomycin- und          314,315
Nigericin-Gruppe mit Alkalikationen, sowie ionenspezifischer Transport in
Modellmembranen. Doctoral Dissertation, Eidgenössische Technische Hoch-
schule, Zürich, Switzerland.

Wipf, H.-K., Olivier, A., and Simon, W. (1970): Mechanismus und Selektivität des
Alkali-Ionentransportes in Modell-Membranen in Gegenwart des Antibioticums
Valinomycin. *Helv. Chim. Acta* 53:1605-1608.

Wipf, H.-K., Pache, W., Jordan, P., Zähner, H., Keller-Schierlein, W., and Simon, W.
(1969): Mechanism of alkali cation transport in bulk membranes using macrotetrolide antibiotics. *Biochem. Biophys. Res. Commun.* 36:387-393.

Wipf, H.-K., Pioda, L.A.R., Stefanac, Z., and Simon, W. (1968): Komplexe von          314
Enniatinen und anderen Antibiotica mit Alkalimetall-Ionen. *Helv. Chim. Acta*
51:377-381.

Wipf, H.-K. and Simon, W. (1969): Selective $K^+$ transport through synthetic
membranes using antibiotics in a potential gradient. *Biochem. Biophys. Res.*
*Commun.* 34:707-711.

Wipf, H.-K. and Simon, W. (1970): Modelle für Kopplungsmechanismus und          317
Trägerinduzierten Alkaliionentransport in Mitochondrienmembranen. *Helv.*
*Chim. Acta* 53:1732-1740.

Woodcock, C.L.F. and Fernandez-Moran, H. (1968): Electron microscopy of DNA          392
conformations in spinach chloroplasts. *J. Molec. Biol.* 31:627-631.

Zankel, K.L., Burke, P.V., and Delbrück, M. (1967): Absorption and screening in
*Phycomyces. J. Gen. Physiol.* 50:1893-1906.

Page

Zhabotinsky, A.M. (1967): *In: Oscillatory Processes in Biological and Chemical*     410,411
    *Systems.* (Symposium in Pushtchino-na-Oke, Acad. Sci. USSR, March 21-26,
    1966.) Moscow: Nauka, pp. 149, 181, 199, 252.

Zwolinski, B.J., Eyring, H., and Reese, C.E. (1949): Diffusion and membrane     361
    permeability. *J. Phys. Colloid Chem.* 53:1426-1453.

# INDEX

# M Y E L I N

A report based on an NRP Work Session
held June 7-9, 1970

by

**Lewis C. Mokrasch***
Neurosciences Research Program
Brookline, Massachusetts

**Richard S. Bear**
Department of Anatomy
University of North Carolina School of Medicine
Chapel Hill, North Carolina

**Francis O. Schmitt**
Neurosciences Research Program
Brookline, Massachusetts

Ava B. Nash
Yvonne M. Homsy
NRP Writer-Editors

---

*Present address, Department of Biochemistry, Louisiana State University Medical Center, New Orleans, Louisiana.

## CONTENTS

## LIST OF PARTICIPANTS

Dr. Richard S. Bear
Department of Anatomy
University of North Carolina
   School of Medicine
Chapel Hill, North Carolina 27514

Dr. Allen E. Blaurock
Cardiovascular Research Institute
University of California Medical Center
Parnassus Avenue
San Francisco, California 94122

Dr. Daniel Branton
Department of Botany
University of California
Berkeley, California 94027

Dr. Richard P. Bunge
Department of Anatomy
Columbia University
College of Physicians and Surgeons
630 West 168th Street
New York, New York 10032

Dr. Donald L.D. Caspar
Children's Cancer Research Foundation
35 Binney Street
Boston, Massachusetts 02115

Dr. Ronald Chandross
Laboratory for Reproductive Biology
University of North Carolina
   School of Medicine
111 Swing Building
Chapel Hill, North Carolina 27514

Dr. Alan N. Davison
Department of Biochemistry
Charing Cross Hospital Medical School
62 Chandos Place
London, WC2, England

Dr. Edwin H. Eylar
Department of Experimental Biology
Merck Institute for Therapeutic
   Research
Rahway, New Jersey 07065

Dr. Jordi Folch-Pi
McLean Hospital
Belmont, Massachusetts 02178

Dr. Asao Hirano
Department of Neuropathology
Montefiore Hospital and Medical Center
111 East 210th Street
Bronx, New York 10467

Dr. Vittorio Luzzati
Centre de Génétique Moléculaire
Centre National de la Recherche
   Scientifique
91, Gif-sur-Yvette, France

Dr. Donald M. MacKay
Department of Communication
The University of Keele
Keele, Staffordshire, England

Dr. Paul Mandel
Centre National de la Recherche
   Scientifique
11 Rue Humann
67, Strasbourg, France

Dr. Lewis C. Mokrasch
Neurosciences Research Program
280 Newton Street
Brookline, Massachusetts 02146

Dr. William T. Norton
Department of Neurology
Albert Einstein College of Medicine
1300 Morris Park Avenue
Bronx, New York 10461

Dr. John S. O'Brien*
Department of Neurosciences
University of California, San Diego
    School of Medicine
La Jolla, California 92037

Dr. J. David Robertson
Department of Anatomy
Duke University
Durham, North Carolina 27706

Dr. Francis O. Schmitt
Neurosciences Research Program
280 Newton Street
Brookline, Massachusetts 02146

Dr. Marion E. Smith
Neurology Service
Veterans Administration Hospital
3801 Miranda Avenue
Palo Alto, California 94304

Dr. Betty G. Uzman
Children's Cancer Research Foundation
35 Binney Street
Boston, Massachusetts 02115

Dr. Henry deF. Webster
Laboratory of Neuropathology
    and Neuroanatomical Sciences
National Institute of Neurological
    Diseases and Stroke
National Institutes of Health
Bethesda, Maryland 20014

Dr. Frederic G. Worden
Neurosciences Research Program
280 Newton Street
Brookline, Massachusetts 02146

*Unable to attend the Work Session because of illness.
Note: NRP Work Session summaries are reviewed and revised by participants prior to publication.

# I. HISTORICAL INTRODUCTION
L. C. Mokrasch

### Myelin Morphology

Myelin is one of the most conspicuous components of nervous systems. Ever since the microscopic anatomy of nervous systems first attracted attention, myelin has been studied by morphologists more than most other nerve components.

The term *myelin* (German, *Mark*) was coined by Virchow (1854) to describe a lipid-rich structure that in the brain was a medullary rather than a cortical entity. At the level of the individual nerve fiber, because the lipid-rich material is in the form of a sheath around the axon, it was inappropriately called the *myelin sheath* (German, *Mark-Scheide*); the myelin is of course a cortical rather than a medullary structure on the individual fiber. This incorrect terminology has persisted to this day and, indeed, the description of nerve fibers as "medullated" or "unmedullated" persisted despite the fact that the correct configuration of myelin was recognized by Virchow himself who observed the tubular structure of myelin and of *myelin forms* that emerge from nerve fibers cut or macerated in water.

Evidence that myelin was a highly ordered structure was provided by the work of Valentin (1862) who examined myelinated fibers under polarized light. However, it was not until later in the nineteenth century that experimental and theoretical studies on the paracrystalline state of lipid aggregates provided a rigorous basis for the conclusions drawn by Valentin. The study of myelin by histologists and anatomists was advanced by the publication of Ranvier (1889) of his technique for osmium staining. The conspicuous black staining of myelin led to the sharp distinction between "myelinated" fibers in which black staining material was visible and "unmyelinated" fibers in which little or no myelin could be observed. With the work of Lehman (1911, 1918), Friedel (1922), and Schmidt (1924, 1936), it became evident that nerve myelin and myelin forms made of pure chemicals (such as unsaturated lipids dispersed in water) are paracrystalline (smectic or smectogenic) in nature.

Evidence was soon forthcoming that the membrane that invested the "nonmyelinated" nerves was very similar to the myelin membrane of other nerves, differing from it mainly in the aggregate thickness of the investment (Friedlaender, 1889; Ambronn, 1890). Apáthy (1889) questioned the validity of describing nerves as myelinated or unmyelinated. Observations also disclosed that the lipid component of the sheath has a sign opposite to that of the protein, and the resulting anisotropy depends upon the contribution of each. Ambronn (1890) found that lipid solvents, such as alcohol, reverse the sign of the uniaxial double refraction from positive with respect to the radial direction to negative in myelinated (mostly vertebrate) nerve fibers, whereupon the optical properties become qualitatively like those of the so-called unmyelinated fibers of invertebrates. On the other hand, reversal of the normally negative sign of invertebrate fibers to positive was achieved by Göthlin (1913) through application of reagents such as glycerine (the *metatropic* reaction).

The studies of Göthlin (1913) clearly showed that myelin was not characteristic of vertebrate nerves alone, but could be found also in invertebrate nerve trunks. He suggested that to classify nerves as myelinated and unmyelinated was misleading. Instead, he classified nerves on the basis of their optical properties into four categories distinguished by the ease with which birefringence could be demonstrated: (1) myelotropic fibers, whose sheath double refraction is dominated by lipids; (2) metatropic, whose normal optical state is that typical of the protein constituents but which can be reversed to indicate lipid presence by glycerine application; (3) the proteotropic condition, wherein the lipid is insufficient to be demonstrated optically and in which the protein predominates optically; and (4) the atropic fibers, in which no birefringence is seen, possibly because of the small size of the fibers and because the lipid and the protein contributions to birefringence cancel each other. Göthlin showed that the birefringence of nerve fibers was principally related to the organization of the lipid portion of the myelin in most cases, and that sometimes, especially in the smallest vertebrate fibers, the C-fibers, a protein-related birefringence could be described. As knowledge accumulated about the paracrystalline (or "liquid-crystal") state and as the related concepts were further refined through the efforts of Lehman (1911, 1918) and Friedel (1922), the stage was set for a series of conceptual and technical breakthroughs in the understanding of the molecular structure of myelin and the myelin sheath.

It became possible to interpret the findings of earlier workers in terms of molecular orientation and shape of colloidal particles or of phases of myelin through theoretical developments by Wiener (1912). He was able to calculate the structural consequences of these factors using the observed optical anisotropy.

Building upon early investigations with polarized light, Schmidt (1936) proposed a model for the myelin sheath that foreshadowed subsequent hypotheses. Although devised specifically for myelin, his model resembles that proposed in 1935 for cell membranes generally by Danielli and Davson. Schmidt thought of the myelin sheath as made up of concentric cylindrical layers of protein between double layers of lipid, normals to both sets being radial. Working at the same time, Chinn and Schmitt (1937) reached a similar conclusion by a different route. Schmitt and Bear (1939) showed further that the metatropic reversal that Göthlin had found in thin invertebrate nerve sheaths was actually a means by which lipid intrinsic double refraction overcame the form double refraction; Göthlin had thought it the result of improved lipid orientation caused by osmotic stresses. Using a variety of fiber sizes from both vertebrates and invertebrates, Schmitt and Bear showed that nerve sheaths exhibit a smooth rather than an abrupt gradation in the myelotropic against proteotropic structural manifestations.

About the same time, one of the most potent tools of the crystallographer was being applied effectively to the study of nerve membranes. Since the discovery of X-ray diffraction by inorganic crystals by Friedrich and his colleagues (1913) and its development as a tool for deducing the internal structure of crystals by Bragg (1914a,b), its application to biological materials awaited the development of special techniques. The pioneering application of X-ray diffraction to the study of nerves by Schmitt, Bear, and Clark(1935a,b) supported the conclusions drawn from optical polarization studies and in addition provided data on the actual molecular dimensions of the smectic laminae.

Another decade passed before the most powerful tool for the revelation of biological ultrastructure, the electron microscope, contributed to our present detailed understanding of myelin structure. It is curious how in the most advanced tool the new and the old combine; osmium tetroxide, which served the light microscopists so well for nearly a century, would also serve the electron microscopists because the heavy atomic nucleus of osmium makes osmicated structures electron-opaque. Sjöstrand (1949) was able to confirm laminar structures in nerve tissue by electron micrographs of preparations stained with osmium. Soon the lamellar structure of myelin was clearly and convincingly visualized in a variety of locations (Fernández-Morán, 1950a,b; Sjöstrand, 1950, 1953).

One important aspect of the myelin lamellae remained unclear through these early investigations. The techniques used to define myelin structure prior to the advent of electron microscopy had provided no evidence for any configuration other than *concentric* lamellae (Schmitt, 1936, 1944, 1950b). This view persisted during the time when electron microscopy was newly applied to myelin (Fernández-Morán, 1950a,b, 1952). The baffling question of how a relatively thick layer of concentric membranes could be deposited and whence it came was answered simply and brilliantly by Geren who suggested, with supporting evidence, that myelin is a spiral, not concentric, arrangement of layers derived directly from the satellite cell (Geren, 1954; Geren and Schmitt, 1954).

During the second quarter of this century, the morphology of membrane systems in terms of molecular arrangements received increasing consideration. If membranes were to be functional as well as structural units, then the geometric interrelations of the constituent molecules would acquire special significance. It was during this time that the classic model for biological membranes was developed (Danielli and Davson, 1935; Davson and Danielli, 1952), a model that was consistent with the polarization optical data and that was proved structurally correct by X-ray diffraction studies of nerve myelin (Schmitt, Bear, and Palmer, 1941) and of lipid-protein systems (Palmer, Schmitt, and Chargaff, 1941).

Subsequent to the application of increasingly powerful tools to its study, the structure of myelin has become well known and acknowledged. Formerly, controversy existed about the gross structure of myelin, such as whether the laminae are concentric in some cases and spiral in others, but current controversies concern the fine structure of myelin and whether it is a typical membrane like all others. As of this writing, the weight of evidence is forcing near unanimity about the structure of myelin, but new controversies have erupted concerning the localization within the membrane of some of the conspicuous myelin constituents. The most recent flux of thought about the structure of myelin will be described in detail in the section written by Dr. Bear. Suffice it to say that the dynamism associated with the study of myelin with respect to its composition and structure appropriately reflects the dynamism of the myelin itself in vivo.

## Myelin Composition

In the history of the study of brain chemistry, examination of the composition of myelin is recent. Tribute must be paid to the earliest investigators whose methods, crude by today's criteria, were able to provide the first identification of brain constituents. By 1854 von Bibra had established that the lipid composition of the brain changes with age (von Bibra, 1854). The methods of extraction used to examine brain lipids were nonquantitative until the work of Folch et al. (1951). Nevertheless, the important first studies with incomplete extractions made it clear that there were a number of lipids which were found in the brain in characteristic composition and that these constituents varied with age and in pathological conditions. A sort of coherence in the many studies on the lipid composition of the brain was devised by Thudicum in his classic treatise (1884) on the composition of the brain. There was, however, no clear concept relating to specific localizations of the lipids in myelin or in any other part of the brain.

The study of the protein composition of the brain was hampered in the beginning by two factors. First, the residue remaining after a vigorous extraction of lipids is not a promising material for the study of proteins, and second, the methods of protein chemistry were not yet well advanced in the nineteenth century. Nevertheless, an attempt was made by Halliburton (1894) to separate and partially characterize several proteins from brain. A somewhat wider study was made more than twenty years later by McGregor (1917), but the isolation of specific proteins from the brain never did receive the degree of attention that was given to other tissues in the early years of this century.

The unpromising material left after the extraction of brain lipids actually was the source for the first isolation of a myelin-specific protein. In 1877 Ewald and Kühne reported on a proteinaceous material remaining after exhaustive treatment of de-lipidated brain with proteolytic enzymes. Because the material was refractory to proteolysis, it was named "neurokeratin." Later, the work of Folch-Pi and his co-workers led to the conclusion that their "trypsin-resistant protein residue" was identical to neurokeratin and actually was a complex of denatured proteolipid protein with polyphosphoinositides (Le Baron and Folch-Pi, 1956; Folch-Pi, 1966).

Before it was possible to isolate myelin, a direct approach to the study of its composition was not feasible. There were, however, two useful means of inferring the composition of myelin. One was a systematic comparison of the composition of white matter and gray matter, and the other was to follow the composition of the brain or its white matter as a function of the degree of myelination during the development of the brain. The latter strategem was used by Koch and his collaborators in the first comprehensive attempt to define the composition of myelin. Working with material from rat, sheep, pig, and human brains, Koch was able to define three periods of development in terms of brain composition. In the rat, the first period extends to about 10 days postnatally and is characterized by a high water content, a low lipid content, and the absence of histologically demonstrable myelin. The second period in the rat continues until about 40 days after birth and is characterized by a reduction in water content and the accumulation of certain lipids and proteins. In the third stage characteristic of the mature animal, the chemical composition of myelin becomes relatively constant (Koch, 1904, 1905, 1907; Koch and Upson, 1909).

Somewhat later, MacArthur and Doisy (1919) carried on similar studies in more detail. These studies on the human brain laid the groundwork for a later attack on the actual composition of myelin. Using isotopes, Waelsch et al. (1940, 1941) reported results which demonstrated that lipids assumed to be present in myelin were indeed deposited during myelination and persisted thereafter.

Four decades after the pioneering work of Koch and associates, a comprehensive report appeared by Brante (1949) on the composition of the brain from which clear inferences were made about the actual composition of the myelin sheath. Brante identified cholesterol, cerebrosides, and sphingomyelin as "sheath-typical" lipids. Johnson et al. (1950) characterized this group a little more strongly as "myelin lipids."

The stage was set for a new series of advances in research on the chemistry of myelin when inositol phosphatides (Folch, 1951) and proteolipids (Folch-Pi, 1955) were characterized as being myelin constituents. The definition of myelin composition from changes observed in white matter is satisfactory only for its most characteristic components. Minor constituents could be only crudely approximated by this procedure.

Several groups, using variations on the classic Schneider-Hogeboom (1952) subcellular fractionation, were able to obtain small amounts of relatively pure myelin from animal brains (Patterson and Finean, 1961; Mandel et al., 1961; August et al., 1961). Somewhat later, the procedure of density-gradient centrifugation was applied successfully to the separation of brain subcellular particles, including myelin (De Robertis et al., 1962).

Clearly, the crucial advance in the study of the composition of isolated myelin, and later its metabolism, was the now-classic procedure of Autilio et al. (1964) by means of which mammalian myelin could be obtained on a scale and with a purity long anticipated by myelinologists.

## II. BIOLOGICAL CHEMISTRY AND DYNAMICS OF MYELIN
### L. C. Mokrasch

### Composition of Myelin

In recent years there has been increasing interest and research in the structure and function of biological membranes. New techniques and theories have led to a proliferation of fresh ideas about membrane structure. The NRP has a special interest in membranes, and has held a Work Session on brain cell microenvironment (Schmitt and Samson, 1969) and another on carriers and specificity in membranes (Eigen and De Maeyer, 1971). In this section, a discussion of the gross aspects and of the constituent proteins and lipids of myelin will be presented.

#### Gross Aspects of Myelin

At the outset of the Work Session, one of the major points raised by the participants was the nature of myelin and how it differs from cell membranes generally. Schmitt asked, "What is the difference between myelin and stacked glial membranes?" Although central nervous system (CNS) myelin is morphologically an extension of the glial cell plasma membrane, there is no evidence that the glial membrane and myelin are chemically identical. Mokrasch cautioned that myelin should not be discussed as a typical membrane, that myelin proteins may be constantly undergoing conformational changes, changes in plasticity, and allosteric modification. A similar doubt that myelin is indeed a model membrane was voiced by Bear, who noted that for myelin there is no X-ray diffraction evidence for the machinery "real" membranes are known to have. Luzzati wondered whether some knowledge about myelin could be obtained from the study of simple systems. He stressed that most discussions about membranes are not carried on in operational terms and that it would be useful to agree on problems that can be tackled experimentally.

Although the concept generally held by investigators is that all biological membranes are lipid-protein complexes, most membranes are viewed as having distinctive compositions (O'Brien, 1967; Siakotos et al., 1969; Korn, 1969a,b). For purposes of this Work Session, only a broad description of the composition of myelin and other membranes is presented, and for more detailed data the reader is referred to reviews by Mokrasch (1969), Norton (1971a), and Davison and Peters (1970).

The chemical distinctness of myelin from other membrane systems is tabulated in Table 1 in terms of simple chemical estimations. It may be seen that myelin isolated from mammalian central or peripheral nervous tissue has low ratios of nitrogen to phosphorus (atom to atom), and of nitrogen to carbohydrate (atom to mole), compared to other membranes. The nitrogen:phosphorus ratio varies from 4 to 5 for myelin to 11, the next closest ratio, for membranes of erythrocyte ghosts and cerebral microsomes. The nitrogen:cholesterol ratio varies similarly from 4.4 to 5.4 for myelin to values 3 to 20 times higher for other membranes. Myelin of the central nervous system (central myelin) seems to be distinguished from other membranous systems, including myelin of the peripheral nervous system (peripheral myelin), by having a higher glycolipid than cholesterol content.

A comparison of myelin from different sources in terms of its percentage composition is given in Table 2. It is noteworthy that myelin from higher taxonomic groups is somewhat poorer in protein and slightly richer in lipids.

Generally, the composition of myelin is found to vary when the following comparisons are made:

1. Myelin of one species versus that of another.
2. Central myelin versus peripheral myelin in the same species.
3. Brain versus spinal cord myelin in the same species.
4. Myelin with a higher density versus that with a lower density in a preparative sucrose density gradient.
5. Young versus old myelin in the same species.
6. The reporting of myelin of the same type by different authors.

Some of these variations are minor and depend upon the analytical methods used and the condition of the donor animal. Other differences clearly depend upon the preparative procedure; for example, delays in the processing of the raw tissue, particularly spinal cord, may result in a significant autolytic loss of the basic protein. It would appear that a physical function, such as the excitability of a nerve fiber, does not necessarily imply the chemical or metabolic integrity of the myelin with which the functioning nerve fiber is invested. A peripheral nerve, such as frog sciatic, may continue to conduct impulses in particular fiber groups (e.g., A fibers) for several weeks if kept in cold Ringer solution between tests, even though a large fraction of the fibers may

TABLE 1

Simple Chemical Estimations of Composition of Myelin and Other Membranes
(Molar or Atomic Ratios Relative to Cholesterol) [Mokrasch]

| Source | Cholesterol | Nitrogen | Phosphorus | Carbohydrate | Investigators |
|---|---|---|---|---|---|
| Human CNS myelin | 1 | 4.4 | 0.9 | 0.48 | O'Brien and Sampson, 1965a |
| Peripheral nervous system bovine myelin | 1 | 5.4 | 1.2 | 0.27 | O'Brien et al., 1967 |
| Rat CNS myelin | 1 | 5.3 | 1.4 | 0.48 | Seminario et al., 1964 |
| Rat brain microsomes | 1 | 20 | 1.8 | 0.19 | Seminario et al., 1964 |
| Rat brain mitochondria | 1 | 61 | 3.0 | 0.25 | Seminario et al., 1964 |
| Rat synaptic endings | 1 | 83 | 2.5 | 0.25 | Seminario et al., 1964 |
| Human red blood cell ghosts | 1 | 12 | 1.1 | 0.0 | Seminario et al., 1964 |

TABLE 2

Gross Composition of Isolated Myelin
In Percent by Weight of Each Component of Dry Preparation  [Mokrasch]

| Source | Protein | Sterol | Phospholipid | Glycolipid | Investigators |
|---|---|---|---|---|---|
| Human white matter | 22 | 19 | 37 | 20 | O'Brien and Sampson, 1965c |
| Bovine white matter | 25 | 22 | 32 | 20 | Autilio et al., 1964 |
| Bovine dorsal roots | 24* | 13 | 44 | 19 | O'Brien et al., 1967 |
| Ovine white matter | 30 | 23* | 26 | 22 | Korey et al., 1958 |
| Rabbit whole brain | 38 | 17 | 31 | 16 | Cuzner et al., 1965 |
| Rat whole brain | 48 | 11 | 29 | 12 | Seminario et al., 1964 |

*Estimated by difference.

have degenerated and stopped conducting. Chemical analysis of such a "still-conducting" nerve would give highly abnormal results.

In the following paragraphs, the major constituents of myelin (phospholipids, glycolipids, sterols, and proteins) are described. The chief glyceryl lipid in myelin is the ethanolamine phospholipid, mainly phosphatidal ethanolamine. Among the other glyceryl phospholipids, the choline-containing lipid is usually the next most abundant, and generally its plasmalogen content is near 10%. The other lipids, serine and inositol glyceryl phospholipids, have a small plasmalogen content and are present in amounts that are more variable than those of the ethanolamine phospholipids.

Of the lipid-bound fatty acids, oleic acid is the predominant member in all classes of glycerophospholipids (O'Brien and Sampson, 1965a; O'Brien et al., 1967). In contrast to the linear carbon chains of the fatty acids, there is a significant content of aldehydes tentatively identified as having branched chains in the plasmalogens (O'Brien and Sampson, 1965a). The predominant chain length of the aldehydes present in the plasmalogens is 18 carbons.

Sphingomyelin is one of the few cerebral constituents that is relatively concentrated in myelin (Nussbaum et al., 1963; Norton and Autilio, 1966), and the sphingomyelin content of peripheral myelin is about twice that of central myelin (Evans and Finean, 1965; Horrocks, 1967; O'Brien et al., 1967).

There is a difference between the sphingomyelin that is relatively concentrated in myelin and that found in other organelles. The two major sphingomyelins are the stearic acid type, which is lowest in rat and human myelin, and the nervonic acid type, which is more concentrated in myelin than in other organelles. There is also a species difference in the degree of sphingomyelin concentration in myelin; that is, 40% of rat brain sphingomyelin is located in myelin, whereas about 70% of bovine brain sphingomyelin is in myelin.

The presence of gangliosides in myelin as a normal constituent is supported by studies on myelin-deficient mutant mice that are also deficient in the monosialoganglioside (GM-1) type of ganglioside. The values reported for gangliosides are more variable than those of any other constituent. The ganglioside pattern of human myelin is more complex than that of the rat (Norton et al., 1966; Suzuki et al., 1967). About 50 $\mu$g of N-acetylneuraminic acid per 100 mg of myelin, or about 0.15% ganglioside dry weight in myelin, primarily of the GM-1 type, are found.

Central myelin contains more than twice as much cerebroside by weight as sphingomyelin, but the two classes are nearly equal in peripheral myelin. Sulfatide in myelin amounts to about one-sixth of the total glycolipid.

The fatty acid content of sphingolipids differs from that of glyceryl lipids in having more unsaturated and longer chain fatty acids, the principal acid being nervonic acid (Gerstl et al., 1967). The fatty acid patterns of cerebrosides and sulfatides are nearly identical, with the exception of the hydroxy fatty acids that are about 38% of the total in the young human and about 80% in the brain of a middle-aged person (O'Brien and Sampson, 1965a).

The principal base among sphingolipids is the $C_{24}$-sphingosine with smaller amounts of the $C_{18}$ base (Pilz and Mehl, 1966). Smaller amounts of dihydroxysphingenine and $C_{16}$- and $C_{20}$-sphingenine are also found (Schwarz et al., 1967).

Sterols are relatively concentrated in the myelin of the nervous system. About 80% of the cholesterol in the white matter of ox brain is localized in the myelin (Norton and Autilio, 1966). Although not the only sterol in myelin, cholesterol is the major component. Desmosterol, which has been shown to be a myelin component (Smith et al., 1967), is more abundant in immature central myelin and decreases with age in rats from about 10% of the total sterols to traces in the mature animal. It does persist in the spinal cord as 1% to 3% of the total sterols in mature rats.

Myelin contains at least three major protein components that have been loosely described as proteolipid protein (Autilio, 1966; Cotman and Mahler, 1967). These are the Folch-Pi-Lees (FLPL), basic (BP), and Wolfgram (WPRL) proteins. It is probable, however, that only one of these has all the properties of a classical proteolipid protein (Folch-Pi, 1959).

The lowest molecular weight protein is a basic protein, which is the antigen for experimental allergic encephalomyelitis (Kies et al., 1965; Martenson and LeBaron, 1966). According to Eylar (1971), in 12 animal species there seems to be only a single basic protein in central myelin, but in the rat two basic proteins are found. Two distinct basic proteins have been observed in myelin from the peripheral nervous system.

The use of markers for the identification of myelin membranes is made difficult for two principal reasons: (1) certain conspicuous myelin components, such as proteolipid, cyclic AMP phosphohydrolase, and triphosphoinositide (TPI), are concentrated but not exclusively localized in myelin; (2) the question of compositional change during isolation must be unequivocally settled before any component can be accepted as a marker. It was pointed out by Mokrasch that the proteins found in myelin are specific for myelin; however, they lack enzymic activity that can be easily assayed. Moreover, proteins with very similar physical and chemical properties are found in other membranous organelles. At present, the biological assays for such proteins lack the speed and precision characteristic of a good chemical assay.

Although no phospholipid is specific for myelin, the relative abundance of phospholipid representing myelin membrane in the brain serves as a rough index of the degree of myelination there. Similarly, the incorporation of $^{32}$P into brain lipids serves as an index of the rate at which myelination progresses. Not all the lipids share in this turnover during myelination, however; for example, the specific activity of sphingomyelin labeled during myelination shows a distinct maximum when the brain weight of a rat is 1.0 g. On the contrary, the specific activity of phosphatidyl choline shows no corresponding peak of incorporation. Compared to other phospholipids, the triphosphoinositide content of myelin is quite low, but its phosphorus turnover is higher than that of any other lipid, especially during myelination.

About 1% to 1.5% of the total lipids in myelin are the minor constituents, glyceryl galactosides (acyl, alkyl ether, and diacyl) and phrenosine esters. Small amounts of nucleic acids with characteristic base ratios have been detected in goldfish Mauthner nerve myelin (Edström, 1964), and the ratios may be changed by nervous activity according to Jakoubek and Edström (1965).

### Proteins

The myelin protein with the longest history of critical investigation is that discovered by Folch-Pi and Lees (1951). Evidence is abundant that this "proteolipid" protein isolated from bovine white matter or central myelin is a monodisperse protein with a probable molecular weight of 34,000. Similar proteins have been isolated from other nervous tissues and membranous organelles, but they are never present in as great quantity as in central myelin (Table 3). In addition, other tissue proteins share some of the unusual properties of the Folch-Pi-Lees proteolipid protein, but not to a degree that would make separations or characterizations difficult.

TABLE 3

Myelin Proteins [Folch-Pi]

| Source | Percents of Total Protein | | | Investigators |
| | Proteolipid | Basic | Wolfgram | |
| --- | --- | --- | --- | --- |
| CNS | 50 | 30 | 20 | Wolfgram and Kotorii, 1968a |
| PNS | 23 | 77 | 0 | Wolfgram and Kotorii, 1968b |
| PNS | 23 | 21 | 55 | Eng et al., 1968 |

*Proteolipid Protein*

Folch-Pi described the properties that characterize this proteo-lipid protein as follows:

1. It is resistant to proteolysis.

2. It is soluble in either aqueous solvents or chloroform-methanol mixtures.

3. It is unstable in the presence of moderate salt concentrations or mild alkalinity.

4. It has a relative deficiency of acidic and basic amino acids.

After years of frustration owing to the fact that the classic purification methods of protein chemistry were inapplicable to the Folch-Pi-Lees proteolipid protein, a number of procedures were devised that permitted a fairly easy isolation of the pure native protein (Tenenbaum and Folch-Pi, 1966; Mokrasch, 1967, 1972). In addition, the quick isolation of crude preparations of the three major myelin protein fractions is now possible (Eng et al., 1968).* The amino acid compositions of the protein portions of these crude fractions are very close to those of the constituent proteins purified by other means (Table 4).

*Also, J. Folch-Pi, manuscript in preparation.

TABLE 4

Amino Acid Composition of the Three Myelin Fractions (mole %)
[Folch-Pi]

| Amino acid | 2:1 chloroform-methanol, insoluble protein | Wolfgram proteolipid* | Basic protein from myelin | Basic protein from white matter† | Trypsin-resistant protein from myelin | Proteolipid protein‡ |
|---|---|---|---|---|---|---|
| Lysine | 6.87 | 6.95 | 7.79 | 7.56 | 4.25 | 4.3 |
| Histidine | 2.06 | 2.29 | 5.04 | 5.67 | 2.24 | 1.9 |
| Arginine | 6.14 | 5.83 | 9.47 | 10.14 | 2.01 | 2.6 |
| Aspartic acid | 9.47 | 9.90 | 6.71 | 6.99 | 4.96 | 4.2 |
| Threonine | 5.29 | 5.16 | 3.71 | 4.15 | 8.98 | 8.5 |
| Serine | 6.00 | 5.84 | 9.46 | 9.71 | 6.38 | 5.4 |
| Glutamic acid | 14.27 | 12.95 | 7.63 | 6.37 | 6.85 | 6.0 |
| Proline | 5.00 | 4.64 | 7.24 | 7.44 | 3.07 | 2.9 |
| Glycine | 7.16 | 8.00 | 15.49 | 15.08 | 11.82 | 10.3 |
| Alanine | 8.96 | 8.45 | 8.90 | 8.81 | 12.06 | 12.5 |
| Half cystine | 1.09 | 1.01 | 0.00 | 0.00 | 4.14 | 4.2 |
| Valine | 5.56 | 5.84 | 1.38 | 1.54 | 6.62 | 6.9 |
| Methionine | 1.77 | 2.13 | 1.28 | 1.16 | 0.94 | 1.7 |
| Isoleucine | 4.94 | 4.27 | 1.57 | 1.76 | 4.61 | 4.9 |
| Leucine | 8.81 | 9.63 | 6.02 | 6.35 | 10.16 | 11.1 |
| Tyrosine | 2.86 | 2.87 | 2.79 | 2.49 | 4.02 | 4.7 |
| Phenylalanine | 3.78 | 4.22 | 5.36 | 4.78 | 6.85 | 7.8 |

*Wolfgram (1966).
†Martenson and LeBaron (1966).
‡Tenenbaum and Folch-Pi (1966).

According to Mokrasch (1967), the purified apoprotein of the Folch-Pi-Lees proteolipid can be obtained by mild treatment of a washed total lipid extract of myelin (or other tissue fractions). The protein thus obtained from bovine white matter or myelin isolated from it behaves as a monodisperse protein by a number of physical tests including chromatography on a wide variety of supports, gel electrophoresis, and salt precipitation (Mokrasch, 1972). A systematic study, using a variety of animals, of proteolipid protein of myelin, both central and peripheral, has not yet been reported.

The unusual resistance of proteolipid protein to proteolysis and its amino acid composition have led to the identification of the protein with neurokeratin and similar preparations of denatured brain proteins (Uzman, 1958; Uzman and Rosen, 1958; Folch-Pi, 1959). Recently, however, it has been possible to devise a dispersing medium that facilitates the attack of proteases (Lees and Burnham, 1967), so the overdue study of the primary structure of the Folch-Pi-Lees protein may soon begin after years of abortive attempts.

Mokrasch reported how the Folch-Pi-Lees protein from any source, when freed of adventitious lipids, can be converted to a water-soluble form (Sherman and Folch-Pi, 1970; Mokrasch, 1972). Provided that the ionic strength of the solution is kept low, the pH below 8, and all the methanol of the original chloroform-methanol solution is removed, very stable protein solutions can be easily obtained. If a two-phase system should be made by adding chloroform and methanol to the aqueous solution of the proteolipid protein, the protein partitions in favor of the aqueous phase (Table 5). If, however, an acidic lipid is added to the mixture, the protein-lipid complex is found in the chloroform-rich layer. This behavior seems to reflect a conformational mobility by means of which the protein can assume stable configurations in media of markedly differing properties and in association with lipids. Such conformational flexibility seems well suited to a protein in an extended conformation in a dynamic membrane like myelin.

*Basic Protein*

After years of conflicting reports about the identity and number of basic proteins present in myelin, Eylar stated that reliable methods have now been devised to isolate the protein in pure and stable form (Nakao et al., 1966a,b; Eylar et al., 1969; Martenson et al., 1970). This protein, also called A1 protein, is the principal antigenic component of normal myelin.

TABLE 5

Partition Properties of Proteolipid Protein;
Effect of a Model Acidic Lipid [Mokrasch, 1971]

| Preparation | Addition | Partition Coefficient* |
|---|---|---|
| Protein in 1:1 CHCl$_3$ :CH$_3$OH | Control | 2.08 |
|  | Cardiolipin, 1 mg/ml | 0.31 |
| Protein in H$_2$O | Control | 39.50 |
|  | Cardiolipin, 1 mg/ml | 0.88 |

*(Mg per ml in upper phase)/(mg per ml in lower phase). Final solvent ratios: CHCl$_3$ :CH$_3$OH:H$_2$O, 5:5:2 (volume per volume).

The basic protein isolated from bovine myelin is homogeneous by gel electrophoresis, immunoelectrophoresis, and ultracentrifugation. It amounts to about 30% of myelin protein and probably plays a major role as a structural protein of myelin. The amino acid composition of the basic protein from bovine central myelin is given in Tables 4 and 6, and the total amino acid sequence has been reported (Eylar, 1970). Unlike the case for histones, the basic amino acids are distributed evenly throughout its sequence.

The protein is not a large one; it has about 170 amino acid residues: one tryptophan, two methionines, a high glycine content, and 25% of the total are the three basic amino acids. It could behave, therefore, as a polycation with a net positive charge of 25 at physiological pH, and it has an isoelectric point of 12. Its average molecular weight is 18,400. Other properties of the protein are summarized in Table 7.

Eylar emphasized that the protein is not a phosphoprotein or glycoprotein and is not cross-linked. Possibly owing to the comparatively high number of positive charges and the coulombic repulsion of those charges, the intrinsic viscosity suggests that the protein is fully

TABLE 6

Amino Acid Content (in Moles Percent) of Sciatic Nerve and CNS Basic Protein
[Oshiro and Eylar, 1970]

| Amino Acid | Bovine A1 Protein | Moles Percent P2 Protein | P3 Protein |
|---|---|---|---|
| Lysine | 9 | 10 | 10 |
| Histidine | 6 | 3 | 2 |
| Arginine | 11 | 6 | 9 |
| Aspartic acid | 6 | 9 | 10 |
| Threonine | 4 | 6 | 6 |
| Serine | 3 | 6 | 7 |
| Glutamic acid | 9 | 9 | 11 |
| Proline | 4 | 6 | 5 |
| Glycine | 14 | 10 | 9 |
| Alanine | 8 | 8 | 7 |
| Valine | 6 | 6 | 5 |
| Methionine | 1 | 2 | 2 |
| Isoleucine | 3 | 4 | 5 |
| Leucine | 6 | 8 | 8 |
| Tyrosine | 2 | 2 | 2 |
| Phenylalanine | 4 | 4 | 3 |
| Tryptophan | 1 | 1 | 1 |
| Molecular weight | 18,400 | 11,000 | 6,000 |

unfolded. This is reasonable for a protein with a high proline content and is confirmed by optical rotatory dispersion (ORD) and circular dichroism (CD) data suggesting no helicity or $\beta$-structure (Table 7). According to Eylar, the protein is comparatively stable: boiling a solution of the protein does not diminish its experimental allergic encephalomyelitis (EAE)-inducing ability or its interaction with antibody (Eylar and Thompson, 1969).

In Eylar's view, the properties described so far seem to make this protein uniquely suited for its role as a lipid-protein membrane

TABLE 7

Chemical Properties of Bovine A1 Protein
[Eylar and Thompson, 1969; Oshiro and Eylar, 1970]

| | |
|---|---|
| Average molecular weight | 18,400 |
| Intrinsic viscosity | 9.3 ml/g |
| Axial ratio (unhydrated) | 9.6 |
| Optical rotatory dispersion | no helical or $\beta$-structure |
| Sedimentation coefficient | 1.4-1.7S |
| N-terminal residue | N-acetylalanine |
| C-terminal residue | arginine |
| Phosphorus | none |
| Cystine or cysteine | none |
| Hexoses, hexosamine, sialic acid | none |

constituent. It is not found in other cellular structures, although there is an analogous protein or proteins in peripheral myelin. It is possible that during myelinogenesis the formation of this protein is the trigger that initiates the formation of myelin. Perhaps the presence of this protein in the satellite cell cytoplasm perturbs the cytoplasmic membrane and the result is myelin.

A number of basic proteins have been separated from peripheral myelin. Whether these are authentic components of peripheral myelin or are derived in part as autolytic artifacts is not yet clear. In any case, there are two EAE-inducing peptides that can be isolated from rabbit peripheral nerve; they are designated P2 and P3. The P2 protein appears capable of inducing experimental allergic neuritis (EAN) in monkeys as well.

The amino acid composition of the bovine CNS antigen compared with rabbit P2 and P3 is shown in Table 6. P2 has a molecular weight of 11,000, and P3, 6,000. P2 and P3 have very similar amino acid compositions and P3 could be a degradation product of P2. Both P2 and P3 have tryptophan and an abundance of glycine. Recent studies* indicate that the A1 protein and the P2 protein have homologous regions of their polypeptide chain where the amino acid sequence is identical. Since the homologous region includes the tryptophan region, it is obvious why the P2 protein is capable of inducing allergic encephalomyelitis.

### Wolfgram Protein

The third type of myelin protein was first described by Wolfgram (1966) and is frequently designated as the "Wolfgram protein." It differs from the other two types by being extractable from myelin with acidified chloroform-methanol mixtures, precipitable therefrom by neutralization of the solvent mixture, and by being relatively rich in dicarboxylic amino acids. For some time it was questioned whether this was an authentic myelin constituent, but there seems to be sufficient evidence now to regard it as such.

Unfortunately, as viewed by Mokrasch, the Wolfgram protein has not been studied as a purified entity but has been observed as a residue insoluble in chloroform-methanol mixtures to which no acid has been added, or as a third component identifiable in gel electrophoresis studies (Mehl and Wolfgram, 1969). Other than the characteristics mentioned, this protein does qualify as a proteolipid type because of its solubility in organic solvents. Unlike the FLPL protein, however, the Wolfgram type is susceptible to attack by trypsin. It is clear that if the structures and properties of the FLPL proteins and of the basic proteins are important in the understanding of myelin structure and function, the characteristics of the Wolfgram type are also important.

## Dynamics and Experimental Alterations

### Myelinogenesis

#### Chemical Aspects: A.N. Davison

It was reported by Banik, Blunt, and Davison (1968) that if the optic nerve of a developing kitten is stained with sudan black, one can observe the first appearance of myelin structure when the animal is about 10 days of age. In the electron micrograph myelin is also seen,

*E.H. Eylar, unpublished data.

but the usual myelin osmiophilia does not appear until about 25 days of age. Wolman (1957), studying the development of rat brain, similarly showed that sudanophilia precedes osmiophilia. Banik et al. (1968) then attempted to use a biochemical test, the thin layer chromatography of extracted brain lipids, to examine this anomaly. Cat optic nerve, at various postnatal periods, was extracted for lipids. When equal amounts of lipid from 16-day-old optic nerve and from an adult cat optic nerve were applied to a thin-layer sheet, developed, and visualized, most of the lipids were present in both cases, but the cerebrosides were relatively deficient in the kitten optic nerve.

Similarly, when myelin is isolated from rat brain at various postnatal ages and analyzed for lipids, one sees a progressive increase in cholesterol and especially in cerebrosides (Table 8), and a corresponding decrease in phospholipids, particularly lecithin (Cuzner and Davison, 1968).

TABLE 8

Myelin Composition of Rat Brain at Various Ages
(Moles Percent Lipid) [Cuzner and Davison, 1968]

| Lipid | 18 Days | 20 Days | 25 Days | Adult |
|---|---|---|---|---|
| Cholesterol | 36 | 41 | 44 | 45 |
| Cerebroside | 8 | 8 | 15 | 13 |
| Total phospholipid | 56 | 51 | 41 | 42 |
| Total plasmalogen | 9 | 13 | 15 | 16 |

It has been possible, according to Norton*, Banik and Davison (1969), and Agrawal et al. (1970a), to separate a myelin-like fraction from the crude mitochondrial fraction of immature rat brain. When myelin is isolated from a sucrose density gradient and exposed to

*W.T. Norton, unpublished data.

osmotic shock, a pellet can be obtained from centrifugation at 13,000 x $g$ for 10 minutes that is a typical myelin preparation. The cloudy supernatant yields another pellet after centrifugation at 78,000 x $g$ for 1 hour that is the myelin-like membranous fraction.

The composition of myelin and myelin-like fractions differs in a manner similar to that noted in mature and immature myelin. Davison reports that relative to cholesterol, the myelin-like fraction has 150% as much phospholipid and 15% as much galactolipid on a molar basis as normal myelin. In these respects the myelin-like fraction also resembles plasma membrane, which normally is richer in phospholipids, especially phosphatidyl choline, and has a smaller content of cerebrosides. When a myelin-like preparation is examined under the electron microscope, structures are seen that resemble vesicles from plasma membranes or microsomes. Cross-contamination is not likely because labeling experiments in which microsomes and plasma membranes were made radioactive and then mixed with the myelin-like fraction showed no transfer of label to the myelin-like fraction.* The ganglioside content of myelin-like membrane fractions shows a resemblance to that of synaptic membranes, being much higher than that of regular myelin isolated from the 16-day-old rat or the adult rat (Agrawal et al. 1970).

Davison discussed other experiments in which biochemical techniques are used to study myelin composition. For example, the use of enzyme markers reduces the likelihood that the myelin-like fraction would be grossly contaminated with mitochondrial, nerve ending, microsomal, or plasma membrane particles. The 2′,3′-cyclic AMP 3′-phosphohydrolase activity (Table 9) is closer to that of normal myelin than of any other structure (Banik and Davison, 1969). Similarly, the aminopeptidase acitivity is close to that of normal myelin.

Gel electrophoresis under a variety of conditions shows that the myelin-like fraction contains very little basic protein; therefore, assuming that the enzyme is a reliable test for myelin, the phosphohydrolase activity cannot be due to a gross contamination with normal myelin. In fact, the basic proteins detected in this system have been shown by other tests to be histones and not the basic protein characteristic of myelin (Agrawal et al., 1970). On the other hand, there is a substantial protein fraction clearly identifiable with the proteolipid protein of normal mature myelin.

---

*W.T. Norton and A.N. Davison, unpublished data.

TABLE 9

Enzyme Profiles of 15-Day-Old Rat Brain
[Banik and Davison, 1969]

| Enzyme | Crude Myelin | Myelin-like | Other Typical Fractions |
|---|---|---|---|
| 2′,3′-Cyclic nucleotide 3′-phosphohydrolase | 128 | 54 | 202 Myelin |
| Total ATPase | 11 | 10 | 34 Microsome |
| Acetyl cholinesterase | 2.5 | 1 | 9 Nerve ending particles |
| Succinic dehydrogenase | 0.5 | 0.4 | 36 Mitochondria |

Results are expressed as $\mu$moles substrate/hr/mg protein

When $^{14}$C-acetate is incorporated into myelin and myelin-like fractions of 16-day-old rat brain in vivo, there is no significant difference between the labeling of the myelin and the myelin-like fractions after either 10 minutes or 24 hours (Agrawal et al., 1970). Similarly, if $^{14}$C-leucine is administered to an 11-day-old rat, the myelin-like fractions show a degree of incorporation remarkably similar to purified myelin at various periods postinjection.

A proposed mechanism for myelination based on these observations, and one that is supported by Hirano and by Webster, was presented by Davison (Figure 1). During early development of the myelin sheath, there are initially a number of loose windings around the axon that are filled with satellite cell cytoplasm. Then compaction occurs when mature myelin is formed by the synthesis of the basic protein and its accumulation at the place where the major dense line of the electron micrograph begins. Somehow, the appearance of basic protein and cerebroside on the cytoplasmic side of the satellite cell triggers the compaction. Consistent with this proposal is Suzuki's observation that the ganglioside composition of myelin changes throughout the life of the animal (Suzuki et al., 1967, 1968).

Referring to the earliest stages of myelination, Folch-Pi noted that the proposed existence of a myelin precursor is supported by the

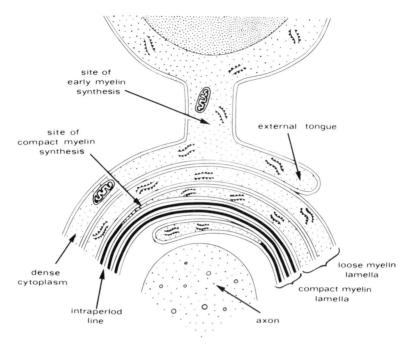

Figure 1. Schematic diagram showing "early" and mature myelin sheaths in the developing central nervous system. [Davison]

fact that neurokeratin and TPI, which are fairly specific myelin constituents, are found in 1- or 2-day-old rats, well before any significant amount of regular myelin can be found. In addition, these substances increase at a significant rate before myelination really begins (Folch-Pi, 1955).

Additional evidence for the existence of a myelin precursor, which may be the satellite cell plasma membrane itself, comes from the study of myelin-deficient mutants by Mandel (Nussbaum et al., 1968,1969). He noted that when there is a deficiency of cerebrosides preventing normal myelination, other membranous components such as protein are degraded.

Concerning the decrease in neosynthesis of phosphatidyl choline during maturation of myelin, experiments carried out by Mandel and his collaborators* suggest that there is a regulation of phosphatidyl

*L. Freysz and P. Mandel, unpublished data.

choline synthesis by substrate inhibition; for example, the inhibition of phosphatidyl choline diglyceride transferase by lecithin does not occur in chick embryo brain extracts; however, when extracts of brain of hatched chick are used, strong inhibition (50%) is obtained under similar conditions. When myelin formation begins, inhibition of phosphatidyl choline synthesis begins. This change and perhaps the inhibited activity of phospholipases may explain the changes that occur in myelin composition with respect to other phospholipids.

The content of basic protein in the "early myelin" is very low. There is evidence of a nonuniform protein content in isolated myelin that is consistent with the notion of a permissible variation in myelin composition. Mokrasch suggests that a continuum of compositions for myelin is possible, beginning with some "threshold" composition, (perhaps similar to that of the plasma membrane of the Schwann cell or glial cell with its large phosphatidyl choline component) at which time the myelin structure becomes stable or coherent, and continuing through a gradation of compositions characteristic of more mature myelin.

*Structural Aspects: H. deF. Webster*

A study to establish the temporal changes in the geometry of myelin and Schwann cell membranes in rat sciatic nerve was recently completed by Webster (1971). Except for Webster's study and some earlier work by Uzman and Nogueira-Graf (1957) and Robertson (1962), little has been done to approach this interesting problem. Of particular interest is the estimation of the increase in membrane area during myelin deposition. This and other structural changes were discussed by Webster and are described in the following paragraphs.

At birth, myelin sheaths are absent in the marginal bundle of the posterior tibial fascicle and only a few are present at age 3 days. At 7 days there are many myelin sheaths, and at 16 days they are larger and more numerous (Figure 2, A, B, C, D). Serial sectioning permits three-dimensional reconstructions of a nerve fiber's myelin sheath at intervals during development. Initially, clusters of small axons (less than $0.8 \mu$) are surrounded by a Schwann cell with a peripheral nucleus. Those axons destined to be myelinated are segregated from the remainder and provided with their own satellite cells by mitosis.

During the formation of the first few spiral turns around an axon, the membrane growth is exponential. At different levels along the same Schwann cell, the mesaxon's length and configuration vary. If the

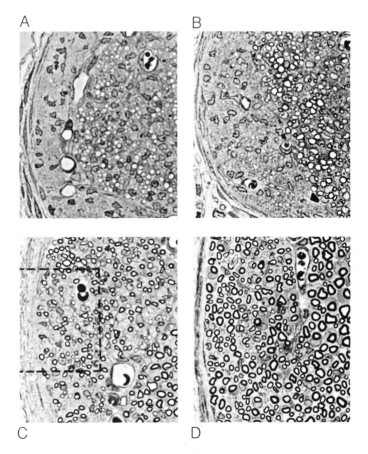

Figure 2. Transverse sections at the same magnification of the marginal bundle in rat sciatic nerves. The perineurium of the posterior tibial fascicle is to the left. At birth (A) there are no compact sheaths, and at age 3 days (B) there is only one; at age 7 days (C) many are apparent. The myelin sheaths are more numerous and larger at age 16 days (D). x 500. [Webster, 1971]

membrane is relatively stationary near the external mesaxon's origin, and if it is free to rotate about the axon, the spiraling as well as concurrent transverse growth would occur naturally if the growth of the membrane is uniform throughout. Sometimes a number of over-lapping surface projections of the membrane give the appearance of a multitude of mesaxons. This is evidence that membrane synthesis occurs diffusely before compact myelin is formed. It is possible that the synthesis of myelin constituents proceeds at an exceedingly rapid rate before the orderly assembly of these constituents into myelin mem-brane. Subsequently, the imposition of order in the assembly of the membrane could be what controls the synthesis of myelin constituents.

According to both Webster and Robertson the position of the inner mesaxon varies initially along the length of the cell. It is frequently seen as a somewhat elongated irregular spiral (Figure 3). The

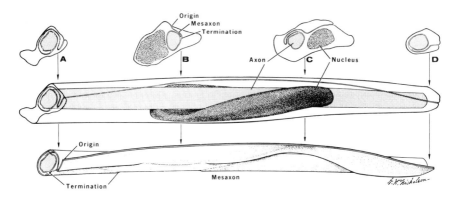

Figure 3. Three-dimensional representations of a Schwann cell with the position of the four transverse levels indicated by arrows. The helical nucleus and relatively constant position of the mesaxon's surface origin are included in the upper figure. The lower one shows the mesaxon as a membrane sheet; its contour and transverse length vary. [Webster, 1971]

relation of the Schwann cell to adjacent axons is also nonuniform. Over short distances a few axons often seem to be carried on the surface of the Schwann cell. They then separate from it to join other clusters of axons or another Schwann cell. It appears that when the first spiral turn is fairly well formed, a specificity is induced in the membrane that prevents a number of axons from being invested by the same myelin sheath. As myelination proceeds, a loose spiral is formed. Further growth leads to the formation of a compact sheath with 4 to 6 layers that are separated by longitudinal ridges of cytoplasm near the origin of the external mesaxon.

Webster showed that as the spiral becomes larger and more compact in myelin development, it also seems to become more orderly. In contrast, when the sheaths are small there is an impressive variety to their shapes and structures, even within the same sheath observed at different levels. In the same section, the sheath may be perfectly regular in nearly all its circumference and still have at one point protrusion of separated layers with included satellite cell cytoplasm.

Occasionally, lamellar structures not connected to the myelin sheath are seen within the satellite cell, but this feature is observed less frequently as maturation progresses. The relation of these cytoplasmic lamellar structures to myelin or its formation is obscure. Their presence

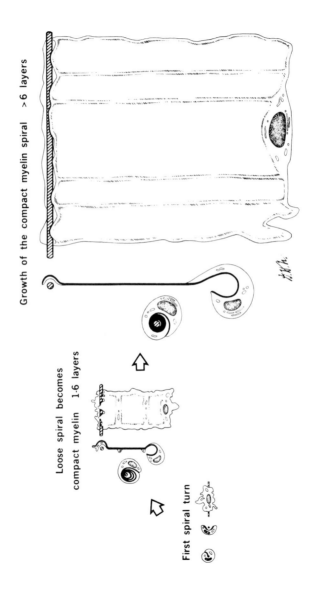

Figure 4. A diagram of myelin formation in a Schwann cell that shows how the location and relative size of the myelin membrane's cytoplasmic interfaces change during the sheath's growth. The Schwann cell's appearance in cross sections is supplemented by transverse and face views of the "unrolled" cell. (To facilitate comparison, the three stages are drawn at the same magnification.) [Webster, 1971]

probably reflects the rapid rate of membrane synthesis by Schwann cells during myelination because they are much less commonly observed in mature nerves.

Even when there are about twenty compact laminae in the spiral, there may be some very loose and separated membranes at the periphery. Their presence, along with the variations in sheath contour that remain in large adult fibers (Webster and Spiro, 1960), make it unlikely that myelination occurs by any simple spinning or winding mechanism.

When the myelin spiral is visualized in an unwound configuration, the changing relationships of Schwann cell cytoplasm and myelin membrane are seen more clearly (Figure 4). During the transition from a loose to a compact spiral, longitudinal ridges of cytoplasm are continuous with those at the nodal ends. These ridges contain numerous organelles such as mitochondria, Golgi elements, endoplasmic reticulum, etc. Probably these cytoplasmic ridges and their organelles permit a vigorous exchange of materials between the membranes and the cytoplasm during rapid sheath growth. However, the greatest increase of myelin membrane area occurs when the sheath is compact and contains more than 6 layers (Figure 4). During this phase the growth rate declines, remodeling is probably more prominent, and the large longitudinal cytoplasmic ridges are transformed into the Schmidt-Lanterman clefts. These strips of cytoplasm are perpendicular to the nerve fiber's axis, they are much smaller, and they do not contain organelles.

To quantitate the real growth of membrane, some calculations were made from the serial electron micrographs of a single Schwann cell segment (Webster, 1971). The thickness of the myelin spiral is measured from a cross section of the structure, and the length of the internodal section of the Schwann cell is derived from the serial section information. If circularity of the windings is assumed and irregularities are neglected, the total membrane areas can be calculated (Figure 5). Whereas the growth of the Schwann cell surface membrane (external plus axonal) proceeds roughly linearly with respect to the bundle diameter, the membrane area of the myelin winding is related logarithmically or exponentially to the diameter. The maximal growth rate of the myelin membrane is its initial value, and the rate decreases with time (Figure 6). When comparison of myelin membrane area to Schwann cell surface membrane area is made, the area of the former is very small in 1-day-old rats. This changes quickly, however, and after 9

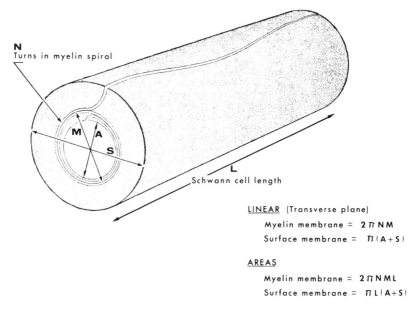

Figure 5. Approximate dimensions of Schwann cell membranes during myelination. [Webster]

days of age the myelin area may be 100 times larger than that of the satellite surface membrane (Figure 7).

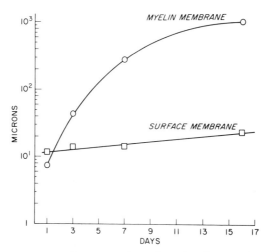

Figure 6. Approximate linear dimensions of myelin and surface membranes in transversely sectioned Schwann cells during myelination. [Webster, 1971]

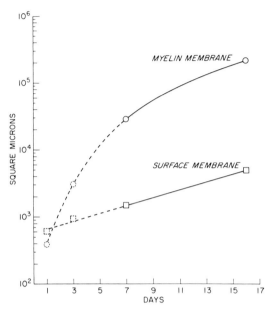

Figure 7. Approximate areas of Schwann cells' myelin and surface membranes during myelination. [Webster]

In Webster's opinion, any proposed mechanism for the initial phase of myelinogenesis must take into account that the rate of increase in area is proportional to the area already present. This view does not seem to allow for any mechanism by which membrane growth occurs at a specific site, such as the internal or external mesaxon, but rather that the membrane grows as a whole and that insertions into the membrane at any internal locus must be more facile than has been suspected hitherto. After the compact spiral is formed, both the growth rate and the relative area exposed to cytoplasm decrease. This area (outer and inner myelin lamellae, Schmidt-Lanterman clefts) may be sufficient for insertion of new membrane material and remodeling until the adult dimensions are achieved.

Concerning the initiation of myelination, Bunge offered the speculation that when an axon is ready to be myelinated, the satellite cell membrane properties are altered and a complementary surface charge mosaic causes it to stick to itself. Once begun, the myelination continues by this same process. Moreover, it is possible that there is a chemical difference between the earliest membranes at the time of contact and the later mature windings. At the earliest stage all the membranous systems are growing actively. When the Schwann cell grows and divides, it grows lengthwise and moves away from its parent or twin cell and establishes an intact and complete basal lamina. Following this, according to Webster, the membrane growth pattern becomes more orderly, and it is possible that restrictions imposed by the presence of neighboring cells or collagen force the membrane growth pattern to shift inward, resulting in a spiral growth.

In terms of nerve conduction, there are some areas concerning the role of myelin and related structures that need further study and clarification. It is clear that the impulse is propagated more rapidly in heavily myelinated than in poorly myelinated or unmyelinated axons. As reported by Robertson (1960c), in saltatory conduction only about 10% of the current flows through the myelin sheath, and the rest flows through the nodes. Thus, the precise role of the Schwann cell or the changes that occur within it during conduction are unknown. According to Bunge, one possibility is that the Schwann cell cytoplasm may form the "return" part of the circuit for the electrical current generated in conduction. This could be established only by intracellular recordings from myelin-related cells; this type of recording has not yet been accomplished. Robertson believes that the "return" is entirely extracellular.

## Myelin Stability and Turnover

*Chemical Aspects: B.G. Uzman and M.E. Smith*

It has been known for several years that $^{14}$C-labeled cholesterol given to suckling animals can be found later in both the CNS and peripheral nervous system (PNS) myelin of the mature animal (Davison et al., 1958, 1959; Dobbing, 1963; and Cuzner et al., 1965). The study of cholesterol turnover reported by Uzman and colleagues (Hedley-Whyte et al., 1968) was done chiefly by means of autoradiography using tritiated cholesterol. Preliminary studies here established the stability of 1,2-$^3$H cholesterol because after injection of cholesterol labeled with both $^3$H (1,2 H position) and $^{14}$C (4 C position), the ratio of the two isotopes remained constant in mice during a 90-day interval.

Although the preparation of the tissue for sectioning removed about 90% of the cholesterol, chromatographic studies at each stage of the process verified that some cholesterol remained in the tissue and that the isotopic compound was cholesterol. Using a rapid dehydration process (Hedley-Whyte and Uzman, 1968), it was possible to increase the retention of cholesterol in the tissue to 60%. The penetration of intraperitoneally injected labeled cholesterol into the nervous system occurs rapidly (Figure 8).

Because the cholesterol is not localized exclusively within the myelin, the Salpeter method of grain density measurement was used to distinguish between the myelin and extramyelin pools of labeled cholesterol. The Salpeter method is one in which the distribution of label is referred to that of a grid of points superimposed on the autoradiograph. The density of grains per point over any particular tissue component establishes radioactivity per area and can be directly compared with density over any other tissue component; this method avoids the common error of assuming that a tissue component over which there is the greatest number of grains is more active—it may just be more abundant. By comparing the density of grains over the myelin section to those outside the myelin, it was established that the

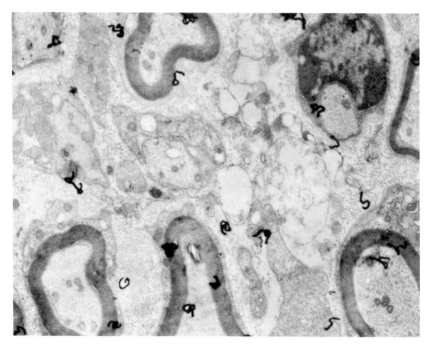

Figure 8. Electron microscope autoradiograph of sciatic nerve of 5-day-old mouse 3 hours after intraperitoneal injection of cholesterol-1,2-$^3$H, exposed 11 months to Ilford L4. Note grains over myelin and also over unmyelinated fibers as well as over tissues more than 0.28 $\mu$ away from the myelin sheath. x 13,000. [Hedley-Whyte et al., 1969]

radioactive cholesterol is relatively concentrated within the myelin. It seems clear, therefore, that the cholesterol is not largely redistributed during the dehydration and infiltration processes.

To determine whether the cholesterol is selectively localized in some portion of the myelin sheath, EM autoradiographs were analyzed by dividing the myelin into regions (inner or outer halves or thirds) and counting the grains over each region (Hedley-Whyte et al., 1969). The results showed that when a 5-day-old mouse is injected with radioactive cholesterol, there is no selective localization of the label at 3 hours, 1 day, or 46 days postinjection. The interpretation is that either the cholesterol is not inserted predominantly in a particular locus, or that, if it is, it rapidly redistributes itself after incorporation within the myelin.

The stability of cholesterol in myelin was further examined in regeneration experiments (Rawlins et al., 1970). If the mouse sciatic nerve is sectioned at its middle branch about 2 mm from the branch point, the adjacent branches can serve as a control against which the regeneration of the middle section can be compared. When an 8-day-old mouse was injected, the nerve sectioned at 43 days, and the animal sacrificed at 73 days, the density of radioactivity of the regenerated myelin was as great as that of normal nerve myelin. This suggests that a pool of bound cholesterol available for reutilization exists in the nerve in some locus other than the myelin sheath. The fact that brain and peripheral nerve retain labeled cholesterol for long periods but that viscera (heart, kidney) and skeletal muscle do not is further evidence of binding of labeled cholesterol within the nervous system; the localization of radioactive cholesterol in extramyelin compartments and its availability for regeneration all suggest a "closed" circulation of cholesterol in the nervous system.

When Smith (1967) injected a 16- to 18-day-old rat with $^{14}$C-acetate and isolated the myelin, she found all the lipids to be labeled. Initially, the specific activity of the labeled myelin falls rapidly owing to dilution by the continued myelination. There is, nevertheless, a considerable disparity in the dilution or decrease in specific activity when the myelin components are compared. Since the apparent turnover of the sulfatide was negligible, it was used as an index of comparison for the other lipids. Phosphatidyl inositol is by far the most metabolically labile of the lipids. While lecithin shows a fair degree of lability, like the sulfatide, the sphingolipids and ethanolamine glyceryl phospholipids are comparatively stable.

Similarly, when 3-month-old rats are injected with $^{14}$C-glucose, the results appear to be the same as those in the other studies except that the phosphatidyl serine seems relatively stable (Smith, 1968). This may imply a difference in turnover of phosphatidyl serine between young and older animals. When spinal cord slices from a 40-day-old rat are incubated with $^{14}$C-glucose, the myelin lipids are labeled with a pattern that accords with the in vivo turnover data (Smith 1969a). It is likely that some of the differences in chain length of the fatty acyl groups are responsible for the differences in turnover rates among the various lipids; O'Brien feels that the correlation is good (O'Brien, 1967). Smith noted that the molar proportions of stable and labile lipids are not constant with age, although the ratio of cholesterol to the other stable lipids is relatively constant (Table 10). The relative stability of certain lipids may be a reflection of reutilization of the lipid substituents.

Smith (1968) reported that compared to lipids the proteins of myelin appear to have a significantly greater turnover. Davison reported similar findings in the basic, but not in the proteolipid, proteins of myelin (Davison, 1961; Wood and King, 1971).* Their turnover rates are very similar, in terms of the decay portion of the curve, to a half-life of about 35 days. The Wolfgram protein is most labeled initially when $^{14}$C-glucose is administered. When $^{14}$C-leucine is administered, the half-lives of the proteins appear to be in good agreement with those obtained using glucose. The relatively greater metabolic lability of the proteins could possibly reflect their more accessible location on the membrane relative to the metabolic machinery of the cytoplasm. How this postulation could apply to the inositol phosphatides is less clear unless, in fact, they are not among the other lipids between the protein layers. The reutilization of the protein degradation products could also be less conservative than that of the lipids. There is evidence that one peptidase, leucine aminopeptidase, may be firmly bound to myelin (Adams et al., 1963; Beck et al., 1968). Whether this or any other enzyme is truly present in myelin or whether it is absorbed by myelin after being released from lysosomes is not yet clear.

The same pattern of protein turnover seen in the in vivo studies mentioned above can also be seen in the in vitro experiments carried out in Smith's laboratory (Smith and Hasinoff, 1971). Using $2 \times 10^{-3}$ M cycloheximide they obtained up to 95% inhibition of protein synthesis in both brain and spinal cord slices. The incorporation of acetate into

*Also N.C. Agrawal, A.H. Bone, R.F. Mitchell and A.N. Davison, unpublished data.

TABLE 10

Molar Ratios of Rat Brain Myelin Lipids
Relative to Cholesterol at Various Phases of Development
[Eng and Smith, 1966]

| Lipid | Myelin Prepared by Discontinuous Gradient Days After Birth | | | | |
|---|---|---|---|---|---|
| | 9 | 15 | 19 | 23 | 90 |
| Cholesterol | 1.00 | 1.00 | 1.00 | 1.00 | 1.00 |
| Stable Lipids | 1.18 | 1.00 | 0.99 | 1.03 | 1.07 |
| Labile Lipids | 1.26 | 0.81 | 0.66 | 0.68 | 0.43 |

| | Myelin Prepared by Continuous Gradient Days After Birth | | | | |
|---|---|---|---|---|---|
| | 15 | 16 | 18 | 22 | 90 |
| Cholesterol | 1.00 | 1.00 | 1.00 | 1.00 | 1.00 |
| Stable Lipids | 1.06 | 1.03 | 1.05 | 1.05 | 1.09 |
| Labile Lipids | 0.66 | 0.59 | 0.59 | 0.61 | 0.44 |

lipids is not inhibited by cycloheximide. Therefore, it appears that the two synthetic systems operate independently. This would be consistent with the evidence that the protein and lipid compartments in membrane also have dissimilar turnover characteristics. Nevertheless, it is true that if either compartment is seriously disturbed, the integrity of the myelin membrane becomes untenable. Smith believes that the cholesterol in the diet may, under certain conditions, provide a greater percentage of myelin cholesterol, perhaps even more than what is synthesized in situ, especially at younger ages.

Experiments using inhibitors of cholesterol synthesis have shown that under certain conditions, large amounts of precursors of cholesterol can be incorporated into the myelin in place of cholesterol.

Thus, in rats fed 20,25-diazocholesterol during the period of rapid myelination, up to 50% of the myelin sterol consisted of desmosterol at 21 days of age, while 7-dehydrocholesterol comprised at least one-third of the myelin sterol in rats administered AY 9944 (Fumagalli et al., 1969; Smith et al., 1970). By 60 days of age the 7-dehydro-cholesterol may have been reduced in situ, or it may have exchanged with cholesterol. The exchange concept would seem to explain the remodeling that is believed to take place in the early myelin composition.

Experiments with 7-dehydrocholesterol suggest that it is incorporated very early into myelin or myelin-like structures, but that it also disappears quickly. Therefore, even in the youngest animals the machinery for synthesizing myelin membrane must be active. Using labeled 7-dehydrocholesterol, however, Banik and Davison (1971) showed evidence for the reduction of the dehydro analog to cholesterol but the sterol remains in the myelin after its original incorporation. Paradoxically, however, the 7-dehydrocholesterol reductase is not present in myelin. Thus, it seems that a rapid exchange of myelin constituents occurs that permits the 7-dehydrocholesterol to be incorporated initially, then to leave the myelin membrane to be reduced and later returned to the myelin. This exchange concept would seem to explain the remodeling that takes place in the early myelin composition. Mokrasch pointed out that a similar exchange of membrane components has been observed between microsome and mitochondrion.

*Mechanical Aspects*

The role of the myelin and related membranes in axoplasmic flow and fast translocation is not clear. It seems that a slow alteration in the cross-sectional dimensions is possible if the myelin lamellae are able to slide past each other. According to Bunge, both Schwann cells and oligodendrocytes exhibit pulsatile movements in tissue culture (Murray, 1965). In addition to the moving extensions of the satellite cell related to the myelin winding, there is considerable activity of the myelin sheath in the vicinity of the Schmidt-Lanterman clefts. Movement of Schmidt-Lanterman clefts has been observed also in *Xenopus* nerves by Singer and Bryant (1969). It is possible that the physical dynamism that has been postulated and observed under certain circumstances may be related to the metabolic dynamism of the myelin sheath. If the activity has a real functional role, then there must be a need for a continuing influx of energy. According to Robertson, in another laminar structure, the retinal rod, there is evidence of periodic changes of birefringence that may be related to the energy supply to the membrane system (Robertson, 1966a). Corresponding alterations have not been observed by Bunge in myelin; in fact, in tissue culture in which the myelinated neurons have been killed by cyanide, the myelin persists for several days before gross alterations can be seen.

According to Luzzati, smectic lipid-protein structures generally contain some unsaturated lipid fatty acids, the presence of which favors the fluidity within the planes of the smectic layers; what relationship this bears to the function of myelin is not yet clear. Banik and Davison (1969) have noted that during maturation and aging the fatty acid composition of myelin lipids does change. In the earliest structures resembling myelin, the fatty acids are somewhat shorter and less unsaturated on the average. In myelin from the mature nervous systems, there is relatively more unsaturation and more α-hydroxyl-ation. Luzzati suggested that this may possibly make the myelin membrane less "liquid" in the same way that cholesterol is regarded as a "stiffening" component of the membrane. Such loss in "liquidity" could impose constraints upon the translocation of lipid molecules within the membrane or upon the motion of one lipid-protein layer relative to its neighbor.

Hirano presented the possibility that the paranodal transverse bands outline a channel providing access from the extracellular space to the periaxonal space. For example, when lanthanum is applied to the brain, it appears in the periaxonal space (Hirano and Dembitzer, 1969). This appearance can be accounted for by its penetration through the spaces between the myelin laminae or by the helical pathway between the transverse bands that mark a line of contact between the axon and the lateral edges of the myelin winding. On the other hand, according to Caspar, lanthanum also disorders the myelin sheath in a peculiar way that may also account for its penetration into the periaxonal space. Nevertheless, Hirano emphasized that the spiral pathway may be a real means of access to the axon and that this conclusion is independent of the lanthanum tracer studies. The conclusion is based instead on observation of the normal morphology. The transverse bands are always present at the lateral loop-axon interfaces in normal mature myelin. Furthermore, they are absent from the axonal surfaces between lateral loops. The transverse bands are, therefore, parallel to the the lateral loops or else they would, at least sometimes, appear between adjacent lateral loops. Since the lateral loops are themselves sections through a single helically wound cytoplasmic rim so, too, must the transverse bands constitute sections through a series of parallel helical densities (Figure 9 in Hirano and Dembitzer, 1969). The spaces between the transverse bands have never been reported as occluded and so constitute a series of open channels at least 200 A wide that wind their way around the axon from the extracellular space at the node of Ranvier all the way to the periaxonal space.

The role of the nodal seal in the facilitation of axonal conduction is unclear, and Bunge suggested that saltatory conduction may be possible without transverse band seals. He wondered how to explain the finding that one oligodendrocyte can invest many (several dozen?) axons. Could this be a means of communication between axons? Robertson commented that there is no clear difference between myelin related to motor and to sensory axons, and in serial sections he (1957) finds that the direction of the spiral in sciatic nerve is random whether the axon is motor or sensory. In sections through a node, he has observed that the spiral reverses half the time when the node is traversed from one Schwann cell to the next.

Bunge indicated that in CNS myelin the spirality also seems to be random. In myelination the oligodendrocyte is hypertrophied (Hardesty, 1905) compared to its later size. A single oligodendrocyte can invest about 15 axons (Peters and Vaughn, 1970) and perhaps as many as 30. An oligodendrocyte actively involved in maintaining this many myelin segments must be doing some very special work. Similarly, an axon of a motor nerve in the gluteus maximus may divide into 1,200 endings. This would be a corresponding kind of complexity.

It is clear from earlier experiments that the biosynthesis of a characteristic myelin constituent (i.e., proteolipid protein) proceeds in preparations from rat brain most vigorously during myelination (Mokrasch and Manner, 1963). In the same study it was shown that the incorporation of amino acids into the proteolipid protein never diminished to a rate consistent with the view, widely held at that time, that the proteins of myelin, once formed, were not metabolically active. Other aspects of the in vitro system, such as its sensitivity to inhibition by added $Cu^{2+}$ and the incorporation of palmitate into the proteolipid lipids, were in good agreement with earlier in vivo observations.

Since this system may exist in a natal, immature, or precursor form (e.g., "premyelin" or "myelin precursor") before the fully developed system is consolidated, and since it may or may not synthesize its own constituents, Mokrasch thinks it would be difficult to make a conceptual framework by analogy between it and what is known about eukaryotic protoplasmic structures. He believes that it is quite clear from the work of several laboratories that myelin, to the extent that it makes its own constituents, i.e., synthesizes lipids (Smith, 1967; Mandel and Nussbaum, 1966) and synthesizes proteins (Mokrasch, 1966; Tolani and Mokrasch, 1967), does so at a rate much lower than that of other protoplasmic structures.

### Enzymes

Although it has been possible to demonstrate the incorporation of precursors into myelin lipoprotein as Mandel has done with $^{32}$P (Mandel and Nussbaum, 1966), and that enzymatic activities, for example, 2',3'-cyclic nucleotide 3'-phosphohydrolase, are localized in the myelin fraction (Kurihara et al., 1969, 1970, 1971; Kurihara and Tsukada, 1967, 1968), myelin is generally regarded as a singularly inert tissue. Some lipid classes are more inert than others; but certain lipid portions exhibit a turnover that can be stimulated. However, most other metabolic activities seem to be absent from myelin (Adams et al., 1963).

Norton commented that discussions on whether myelin is enzymically competent usually center about the relative distribution of its enzymes and the intensity of their activity. An investigator who accepts the notion that myelin is fairly inert will consider any enzyme activity as being surprisingly high, whereas an investigator comparing myelin to mitochondria will describe it as being fairly inert. Thus, experiments involving localization of enzymes in myelin, either by isolation or by histochemical techniques, carry the burden of proving that the well-known artifacts of localization are satisfactorily minimized.

Another important concept that arose during the discussion on the metabolic competence of myelin, the concept that found general acceptance among the participants, is that the structure need neither synthesize nor repair its own constituents. Components synthesized elsewhere in the cell could join the myelin membrane in a spontaneous self-assembly process regulated by the relative proportions of other constituents already present. Similarly, constituents modified by oxidation or by scission in the membrane could be sterically unsuitable for continued residence in the membrane and so be extruded.

Among the enzymes for which various myelin preparations have been examined by Mandel and his collaborators are the glycolytic and respiratory enzymes. Generally, enzyme activities from either of these sets have been found to be lacking. This finding is consistent with the absence or questionable presence of glycolytic intermediates in myelin. Mokrasch remarked that a recent report suggests that some glycolytic enzymes can be found associated with myelin from rabbit spinal roots (Miani et al., 1969). If more of such reports regarding myelin from different sources and procedures accumulate, then our long-standing view about the metabolic activity of myelin may have to be revised. In any case, a short catalogue of well-documented enzymatic activities that have been found either in or associated with myelin is presented in Table 11.

Because cyclic $2',3'$-AMP is absent from myelin, a role for cyclic AMP phosphohydrolase in myelin seems doubtful, according to Mandel. Similarly, adenyl cyclase is absent from myelin. Actually, cyclic AMP phosphohydrolase was discovered in myelin when $3',5'$-cyclic AMP (Drummond et al., 1962) and ribonuclease activity (Mandel) were being sought. It may be that the enzymatic activity measured as phospho-hydrolase is a function subsidiary to some other undisclosed activity.

Despite these reservations, Mandel feels that the use of cyclic AMP phosphohydrolase assay is still a good tentative choice as a myelin marker. Investigations carried out in his laboratory (Kurihara et al., 1969, 1970, 1971) on myelin-deficient mice, the so-called "Jimpy" and "Quaking" mice, which are deficient in both myelin and phospho-hydrolase, support its use as a myelin marker. This enzyme was described in the CNS by Drummond and his co-workers (1962) and later identified in the myelin fraction by Kurihara and Tsukada (1967, 1968). The specific activity of purified myelin for this enzyme is about 7 times higher than that of the homogenate. Microsomes have been shown to have about the same specific activity as the homogenate (Table 12; Kurihara et al., 1971)

The use of negative markers, enzymes that are localized in nonmyelin structures, is a means of verifying the relative paucity of other organelles in the myelin preparation (Agrawal et al., 1970a)*.

### Pathology

*Mutants: P. Mandel*

There are a number of basic questions concerning the instability of myelin in disease: What changes occur in the lipid protein complex to

*Also T. Waehneldt, unpublished data.

TABLE 11

Enzyme Activities Detected in Myelin
[Mokrasch]

| Enzyme | Marker For | Myelin Activity Percent of Total | Investigators |
|---|---|---|---|
| 2',3'-Cyclic AMP phosphohydrolase | Myelin (?) | 60 | Kurihara and Tsukada, 1967, 1968 |
| Leucine aminopeptidase | ? | 17 | Adams et al., 1963 |
| Alanyl-$\beta$-naphthylamidase | ? | 15 | Beck et al., 1968 |
| Phosphatidate phosphohydrolase | Membranes (?) | 12 | Salway et al., 1967 |
| ATPase (Na$^+$, K$^+$) | Synaptosomes and membranes | 2 | Adams et al., 1963 |
| Acid phosphohydrolase | Lysosomes | 3 | Adams et al., 1963 |
| Glucose-6-phosphate-NADP dehydrogenase | Cytoplasm | 3 | Adams et al., 1963 |
| Cytochrome oxidase | Mitochondria | 2.5 | Waksman et al., 1968 |
| Alkaline phosphohydrolase | Vascular endothelium | 3 | Adams et al., 1963 |
| Glutamic decarboxylase | Synaptosomes | 2 | Adams et al., 1963 |
| Succinic dehydrogenase | Mitochondria | 3 | Adams et al., 1963 |
| $\gamma$-Aminobutyric-$\alpha$-ketoglutaric transaminase | Mitochondria | 1 | Waksman et al., 1968 |
| Lactate dehydrogenase | Cytoplasm | 1 | Salway et al., 1967 |

TABLE 12

2',3'-Cyclic AMP-3'-phosphohydrolase in Mouse Brain
[Kurihara, Nussbaum, and Mandel]

| Fraction | Specific Activity (U/mg Protein) | |
|---|---|---|
| | Normal | Jimpy |
| Homogenate | 1.94 | 0.255 |
| Crude myelin | 7.98 | 0.858 |
| 1 x Purified myelin | 9.57 | 1.03 |
| 2 x Purified myelin | 13.1 | 1.3 |
| Microsomes: | | |
| Recentrifuged pellet | 1.64 | 0.29 |
| Recentrifuged supernatant | 1.32 | 0.26 |

cause a less stable structure? Are there lysosomal changes, or is there a deficiency in the biosynthesis of the components? Why do some toxic agents induce central demyelination and others peripheral demyelination? What causes myelin abnormalities in genetically induced or infection-induced demyelinating conditions? Is myelin regeneration part of its normal metabolic turnover, or is it a process related to its original deposition? Does myelin have hydrolytic enzymes for its own destruction? These are intriguing questions, and some of them are now being studied.

There are several simple forms of natural demyelinating states that are useful for study, for example, Wallerian degeneration, neurological mutants, and multiple sclerosis. It is expected that the chemical study of these conditions will eventually reveal the defect and the mechanism responsible for demyelination. Presumably, disorders in the enzymic machinery and/or in the substrate availabilities are important in causing the pathology.

In addition to these demyelinating states, there are several more that are amenable to experimental induction and manipulation. These are the demyelinating states induced by or resulting from nutritional deficiencies, intoxications, genetic aberrations, autoimmune phenomena, circulatory defects, and cerebral edema. In the following discussion, Mandel focuses attention primarily upon myelin defects related to genetic factors. He points out that, from experiments on the hybridization of rat brain cytoplasmic RNA and of whole-cell RNA with rat brain DNA, an approximation to the possible number of RNA molecules that can be transcribed and of proteins that can be made by brain cells can be obtained. Since ribosomal RNA can hybridize with 0.15% of brain DNA, Mandel suggests that there are about 6,000 cistrons* for ribosomal RNA in brain (Deshmukh et al., 1970). The dRNA hybridizes with about 1.2% of brain DNA. This amount of DNA can code between 30,000 and 300,000 different proteins with molecular weights in the range $10^4$ to $10^5$. These calculations indicate a high degree of protein and enzyme redundancy, i.e., a possible abundance of isoenzymes or a degree of molecular polymorphism that could provide a spectrum of enzyme activities. Therefore, when a mutation occurs, there may not be an absolute loss of a certain enzyme activity, but rather an alteration in the affected protein's structure that changes the enzyme activity in some significant way. For example, in Tay-Sachs disease, the activities of $\beta$-hexosaminidase are altered in a manner that shows that the isoenzyme A and B hexosaminidase ratio and the total enzyme activity are both altered (Okada and O'Brien, 1969). Similarly, Fabry's disease is an example of a defective gene that produces a deficiency of two enzymes: 3-hexosyliceramide galactosyl hydrolase and dihexosyliceramide galactosyl hydrolase (Wolfe et al., 1970; Brady, 1970).

It is known from the work of Sidman and his colleagues (1964) and other investigators that there are neurological mutants among mice which are myelin-deficient. Two such mutants are the Quaking and Jimpy mice mentioned above. Except for cerebrosides and sulfatides, the lipid compositions of Jimpy myelin and normal myelin are similar (Nussbaum et al., 1970). However, the cerebrosides in the myelin of the mutant brain are markedly reduced in quantity (Table 13). Similarly, the variation of myelin components with maturation is another characteristic of the difference between normal and mutant myelin (Nussbaum et al., 1969). There is a large increase in the galactolipid of normal myelin, whereas the galactolipid content of Jimpy myelin

*A cistron is a section of DNA that is transcribed as a functional genetic unit and that corresponds either to messenger RNA coding for a peptide chain or to another RNA molecule, such as tRNA, rRNA, 5S RNA.

TABLE 13

Lipid Composition of Myelin and Microsomal
Fraction from Normal and Jimpy Brain
[Nussbaum et al., 1970]

| Phospholipid | (Percent total lipid phosphorus) | | | |
| | In Myelin | | In Microsomes | |
| | Normal | Jimpy | Normal | Jimpy |
| --- | --- | --- | --- | --- |
| Phosphatidyl ethanolamine | 43.7 | 36.3 | 33.9 | 32.4 |
| Phosphatidyl choline | 34.4 | 39.2 | 41.7 | 44.1 |
| Phosphatidyl serine | 11.0 | 14.5 | 12.8 | 12.4 |
| Sphingomyelin | 4.7 | 4.5 | 3.5 | 3.9 |
| Phosphatidyl inositol | 2.8 | 3.1 | 4.5 | 4.7 |
| Phosphatidic acid | 1.4 | 0.6 | 0.6 | 0.7 |

| Galactolipid | (Percent total lipid galactose*) | | | |
| --- | --- | --- | --- | --- |
| Cerebroside | 83.0 | 75.0 | 76.3 | 39.3 |
| Sulfatide | 17.0 | 25.0 | 23.7 | 60.7 |
| Galactose/phosphorus M ratio | 0.316 | 0.003 | 0.026 | 0.0014 |

*Cerebrosides plus sulfatides.

remains vanishingly small. In addition, there is a decrease in protein after about 22 days that may possibly be due to a removal of a myelin-precursor type of protein without its being replaced. The reduction of cholesterol and phospholipids is much less pronounced than that of cerebrosides and sulfatides. The distribution of phospholipids shows only a relative decrease of phosphatidyl ethanolamine and plasmalogen. There are several points at which glycolipid biosynthesis can be blocked in its later stages: at the acylation of sphingosine or at the glycosylation of sphingosine or ceramide (Figure 9). Except for the ganglioside fraction Gm-1, the ganglioside content of Jimpy brain is normal. This situation implies that $C_{16}$-$C_{18}$ fatty acid ceramide formation is not the site of the defect.

Uridine diphosphate (UDP) galactose-sphingosine galactosyl transferase was tested in both normal and mutant mice (Neskovic et al., 1969, 1970b). In normal mice, this enzyme resembles an induced

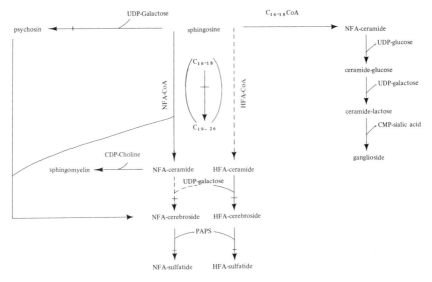

Figure 9. Pathways of sphingolipid biosynthesis. The unbroken line represents demonstrated steps, and the dashed line probable steps. The sites of metabolic error in Quaking and Jimpy mice are indicated. [Neskovic et al., 1969, 1970a,b]

CoA = coenzyme A
NFA = monohydroxy fatty acids
CMP = cytidyl monophosphate
HFA = hydroxy fatty acids
PAPS = 3'-phosphoadenosyl-5'-phosphosulfate
CDP = cytidine disphosphate

enzyme, its maximum activity corresponding to the period at which the maximum rate of myelination proceeds. In the Jimpy mutant, there is no significant increase in the low activity of this enzyme at any time, whereas in the Quaking mutant there is a small increase in the enzyme but still much less than that in the normal mouse.

One possible reason for the decreased deposition of glycolipid in mutant myelin could be the action of an inhibitor. In vitro experiments by Mandel showed that there is nothing in mutant brain that inhibits glycolipid biosynthesis in normal brain preparations when the two are mixed (Neskovic et al., 1969).

When $^{14}$C-galactose is injected into Jimpy brain, there is a greatly depressed incorporation of the label into galactocerebrosides, glucocerebrosides, sulfatides, and glycerogalactolipids. On the other hand, there is no significant decrease in the incorporation of $^{14}$C-galactose into the gangliosides of Jimpy mice. However, activity of UDP-galactose-sphingosine transferase is also deficient in the Jimpy mouse (Neskovic et al., 1970b). Like the enzyme making psychosin from sphingosine, this enzyme normally increases to a maximum activity during the course of myelination; but it does not do so in Jimpy brain.*

*N.M. Neskovic, J.L. Nussbaum and P. Mandel, unpublished data.

Kandutsch and Saucier (1969) have recently shown that hydroxymethylglutaryl-coenzyme A reductase activity is considerably lower and decreases faster in Jimpy brain between 10 and 19 days.

In experiments carried out by Mandel and his co-workers (Kurihara et al., 1970, 1971), the activity of cyclic AMP 3'-phosphohydrolase was found to be deficient in Quaking and Jimpy mutants. In the Jimpy mutant, the specific activity in myelin is about 1/10 that in normal myelin and about 1/6 that in microsomes and in the soluble fraction of brain from normal animals (Table 12). The enzyme is also markedly less active in the spinal cords of Jimpy mice (15% to 20% of normal activity). The same enzymatic activity is reduced in the brain and spinal cord of Quaking mice (respectively, 25% and 50% to 60% of normal activity).

Generally, the fatty acids found in sphingomyelin and ganglioside are shorter than those in cerebroside and sulfatide. Moreover, the long-chain content of the fatty acid ($C_{19}$ to $C_{24:1}$) of the sphingolipids in Jimpy myelin is greatly reduced (Nussbaum et al., 1971). This finding suggests that the process of chain elongation is retarded or that the fatty acyl transferases have been affected by the genetic defect. In sphingomyelin from mouse brain microsomes, on the other hand, there is no difference between the fatty acid chain length of the mutant and that of the normal.

In the mutant mice discussed here, the pathology seems to be restricted to the CNS. Consistent with this is the observation that the activity of cyclic AMP phosphohydrolase is normal in the peripheral nerve of Jimpy and Quaking mice but is reduced in mutant spinal cord (Kurihara et al., 1970). Similarly, the UDP-galactose-sphingosine transferase is normal in the peripheral nerve of Jimpy (Neskovic et al., 1970a).

The study of mouse mutants shows clearly that the controls of myelination in the CNS and the PNS are genetically independent. Several enzymatic deficiencies are the result of the genetic defect in Jimpy and in Quaking. A number of possibilities can be envisaged to explain this multienzymatic defect:

1. A mutation of a regulatory gene or inducer that controls the action of several enzymes involved in synthesizing activities;

2. An impairment in the transcription of the first enzyme in a sequential DNA transcription;

3. A mutational defect in oligodendroglia differentiation or maturation;

4. A mutation of a factor responsible for the migration of oligodendroglia to the axons, thus impairing oligodendrocyte-axon contact. This contact may be necessary for the initiation of the biochemical phenomena associated with myelination.

Data suggesting the involvement of degradative enzymes in the mouse mutants have also been reported (Bowen and Radin, 1969; Kurtz and Kanfer, 1970; Farkas et al., 1970; Deshmukh et al., 1970). Histologically, there seems to be a difference in the oligodendroglial pattern of the mutant mouse brains. Instead of the normal chain-like arrangement, the mutant oligodendrocytes are less orderly (Farkas et al., 1970).

### Edema Damage: A. Hirano

Any toxic agent applied locally to the brain produces an edema; somehow the fluid escapes from the vascular endothelium and invades the neuropil. The edema fluid spreads into the white matter and, as has been shown by Hirano and his collaborators (Hirano et al., 1964; Hirano, 1969), serves to separate myelinated fibers or small groups of myelinated fibers. In spite of the presence of large amounts of extracellular fluid, the individual myelinated fibers apparently retain their morphological integrity; that is, the fiber diameter and interlamellar distances appear unchanged (Figure 10). The separation of individual fibers facilitates analysis of the structure of the myelin sheath.

Several weeks after treatment with any one of a variety of toxic agents, one can find certain configurations ordinarily not visible. These include elongation of the outer loop and the presence of various organelles such as mitochondria, dense bodies, and numerous microtubules in any of the cytoplasmic areas of the sheath (Hirano et al., 1966). In addition, isolated islands of cytoplasm resulting from unfused portions of the myelin-forming cell may be observed anywhere in the sheath but especially near the outer loop (Figure 11). Similarly, apparently separate cell processes have been observed in the periaxonal space. Sometimes these processes may constitute an additional myelin sheath (Hirano and Dembitzer, 1967).

Certain toxic agents appear to result in characteristic changes not seen so commonly after the application of most other agents. For example, systemic intoxication with cyanide results in, among other changes, a tremendous expansion of the axonal diameter accompanied by a concomitant enlargement of the outer diameter of the entire myelinated fiber (Hirano et al., 1967). While the interlamellar distances

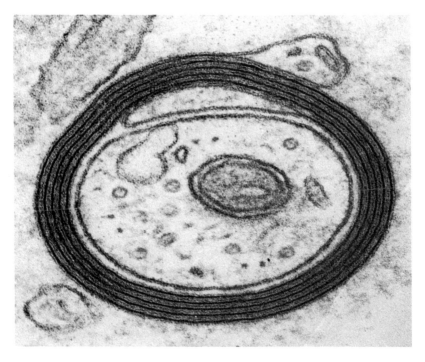

Figure 10. A cross section of a myelinated axon in a severely edematous area of an experimentally treated brain. x 175,000. [Hirano, 1969]

in such altered fibers apparently remain unchanged, the number of lamellae decreases. Hirano believes that these alterations of the myelin sheath arise by the slippage of the lamellae past one another at the major dense line, much as the mainspring of a clock unwinds (Hirano and Dembitzer, 1967). A similar mechanism may apply when the myelin sheaths adjust to the enormous, apparently empty, intramyelinic splits of the intraperiod line that result from triethyltin intoxication (Aleu et al., 1963; Hirano et al., 1968b). Silver nitrate implantation has, on rare occasions, resulted in an opening of the external mesaxon permitting the infiltration of edema fluid at the level of the intraperiod line (Hirano et al., 1966). After silver nitrate intoxication, the infiltration penetrated for only a few turns of the spiral, but more recently complete penetration all the way to the periaxonal spaces has been observed in myelinated fibers of peripheral nerve tissue cultures exposed to sera of patients with Guillain-Barré syndrome (Hirano et al., 1971). Lampert and Carpenter (1965) have

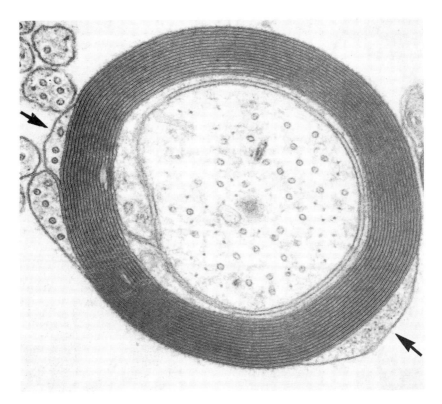

Figure 11. A cross section of a myelinated axon in an edematous area of a brain treated with cyanide four months previously. Two apparently isolated cytoplasmic islands formed from unfused portions of the myelinating cell are visible at the arrows. x 100,000. [Hirano et al., 1968a]

previously reported similar changes in central nerve tissue associated with EAE.

From experiments on the induction and reversal of edema as it affects myelin, it was proposed that one possibility for the reformation of the tight myelin spiral is that the water between the laminae is actively pumped out into the extracellular space. It may be that the extrusion of water also has a role in the formation of compact myelin during the early stages of myelination. However, it is noteworthy that the recovery of myelin from the edematous condition leads to a somewhat disordered membrane structure unlike the preedematous state. It is possible that if myelin can exist and function in an abnormally disordered state, then perhaps a precise molecular configuration of myelin constituents is not absolutely necessary.

*Myelin Degeneration in Tissue Culture: R.P. Bunge*

Bunge commented that an important advantage of studying myelin in cultures is that it is readily visible; alterations and repair are easy to observe. This is particularly true for peripheral nerve; for example, cells derived from the dorsal ganglion of rats (sensory neurons) are used for tissue culture. (See culture stained with osmium in Figure 12.)

Concerning the stability of myelin in culture, Bunge thinks that it is possible to maintain myelin in culture for a period representing about one-third of the lifetime of the rat from which the tissue was obtained. Actually, cultures have been maintained for as long as a year. Central myelin, unfortunately, is not so stable in culture. Figure 13 indicates some of the types of degeneration that have been studied in nerve tissue culture.

When the axon is lost, myelin degenerates to a series of ovoids that are progressively broken down (Figure 13). Diphtherial toxin induces a segmental demyelination; not all the myelin along an axon is affected – only random segments are destroyed, but the Schwann cell itself is uninjured even when its myelin is destroyed. According to Peterson and Murray (1965), the Schwann cell seems to dispose of the degenerated myelin and, having done so, begins preparing new myelin. It becomes an interesting question why a particular myelin segment is more susceptible than others.

Cyanide-induced myelin degeneration involves the death of both the neuron and the Schwann cell (Masurovsky and Bunge, 1971). In this case, however, myelin ovoids generally do not form, and the general tubular shape of myelin remains. When the myelinating cells are exposed to 0.01 M cyanide for 24 hours in tissue culture, the nerve cells degenerate and the myelin swells, but it does not generally break up into balls or vesicles (Figure 13). Three days later the myelin lamellae begin to separate along the interperiod line, and later the major dense line begins to separate into structures 90 A thick. This breakdown process takes place in the absence of any cellular activity and is perhaps an exhibition of a native instability of the myelin membrane.

Bunge also noted that myelin can be damaged experimentally by "glycerol reversal" and by trypsin treatment. When a 15% glycerol

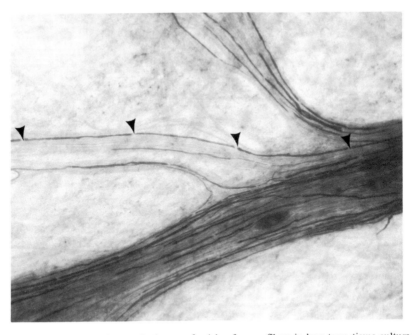

Figure 12. This photomicrograph shows a fascicle of nerve fibers in long-term tissue culture. Many of the fibers are myelinated. One individual myelinated fiber is marked with a series of arrows indicating the sites of nodes that separate the individual segments of myelin. These four arrows thus delineate three individual myelin segments. This type of tissue culture preparation containing peripheral myelin is useful in the study of myelin damage and repair. From a rat dorsal root ganglion which had matured in tissue culture for several months, and was then fixed in $OsO_4$ and stained with sudan black. x 200. [Bunge]

solution in mammalian Ringer is applied to nerves in culture, the nerves shrink somewhat at first, but after equilibrium is established, if the culture medium is reversed to normal, the excess glycerol inside the fiber causes it to swell and the myelin acquires a beaded appearance. Although the changes in the appearance of the myelin are drastic, the nodes remain relatively intact and breakdown of the sheath does not occur. The myelin sheath is split in many places apparently at the interperiod line, and the axon is irregularly distended. In some respects, this is similar to the edema damage caused by triethyltin as described by Hirano. However, after 3 weeks the myelin sheath has largely recovered its normal structure and very few sheaths appear to have been destroyed.

Bunge (1970a) reported that when trypsin is applied to a mature culture, the damage to the myelin is severe. Trypsin in 0.2%

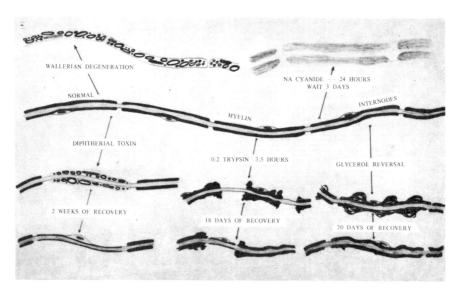

Figure 13. This diagram indicates some of the types of myelin damage or breakdown (and repair) that have been studied in peripheral nerve tissue cultures. The normal myelin internodes break down in different patterns in response to axon loss as in Wallerian degeneration, in response to cyanide-induced necrosis, to diphtheria toxin application, to trypsin application, or to glycerination reversal. In some of these conditions a period of repair follows the initial damage. For details, see text. [Bunge]

concentration causes the myelin to pull away from the nodes and undergo a considerable distortion in 3 hours. The neurons are not affected seriously by this treatment, and the Wallerian pattern of myelin degeneration is not generally observed. After 3 weeks the myelin has not broken down, but in some cases the myelin has pulled back from the node and a new Schwann cell may have moved into the area to invest the bare axon. In other instances, the myelin from an adjacent Schwann cell seems to be moving back toward the exposed node.

    When a real breakdown in the myelin sheath is induced, as in diphtherial toxin application, without causing the death of the Schwann cell, the myelin-related Schwann cell cleans up the myelin debris in about a week. Remyelination may then occur.

*Experimental Allergic Encephalomyelitis (EAE): E.H. Eylar*

Eylar considers that among the membrane proteins the basic protein from myelin has some unusually good features: it can be obtained in a pure form in comparatively large quantities, it is easily soluble in aqueous systems, and it has a measurable biological activity in being the EAE antigen. Experimental allergic encephalomyelitis is commonly regarded as a model for human demyelinating diseases, but a clear indication of its immunological similarity to multiple sclerosis is lacking.

When the basic protein is mixed with the complete Freund's adjuvant and injected into the footpads of guinea pigs, it elicits an immunological response from the macrophages and lymphocytes in the lymph nodes (Eylar et al., 1969). After about 5 days these cells begin to accumulate in the brain, and after 10 days their action on myelin is evident by neurological signs and, finally, by death. It is difficult to induce EAE in the rabbit with bovine central myelin basic protein and even more difficult to do so in the rat.

The characteristic lesion in guinea pig brain is a monocyte perivascular invasion that is a positive histological sign for the early stage of EAE (Eylar et al., 1969). According to Eylar (1971), the invasion seems to be not a process involving antibodies as such, but rather a hypersensitization process in which the lymphocytes find the basic protein (to which they have been sensitized) in the central myelin. Presumably, when the cells reach the myelin they release macrophage-inducing factors that attract macrophages to the sites. These theories have not been substantiated by evidence as yet.

The skin test, an appearance of redness, is a classic test for delayed sensitivity that can be used to assay the degree to which the animals are affected (Eylar et al., 1969). Five to 10 days after the original injection of the basic protein into the footpads, 5 to 50 $\mu$g of the same protein are injected into the skin. The reaction is delayed because it takes about 24 hours for this redness to appear. One can correlate the size of the skin lesion in the skin test with the number of lesions found in the brain. These findings imply that the same cell is responsible for both kinds of lesion.

Eylar has also noted that exposure of the basic protein to 8 M urea fails to change its EAE-inducing properties (Eylar and Thompson, 1969); therefore, it would seem that the activity of this protein is not due to a special tertiary structure or combination of allosteric isomers.

In fact, it seems probable that only its primary structure is responsible for its activity.

The biological activity of the basic protein is present in some of the peptide fragments resulting from the action of pepsin or trypsin (Hashim and Eylar, 1969). The EAE activity of the mixture of peptic peptides is less than that of the native protein; however, when the peptides are separated, there is no large loss of activity in the active peptides tested. Two peptic peptides, both containing the single tryptophan, can be isolated (Table 14). One tryptic peptide, which is active, has also been isolated. The isolation of these highly active peptides confirms the hypothesis that the biological activity of the

TABLE 14

Sequence and Encephalitogenicity of Peptides
Related to the Putative Disease-Inducing Site
[Westall et al., 1971]

| Peptide | Encephalitogenic | Sequence | | | | | | | | | |
|---|---|---|---|---|---|---|---|---|---|---|---|
| T27 (Tryptic) | + | | | Phe | Ser | Trp | Gly | Ala | Glu | Gly | Gln | Lys |
| E (Peptic) | + | Ser | Arg | Phe | Ser | Trp | Gly | Ala | Glu | Gly Pro | Gln Gly | Lys Phe |
| S1 (Synthetic) | + | Ser | Arg | Phe | Ser | Trp | Gly | Ala | Glu | Gly | Gln | Lys |
| S2 (Synthetic) | – | Ser | Arg | Phe | Ser | Trp | Gly | Ala | Glu | Gly | Gln | |
| S3 (Synthetic) | + | Ser | Arg | Phe | Ser | Trp | Gly | Ala | Glu | Gly | Gln | [Arg] |
| S4 (Synthetic) | – | Ser | Arg | Phe | Ser | Trp | Gly | Ala | Glu | Gly | Gln | [Ile] |
| S5 (Synthetic) | – | Ser | Arg | Phe | Ser | Trp | Gly | Ala | Glu | Gly | [Ile] | Lys |
| S6 (Synthetic) | + | Ser | Arg | Phe | Ser | Trp | Gly | Ala | [Ile] | Gly | Gln | Lys |
| S7 (Synthetic) | – | Ser | Arg | Phe | Ser | [Phe] | Gly | Ala | Glu | Gly | Gln | Lys |
| S8 (Synthetic) | – | Ser | Arg | Phe | Ser | [Val] | Gly | Ala | Glu | Gly | Gln | Lys |
| S9 (Synthetic) | + | Ser | Arg | Phe | [Ala] | Trp | Gly | Ala | Glu | Gly | Gln | Lys |
| S10 (Synthetic) | + | Ser | Arg | [Val] | Ser | Trp | Gly | Ala | Glu | Gly | Gln | Lys |

| Required Sequence | | --------- Trp -------------------- Gln Lys |
|---|---|---|

parent protein is due to an amino acid sequence rather than to any structure of higher order.

If the biological activity of the protein is conferred by its primary structure, then there should be either a crucial amino acid or a combination of amino acids that is responsible. A number of peptides were prepared by the technique of Merrifield (1963) using the solid state peptide synthetic technique (Westall et al., 1971). Of the peptides synthesized (Table 14), those lacking the tryptophan, the terminal basic amino acid (arginine can substitute for lysine and is carboxy-terminal in the human peptide), and the penultimate glutamine are inactive. Another requirement is that there must be at least two amino acids preceding the tryptophan; otherwise, most of the other amino acids can be replaced with glycine and the resulting peptide is active.

Eylar feels that the presence of an intact tryptophan residue in the original protein is crucial. If the tryptophan is modified by coupling it to 2-hydroxy-5-nitrobenzyl bromide (Barman and Koshland, 1967), the resulting product is inactive in inducing neuropathy even at doses 150 times larger than the usual effective dose of 1 $\mu$g. The modified protein has unchanged antibody-combining ability as measured by the passive hemagglutination-inhibition test (Eylar, 1971). Corresponding to EAE, EAN is produced when an animal is injected with a preparation of peripheral nerve (Arnason, 1971). If, for example, an emulsion of sciatic nerve is used, after 10 to 20 days the animal becomes paralyzed and the histological picture resembles that described for EAE, except that in the latter case the perivascular invasion of lymphocytes occurs throughout the peripheral nervous system. In about 50% of the cases, the symptoms resemble those of the Guillain-Barré syndrome, which in the human is thought to develop after a virus infection and which has a mortality rate of 30%.

Ordinarily, extensive demyelination, which does not normally occur in EAE when the symptoms appear, does so at a much later stage. Accordingly, the correspondence between demyelination and paralysis in EAE is not clear. On the other hand, when peripheral nerve is injected, the occurrence of symptoms and peripheral demyelination is simultaneous. Occasionally, neuronal and axonal lesions are also observed, but their appearance depends to some extent on the preparation used and the animal tested. The involvement of the autonomic nervous system in these diseases is not clear. It has been shown (Arnason, 1971) that the use of whole tissue can produce demyelinating antisera, but the purified basic protein does not do so. Arnason (1971) finds that lymph cells from sensitized lymph nodes will cause demyelination of sciatic nerve cultures.

Eylar reports that the two basic protein fractions from peripheral nerve can induce paralysis. P2 and P3 cause lesions in the CNS as well as in the PNS, even at doses as small as 10 $\mu$g. Furthermore, they are immunologically similar to the CNS antigen since they combine with the same antibody. In his opinion, an important consideration is that in order to produce demyelinating antibodies, the whole tissue must be injected. It has been found that even though 20% of normal persons carry antibodies to myelin, they show no symptoms.

In most EAE and EAN lesions, axonal destruction can be demonstrated microscopically. However, a severe clinical deficit occurs when the histology shows nothing more than the presence of lymphocytes. In the attack on myelin due to the allergic challenge, the myelin separates at the interperiod line and not at the major dense line. Webster cautioned that there is no clear relationship between the severity of the symptoms, their progression, and the number or extent of demyelinating lesions. Similarly, Eylar commented that in monkeys and rabbits there is a cyclic aspect to the symptoms of EAE. On the other hand, in EAN the symptoms appear after the insult, and if remission occurs, it is permanent. A clinical picture similar to that of EAE in animals and multiple sclerosis in humans is seen in the reaction to rabies vaccine. Over a series of months, a cyclic syndrome was observed in patients who finally died of the disease. This was the result of a single immunological insult (Shiraki and Otani, 1959).

Mandel added that the role of myelin in the facilitation of conduction is seen in these immunological demyelinations and also in mutant mice (Jimpy); nerve conduction continues, but more slowly.

### Isolation of Myelin and Its Related Structures

*General Principles: W.T. Norton*

As described by Norton (1971b), the first step in the isolation of myelin and related structures is usually a homogenization of the tissue in sucrose solutions. Presumably this step causes the myelin to disaggregate into vesicles that are subsequently isolated by centrifugal procedures that depend upon the density or the size of the vesicles. The original method of Autilio and her co-workers (1964) leads to a loss of myelin, particularly that fraction which has a density greater than that of 0.656 M sucrose. However, the method was developed for bovine white matter and is probably not directly applicable to whole rat brain.

A more useful approach is to make a more dilute homogenate of brain and to layer it over a 0.85 M sucrose solution for centrifugation (Norton, 1971b). This gradient gives a complete recovery of myelin with whatever range of densities it may have. The material floating on the sucrose gradient after centrifugation is collected and shocked osmotically in water to remove nonmyelin constituents. Finally, the preparation is isolated after sedimentation on another sucrose gradient. The myelin recovered from the first sucrose gradient and water shock procedure gives a distribution on a continuous CsCl gradient (0.3 to 1.3 M), with a single major peak and some tailing in the denser part of the gradient. On the other hand, a crude preparation of myelin gives a bimodal distribution with the heavier component corresponding to the "early myelin" fraction. A similar distribution has been observed by Greenfield, Norton, and Morell (1971) when myelin is prepared from the Quaking mouse, in which case the myelin resembles immature myelin. Norton noted that if sucrose with salt is employed in the original homogenization, myelinated axons, and not free myelin, are fairly easily isolated and the myelin can be removed from the axons by appropriate treatment (Norton and Turnbull, 1970; De Vries and Norton, 1971). Isolated axons are deficient in microtubules, but neurofilaments persist.

Norton has also isolated oligodendroglia from bovine white matter and shown them to have processes attached (Table 15). These

TABLE 15

Properties of Bovine Oligodendroglia
[Norton, Poduslo, and Raine, 1971]

---

Yield: 11-14 × $10^6$ cells/g white matter (0.25-0.30 mg dry weight)

Weight per cell: 22-45 picograms

Diameter ~ 8 $\mu$

Protein ~ 50%

Lipid ~ 30%

Lipid composition
| | |
|---|---|
| cholesterol | 14.1% of total lipid |
| galactolipid | 9.9% of total lipid |
| phospholipid | 62.2% of total lipid |

---

oligodendroglia were isolated after trypsinizing bovine white matter
with a medium containing 0.1% trypsin, 10% hexose, and 0.1 M KCl at
pH 6 (Norton et al., 1971; Poduslo and Norton, 1971). The disaggregated tissue is passed through a nylon or stainless steel screen (200
mesh) and the resulting suspension is applied to a sucrose step gradient
for centrifugation. The oligodendrocytes ($\sim$ 90% pure) sediment to the
1.55 M sucrose layer, yielding about 7 mg/20 g of white matter.
Astrocytes are present in other fractions. The electron micrographic
morphology is good; however, the viability is unknown. Most of the
processes are removed from cells isolated from bovine white matter.

### Criteria of Myelin Purity

According to Norton, to verify that a membrane fraction is
myelin, the use of electron microscopy is valuable but not sufficient
and chemical criteria are necessary. In Mokrasch's opinion, the solubility of myelin in chloroform-methanol mixtures is a weak criterion
because of artifactual alterations that may either increase or decrease
the solubility of myelin. Moreover, the previous history of the preparation seems to be of importance. He added that, like other membrane
systems, myelin is regarded as being insoluble in water or in most
aqueous systems. Myelin is osmotically responsive to the aqueous
environment, and certain ions cause it to lose lipids (Wolman, 1965;
Wolman and Weiner, 1965). It is possible to solubilize myelin in
aqueous systems by using detergents. Organic solvents also remove most
or all of the lipids from myelin along with varying amounts of protein
(Rumsby and Finean, 1966a,b,c). The most useful solvent for the
extraction of all components of myelin is a mixture of chloroform and
methanol (Folch and Lees, 1951; Wolfgram and Rose, 1961), and,
depending on the aqueous milieu of the myelin, it may be completely
soluble or only partially soluble in chloroform-methanol (CM) mixtures. The effective solubilization of peripheral myelin depends upon
the presence of water (Wolfgram and Rose, 1961), whereas its presence
is less important for the solubilization of central myelin. Folch-Pi
reports that myelin is also soluble in a phenol-formic-acid-water system,
which is useful for disc gel electrophoresis (Mehl and Wolfgram, 1969).
Myelin isolated by the procedure described by Norton (1971a)
was analyzed for a number of constituents presumed to be con-
taminants (Table 16). Generally, the presence of nucleic acid, ganglioside, and adenosinetriphosphatase was reduced to low values compared
to those in other membranes, but never to zero. Concerning the criteria

TABLE 16

Criteria of Purity of Rat Brain Myelin in Percent
[Norton, 1971a]

| | | | | Age, Days | | | |
|---|---|---|---|---|---|---|---|
| | 15 | 20 | 30 | 60 | 144 | 190 | 425 |
| Nucleic acid | 0.26 | 0.58 | 0.24 | 0.20 | 0.17 | 0.18 | 0.13 |
| Recovery of nucleic acid | 0.15 | 1.05 | 0.93 | 1.23 | 1.35 | 1.72 | 1.96 |
| Total ATPase, $\mu$mole P/mg/hr | 0.81 | 2.17 | 0.84 | 0.32 | 0.29 | -- | -- |
| Recovery of ATPase | 0.17 | 1.06 | 0.56 | 0.26 | 0.28 | -- | -- |
| Total NANA, $\mu$g/100 mg | 48.7 | 44.7 | 39.6 | 39.5 | 35.6 | 51.7 | 68.0 |
| Recovery of NANA | 0.24 | 0.55 | 0.88 | 1.18 | 1.43 | 2.48 | 4.57 |

of purity, Davison and Norton agreed that the problem is particularly acute when myelin is isolated from the developing brain where the composition (hence properties) of myelin is changing constantly (Table 16). In sum, all the criteria taken together give the best indication of purity.

Norton commented that, curiously, the composition of axons is rather similar to that of myelin (Table 17). Since cerebroside is found in axons and since it is deposited during myelination, it seems that the composition of axons changes along with that of myelin. Similarly, the thin-layer chromatographic patterns of lipids from isolated axons and myelin are very much alike. Cerebrosides, similar to those found in myelin, are among the lipid components of the oligodendrocytes but have not been found in neurons or astrocytes; whether these are located in the axonal membrane or not is uncertain. Nevertheless, this reinforces the relation of the isolated cells to the myelin that they are presumed to produce.

TABLE 17

Lipid Composition of Bovine Axons Compared to Nonmyelin and Myelin
(Expressed as mg/mg Phospholipid)
[Norton and Turnbull, 1970]

| Lipid | Axons | Myelin-free Fraction | Myelin |
|---|---|---|---|
| Cholesterol | 0.33 | 0.70 | 0.65 |
| Galactolipid | 0.35 | 0.55 | 0.68 |
| Phospholipid | 1.00 | 1.00 | 1.00 |
| Ethanolamine lipids* | 0.31 | 0.28 | 0.40 |
| Choline lipids* | 0.33 | 0.30 | 0.25 |
| Sphingomyelin | 0.13 | 0.23 | 0.17 |
| Serine plus inositol lipids* | 0.16 | 0.15 | 0.17 |

*Glycerophospholipids.

## III. THE STRUCTURE OF THE MYELIN SHEATH
### R. S. Bear

## Optical Studies

### Microscopy

Except for obvious gross discontinuities at the nodes of Ranvier and occasional indications of the Schmidt-Lanterman incisures, fresh native nerve fibers show no visible microscopic inhomogeneities in their myelin sheaths, despite the many artifactual structures that appear in fixed and stained preparations. After more than a century of investigation, these circumstances are fairly well understood in the light of modern conceptions of the structure of the myelin sheath, as will appear below.

The homogeneous appearance of myelin in light microscopy is now recognized to be a consequence of its fluid crystalline organization. The constituents are arranged radially in compartments whose dimensions are small in comparison to the wavelength of visible light; hence, they remain unresolved under light microscopy, requiring the shorter X-rays or electron waves for resolution.

### Double Refraction

Nevertheless, observations in visible light did allow early investigators to recognize some of the essential qualitative aspects of myelin structure. Because the predominant lipid and protein constituents are arranged in definite orientation, the optical properties of myelin are distinctly not isotropic as shown by observations in polarized light. As a result of the investigations, notably of Valentin, Klebs, and von Ebner during the latter half of the nineteenth century, it is known that the myelin sheath, under native conditions, is optically described as being positive uniaxial, with the optic axis oriented radially. This means clearly that the constituents are organized with randomness about every radial direction in the sheath but with some sort of nonrandom distribution along the radii. (For a review of the early data see Schmitt and Bear, 1937, and Schmidt, 1936.)

During this period, observations also disclosed that the anisotropy of the sheath is oppositely determined by the lipid and protein components. Ambronn (1890) found that lipid solvents such as alcohol reverse the sign of the uniaxial double refraction from positive (with respect to the radial direction) to negative in myelinated (mostly vertebrate) nerve fibers, whereupon the optical properties become qualitatively like those of the so-called unmyelinated fibers of invertebrates. On the other hand, reversal of the normally negative sign of invertebrate fibers to positive was achieved by Göthlin (1913) through application of reagents such as glycerine. Accordingly, it became possible to distinguish the following: myelotropic fibers whose sheath double refraction is dominated by lipids; metatropic ones, whose normal optical state is that typical of the protein constituents but which can be reversed to indicate lipid presence by glycerine application; and the proteotropic condition, wherein the lipid is insufficient to be demonstrated optically.

The possibility of interpreting these findings in terms of molecular orientation and colloidal particle or phase shape resulted from the theoretical development by Wiener (1912) in which the structural consequences of these factors in terms of the observed optical anisotropy became possible. Wiener distinguished the "form double refraction," exhibited by oriented anisodiametric particles (rods, disks, lamellae, etc.) that might themselves be internally isotropic but are surrounded by media of different refractive indices, from the "intrinsic (crystalline or micellar) double refraction" possessed by the molecules or particles themselves. The two can be distinguished experimentally by immersion experiments, in which the particle environment is varied; when the environment (solvent) and particle (solute) refractive indices are identical, form double refraction disappears, leaving only the residual intrinsic double refraction of the dispersed component.

Wiener's treatment was quite general and capable of covering a wide range of circumstances, but it emphasized two particular cases: the oriented rodlet situation, most like that of fibrous systems, in which the form double refraction is positive with respect to the optic axis that coincides with the predominant orientation of the rod axes when particle and solvent refractive indices are not equal; and the platelet or lamellar case in which the thin dimensions of discs or sheets are oriented predominantly in a given direction, always yielding a negative sign relative to this direction when solvent and particle indices are unequal. Intrinsic particle double refraction is expected to be positive relative to the axis of greatest electrical polarizability (at optical frequencies) of the particle. Polarizability is normally greatest along the predominant orientation of the main chain bonds in long molecules, unless highly polarizable side chains overcome this.

It is sometimes said that the intrinsic double refraction of the lipids dominates in the normally observed myelotropy of medullated nerves, opposing and overcoming the form double refraction and weak intrinsic double refraction contributed by the protein. Thus, the lipid molecules are expected to have their long axes radially oriented, contributing strong positive double refraction relative to the radial optic axis. It may be noted, however, that the lamellar interleaving of protein and lipid phases (as developed in more detail from the subsequent X-ray diffraction and electron microscopic studies) provides an ideal system for producing the lamellar type of form diffraction of Wiener though the magnitude of this contribution cannot be demonstrated reliably by immersion experiments because of the lability of the myelin to foreign solvents and, indeed, the lack of a solvent phase in the usual sense. Thus we can say that the lipid molecular orientation overcomes both the overall form anisotropy of the sheath as well as the contributions of intrinsic protein asymmetry.

Investigations of the early period by means of polarized light culminated in the prophetic proposals of a model for the myelin sheath by Schmidt (1936), who envisaged concentric cylindrical layers of protein intercalated between lipid double layers, with normals to both sets of layers being radial (Figure 14). Simultaneously, Chinn (1937) in

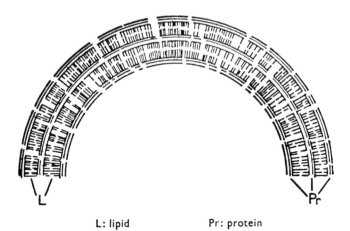

L: lipid          Pr: protein

Figure 14. Cross section of myelin layers as proposed by Schmidt (1936) from polarized light investigations.

Schmitt's laboratory came to a similar conclusion on other grounds. Schmitt and Bear (1939) extended the generality of this view by showing the metatropic reaction of Göthlin was in reality a means by which lipid positive intrinsic double refraction overcame the negative form double refraction in thin invertebrate nerve sheaths, rather than being the result of improved lipid orientation caused by osmotic stresses (as Göthlin believed). Schmitt and Bear also showed that nerve sheaths over a wide spectrum of vertebrate and invertebrate sources and fiber sizes show a continuous rather than an abrupt gradation in the myelotropic versus proteotropic structural manifestations. The Schmidt model, made especially for myelin, resembled a model proposed a little earlier by Danielli and Davson (1935) for membranes generally.

## X-Ray Diffraction Studies of Living Nerve

### Wide-Angle Diffraction: F. O. Schmitt

When a collimated pencil of monochromatic X-rays (commonly the characteristic CuKα X-radiation in studies of organic structures) is passed through a nerve (optimum diameter about 1.0 mm), a diffraction pattern can be registered photographically. The pattern obtained depends on the atomic and molecular structure of the fiber. Because of a reciprocal relation that exists between the forward angles relative to the initial beam at which diffraction intensity maxima occur and the structural spacings that cause the diffraction, the observations at wide angles relate to the finest details of structure.

Use of this method with nerve dates from early exploratory studies in the 1920's, but the first significant investigation was that of Boehm (1933). He observed three diffractions from fresh nerves (Figure 15): a radially very diffuse, tangentially unoriented halo corresponding to a spacing of about 3.3 A, ascribable to the liquid water present; a meridionally accentuated, moderately diffuse ring of spacing 4.8 A; and an equatorial spot of spacing 17 A. Boehm correctly believed the 4.8 A ring to be derived from the myelin sheath but thought that the 17 A reflection came from adventitious connective tissue because it behaved like a collagen equatorial diffraction in decreasing to 11 A upon drying. Although Boehm may have been partially correct about the equatorial reflection, subsequent comparative studies by Schmitt and his co-workers (1935b) with nerves of widely varying amounts of associated

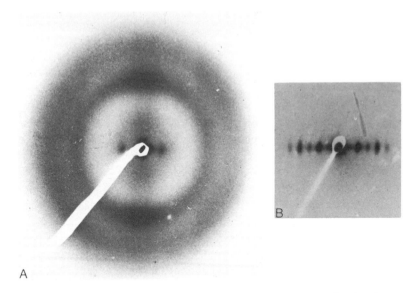

A

Figure 15. (A) Wide-angle X-ray diagram of fresh bull frog motor root, showing the outer diffuse halo due to moisture present, the inner somewhat narrower and axially accentuated halo due to lipid paraffin chains, and the 15.5 A equatorial spots that show the 11th order of the 171 A radial myelin period. (B) Small-angle X-ray diagram showing the first five orders of the 171 A radial period. [From the studies of Schmitt, Bear, and Clark, 1935]

connective tissue revealed a persistent reflection at 15.5 A. This continues to be recognized as being due to the myelin sheath, representing the 11th diffraction order of the large radial myelin period of 171 A (in amphibian peripheral nerve), considered further below.

It appears to this day that the sole diffraction by myelin at the wide-angle spacing range (spacings less than 10 A) is the 4.6 to 4.8 A diffuse ring. It is, nevertheless, a significant diffraction that permits the following conclusions:

1. *The radial diffuseness* of the ring identifies it as having been developed from a liquid component of the myelin sheath. Crystalline powder diffractions are rings of much sharper radial concentration.

2. *The radius* of the ring, measured at angles corresponding to Bragg law spacings of about 4.7 A, fairly conclusively identifies the source as being the hydrocarbon chains of the lipid molecules. At the same time, a smaller contribution from the myelin protein cannot be excluded because proteins regularly show diffuse diffraction at 4 to 5 A. Studies of liquid hydrocarbons have revealed that it is improper to use the Bragg law in such cases, and that this typical diffraction relates

to nearest neighbor intercarbon-atom separations, corresponding to hydrocarbon chains with about 5.0 A separations between chain axes (Warren, 1933).

3. *The meridional accentuation* of the ring intensity can be accounted for by assuming that the hydrocarbon chains are oriented predominantly radially in concentric cylindrical layers of the myelin sheath. Schmitt and his collaborators (1935b) calculated the intensity distribution around the ring expected of a perfectly oriented system of the kind just described. Because the observed meridional accentuation was not so great as that ideally expected, it was suggested that a degree of departure from perfect radial chain orientation occurs, which is consistent with the concept of fluidity of hydrocarbon layers.

**Small-Angle Diffraction: A. E. Blaurock, D. L. D. Caspar, and F. O. Schmitt**

When the incident beam's collimation is made finer (less angular spread) and the specimen-to-film distance increased, it is possible to explore very small angles near the undeviated beam. Then, as Schmitt and his co-workers (1935b) discovered, one encounters on the equatorial axis of the pattern a series of equally spaced reflections, each separated from neighbors by about half a degree of angle corresponding to a relatively large period radially oriented in the myelin sheath (Figure 15B). The ensuing conclusions follow from this equatorial series:

1. The equatorial reflections are diffraction orders of a period of 171 A in amphibian frog motor roots and sciatics, which at that time had most frequently been studied. A recent and extensive survey of variations in this period has been provided by Blaurock and Worthington (1969) who observed relatively larger periods in frog, rat, and chicken sciatic nerves (171, 176, and 182 A, respectively) than in "central nervous system myelin" as exemplified by frog optic nerves and spinal cord, rat and chicken optic nerves, and beef brain white matter (154, 153, 159, 155, and 157 A, respectively). In addition they report the spacings of lateral line, spinal cord, and optic nerves of one variety of fish (*Tilapia mossambica*) to be 159, 156, and 156 A respectively, but also note that this more intense series is accompanied by a weaker series related to spacings of 182 to 184 A in the same fish nerves. A similarly broad survey has been made by Höglund and Ringertz (1961) with much the same results, although the weaker

spacing in fish nerves was not observed. Mammalian (dog and cat) spinal roots were found to have a spacing of 184 ± 5 A in the early studies of Schmitt and collaborators (1941). The above figures give some suggestions of clustering about the values of 156, 173, and 183 A, rather than being distributed continuously through the entire range. The interesting cases of two distinguishable periods in individual fish nerves give further indication that the myelin radial structure may vary in discrete increments. No large radial periods have been observed for invertebrate nerves, probably because of the difficulty of obtaining nerves with sufficient thickness and concentration of myelinated sheaths.

2. The limited number of members of the diffraction order series that are observed indicates the minimum size of structural detail that is available in the small-angle information. If the $h$th order is the highest one detected, the resolution is about $d/2h$, where $d$ is the period concerned. Most investigators have readily observed the first five orders and the eleventh order. Blaurock and Worthington (1969) indicate that in frog, rat, and chicken sciatic nerves, the eleventh order terminates the series to yield a resolution of about 8 A, while in the optic nerves of the same animals, the tenth order is the terminal one although, because of the smaller periods, the resolution still remains 8 A. In fish nerves the resolution is somewhat better, decreasing to 7.5 A, because an eleventh order is terminal despite the reduced period. Of course these terminal orders depend somewhat upon the ability of the X-ray apparatus used to reveal weak orders. Still more recent results with frog sciatic nerve as reported by Blaurock at the Work Session (also Caspar and Kirschner, 1971) have extended the data to $h=15$, $d/2h=5.7$ A, and Blaurock and Wilkins (see discussion below) have observed diffuse equatorial diffraction bands out to an angle corresponding to a 9 A spacing or a 4.5 A resolution. This represents a resolution limit that will be difficult to surpass; hence, atomic detail will not be achieved readily, if, indeed, the complexity and variability of the lipid mixtures involved would be expected to permit better resolution.

3. The relative intensities of the various diffraction orders reveal details of the distribution of matter within the radial period. The full realization of this goal of diffraction studies is not readily achieved, and its final solution will be discussed further, after auxiliary information provided by electron microscopy has been surveyed. Nevertheless, certain qualitative facts were noteworthy from the beginning of the diffraction studies. The second and fourth orders dominate the

pattern normally, being typically from 4 to 50 times as strong as the odd orders with indices 1, 3, and 5. The second order dominates all others in every case, but in fish nerves the fourth order intensity declines to approximate equality with the third. This preeminence of the lower even orders over the odd ones indicates that the radial period of the myelin is approximately halved, i.e., the first and second halves of the period are similar; if all odd orders were missing, the halving would be complete and the period would be half that actually observed.

## X-Ray Diffraction Studies of Related Structures

### Variations from Normal Structure

The X-ray diffraction studies pertinent to the structure of myelin are not limited to applications to living intact nerve, but much information of value has been obtained from investigations of nerves subjected to abnormal conditions and from model systems of nerve lipids, lipid mixtures, and other combinations involving substances similar to those found in myelin. In the following paragraphs are brief reviews of the facts bearing on the structures of myelin sheaths.

1. The simplest modification of nerve myelin is achieved by the process of drying (Schmitt et al., 1935b; Elkes and Finean, 1949). The ultimate result (Bear et al., 1941) is an irreversible dissociation into several phases: one remains reasonably well oriented and corresponds to a contracted myelin residue ($d$=144 A in frog sciatic, 158 in mammalian spinal roots); orders observed are 1, 2, and possibly 4; a less well oriented single diffraction ($d$=60 to 67 A), probably ascribable to a phase rich in sphingomyelin and cerebrosides as judged from results with these as pure substances and mixtures; a very poorly oriented, near ring ($d$=42 to 45 A), also observed in phospholipid; and a moderately well oriented arc ($d$=34 A) that can be duplicated with pure cholesterol. The separateness of these phases is apparent from their differing degrees of orientation, and their likely constitution is indicated by duplication from isolated lipids and lipid mixtures. It is apparent that in the native hydrated state the lipids, despite disparate molecular dimensions, are able to unite in a single mixed lipid smectic phase. Upon removal of water, these disparities become serious enough to favor separation of

phases, segregating more compatible molecules. A "myelin skeleton" remains, however, in which the original 171 to 184 A radial periods have been decreased by about 26 A. The second order is still dominant in the skeletal residue, indicating that the structure retains the near-halving discussed above.

2. Other studies have compared the effects of temperature and solvents on the small-angle diffraction of nerve and total lipid extracts (Elkes and Finean, 1953a,b). Although the details are too complex to repeat here, the general conclusions resemble those disclosed by the drying experiments in indicating the sensitivity of the mixed lipid phase of normal myelin to separation of individual phases in unnatural environments and the persistence of the original skeletal structure except when exhaustive lipid extraction has been achieved. Some treatments produce expansion of the myelin period: at certain intermediate stages of solvent application the normal 171 A period can rise as high as 183 A (lipid swelling?) and a similar increase accompanies freezing at $0°$ to $-2°$ C. The native structure resists temperature increase to about $55°C$ but deteriorates above that. Swelling in hypotonic aqueous solutions has provided valuable insight into myelin structure, but this aspect is more appropriately treated in subsequent sections. Total lipid extracts yield phases similar to those of disintegrated myelin but lack the large spacing "skeletal myelin" phase present in treated nerves.

**Synthetic Systems: V. Luzzati**

A considerable body of information about the extensive polymorphic capabilities of lipid-water systems has been developed, notably by Luzzati (1968) among others. Attempts to summarize some of the most significant aspects of this polymorphism relative to myelin sheath structure are made in the following paragraphs.

1. Lamellar systems are commonly observed; a frequently encountered one, $L\alpha$, consists of a two-dimensional arrangement providing the total bilayer surfaces of constant thickness in which the hydrocarbon "tails" of the lipid are oriented into a fluid layer and the polar "heads" sit at the interface. Thus, systems of long-range order are provided despite short-range disorder of the paraffin chains. A similar structure, $L\beta$, contains smaller amounts of water. The paraffin chains are rigid and hexagonally packed, although some degree of disorder

permits accommodation of differences in tail lengths and degree of saturation that may be present. Alternate L$\alpha$ and L$\beta$ bilayers can form L$\gamma$, whose structure periodicity is larger than either constituent. A relatively rare L$\delta$ lamellar structure has the cross sections of paraffin chains forming a square array of 4.7 A on a side. The paraffin chains are believed to be helically coiled and not restricted as to rotation about their long axes.

2. The domains of lamellar structures in phase diagrams of pertinent lipid-water systems are of interest. For example, in the high-temperature, high-lipid region of the aqueous egg lecithin diagram, L$\alpha$ occurs at water contents above 4 to 45% and at room temperature to 100°C, and L$\beta$ is found at a slightly lower moisture content and at temperatures below 50°C (Luzzati et al., 1968b). This system, like other similar ones with polar heads lacking net charge, incorporates relatively little water in forming L$\alpha$. This contrasts with the beef-heart mitochondrial lipid-water situation, where the L$\alpha$ phase dominates the phase diagram from concentrated mixtures (weight fraction of lipid 0.8) to very dilute cases (below 0.2 lipid) and at temperatures above about $-10$°C, and huge amounts of water are incorporated. Direct observation of the effects of adding cationic or anionic lipids to egg lecithin have dramatically demonstrated increase in the thickness of L$\alpha$ water layers to more than 250 A, (Gulik-Krzywicki et al., 1969b). Of interest for the present considerations, the L$\alpha$ phases exhibit simple X-ray patterns like those of nerve myelin: a small-angle series related to the large interlamellar spacing and a diffuse wide-angle diffraction of spacing, about 4.5 A, characteristic of paraffin chains.

3. Lipids are not always required to form bilayer systems, however. For example, Luzzati (1968; Luzzati et al., 1969) has distinguished structures formed from "rods" that are built either with polar heads peripheral and paraffin tails central (Type I) or vice versa (Type II), in addition to lamellar phases of the bilayer type. When rods are parallel but packed in hexagonal array, one obtains an H (for hexagonal) structure. In others, the rods emerge from connecting points in 3's or 4's to form three-dimensional lattices with tetragonal (T), rhombohedral (R), or cubic (Q) crystallographic symmetry. Some of these, such as T and R, are stacks of four- or three-connected layers of rods, respectively, which as individual layers can be said to provide models for lamellae. Ribbon-like aggregates having uniform clusters of polar head double rows with fluid paraffin chains occupying the space between provide rectangular or oblique net cells. It should be noted,

however, that in egg lecithin-water and beef-heart mitochondrial lipid-water phase diagrams, the nonlamellar structures, when they occur, are found under more anhydrous conditions and at generally more elevated temperatures than those that produce Lα; on the contrary, the hexagonal structure is predominant in lysolecithin. It seems unlikely that these structures are significant for nerve myelin structure in the fresh state. It is possible, however, that the rather cursory interpretations of the past that have assumed that the small-angle diffractions of dried pure or mixed lipids, or that the phases separating out from myelin in dried or altered nerve, measure simply the thickness of the pertinent bilayers should be reexamined.

### Protein-Lipid Combinations

Nerve myelin structure differs in two material ways from those disclosed above for the lipid-water Lα lamellar structures and resembles the Lγ phase: (1) the magnitude of the myelin period is more than twice that which has been observed for any of the lipid-water or lipid-anhydrous phases except in the instances of great swelling and in the Lγ phase; and (2) the near halving of the myelin period as indicated by the dominance of the second and fourth orders (see above), which is not seen in artificial systems (with the exception of the Lγ phase), even in cases of great swelling (see Gulik-Krzywicki et al., 1967). Noting facts of this kind and also taking into account the optical evidence for the presence of oriented protein in the myelin sheath, Schmitt and his collaborators (1941) concluded that the myelin sheath period includes *two* lipid bilayers, with polar faces at the edges and between them carrying protein or other polar substances and water in layers with alternating material differences in composition and/or structure. Without these alternating polar layer differences, the myelin period would be truly halved, and the odd orders would be completely missing.

Studies on artificial lamellar structures carrying protein as well as lipid and water are few. Palmer and his co-workers (1941) studied complexes of cephalin with histone and globin and found that dry cephalin bilayers exhibit 43.8 A thickness, and that the addition of histone increased this spacing to 56 to 59 A and that of globin to 50 to 52 A. It had previously been observed (Palmer and Schmitt, 1941) that cephalin with water alone is capable of yielding swollen bilayers as great as 150 A thick (total nerve lipid also behaves similarly). The cephalin, presumably because of its content of phosphatidylserine (negatively charged polar heads at neutral pH's), appeared to be the chief component of nerve lipid contributing to its swelling properties. Swelling was less extensive in Ringer solution, and $Ca^{2+}$ ions were particularly effective in collapsing the water layers by discharge of the bilayer negative charges. The proteins apparently can play a similar role in preventing or reducing swelling. The relative thinness of the additional spacing increments provided by the proteins in these cases (6 to 15 A) suggested that unrolled polypeptide chains were present.

Models of lamellar systems with globular protein layers interspersed between lipid bilayers have been provided by Luzzati and his collaborators (Gulik-Krzywicki et al., 1969a) from studies involving optical methods (spectrophotometric and circular dichroism) as well as X-ray diffraction. In ferricytochrome c-phosphatidyl inositol-water and ferricytochrome c-cardiolipin-water systems, two proteolipid phases were found: "phase I" has one globular protein layer about 25 A thick interspersed between bilayers of lipid 43 A thick, yielding a total thickness of about 68 A (anhydrous condition); "phase II" inserts two layers of protein globules to build up a period of about 91 A. The protein molecules are quite closely packed and do not appear to have been materially altered in conformation from the freely dispersed globular condition. The lysozyme-cardiolipin-water system presents phases similar in structure to those of phase I above, in that single layers of globular protein are interleaved with lipid bilayers, but the dimensions, hydration, and compactness vary.

None of the synthetic models duplicate nerve myelin's properties of double bilayer and alternating protein-aqueous layer structure; the reasons for this are "biological" and are discussed below under Electron Microscopy.

## Electron Microscopy of Myelin

Shortly after the early models of the structure of the myelin sheath were derived from optical and X-ray diffraction studies, electron microscopy became generally available as a tool for direct visualization of submicroscopic structure. Unlike the optical and X-ray methods, however, the EM procedures do not permit observations of living tissues and must rely upon the residual artifacts remaining after fixation, staining, and embedding, or other procedures required to prepare the specimen for observation in a vacuum under bombardment by an electron beam. Nevertheless, properly interpreted electron micrographs have yielded additional knowledge of myelin structure that has proved indispensable to further full utilization of the diffraction data. The significant specifically structural contributions from electron micros-copy are set forth in the following sections.

## Myelin Structure in Stained Preparations

Preparation of nerves for examination of myelin sheaths is commonly done as follows: fixation in glutaraldehyde, additional fixation and staining with neutral buffered $OsO_4$ or $KMnO_4$, dehy-dration through an alcohol series, and preparation for sectioning by freezing or by gelatin, methacrylate, or epon embedding. The resultant electron micrographs readily disclose the lamellar nature of myelin: layers are piled concentrically about the axon with surfaces normal to radii, just as had been suspected from the earlier optical and X-ray analyses. The results differ in detail according to the procedural variations (Fernández-Morán and Finean, 1957), but in general the structure appears as a set of parallel, very densely stained lamellae with a less densely stained band halfway between separating lighter regions. The period or distance between nearest "major dense lines" is often considerably less than the native fresh period (171 A in frog sciatic, for example) and can decline even below that of air-dried material (144 A) to as low as about 120 A. The lighter "intraperiod" lines vary materially with conditions, being relatively well developed after $KMnO_4$ fixation. Fernández-Morán and Finean (1957) have compared changes in sheath structures at various stages of the preparative process by means of small-angle X-ray diffraction to trace the spacing and band density changes from the original state to that finally observed in the EM. For example, the spacing contraction occurs mostly during fixation and dehydration, with some expansion during embedding, and final further contraction in the microscope. When the major dense line is much more prominent than the intraperiod line, the first order diffraction is dominant on the X-ray pattern, as expected, and elevation of the intraperiod density is correspondingly represented in the dif-fraction pattern by an increase in the second-order intensity. Although the EM appearance of density distribution through the radial myelin period is somewhat at variance with the distribution of matter indicated by X-ray data on fresh nerve, the two views do agree on the general lamellar nature and magnitude of the period as well as its division into approximately equivalent halves.

**Formation of Myelin Sheath: B. G. Uzman, J. D. Robertson, H. deF. Webster, and R. P. Bunge**

A major advance occurred when Geren (1954) studied the formation and development of myelin around the axons of chick embryos. A Schwann cell engulfs a nodal segment of the axon that then has about it, in addition to its own plasma membrane, a Schwann cell membrane that remains attached by a pair of membranes (mesaxon) to the external Schwann cell membrane as shown in Figures 16 and 19A. Addition of membrane material results in a spiral wrapping of the membrane pair about the axon to form the eventually compact, layered myelin sheath. Throughout the process, the connections of the membrane halves of the membrane pair to the Schwann cell surface

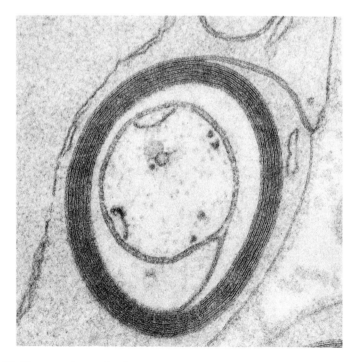

Figure 16. Electron micrograph of young mouse sciatic nerve fiber with developing myelin sheath. Compare details with Figure 19 A. x 160 000. [Robertson, 1966b]

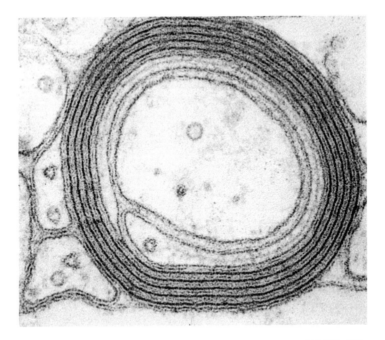

Figure 17. A myelinated axon from the cerebral white matter of a rat. x 166,000. [Hirano and Dembitzer, 1967]

remain traceable. Consequently, it is possible to identify the major dense line of the myelin period with the apposition of membrane surfaces continuous with the cytoplasmic side of the Schwann cell membrane, while the intraperiod line represents apposition of membrane surfaces continuous with the external surface of the Schwann cell.

The significant facts thus emerging are that the myelin sheath, because of the manner in which it is formed, has the following characteristics: (1) a radial period that is twice that of a cell membrane (thus explaining why the X-ray information required interpretation in terms of a double bilayer of lipid with differing polar layers alternately interleaved, corresponding to the different cytoplasmic and extracellular connections); and (2) planes of symmetry at both the major dense (cytoplasmic) and intraperiod (external) levels, a circumstance that, as will appear below, greatly facilitated the eventual full interpretation of the small-angle diffraction data.

This view of the development of myelination has been amply confirmed for peripheral nerve (e.g., Robertson, 1955) and extended to

myelination in the CNS (Maturana, 1960; Peters, 1960) with modifications appropriate to the different cells, oligodendrocytes instead of
Schwann cells, which are active in myelination of the central fibers. In
these cases, processes extend from a glial cell to endow several axons
with spirals of myelin layers as shown in Figure 17. The resultant
pattern of membrane pairs with cytoplasmic and external cell connections (Figure 18) is the same as that in peripheral nerve (Bunge,
1968; Hirano and Dembitzer, 1967).

**Stain Identification of Polar Layers**

The morphological relationships of the stained dense lines in
their connection with external and cytoplasmic membrane surfaces
would lead to the conclusion that the stains are emphasizing the
proteinaceous or polar membrane parts, because models such as that of
Danielli and Davson (1935) have generally placed the protein components at the polar external surfaces of lipid bilayers. However, osmic
staining in particular has presented some difficulties in the acceptance
of this view because of the reputed ability of osmium tetroxide to stain
lipids preferentially, presumably at hydrocarbon chain unsaturation
points. After some confusion it now appears to be agreed that the stain
identifies the polar layer; even in bilayers alone the polar heads are
stained (Stoeckenius, 1962). A particularly simple demonstration of
this was given by Robertson (1960a,c, 1961c), who showed that
bilayers of egg cephalin, which had been floated apart by hydration,
take on stain at the external surfaces that had been in aqueous
environment before EM examination. It may therefore be said that the
electron optical studies show a structuring of the myelin membrane pair
that indicates the alternating polar layers with unstaining hydrocarbon
chains of the bilayers between them, in agreement with the suggestions
from optical and X-ray diffraction information.

**Penetration Channels of Myelin: J.D. Robertson and A. Hirano**

The EM studies have also provided useful knowledge about
channels through or along which penetration of the myelin can be
accomplished. A most useful case is the demonstration by Robertson
(1958a) that in hypotonic solutions the membrane pairs split along the
externally connected dense line to permit influx of aqueous solutions

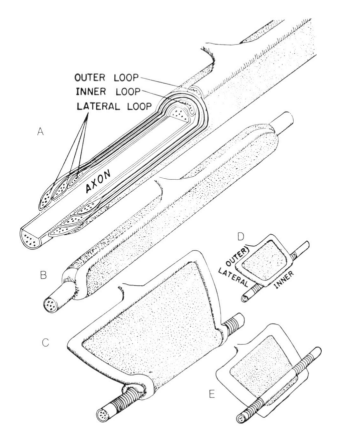

Figure 18. Diagram of the shovel-shaped process of an oligodendrocyte from the CNS and its configuration when wrapped around an axon. In (A) sections are cut to disclose the outer, inner, and lateral loops where oligodendrocyte cytoplasm is enclosed by membrane; (B) shows an intact single node, and (C), (D), and (E) illustrate a hypothetical unwrapping of a mature myelin sheath resulting in a shovel-like oligodendrocytic process. Note the transverse axonal bands near the nodes, which provide a helical path into the periaxonal space. [Modified after Bunge et al., 1961 by Hirano and Dembitzer, 1967]

(Figure 19). X-ray diffraction studies have also shown that the intercalated aqueous layers can be quite regular in thickness and large, permitting observation of small-angle diffraction series of spacings over 400 A; this possibility has been exploited to aid in interpretation of the X-ray information, as discussed below.

The externally connected dense line is also the path of other sheath penetrations; for example, lanthanum nitrate can move into even the innermost sheath layers (next to the axon) through channels

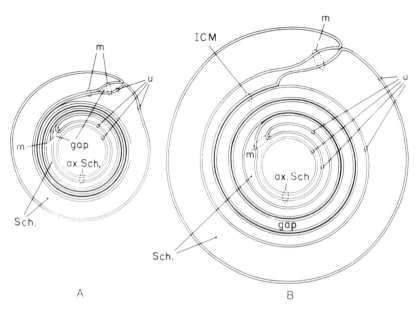

Figure 19. Diagram showing the relationships between the Schwann cell (Sch) and axon membranes; m is the outer and inner mesaxon; ax. Sch. is the axon Schwann membrane pair and u is the unit membrane. Gaps between the pair members are found at the outer and inner mesaxons though elsewhere the mesaxon pair is normally closely applied as in A. Hypotonic solutions cause parting of the mesaxons as in B, though the major dense lines where cytoplasmic faces of the membrane are apposed remain together to exhibit internal compound membrane (ICM) pairs. [After Robertson, 1958]

provided by the periaxonal space at the nodes of Ranvier and the externally related myelin levels (Figure 18). Loops of myelin corresponding to incisures are applied to the axon near the node close to "transverse bands" developed on the axon in such a way that a continuous helical path is available into the periaxonal space, along which the lanthanum salt invades the sheath via the externally connected intraperiod lines (Revel and Hamilton, 1969; Hirano and Dembitzer, 1969). On occasion, also, the intraperiod line has been observed to be double (Robertson, 1961a; Napolitano and Scallen, 1969), with a 20 A separation, again suggesting a readiness for parting along this line.

There are a number of circumstantial indications that myelin layers are capable of tangential slippage: for example, the phenomenon of swelling (Robertson, 1958a) implies an ability of the sheath spiral to expand as the total thickness increases severalfold; Hirano and

Dembitzer (1967) observed that when injurious experimental conditions were introduced into the cerebral white matter of rats, they developed enlarged axons that would require some slippage of sheath layers to accommodate the increased diameter. Furthermore, Webster's (1971) finding (also Rosenbluth, 1966) that during myelination the greatest increase in myelin membrane area occurs after the sheath has become a compact lamellar spiral would indicate that when processes of internal accommodation are going on, there is lamellar slippage or intussusceptional addition of new material or both.

The observations do not permit kinetic views of layer slippage but rely on static situations of individual micrographs. The candidate layer for the most ready and frequently observed parting would seem to be the intraperiod line, but the evidence may be circumstantial since this line is directly accessible to externally applied agents that would not have to pass through any membrane. The myelin lamellae do on occasion part at the major dense line; for example, the Schmidt-Lanterman incisures are loops of cytoplasm enclosed in membrane pairs parted at the cytoplasmic, major dense line (Robertson, 1958a,b; Hall and Williams, 1970). Some of the similar but anomalous cytoplasmic lakes developed under the white matter injuries employed by Hirano and Dembitzer (1967) also cause splitting along the major dense line. In all major dense-line-splitting, cytoplasmic enlargements provide pools that can be influenced only by reagents capable of passing through at least one intact membrane.

It is possible by fixation in glutaraldehyde, followed by extraction with lipid solvents (neutral and acidified chloroform-methanol, carbon tetrachloride containing $OsO_4$, ethanol, and acetone), followed by embedding in epoxy, to achieve myelin sheaths in which the laminar period is apparently well preserved, as far as can be detected in the EM, despite removal of 95% of the lipid (Napolitano et al., 1967). Robertson (1960c) showed that the nonlipid layers may be preserved after alcohol ether-phosphotungstic acid (PTA) treatment, but they are freefloating if embedded in methacrylate. This is also true of fibers treated according to Napolitano and his colleagues if methacrylate is used. Collapse of the structure results if water is admitted after the chloroform-methanol treatment. Unfortunately, the protein layers, although seemingly well preserved according to the electron micrographs, are sufficiently disordered so that small-angle X-ray diffraction, which is very sensitive to disorder, has been found incapable of yielding coherent diffraction of protein distribution in the better preparations (Robertson, 1969).

## Electron-Density Distribution

The diffraction of X-rays by matter is determined by the distribution of atom-attached electrons therein. The amplitude of the coherent wave contributed by each atom to the diffraction effects is proportional to its atomic number, as is particularly true at small diffraction angles where forward scatter (close to the direction of the incident beam) depends very little on interference effects between the scattering by the several electrons within each atom. Consequently, analysis of the small-angle diffraction by any system is directed toward working backward from the observed diffraction to achieve a conception of the way in which the electrons are distributed, producing, in effect, the desired "image" of the structure in these terms. In microscopy the objective performs an analogous function by recombining the diffraction effects into a simulation (image) of the original object that brings the light or electron waves back together in the same relationships existing as they emerged from the object. Unfortunately, satisfactory objectives for X-rays cannot be achieved because of the very small wavelength involved ($\lambda=1.54$ A for CuK$\alpha$ radiation), which, however, is essential for the resolution of fine structural detail.

### Phase Problem

When an X-ray beam passes through a periodic structure, it performs a harmonic analysis of the electron distribution. Electrons that are separated by one whole period contribute to the amplitude of the coherent wave progressing toward the first diffraction order; separations of one-half period contribute to the second order, those of one-third period to the third order, and so on. Conversely, given the series of diffraction orders, their observed amplitudes should be capable of being recombined to indicate the original electron distribution. Unfortunately, here one encounters the major "blind spot" of the diffraction method: all ways of recording diffractions measure the intensities only (which are proportional to the squares of amplitudes) and do not disclose the relative time relations (phases) with which the peaks and troughs of the several orders reach the recording device (e.g., film), because of the extremely high frequencies involved. Consequently, the spatial relationship with which the harmonic components derived from the series of diffraction orders should be combined is unknown, and an infinite number of images compatible with the diffraction effects can be formulated by combining the components in all possible relationships.

This difficulty becomes greatly simplified if a center of symmetry (or a plane of symmetry in the case of the lamellar structures, with unidirectional radial periodicity of structure, as in the myelin sheath) is present; then the identity of structure encountered in going in opposite directions from the center of symmetry dictates that the harmonic components should be combined with peaks or troughs in register and not in other possible relationships. Only in this way can symmetry centers be developed. Each component is to be assigned a positive or negative sign. Systematic consideration of all possible images is thus facilitated, but the number of conceivable structures may still be formidable, since, with $n$ orders observed, each of which may be either + or -, the number of possible structures becomes $2^n$. More accurately, the number of different structures is $2^n/2$ because those permutations of signs differing by reversal of all odd-order signs without change of even orders yield structures with origins at a new center of symmetry but otherwise identical. Also, sign permutations in which all signs have been reversed provide structures that are related by inversion of peaks on one into troughs of the other, and vice versa; separate calculations for the members of such pairs are not required, so that another halving of the considerations is possible.

Before the advent of the EM studies, i.e., prior to about 1954 when the manner of myelin development was discovered and the presence of planes of symmetry at both the major dense layer and the minor intraperiod one was rather conclusively demonstrated (as discussed above), it was virtually impossible to attempt radial plots of electron density for myelin using diffraction data. Even though the models proposed might well have been expected to have planes of symmetry, this was in no way confirmed by the optical or diffraction data, or by the electron micrographs with inherently poor resolution.

Even so, the simplifications introduced by the knowledge of symmetry could not immediately be very useful because with six diffraction orders commonly observed (orders with indices 1 through 5, and 11), the possible structures still remained quite numerous ($2^6/2=32$). This difficulty was overcome when it was realized that the myelin layers could be floated apart in an orderly manner so that small-angle diffraction was still observable. It then became possible to assign phases (+ or -) unequivocally, at least to the lower orders, and reduce the number of possible structures to relatively few.

The way in which swelling accomplishes this assignment of phases may be explained as follows: If unit "packages" of structure remain essentially undisturbed as aqueous layers are interspersed to provide a wider periodicity, the potential contribution of each unit package to the diffraction of the total system remains unaffected with respect to the angular distribution of its own interference capabilities.

The uniform solvent layers lack structure and hence are not capable of providing variations in scattered amplitude. The large periodicity of the total swollen structure produces its own interference pattern that permits the scattering amplitude of the unit package to be observed only at angles consistent with the diffraction order positions for the total structure. It is said that the pattern of the swollen system is one in which the curve of scattering amplitude appropriate to the unit package is "sampled" by the interference properties of the total structural period. By varying the period of the swollen structure through a number of degrees of swelling, sufficient samplings are achieved so that the outline of the unit package's scattering amplitude ("transform amplitude" of the crystallographer) can be obtained as shown in Figure 20.

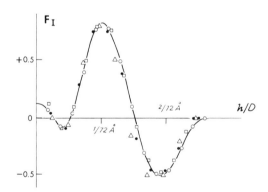

Figure 20. A plot of the amplitudes of myelin diffraction orders $[I_{obs}(h)]^{1/2}$ after correcting for order index ($I_{obs} \times h$) and placing them on the same arbitrary scale, against diffraction angle expressed as inverse A spacings. Data are for frog sciatic nerves in 0.24M sucrose (o), d = 388A; in 0.24M sucrose (△), d = 293A; in 0.24M sucrose with 4mM NaCl (□), d = 373 A; in 0.24M sucrose with 4mM KCl (●), d = 326 A. Zeros are expected at approximately 1/(2 × S), 3/(2 × S), and 5/(2 × S) where S is the membrane pair separation (compare Figure 21 A); hence S according to the data shown here is about 72 A. The continuous curve shows the Fourier transform of the proven bilayer profile [Blaurock, 1971]; the curve differs in detail from a similar curve shown in Worthington and Blaurock (1969b), where the same experimental data are given (compare also Moody, 1963). [After Blaurock, 1971]

A typical unit's transform consists of a plot of a continuous curve of amplitude versus diffraction angle with segments of the curve alternating from positive to negative and back again as the phase of the scatter varies in relation to waves coming from some origin of reference in the structural unit (chosen at a center of symmetry of the unit when present). The sampling process actually measures squares of amplitudes; the experimentally determined curve consists entirely of positive loops that rise abruptly at the angles where the true transform changes sign. The problem then reduces to finding some way to assign a sign (+ or -) to one of the loops from which those of all others are determined by alternation. Even if this is not achievable, there will be only two possible assignments that will yield electron density plots that are relatively inverted, i.e., with peaks of one structure occurring where troughs of the other are found; usually only one of these will provide a structure that is sensible in relation to other facts that are known. The whole process determines regions of diffraction angle that are to be assigned positive or negative signs for recombination of the harmonic components whose amplitudes are proportional to square roots of observed order intensities. These regions apply to both the swollen and unswollen states of the system; thus electron density plots of all states are achievable. By using swelling agents of differing known electron densities, scaling of the plots can be achieved on an absolute basis.

This program was first attempted for the nerve myelin sheath by Finean and Burge (1963) and less completely by Moody (1963). The former investigators presented both X-ray and EM data indicating that the swelling does indeed proceed as required by the theory outlined above, namely, with no significant alteration of the membrane pair unit during swelling to the resolution considered; and they were able to assign - + + - - signs, respectively, to the first five diffraction orders. One of two assignments of Moody was the same, and he provided an argument as to why the first order should be negative based on the greater overall electron density of myelin as compared to the swelling agent. It will be noted that the second- and third-order signs are the same but opposite that of the first, while those of the fourth and fifth are again reversed; this is a consequence of the fact that the several orders of the unswollen myelin sample the membrane pair transform at consecutive loops as 1; 2,3; 4,5; where semicolons indicate loop boundaries.

Subsequently, Worthington and Blaurock (1969b) conducted a somewhat more detailed study involving swelling of frog sciatic nerves in distilled water and sucrose solutions (spacings from 252 A in water to 388 A in 0.24 M sucrose). An essential development in their considerations was to correct the observed intensities by multiplying each by its order index. This correction takes account of the fact that greater volumes of myelin are in a position to diffract to the lower orders than to the higher ones as had been demonstrated by Blaurock and Worthington (1966). Nevertheless, by assuming a bilayer structure to each membrane, the phase signs were again determined to be the same as those in the earlier studies, although the electron density plots would reflect greater emphasis on the finer details contributed by the higher orders. Finally, Blaurock (1971) proved crystallographically this choice of phases without the need to assume bilayer structure.

When electron-density plots are obtained by using the first 6 orders of the unswollen myelin pattern with signs of orders 1 through 5 as given above and order 6 positive, the results are qualitatively like those originally presented by Finean and Burge (1963). There is moderately high density in a region near the origin, which was automatically at the center of symmetry of the membrane pair as it floats free in the swelling media, namely, at the cytoplasmic-connected major dense line of electron microscopy; there is also a similarly high density about the half period, namely, at the intraperiod line or externally connected level where the membrane pairs swell apart. Approximately halfway between these regions are low-density troughs. This general picture (see the higher resolution plots of Figure 23) is to be expected of alternating polar and nonpolar layers, because the latter (with hydrocarbon molecular tails) should have densities of about 0.25 to 0.29 electrons per $A^3$ characteristic of liquid hydrocarbons; water has a density of 0.34 electrons per $A^3$ and a typical protein, e.g., collagen in the anhydrous state, has 0.45 electrons per $A^3$, as is easily calculated from the mass densities and compositions of these materials.

The use of only the first 6 orders achieves a resolution of about 14 A, and it is desirable that the information at higher diffraction angles also be incorporated to improve detail. Attempts to do this with swelling methods run into the difficulty that information beyond angles corresponding to the sixth order do not as unequivocally determine the transform loops, because the essential assumption—that in swelling the myelin pairs remain unaltered in detail—becomes less certain. Thus, small variations in the structure of the membrane pairs can be significant in affecting the corresponding experimental detail.

**Patterson Plot: A.E. Blaurock**

There is another way to examine the X-ray data, and that is by what is known as a Patterson plot, which is effective in extracting information about the structure of the myelin sheath. Whereas the electron-density plots use harmonic amplitudes that are the square roots of observed intensities (after correction; see above) and result in a direct "image" of the structure, the Patterson plot uses amplitudes of the harmonic terms that are direct, corrected, observed intensities; the result is a graph showing, in effect, the frequencies (or probabilities) for specific separations that occur between parts of the structure with similar electron densities. There are therefore no phasing problems connected with the Patterson plots; they produce no direct images. However, interpretation does involve an effort to seek "images."

Patterson plots for swollen myelin have the advantage that when the membrane pairs are floated far enough apart (aqueous layer thicker than the membrane pair), the plot becomes simpler because the region corresponding to electron separations of from zero up to the total thickness of the pair presents information about that single pair (Figure 21). When pairs come closer or are in contact, their inter-relations show up and complicate the situation. Thus, Blaurock (1967, 1971) predicted that the Patterson of myelin would show a prominent peak that directly measures the center-to-center distance between the membranes in a pair and that this pair-separation peak would stand by itself in the Patterson of swollen myelin but would overlap with its mirror image in the Patterson of normal myelin. Basing his argument on simplified strip models,* Worthington (1969) showed that the Patterson of highly swollen myelin should include beyond the origin peak (which always dominates the Patterson because it presents the identity expectation that every electron of the structure falls upon itself) a trough that measures the thickness of the layers, followed by a strong second peak that directly measures half the thickness of the membrane pair. Worthington and Blaurock (1969b) then obtained an experimental Patterson plot for swollen myelin and found that it agreed with expectations; hence, it became clear that the two membranes in a pair are 72 A apart, center-to-center, that the total pair thickness is about 144 to 145 A, and that the individual layer thicknesses are around 20 to 30 A. This was in agreement with estimates based on how strip model parameters should influence experimental swelling curves, thereby determining nodal zero positions and maximum amplitudes. The membrane pair thickness of 144 A can be compared to the same value noted earlier as the experimentally observed period of the myelin

*In these strip models, the polar and apolar layers of the centrosymmetric membrane pair were represented by "strips" of specified constant radial thickness but of different electron densities.

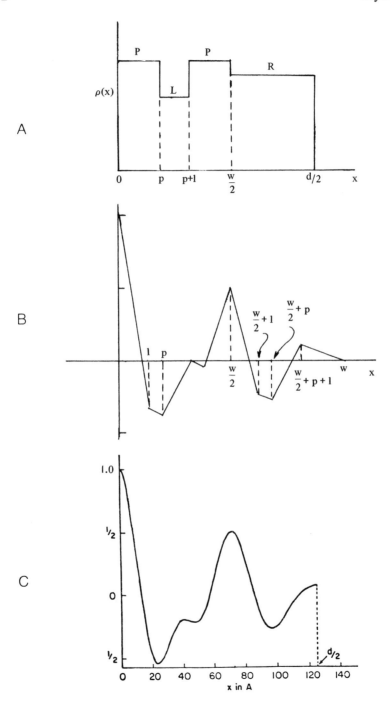

"skeleton" that remains after drying of frog sciatic nerve; the near-identity of the two values is in fact something of a coincidence, since the wet pair evidently contains a certain amount of water.

The Patterson plot for intact unswollen myelin is more difficult to unravel in structural terms. Based on the success of the Patterson interpretation for swollen myelin, an interpretation has, however, been made by Blaurock and Wilkins* for frog sciatic nerves as shown in Figure 22B and C. Using added data achieved at angles corresponding to higher orders out to Bragg spacings of about 9 A (hence providing a resolution of about 4.5 A), these authors include in the analysis not only the strong lower but also the weak higher orders. They include, beyond the first five, orders with indices 6, 7, 8, 10, 11, 13, and 15, certain radial diffuse scattering at spacings of 22, 15, 11 and 9 A. The diffuse series is identified with the second, third, fourth, and fifth submultiples of a spacing of 45 A, which is believed to be a spacing within individual membranes that becomes evident when membrane pairs occasionally develop less than perfect apposition. After proportionally adding the integrated, diffusely scattered energy to the neighboring sharp terms, a Patterson plot was obtained.

Going directly to the eventual interpretation, we may note that the total Patterson for frog sciatic nerve can be accounted for in terms of two major dimensions in addition to the observed 171 A spacing (Figure 22B): (1) the 45 A intramembrane spacing believed to be the distance between the centers of extra-dense layers on both edges of the apolar phase of the bilayer where the electron-dense phosphate groups of the phospholipid heads are located in a lipid bilayer, and (2) a 78 A distance between the centers of the two membranes in a pair. The 45 A interpeak separation is symmetrically placed within a membrane, with

*Unpublished data.

Figure 21. A comparison of a simple strip model for swollen myelin (A) with the corresponding theoretical expectation in a Patterson plot (B) and with an experimentally determined Patterson plot (C). Abscissas are dimensions through the myelin layers: thickness and electron densities, respectively, are p and P for polar (protein containing) strips; l and L for the lipid strips; and (d-w) and R for the interposed aqueous layer, where d is the entire swollen period and w is the membrane pair thickness. Only the first halves of the structure or plot are given because centers of symmetry are present at x = 0 and d/2. Because the individual membrane pairs of the swollen myelin are far enough separated by the aqueous layers, the Patterson of the swollen myelin (C) resembles that expected for a single membrane pair (B). The peak at x = 72 A then measures w/2, providing again an estimate for w (144 A, compare Figure 20). [After Worthington (1969) and Worthington and Blaurock (1969b) with scaling altered so that abscissas are comparable]

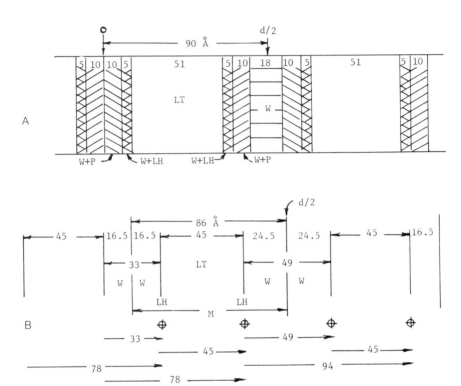

Figure 22. A comparison of the Vandenheuvel (A) and Blaurock-Wilkins (B) models for the nerve myelin period. W indicates major positions of water, P of hydrated protein, LH of lipid heads, LT of lipid tails. The origin (O at top) is the cytoplasmic apposition of membranes; at d/2 is the external apposition. The period accounted for by the Vandenheuvel model is larger (180 A) than in the B-W model (171 A), but the membrane thickness (M = 85.5 A) in the latter

two of the electron-dense layers accordingly lying 16.5 A to each side of the cytoplasmic apposition (45 + (2 x 16.5) = 78 A).

    With the above model, the major features of the Patterson plot for intact myelin are shown in Figure 22B and C. At the origin the four major densities of the membrane pair contribute a strong peak, due to four identity correlations: (1) at 33 A, the single correlation of the major densities across the cytoplasmic apposition of membranes; (2) at 45 A, the two correlations between major densities within each of the two membranes; (3) at 48 A, a single correlation of densities across the external membrane apposition; (4) at 78 A, the correlation between the two membranes in a pair. All these show a correlation not only of the major densities but of the troughs as well. These are the experimentally recognizable details of relative amplitudes and locations up to

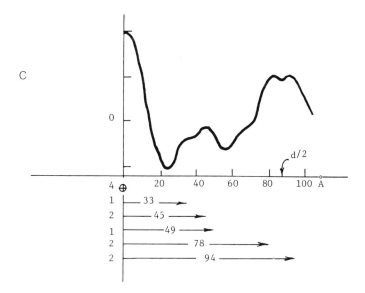

model is less than in the former (81 A). LH positions in the B-W model indicate locations of major peak densities expected where lipid phosphate atoms are present.

Under (B) are indicated all vectors between the LH peaks that are in the forward direction (left to right) and of length up to or slightly greater than d/2. The symbol for an identity is the crossed circle. 22C is the experimental B-W Patterson plot for intact frog sciatic nerve (ordinates in arbitrary units). Below (C) the identities and vectors of (B) are lined up from the origin, with integers indicating the number of each. The vector lengths account well for the positions of major Patterson peaks, while their multiplicities are in fair qualitative agreement with the relative peak heights.

In both models and the Patterson plot the positions with abscissas 0 and d/2 are centers of symmetry where d is the total period. Dimensions are given in A.

[Dimensions in (A) from Vandenheuvel (1965) and data for (B) and (C) from unpublished data of Blaurock and Wilkins put on comparable scales.]

spacings of $d/2 = 85.5$ A, allowing for the fact that (a) resolution of the 45 A and 48 A Patterson peaks is not achievable, and (b) the 78 A peak merges somewhat with the corresponding 93 A peak beyond $d/2$.

## Fourier Plot: A. E. Blaurock and D. L. D. Caspar

Starting with these general features of the Patterson plot, it is possible to project phases for the higher orders out to the fifteenth and to use them to achieve a direct electron-density plot for frog sciatic myelin at good resolution. The result repeats the general features of the lower resolution earlier plots, but accentuates the major peaks bounding the low-density troughs. The major peaks also bound the polar appositional plateaus, the plateau at the externally connected apposition being wider (about 40 A) than that at the cytoplasmic apposition (about 25 A). Rising from the trough depth to the major peaks, the curve carries shoulders, one toward the cytoplasmic apposition being somewhat lower than another toward the external apposition (Figure 23). The shoulders are believed to reflect the presence of the considerable proportion of cholesterol (see below). Thus, except for the shoulder asymmetry and the different widths of polar plateau at cytoplasmic and external surfaces, each membrane is very nearly symmetrical about its center (the trough valley).

Caspar and Kirschner (1971) have carried out a program similar to those outlined above but differing in some essential respects. They have compared rabbit optic and sciatic and frog sciatic nerves (Figure 23B,C, and D). The myelin periods of these sources are 156, 180, and 170 A, respectively, varying widely enough so that each diffraction series samples the transform of the membrane pair in a manner similar to that described above for the swelling experiments. As before, the assumption is implicit that the myelin structures of the sciatic and optic nerves are similar at least with respect to the larger features of structure involved at the lower diffraction orders. The optic nerve myelin is distinguishable in diffraction by the marked emphasis placed on the even second and fourth orders, with weak development of the tenth order, while all odd orders are very weak; it is also distinguishable by the smaller total period. Consequently, one knows immediately that individual membranes of the optic nerve membrane pairs are closely similar and nearly centrosymmetric about their apolar centers. Making comparisons of the optic and sciatic nerves, Caspar and Kirschner (1971) took detailed accounts of the departures from membrane symmetry, to phase the diffractions and provided electron-density plots for the three nerves studied. As expected, the optic nerve plot reveals the near centrosymmetry of individual membranes, though the asymmetric apolar shoulders remain. The greatest development of membrane asymmetry that characterizes the sciatic nerves occurs at the polar plateaus, which are thicker than in the optic nerve, particularly at the external apposition. Trough width in rabbit sciatic is greater than that in frog sciatic nerve, suggesting that lipid hydrocarbon chains may be longer in the rabbit case.

Caspar and Kirschner (1971) have also made additional suggestions as to the meaning of the electron-density plots in terms of chemical structure (Figure 24). The shoulders on the sides of the apolar troughs are believed to indicate an asymmetry in the distribution of the cholesterol molecules; the details of the asymmetry suggest that

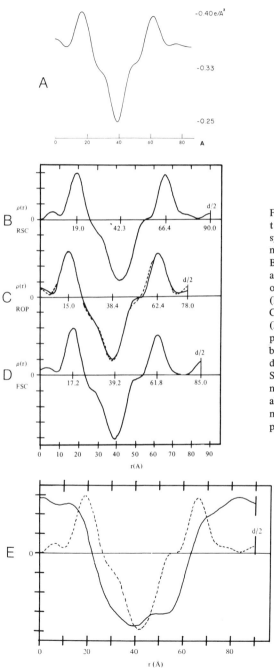

Figure 23. Comparison of electron density plots (Fourier synthesis) for (A) frog sciatic nerve myelin [Unpublished data of Blaurock and Wilkins], (B), (C), and (D) rabbit sciatic (RSC) and optic (ROP) and frog sciatic (FSC) nerves, respectively [After Caspar and Kirschner, 1971]; and (E) neutron scattering density plot for rabbit sciatic nerve equilibrated with $D_2O$. [Unpublished data of Caspar, Kirschner, Schoenborn and Nunes]. Ordinates are in arbitrary units, while abscissas are in A, covering half a membrane pair with origin at the pair center.

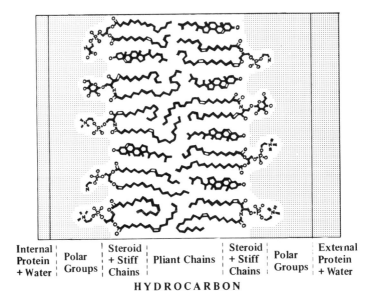

| Internal Protein + Water | Polar Groups | Steroid + Stiff Chains | Pliant Chains | Steroid + Stiff Chains | Polar Groups | External Protein + Water |

**HYDROCARBON**

Figure 24. Diagram showing arrangement of typical lipids of a myelin bilayer relative to aqueous protein layers in a single membrane half of the myelin period. Note that twice as many cholesterol molecules are present toward the external protein as there are toward the internal polar layer. [After Caspar and Kirschner, 1971]

perhaps twice as many cholesterol molecules are present in the halves of the lipid bilayers that are adjacent to the external appositional plateaus as are present toward the cytoplasmic plateaus. Electron-density plots for artificial bilayers made of lipids, including cholesterol, extracted from erythrocytes show similar shoulders, though, by assumption, they are symmetrical in this case (Rand and Luzzati, 1968). Also, Levine (1970) and Levine and Wilkins (1971) have shown that in the profile of aqueous egg lecithin/cholesterol mixtures (1:1 mole ratio), similar shoulders appear that reflect the presence of cholesterol. As discussed below, the asymmetry in myelin lipids may result from an asymmetry of the protein layers.

The nature of the plateaus at both cytoplasmic and external appositions has been confirmed in several ways by the structural studies. Swelling is evidence that aqueous solutions part the external appositions, and this indicates the presence of hydrophilic chemical substances there. Blaurock (1971) reports that the cytoplasmic appositional plateau increases in thickness in frog sciatic nerve by as much as 8 A when salt solutions are applied, also indicating the hydrophilic nature of this plateau.

Perhaps the most dramatic evidence of the hydrophilic character of the appositional plateaus is provided by the studies of Caspar and
his co-workers* in which neutron diffraction was determined with
nerves that had been exposed to heavy water permitting substitution of
exchangeable hydrogens by deuterium (Figure 23E). Although X-rays
are scattered equally and relatively weakly by hydrogen and deuterium
as compared to other elements, the situation is quite different with
neutron waves, which are scattered to about the same degree by all
elements commonly found in organic substances; deuterium particularly becomes on a par with C, O, and P, although it is somewhat less
effective than N. Neutron-scattering density plots, analogous to
electron-density plots, show marked accentuation at the appositional
plateaus, indicating that the exchangeable hydrogen replaced by deuterium is located there, as well as possibly the protein N, also a good
neutron scatterer; the C and unexchangeable H of the hydrocarbon
layers show relatively depressed scatter (indeed, their neutron-scattering
factors are known to be opposed in sign, hence partially antagonistic);
and the major electron-density peaks between troughs and plateaus are
no longer in evidence because the P atoms are not exceptionally good
neutron scatterers. The general interpretations of the chemical distribution of matter through the myelin layers seem well supported by the
neutron studies.

**Correctness of the Proposed Models**

The correctness of the models (electron-density plots) that have
been derived from the small-angle X-ray diffraction studies of myelin
depends most crucially on whether the right phasing of the several
diffraction orders has been achieved. Considerable confidence that this
has been done satisfactorily is derivable from the facts (a) that the
majority of the independent investigations that have considered the
phasing problem have resulted in agreement, at least for the major first
five orders (Moody, 1963; Finean and Burge, 1963; Worthington and
Blaurock, 1969a,b; and Caspar and Kirschner, 1971); and (b) that these
investigations have invoked several different approaches: considerations
based on strip models, on the results of swelling experiments, on the
consequences of the near centrosymmetry of the membrane bilayers,
particularly in CNS (optical) nerves, and on combinations of these. The

*D.L.D. Caspar, D.A. Kirschner, B. Schoenborn, and A. Nunes. Unpublished data on neutron
diffraction analysis.

experimental data, when manipulated according to theoretical expectations, fit into the anticipated behavior so well that they may be said to
confirm the underlying assumptions to approximations suitable for
justification of the procedures used. This is most certainly true for the
first five orders that determine the major features of the final electron-
density plots. The studies that have employed swelling (most recently,
Blaurock, 1971) or have taken advantage of the near-symmetry of the
individual membranes of the membrane pair (Caspar and Kirschner,
1971) depend rather directly upon the diffraction data. Because they
invoke rather indisputable and relatively minor initial assumptions, they
may be said to have determined unequivocally the gross features
regarding the relative positions of dense (polar) and less-dense (apolar)
layers through the membrane pairs.

The finer details introduced by assignment of phases to the
orders beyond the fifth and the introduction of these higher terms into
the electron-density plots are more open to dispute. At the higher
diffraction angles the loops of the membrane pair transform become
less clearly defined and more dependent on minor variations in the
membrane pair structure during swelling or between different types of
nerve (Caspar and Kirschner, 1971). Nevertheless, the electron-density
plots obtained in the two careful independent investigations, which
have utilized the higher terms, are so similar that a reasonable degree of
confidence can be felt in these results. These investigations develop
polarity (Figure 23) in the individual membranes at the lipid parts of
the bilayer as well as at the polar cytoplasmic and external surfaces. On
the other hand, refinements based on strip models tend to assign the
polarity to the polar surfaces alone (King and Worthington, 1971),
possibly a residual effect of an initial assumption of a symmetrical
arrangement of strips to represent each bilayer. The methods that
develop asymmetry of lipid distribution depend less on a priori
assumption of specific models and would seem to be more reliable for
the finer details.

Patterson plots have played a supportive role, as shown in
Figures 21 and 22, in the derivation of myelin models. It must be
noted, however, that the results obtained by this route are less
definitive. The theoretical fact is that any combination of phases used
with observed diffraction amplitudes would produce a model consistent
with the Patterson plot obtained through use of the same diffraction
intensities. Patterson plots are uniquely interpreted in the cases of
crystals containing relatively simpler molecules in which a few

unusually heavy atoms stand out in each unit cell and can be distinguished as more or less isolated components. Then the image-seeking process of Patterson-plot interpretation provides clues as to where the heavy atoms and the atoms of some of their environment may be located in the unit cell so that a start on the structural elucidation is secured. Some conception of the reliability of the structure then emerging is derived from knowledge of what expectations are reasonable for the known chemical structure of the molecules involved. With effectively continuous electron distributions such as are involved in the case of myelin, a priori knowledge of chemical distribution in the structure is only indirectly available; indeed, ideally one would avoid use of Patterson plots entirely in such cases to avoid a degree of circularity of argument.

Not all investigators agree on the phase assignments, even for the first five orders. Akers and Parsons (1970), for example, have concluded that *all* orders including the ones of higher index should be given positive signs. Crystallographic experience generally is against the probability that such a one-sided distribution of signs will exist; usually, more nearly equal numbers of positive and negative signs are encountered in a series of terms for the development of electron-density plots in centrosymmetric structures. Nevertheless, it is possible that an unusual situation is present in myelin.

The Akers-Parsons development is based on diffraction studies of frog sciatic nerves that have been subjected to light incremental staining by $OsO_4$, $PtCl_4$, and $KMnO_4$. These treatments caused all orders (1,2,3,4,5,6,8, and 11 were observed) except the sixth to increase in intensity as the time of exposure to the stain was prolonged. Patterson functions suggested that the stain was entering at the cytoplasmic center of symmetry (origin) and near the external apposition (cell center). A model with Gaussian-type distributions was assumed for the stain distribution at these locations, with parameters introduced for the stain peak heights and widths and for the accurate location of the near-central peak. A computer program, testing all phase permutations as well as variations in stain parameters and focused on minimizing a residual factor that measured the ability of a given variant model to explain the observed intensities, was used to determine the best phase assignments, all of which turned out to be positive.

The resulting electron-density plot differs quantitatively at many points from those shown in Figure 23, but it is particularly distinguished by the introduction of new density peaks midway between the cytoplasmic and external centers of symmetry rising out of the centers of the troughs of Figure 23. No specific compositional distribution is offered to explain this new feature, although it is suggested that proteolipid, rather than lipid alone, might be located at membrane centers.

The complexity of the Akers-Parsons procedure makes critical analysis of its validity difficult. One notes, however, that the use of all positive phases results in the development of an electron-density distribution resembling in shape the Patterson plots given for the same specimens. This is a consequence of the fact that Patterson plots incorporate all diffraction intensities with positive signs, while their density plots use amplitudes (square roots of intensities) in the same way. Since the Patterson plots were initially used to suggest the Akers-Parsons variable model (location of stains), there may be a degree of circularity in the process (for reasons already mentioned), with the computation "zeroing in" to the one of many possible solutions that has been subtly selected initially.

One understands from the diffraction theory that if a computer is given a program allowing it to determine all possible structures that agree with a certain set of $n$ diffraction intensities, it will discover $2^n$ solutions. An arbitrary initial limitation on the models considered will have the likely result of limiting the number of solutions discovered by the computation. Although Akers and Parsons did locate their stain where electron micrographs indicate it to be, namely, at cytoplasmic and external appositions, it is well known that staining at the levels achieved in electron micrographs results in diffraction intensities quite different from those observed under the light staining conditions employed by Akers and Parsons. It follows that there is no satisfactory means for determining a priori how the stain enters under their conditions, so it is difficult to present to a computer a satisfactory initial model for optimization.

Worthington (1970) also has criticized the Akers-Parsons procedures. One concludes that models of the kind presented in the electron-density plots of Figure 23 are based on fundamental grounds, involve a minimum of a priori assumptions, and emerge consistent with the other optical, chemical, and electron-microscopic evidence. Consequently they can be accepted with a considerable degree of confidence.

## The Proteins of Myelin

Although one can visualize the general structure of the nonpolar lipid bilayers to some degree, even to the prevalent locations of phosphate groups of their polar heads, the density plots give very little specific information about the proteins, and possibly also polysaccharides, located in the appositional plateaus. The only indication in diffraction relative to the protein in myelin may be a diffuse meridionally accentuated ring occurring at angles corresponding to a spacing of 12 A reported by Blaurock and Wilkins.* The meridional (axial) accentuation, like that of the 4.7 A hydrocarbon ring, indicates structure in cylindrical planes normal to nerve radii. These investigators believe that the 12 A spacing should be assigned to protein because it was not found in lipid emulsions; a spacing of this general size, though somewhat smaller, was found in other membranes. It is perhaps surprising that the phosphate heads of the phospholipids, whose spacings within tangential layers at the bilayer boundaries could also diffract with axial accentuation, should not be in evidence in view of their domination of many other features of the radial diffraction (sharp as well as diffuse) by myelin, as discussed above.

Finean (1965) also mentions a 10 A ring and a faint, meridionally intensified ring at about 3 A. The 10 A ring may correspond to that described by Blaurock and Wilkins, while the 3 A ring is probably ascribable to the connective tissue present in many peripheral nerves.

### Chemical Studies of Myelin: W. T. Norton

Chemical study of the components of myelin has been greatly facilitated in recent years by the development of methods for the isolation of relatively pure myelin (Autilio et al., 1964). The procedure consists of several cycles of osmotic shock and homogenization of nerve tissue in 0.32 M sucrose, centrifugation over 0.85 M sucrose, collection of material at the interface, and washing and pellet concentration by centrifugation. Centrifugation on a density gradient of sucrose yields two major fractions, termed "light" and "heavy," differing moderately in composition: the heavy one has 5% of chloroform-methanol (CM) insoluble residue, 23% proteolipid protein, and 72% lipid (by dry weight); the light one has less of the CM-insoluble residue and proteolipid protein (1.0% and 21%, respectively) and more lipid (78%). The fractions are believed to be better than 95% pure as judged by

*Unpublished data.

morphology in the electron microscope and by comparisons of the nucleic acid content of myelin fractions and the original tissue source.

In contrast to myelin in the intact nerve tissue, the isolated myelin shows little residue insoluble in the standard CM extractants ($CHCl_3$ : $CH_3OH$ = 2:1 by volume, equimolar for both solvents); some purified myelins dissolve completely. Consequently, these preparations offer ideal starting material for separation and characterization of myelin components.

In the following paragraphs, summaries of the known facts regarding the three major protein-containing components of myelin are presented. Although the total myelin is, in a sense, a lipoprotein complex, the isolated components differ according to the conditions for the removal of lipids from the proteins. Two of the major fractions can be isolated with lipid still attached to the protein, giving the complex solubility properties characteristic of lipids. Consequently, they have been termed proteolipids.

### The Folch-Pi-Lees Proteolipid (FLPL): J. Folch-Pi

This protein is soluble in the standard CM extractant and can be isolated from smaller myelin components by various procedures such as shaking with water, concentration of the FLPL-containing chloroform phase by evaporation to dryness, and removal of excess lipid with ethyl ether and ethanol extraction; dialysis of the CM solution against CM; or chromatography of CM-water emulsions on silicic acid columns (Folch and Lees, 1951; Folch-Pi, 1966). The various preparations differ in the amounts of lipids retained on the protein: the lipids removable by ethyl ether (cholesterol) and ethanol (cerebrosides and some phosphatides) are least firmly bound, presumably by van der Waals forces, to the hydrophobic protein side chains; those removable by silicic acid chromatography, chiefly phosphatidyl serine and phosphatidyl ethanolamine, do carry charged atoms that result in moderately firm ionic linkage to the protein; and, finally, phosphoinositides, which are highly negatively charged, that require acidification of CM solution with HCl for removal. The protein obtained when all lipids are removed is insoluble in organic and aqueous solvents once it is dried but can be obtained in clear solution by dialysis against CM and CM-HCl solvent; finally, after gradual increase in aqueous content of the dialysate, the protein remains dispersed in water (Tenenbaum and Folch-Pi, 1966). CM solutions show optical rotation consistent with a high helix content, but in aqueous environments the helix content disappears. The protein is precipitated at alkaline pH's and redissolves in acid, although irreversible changes occur in alkali or shortly after lyophilization, whereupon acidic urea or CM (pH 3) is required for redispersal. The FLPL protein is attacked by pronase but is resistant to trypsin and pepsin. Thorun and Mehl (1968) measured a molecular weight of 34,000 using gel electrophoresis on an acrylamide density gradient, while the minimum molecular weight consistent with the least abundant amino acid residue is 12,500, raising the possibility that the molecule is trimeric.

### Wolfgram Proteolipid (WPRL)

This proteolipid dissolves, along with FLPL, in acidified CM (HCl, pH 2), but it can be precipitated from solution by adjustment to pH 5 (Wolfgram, 1966). The two proteolipids are also distinguishable with respect to behavior upon application of the detergent-salt mixture, 5% triton X 100, 0.5 M ammonium acetate, in which the WPRL is insoluble while the FLPL is soluble (Eng et al., 1968). The amino acid compositions of the protein moieties are also significantly different (see below) and the protein of WPRL is distinguished by susceptibility to attack by trypsin. Because the WPRL has been known for a comparatively short time, it has not yet received so detailed a characterization as the FLPL.

### Encephalitogenic Basic Protein (EBP): E. H. Eylar

While undoubtedly involved with lipids in myelin, EBP is usually isolated by means that directly produce the essentially lipid-free protein: the nerve tissue is excised and quickly frozen (to avoid autolytic degradation of the protein, which is very susceptible to proteolytic enzymes); defatting is done with CM after lyophilization and blending with methanol at low temperature; acid extraction of the residue (HCl, pH 2.1) yields a supernatant from which a precipitate at pH 7.0 ($NH_4OH$ addition) is separated; the supernatant is passed through a diethylaminoethyl (DEAE)-cellulose column, the eluate from which contains the EBP (Eylar et al., 1969). The undegraded molecule (A1) shows a molecular weight of 16,400 to 18,600 by sedimentation equilibrium, and it has a high intrinsic viscosity (9.3 ml/g), suggesting that it is highly unfolded (Eylar and Thompson, 1969; Oshiro and

Eylar, 1970). More recent studies indicate the presence of 170 residues per molecule, whose total sequence of amino acid was determined and was found to have a molecular weight of 18,400 (Eylar, 1970). About 25% of the residues are basic and the isoelectric point is over 12. There is no glycoprotein, no disulfide cross-links, no phosphorus, and no appreciable helical or $\beta$ structure in solution.

The molecule is thermally stable, retaining its notable EAE antigenicity after 1 hour at 100°C, or 8 hours in 8 M urea, or incubation at pH 10 for 8 hours. Peptic peptides of 14 residues and 22 residues in size retain the antigenic activity for induction of EAE; these peptides have been synthesized and the essential residues defined (Eylar et al., 1970). Modification of a single tryptophan residue abolishes biological activity.

The basic proteins of the CNS and PNS differ in number, composition, and activity (Eylar, 1970). The PNS myelin apparently has two basic proteins or peptides with molecular weights near 6,000 and 11,000. These proteins show some biological similarity to the CNS factor because they induce EAE, but they may also be the factors that induce EAN as well. Aside from the fact that the basic proteins may be exclusively present in myelin (i.e., not in other membranes), considerable biological and medical interest stems from their use as experimental models for human demyelinating diseases (such as multiple sclerosis) and delayed hypersensitivity mechanisms in which animals react to their own proteins when these are displaced from normal sites into the circulation.

### Comparison of Myelin Proteins: J. Folch-Pi

Comparison of the myelin proteins in terms of differences between CNS and PNS amino acid composition can be made by reference to Table 18, which is a composite of analytical data derived from the pertinent information supplied by Wolfgram and Rose (1961), Nakao and his co-workers (1966), Tenenbaum and Folch-Pi (1966), Wolfgram (1966), Eng and his co-workers (1968), Wolfgram and Kotorii (1968a,b), and Eylar and Thompson (1969). The data apply to bovine tissues, for which the most complete information is available. Unfortunately, many analyses do not report amide ammonia, so the averages in Table 18 preserve the proportions of amide to acidic residues originally given by Wolfgram and Rose (1961), Nakao and his co-workers (1966), Wolfgram (1966), and Eylar and Thompson (1969).

TABLE 18

Amino Acid Composition of Myelin Protein (mole percent) (1,3,5,6)*

| Residues | CNS FLPL (1,3, 5,6)* | PNS FLPL (1,6)* | CNS WPRL (4,5)* | PNS WPRL (4,5)* | CNS EBP (2,5, 6,7)* | PNS BP (6)* | Total Myelin Prot-CNS (6)* | Total Myelin Prot-PNS (5)* |
|---|---|---|---|---|---|---|---|---|
| *Nonpolar* | *45.7* | *48.5* | *37.0* | *39.0* | *32.7* | *37.3* | *38.3* | *43.0* |
| Glycine | 10.6 | 8.8 | 8.4 | 9.2 | 14.0 | 9.4 | 10.1 | 9.5 |
| Alanine | 12.0 | 8.9 | 8.4 | 7.6 | 7.9 | 7.1 | 9.7 | 7.4 |
| Valine | 6.9 | 10.3 | 6.2 | 8.7 | 2.5 | 7.6 | 5.4 | 10.3 |
| Leucine | 11.1 | 14.3 | 9.5 | 8.3 | 6.2 | 8.4 | 9.1 | 9.8 |
| Isoleucine | 5.1 | 6.2 | 4.5 | 5.2 | 2.1 | 4.8 | 4.0 | 6.0 |
| *Hydroxyl-Containing* | *14.0* | *13.3* | *12.3* | *12.0* | *13.8* | *15.8* | *15.3* | *13.9* |
| Serine | 5.6 | 7.8 | 6.6 | 6.7 | 9.1 | 7.6 | 9.3 | 7.8 |
| Threonine | 8.4 | 5.5 | 5.7 | 5.3 | 4.7 | 8.2 | 6.0 | 6.1 |
| *Acidic* | *10.3* | *10.4* | *21.0* | *21.5* | *13.4* | *17.7* | *14.6* | *15.1* |
| Aspartic acid | 4.2 | 4.7 | 8.7 | 9.0 | 6.8 | 8.9 | 6.2 | 6.8 |
| Glutamic acid | 6.1 | 5.7 | 12.3 | 12.5 | 6.6 | 8.8 | 8.4 | 8.3 |
| Amides ($NH_3$) | (8.8) | (6.9) | (8.2) | (7.6) | (8.6) | (8.4) | (8.6) | (7.5) |
| *Basic* | *9.3* | *7.4* | *15.0* | *13.1* | *24.2* | *17.4* | *14.6* | *13.6* |
| Lysine | 4.5 | 2.6 | 7.2 | 6.1 | 8.8 | 10.7 | 6.2 | 6.4 |
| Histidine | 2.2 | 2.1 | 2.2 | 2.4 | 5.6 | 1.8 | 3.1 | 2.4 |
| Arginine | 2.6 | 2.7 | 5.6 | 4.6 | 9.8 | 4.9 | 5.3 | 4.8 |
| *Aromatic* | *14.1* | *14.1* | *7.6* | *8.0* | *7.7* | *7.5* | *9.4* | *9.8* |
| Phenylalanine | 8.0 | 7.6 | 4.7 | 4.8 | 4.6 | 4.8 | 5.6 | 5.6 |
| Tyrosine | 4.5 | 4.4 | 2.7 | 2.9 | 2.2 | 1.5 | 2.5 | 3.3 |
| Tryptophan | 1.6 | 2.1 | 0.2 | 0.3 | 0.9 | 1.2 | 1.3 | 0.9 |
| *Sulfur-Containing* | *3.9* | *3.6* | *2.2* | *2.7* | *1.4* | *0.8* | *3.4* | *1.5* |
| Cystine/2 | 2.9 | 1.7 | 0.3 | 1.1 | 0.1 | 0.0 | 2.0 | 0.4 |
| Methionine | 1.0 | 1.9 | 1.9 | 1.6 | 1.3 | 0.8 | 1.4 | 1.1 |
| *Amino Acids* | *2.7* | *2.7* | *4.9* | *3.7* | *6.8* | *3.5* | *4.4* | *3.1* |
| Proline | 2.7 | 2.7 | 4.9 | 3.7 | 6.8 | 3.5 | 4.4 | 3.1 |
| Average Residue Weight | 111 | 108 | 110 | 109 | 108 | 109 | 107 | 108 |
| Acids less amides | −1.5 | −3.5 | −12.8 | −13.9 | −4.8 | −9.3 | −6.0 | −7.6 |
| Bases less histidine | +7.1 | +5.3 | +12.8 | +10.7 | +18.6 | +15.6 | +11.5 | +11.2 |
| *Net Charge* | +5.6 | +1.8 | ± 0.0 | −3.2 | +13.8 | +6.3 | +5.5 | +3.6 |

*Sources: 1. Wolfgram and Rose, 1961.      5. Eng et al., 1968.
          2. Nakao et al., 1966a,b.        6. Wolfgram and Kotorii, 1968a,b.
          3. Tenenbaum and Folch-Pi, 1966.  7. Eylar and Thompson, 1969.
          4. Wolfgram, 1966.

547

The amide for PNS basic protein has not been reported; the estimate in Table 18 takes account of the usual decline in the proportion of amide to acidic residues observed generally in passing from the central to peripheral tissues with the proteolipids.

Eng and his collaborators (1968) determined that bovine CNS myelin contains protein to the extent of about 22% of dry weight, and a little less (19%) is found in bovine sciatic myelin. Of this total myelin protein, 29%, 54%, and 17% are, respectively, derived from basic, Folch-Pi-Lees proteolipid, and Wolfgram proteolipid proteins in CNS, while the corresponding figures for bovine PNS myelin (sciatic) are 21%, 23%, and 55%. The figures indicate the general tendency for FLPL to decline, with replacement by WPRL, when one passes progressively from central to peripheral nerve tissues (Folch-Pi, 1966). The amide figures for total myelin protein in Table 18 are estimates obtained by combining figures for the individual proteins in the proportions suggested by Eng and his co-workers (1968).

The survey of Table 18 shows that while the CNS and PNS versions of a given type of myelin protein vary significantly with respect to specific residues (hence require tissue-specific templates for their biosynthesis), the proportions of the several side-chain categories are quite typical and relatively constant within each kind of protein whatever the tissue origin. Both CNS and PNS FLPL proteins are distinguished by high contents of nonpolar, aromatic, and sulfur-containing side chains, possibly corresponding to their resistance to peptic and tryptic digestion (responsible for the classical neurokeratin formation). Both WPRL proteins are moderately low in hydroxyl-containing side chains and notably strong in acidic and amide side chains. Both basic proteins are low in nonpolar, aromatic, and sulfur-bearing side chains (corresponding to ready attack by proteolytic enzymes), high in basic side chains, and moderately well provided with proline (which is resistant to the formation of regular main-chain conformations of the common types, $\alpha$ and $\beta$).

Despite the unsatisfactory situation relative to amide analyses, an attempt has been made in Table 18 to draw up a charge balance sheet at neutral pH. It then appears, not unexpectedly, that the basic proteins have a large excess of positive charges corresponding to the high isoelectric point. The FLPL proteins are modestly charged on the positive side, whereas the WPRL proteins may be neutral or even modestly negatively charged. The proteolipid solubilities depend on lipid content as well as protein, but it may be noted that the apparent relative net charges of the proteins are in line with the fact that the WPRL precipitates from CM solutions at slightly acid pH, while the FLPL remains soluble in neutral CM solvent. The net charge of the total myelin protein also seems to be definitely positive, which may contribute to the complete solubility of this mixture in neutral CM prior to "cracking" by treatment with acid.

Except for the plentiful basic side-chain components of the basic proteins, comparable with those of histones, the only outstanding characteristic of the three myelin proteins is their high content of nonpolar side chains, which are at high-ranking levels compared to other common proteins. Glycine is unusually prevalent, except in comparison with rather special proteins such as the collagens, elastins, and fibroins. It is notable that the myelin proteins also maintain a good level of hydrophilic side chains. They should be capable of considerable flexibility (because of glycine), and with an appropriate configuration they can present either hydrophilic or hydrophobic surfaces, or both, to neighboring substances. The rather chameleon-like solubility properties of the FLPL protein have been described above.

## Lipids

The lipids of myelin provide the bulk, about three-quarters by weight, of its substance. Table 19 provides a summary of those constituents that occur in more than trace amounts, permitting a comparison of bovine CNS and PNS myelin as well as indicating some of the special features that characterize the acyl chains of the glycerophosphatides and sphingolipids (sphingomyelin and the cerebrosides). Comparable information regarding the acyl groups of bovine CNS lipids is not available, and, instead, the data for human CNS myelin are incorporated in Table 19.

In terms of their content, cholesterol, and particularly the cerebrosides, are unique to nerve myelin among membranous structures (O'Brien, 1967). These constituents, which provide together 52% to 65% of the lipid molecules of myelin, appear to be more prevalent in CNS than in PNS myelin. Glycerophosphatides account for about 30% of the molecules, slightly more in PNS and a little less in CNS myelin. According to O'Brien and his co-workers (1967), there is evidence that in a number of species and in several central and peripheral nerve tissues, a characteristic difference between CNS and PNS is seen in the

TABLE 19

Comparisons of Lipids of Spinal Root (PNS) and Brain (CNS) Myelins

| | Bovine PNS Myelin (a) | Bovine CNS Myelin (b) | Human CNS Myelin (c) | Bovine PNS Acyl Groups (d) Average Number C Atoms | Aldehyde | Unsaturated (Mole % in Lipid) (f) | OH-substitution (f) | Human CNS Acyl Groups (e) Average Number C Atoms | Aldehyde | Unsaturated (Mole % in Lipid) (e) | OH-substitution (g) |
|---|---|---|---|---|---|---|---|---|---|---|---|
| Total Lipids | 75.9 | 75.3 | 78.7 | | | | | | | | |
| | (% dry weight) | | | | | | | | | | |
| | Mole % in Total Lipid | | | | | | | | | | |
| Neutral Lipids | *51.1* | *62.5* | *57.6* | | | | | | | | |
| Cholesterol | 40.9 | 44.5 | 40.4 | | | | | | | | |
| Cerebrosides | 10.2 | 18.0 | 15.7 | 22.1 | | 21.6 | 35.2 | 23.5 | | 46.3 | 55 |
| Ceramide | | | 1.5 | | | | | 20.3 | | 39.2 | tr. |
| Amphoteric Lipids | *38.7* | *29.5* | *33.6* | | | | | | | | |
| Choline GP (h) | 10.0 | 8.5 | 8.4 | 17.2 | | 45.3 | | 17.2 | 1.0 | 53.4 | |
| Ethanolamine GP | 12.5 | 14.7 | 11.8 | 18.0 | 39.3 | 57.1 | | 17.6 | 50.3 | 59.2 | |
| Sphingomyelins | 12.9 | 5.5 | 4.4 | 22.3 | | 34.0 | | 21.3 | | 47.2 | |
| Uncharacterized | 3.3 | 0.8 | 9.0 | | | | | | | | |
| Acidic Lipids | *10.2* | *8.0* | *8.8* | | | | | | | | |
| Serine GP | 6.9 | 5.0 | 5.3 | 17.8 | 14.6 | 64.3 | | 17.9 | 36.5 | 52.6 | |
| Cerebroside Sulfate | 1.3 | 2.5 | 3.5 | 22.6 | | 15.2 | 32.0 | 23.1 | | 51.1 | 20 |
| Inositol GP (i) | 2.0 | 0.5 | | | | | | | | | |
| Average Paraffin Chain: | | | | 17.9 | | 58.9 | | 18.2 | | 64.2 | |

(a) Data of O'Brien et al., 1967.
(b) Data of Norton and Autilio, 1966.
(c) Data of O'Brien and Sampson, 1965a.
(d) Calculated from tables of O'Brien et al., 1967.
(e) Calculated from tables of O'Brien and Sampson, 1965b.
(f) Figures give percent of acyl groups bearing the indicated properties: aldehydogenic functions (plasmalogens), one or more unsaturated carbon-carbon bonds, and hydroxyl substitution (phrenasins). The complements of the figures, to 100, give the percentages carrying the opposite properties: normal acidic functions, saturation, and nonsubstitution. Unsaturated figures include both the normal and aldehydic or OH-substituted chains where pertinent.
(g) OH-substitution values from O'Brien et al., 1964.
(h) GP = glycerophosphatides.
(i) PNS value from O'Brien, 1967.

sphingolipids, in that peripheral myelin contains a higher ratio of sphingomyelin to cerebrosides than does central myelin, though the sum of the two remains about the same.

### Acyl Groups: J. S. O'Brien

Table 19 gives average acyl group C atom numbers as an approximate indication of the relative lengths of the paraffin chains of the typical sphingo- and phospholipid molecules. The former are as much as 4 to 5 carbons longer than the latter, though it may be noted that the total bulk of the sphingolipid nonpolar volume is reduced to near equality with that of the typical phospholipid, because with each acyl group of a sphingosine moiety there is a shorter, singly unsaturated paraffin chain only 15 C atoms in size (not counting 3 C atoms comparable to those in the glyceryl moiety of other lipids) that is not included in the acyl averages.

The glycerophosphatides are the bearers of the most unsaturation. As it is often stated that the characteristic property of myelin-form formation by lipids requires some unsaturation in the paraffin chains, one may look to these as sources for much of the liquidity of the nerve myelin. Sphingomyelin, the cerebrosides, and, of course, cholesterol, are of relatively less significance in this respect, tending instead to add lipid stability (O'Brien, 1965).

The ethanolamine and serine glycerophosphatide fractions (both cephalins) bear the most plasmalogen components, while more than half of the cerebroside acyl groups have OH-substituents (presumably the α-hydroxy substitution in the cerebronic acid component typical of phrenosins). The oxygens of the aldehydogenic chains of the phosphatidal ethanolamines and serines would be poorer hydrogen-bond formers than the ester links of the normal fatty acid-derived acyl groups. On the other hand, the substituted hydroxyls add more hydrogen-bond-forming ability to the cerebrosides. The information concerning these variations could be of significance relative to the junctions between nonpolar and polar regions of the myelin structure.

Average values for acyl group properties are useful in judging the overall bulk properties of the several lipids: for example, as shown above, except for cholesterol it may be said that because average molecular sizes of the paraffin chains for all molecules are not too different, it should be possible to fit them into a nonpolar bilayer of a given thickness equal to twice the average paraffin chain length.

However, longer chains will have to be generally apposed to shorter
ones. Indeed, the chain lengths vary quite considerably in the lipid acyl
groups. O'Brien and his collaborators (1967) provide analyses that show
the following facts about chain lengths, expressed as *n:m*, where *n* is
the number of C atoms and *m* the number of unsaturations. The
principal components in moles percent of the several lipid acyl fractions
of bovine PNS myelin are as follows:

   *Phosphatidal ethanolamines:* (16:0), palmitaldehyde, 19.0%
(18:0) stearaldehyde, 34.2%; and (18:1) octadecenaldehyde, 34.1%.
hyde, 34.1%.
   *Phosphatidyl ethanolamines:* (16:0), palmitic acid, 21.0%
(18:1) oleic acid, 42.0%; but with appreciable amounts of unsaturated
longer chains rising to (22:5).
   *Phosphatidal serines:* (16:0) 25.3%; (18:0), 35.9%; and (18:1),
17.1%.
   *Phosphatidyl serines:* (18:0), stearic acid, 17.6%; (18:1),
63.5%; and again with some very long unsaturated chains.
   *Phosphatidyl cholines:* (16:0), 40.0%; (18:0), 11.0%; (18:1),
36.0%; with a few unsaturated chains as long as (20:4).
   *Sphingomyelins:* (18:0), 12.0%; (22:0), 15.0%; (24:1) nervonic
acid, 31.0%; (24:0) lignoceric acid, 22.0%; and significant proportions
as long as (26:0) and (26:1).
   *Unsubstituted cerebrosides:* (22:0), 10.8%; (24:1), 23.6%; and
(24:0), 33.5%, and significant proportions as long as (26:0) and (26:1).
The sulfates show similar predominant acyl groups.
   *Hydroxy-substituted cerebrosides:* (18:0), 21.2%; (22:0),
13.2%; (24:1), 8.0%; and (24:0), 39.3%; also (25:0) is present, and
significant proportions as long as (26:0) and (26:1). The same principal
acyl groups are present in the hydroxylated sulfates.

   While noting that central-peripheral comparisons are not yet
possible in the above analytical detail for any species other than bovine
(thus bovine comparisons cannot be generalized), O'Brien and his
co-workers (1967) did venture to use the experience of O'Brien and
Sampson (1965a) with human CNS myelin to pick out the CNS-PNS
differences in bovine tissue that may be significant. They noted the
overall similarities in the fatty acids of the several myelins, but pointed
to one specific CNS-PNS difference: peripheral choline GP contains
5.6% of linoleate (18:2), whereas the same central lipid contains less
than 0.5%. This difference, though relatively minor in the amounts
involved, is interesting because linoleate is supplied directly from

dietary sources, and this linoleate comparison may reflect the influence of the blood-brain barrier acting against incorporation of dietary lipid into central myelin. Sphingolipids in the CNS contain more (5% to 20%) of the $C_{25}$ to $C_{26}$ fatty acids than are found in the same PNS lipids (less than 2%). Correspondingly, the PNS lipid has more of (22:0) to balance.

### Electrostatic Relationships Between Lipids and Proteins

It is of interest to compare the ionized groups of myelin lipids with those of myelin proteins. At neutral pH, the acidic lipids shown in Table 19 are a measure of the net negative lipid charge, which is about 8 to 10 moles percent (or a little more, because the polyphosphatidyl inositols may carry several phosphates). Using cholesterol content (about 18% of dry weight in PNS and 21% in CNS of bovine myelin) as a reference,* one finds for PNS myelin that 24 g of protein accompanied by 76 g of lipid would provide a system containing 0.022 moles of amino acid residues to 0.049 moles of cholesterol. Noting from Table 18 that 100 moles of amino acid residues contain about 3.6 moles of positive charges and from Table 19 that 40.9 residues of cholesterol are associated with 10.2 moles of negative charges, one finally obtains a ratio of 0.008 positive protein charges to 0.012 negative lipid charges. Corresponding calculations for CNS myelin yield 0.013 positive against 0.010 negative charges. Because of uncertainties and approximations in the data and calculations, these results cannot be accepted with great assurance, but it is interesting that the PNS net charge is negative while that for CNS is slightly positive. These findings may correlate with the relative swelling properties of lamellae noted in the discussion of X-ray diffraction studies of related structures, i.e., the swelling properties of PNS myelin are those of negatively charged lamellae (see also Electron Microscopy of Myelin, this chapter).

Other lipids, though present in myelin in trace amounts, may have functional and structural significance beyond their scarcity. Notable are the gangliosides which, though sometimes thought to be limited to synaptic nerve components (Lowden and Wolfe, 1964), are believed by some (Suzuki et al., 1968) to be a definite though scarce component of axonal myelin ($50 \mu g$ of N-acetylneuraminic acid (NANA) per 100 mg of myelin compared to 76 mg of total lipid per 100 mg of myelin). The myelin ganglioside is chiefly monosialoganglioside, as compared to polymerized forms found elsewhere. The typical

*Cholesterol has a definite, single molecular weight of 386.64.

myelin ganglioside is, accordingly, a complex molecule consisting of sphingosine with its acyl (typically stearate) and associated 18-carbon monohydroxylated, monoamino, monounsaturated "tail" chain from which extends a sequence of glucose, galactose, N-acetylgalactosamine and sialic acid residues. The sialic acid carries a charged group (–COO–), and sialic acid-containing structures have been featured in recent discussions of membrane surface properties (review by Lehninger, 1968).

## Distribution of Constituents in Myelin

The general distribution of proteins and lipids, even of the specific lipid cholesterol, has been inferred from the electron-density plots of myelin (see section on electron-density distribution). What further can be said of the ways in which various chemical components are built into myelin?

Taking into account all constitutional and structural data then known, Vandenheuvel (1965) made a most ambitious attempt to provide a detailed stochastic model. Starting with the analytical data of Norton and Autilio (1966) for bovine brain myelin, Vandenheuvel calculated that about 20 residue moles of total protein are associated with 5.4 moles of cholesterol and 7.0 moles of other lipid. Finean (1953) had suggested that 1:1 complexes of glycerophosphatide and cholesterol may be present in myelin; this also seemed evident from surface-film studies of phosphatides whose area per molecule is contracted when cholesterol is added in equimolar amounts (van Deenen, 1966). Accordingly, Vandenheuvel (1965) designed models for such complexes, extending them to include sphingolipids. Thus, the 20 moles of protein residues must provide surface for 5.4 moles of complex (at 100 A each), plus 1.6 moles of uncomplexed lipid (at 55 A each) or a little over 30 $A^2$ of lipid per residue. Protein can be spread this thinly only if it is in a $\beta$-like configuration with side chains extending in the surface. The general properties of the myelin proteins described above could readily permit this nonglobular kind of structure, and, indeed, the thinness of space available for each protein layer (about 10 A) indicates it to be fairly well spread out. Vandenheuvel then proposed a model for the myelin layers capable of containing the lipids present and requiring only minor adjustments to agree with the structures subsequently derived from electron-density plots (Figure 22A).

It is difficult, however, to distribute the lipid between halves of the bilayers so that cholesterol is twice as prevalent on one side as on the other if the cholesterol-lipid complexes include all lipids possible. Moreover, the necessity of assuming rigid 1:1 complexes in membranes seems doubtful from the work of Luzzati and his colleagues (for example, Rand and Luzzati, 1968). Their studies favor the view that interactions between lipid and cholesterol molecules are not specific and immobile. However, the insertion of cholesterol molecules so that the steroid nucleus and its hydroxyl will interdigitate with the lipids near their polar heads with the remainder of the nucleus and its hydrocarbon tail in the hydrophobic phase may restrict lipid molecular movement, thus contracting the area per molecule in the membrane surface areas. The detailed model provided was derived from X-ray studies of lamellar structures formed from lipids, including cholesterol, extracted from erythrocytes.

If the general protein conformation is that of pleated $\beta$-chains (Pauling and Corey, 1951), then it follows that the $>$NH and $>$C = O groups of the polypeptide backbones are oriented toward the lipid polar heads or into the aqueous phases, ready for H bond formation. The alternate side chains are displaced slightly toward or away from the lipid phase owing to the backbone pleating; thus, if the protein layers are to present predominantly nonpolar and polar surfaces, the hydrophilic and hydrophobic side chains may be expected to alternate. Amino acid sequence data should reveal the probability of this configuration; the only sequence information now available is that of the basic CNS protein, and it does not notably confirm this view.

Speculation may be made that the basic proteins are concentrated on the cytoplasmic side of each membrane unit. Not only does this location protect the allergenic properties of the protein from exposure to extracellular circulatory space, but it may also be the cause of the strong reaction of osmic acid with cytoplasmic appositions of myelin to form the major dense line of electron microscopy.

Nevertheless, in neither CNS nor PNS myelin is the amount of basic protein sufficient for complete coverage of the cytoplasmic surfaces if one assumes that the cytoplasmic and external surfaces each require about the same amounts of protein. According to the data of Eng and his co-workers (1968), the basic protein could account for only 42% to 58% (i.e., twice their total presence) of the cytoplasmic protein in CNS and PNS myelin respectively. Consequently, the proteolipid proteins will have to be present at both the cytoplasmic and externally connected surfaces.

There are no obvious clues as to how the two proteolipid proteins (FLPL and WPRL) might be distributed between the two surfaces. Wolfgram proteolipid protein at external surfaces would promote the swelling of PNS myelin because of its negative charge. This protein is not present in amounts sufficient to cover the entire external surface in CNS myelin, although it is sufficient in PNS myelin. It may be significant that most studies of myelin swelling have been done with peripheral nerves (sciatics); for example, when Finean and Burge (1963) used rat optic nerve, they found considerable disruption of the myelin, i.e., deterioration without spacing changes, although the small-angle X-ray pattern remained unaffected by long periods of immersion of the nerve in water.

If WPRL protein is predominantly localized at externally connected surfaces, then the FLPL protein would be left as a filler, complementing the role of other proteins. Consequently, it will add positive charges to the cytoplasmic surfaces in addition to those provided by the basic protein, and it may possibly enhance the osmic acid staining. It will be particularly prevalent at the externally connected surfaces of CNS myelin, adding further cause for neutralization of the negative charge and reduction of swelling.

If these speculations regarding protein distribution are realistic, then one can postulate that the cytoplasmic surfaces of myelin protein are always richer in positive charges than are the externally connected ones. One might expect the acidic lipids, and possibly even the phosphate groups of the amphoteric lipids, to gather toward the cytoplasmic surfaces, causing some of the neutral cholesterol and cerebroside molecules to be concentrated at the external surfaces. This may well be the case, at least for cholesterol, as discussed earlier.

If the proteins are to cover the surfaces efficiently, particularly in view of their thin distribution, one might expect that their main chains would run more or less parallel at some definite angle to the fiber axis over wide patches of the surface; to be sure, the amino acid composition indicates that, on the average, the proline content necessitates kinks in the $\beta$-configuration every 20 to 30 residues. The parallel chain patches would, nevertheless, be of sufficient size to yield diffraction evidence of their presence that may be provided by the diffuse 12 A spacing observed by Blaurock and Wilkins.* The size of this spacing is reasonable for parallel chains spaced by the populations

---

*A.E. Blaurock and M.F.H. Wilkins, unpublished data on high resolution profiles of myelin and visual cell disc membranes.

of side chains indicated in Table 18. On the other hand, if this spacing were to be thought of as developed from separations between phosphate groups, the average cross sections expected per phosphorus-containing lipid (55 to 100 $A^2$) would seem to indicate that the spacing would be somewhat smaller than 12 A. Moreover, no approximately 12 A diffuse spacing has been reported from phospholipid-water mixtures. Although these matters require further investigation, at the present time it would appear that the protein chains cover the lipid rather lightly and are oriented, possibly in patches, in a random relationship to fiber axes.

When one attempts to correlate myelin structure with lipid composition (acyl chain lengths, etc.), it soon becomes evident that the chemical analysts favor data-gathering from the larger animals (bovine and human), whereas the diffractionists find it more convenient to work with nerves of smaller animals. Consequently, the two kinds of information are not available for the same tissues; hence, attempts to explain the different myelin periods reported for PNS and CNS myelin and for the same tissues from different species (see X-ray Diffraction Studies of Living Nerve, this chapter) are handicapped.

If further surveys support the suggestion that myelin periods cluster about values of 156 A, 173 A, and 183 A rather than being continuously distributed in this range, it will become important to determine the structural causes of this phenomenon. These increments of about 10 A and 17 A could be due to *three structural factors:* (1) lipid acyl chains present in various myelins possess lengths that show discrete increments; (2) various myelins contain different configurations, constituents, or numbers of protein layers at their polar phases; and (3) some particular balance of ionic forces at externally connected layers results in equilibria at specific swelling states, discrete aqueous layers of different thickness being maintained.

In regard to the first structural factor, variations of period length amounting to 10 A to 17 A would require average acyl chain lengths of individual lipids to change by a quarter of these amounts, or 2.5 A to 4.25 A. This would be accomplished by changing the average number of C–C bonds by two to almost four (each C–C bond projects on the chain axis by about 1.25 A). It can be readily shown from Table 18 that the average acyl or acyl-like paraffin chains of the bovine PNS myelin lipids, including the nonacyl (base) chains of the sphingolipids, contain almost 18 carbon atoms. A similar computation for adult human CNS myelin from the data of O'Brien and Sampson (1965b) and

O'Brien and his co-workers (1964) yields an average chain length corresponding to a little over 18 carbon atoms. The difference in paraffin chain length is less than 2% and is contrary to the observation that CNS myelin periods are less than those of PNS myelin.

Of course not all lipid tails are relatively "straight"; a few branch, but many are bent at an angle (Vandenheuvel, 1965) because of unsaturation. The average percentages (data of Table 19) of chains that carry at least monounsaturation are 59% in bovine PNS and 64% in adult human CNS myelin. This may be a signficant difference, but comment about it has not yet appeared. It is in the right direction for radial contraction of lipid bilayers in CNS myelin, although quantitative evaluation of the amount of contraction to be expected is not easily achieved.

As the lipid tails flex and contract longitudinally, the area per lipid molecule on the protein surfaces should increase; hence, more protein would be required in CNS than in PNS myelin to provide a comparable surface coverage. Data on this point are equivocal; for example (Table 19), because lipid contents of both CNS and PNS bovine myelins are about the same by weight, presumably protein content would not be very different. On the other hand, the analyses of Eng and his co-workers (1968) directly specify figures of 19% and 22% for the amounts of protein by dry weight in bovine PNS and CNS myelins respectively. A radial myelin shrinkage of 10% requires an equal expansion in protein if lipid volume does not change. The 15% increase in protein content of CNS myelin seems well able to cover the radial shrinkage relative to PNS, although the reading of such differences in this way is not very reliable.

The above explanation of unsaturation for the generally shorter radial period of CNS myelin may make reasonable the configurational property of the lipid constituents, but the remaining aspects of the second structural factor cited, i.e., variation of kinds of constituents or numbers of protein layers, do not seem likely. Variations in the amounts of different lipid categories between CNS and PNS myelin (as shown in Table 19) would not be expected to alter average molecular volumes or lengths greatly; the sphingolipids, whose varying amounts of chain lengths are notable, have similar molecular dimensions, and their total amount remains relatively constant. If whole protein layers were added in PNS relative to those in CNS myelin, increments in protein content as high as 50% would result, which is outside the realm of possibility indicated here for protein variations.

That increments could be due to variations in normal equilibrium swelling states between CNS and PNS myelins (the third structural factor) may have some validity. The failure of CNS myelin to swell under artificial conditions, as mentioned above, suggests that this myelin is incapable of swelling even in vivo; hence, the normal state is compact. The electron-density plots provided by Caspar and Kirschner (1971) for two sciatic nerves (rabbit and frog) and one optic nerve (rabbit) indicate that in both sciatic myelins the externally connected appositions are relatively broad, whereas in the optic nerve both cytoplasmic and external appositions are of about equal shorter breadth. The cytoplasmic appositions are also somewhat broader in PNS than in CNS myelin. The lipid troughs of these plots do not markedly show the effects of CNS unsaturation as expected, though in the rat the troughs are moderately narrower for CNS than for PNS myelin. Consequently, of the two possible explanations for the shorter radial periods of CNS myelin, lipid unsaturation and swelling, the latter would appear to be more significant. Neither of these explanations, however, readily explains the per saltum nature of the period changes if, indeed, this characteristic is upheld by broader surveys of myelin sources.

In this connection, it is of interest that Napolitano and Scallen (1969) observed a splitting of the intraperiod line into two components separated by 20 A out of the total 180 A period in peripheral nerves that had been fixed in buffered aldehyde and 0.2% digitonin (to prevent cholesterol and other lipid loss during manipulation). The same splitting was not achieved in CNS (optic) nerve, again suggesting that the CNS myelin is normally more compact and unswollen.

### Water Content in Myelin

One of the more elusive and uncertain quantities concerning myelin constitution is the water content. Indirect estimates in complex nerve tissues are unreliable, and direct measurements with isolated myelin may not reflect the normal condition. Vandenheuvel (1965) estimated that 39% of his structural model, which is essentially designed for PNS myelin, is available for filling by water. Schmitt and his collaborators (1941) suggested a range of 35% to 50% from comparisons based on water content measurements in CNS tissue and squid giant axoplasm. Finean (1957, 1960) has followed the kinetics of nerve drying, along with observation of X-ray and electron micrographic changes during drying, and favors a water content of 40%. A water content of about 50% may be calculated from the average electron density found for the membrane pair by Blaurock (1971). According to these arguments, which indicate that CNS myelin is more compact than PNS myelin, the former should have less water than the latter.

### Role of Carbohydrates in Myelin

It is believed that cells generally carry a surface coat rich in carbohydrate (Rambourg et al., 1966; Schmitt and Samson, 1969). This coat, which may be important in the phenomenon of cell recognition during histogenesis (Moscona, 1962), may be derived from vesicles provided by the dictyosomes of the Golgi apparatus as they add more membrane area to the cell surface and discharge contents to the exterior (Grove et al., 1968; Rambourg et al., 1969). The carbohydrate-rich components are not necessarily a part of the plasma membrane, but their close association might be expected to produce some attachment to the external protein-polar layer.

Although little is known about the details of myelin membrane synthesis, there are some indications that carbohydrate may be present at the externally connected appositions. From their histochemical studies, Wolman and Hestrin-Lerner (1960) concluded that a "myelosaccharide" is associated with myelin. They thought that it was present at the cytoplasmic appositions, but Finean (1960) has argued that it is external. Robertson (1960a,b,c) pointed out that in certain hypertonic situations the normal gap of about 150 A between closely approaching cell membranes collapses, as though a thin micropolysaccharide gel had suddenly undergone syneresis. The particularly good staining by $KMnO_4$ of the intraperiod lines in myelin electron micrographs might be evidence that the apposed external surfaces contain carbohydrate. However, analyses of myelin isolated from bovine spinal roots failed to detect significant amounts of glucosamine and galactosamine (Wolfgram and Kotorii, 1968b), although very small quantities of gangliosides containing aminohexoses are believed to be present in brain myelin (Suzuki et al., 1968), and these could be arrayed externally to the protein coats and bound to them.

At the present time, conclusions of the kind attempted above regarding the particular locations of specific constituents of myelin cannot be reached reliably, although it may be significant that one can now begin to examine such questions.

## Comparison of Myelin Structure With Other Biological Membranes

Comparisons of nerve myelin with other biological membranes quickly reveal that myelin is unique and has possibly evolved for the special purposes of preventing current losses from the axon and of speeding propagation by nodal saltatory conduction. Myelin contains unusually high proportions of lipids (75%), whereas other membranes rarely have more than 50% (O'Brien et al., 1967). Correspondingly, myelin contains less protein. The latter fact is consistent with the general impression that myelin is less metabolically active than other membranous systems. Adams, Davison, and Gregson (1963) tested CNS and PNS myelins by 24 histochemical methods for enzyme activity and found no oxidative capacity and relatively little hydrolytic ability (some proteinase and moderate alkaline phosphatase activity).

### Structural Evidence: D. Branton and A. E. Blaurock

Electron microscopic examination of freeze-fracture preparations, in which membranes fracture in the plane of the lipid bilayer, reveal that myelin fracture faces are smooth as expected for hydrocarbon faces while other membranes (of retinal rods, chloroplasts, red cells, mitochondria) reveal particles, a cobblestone appearance, or septal ridges, suggesting an invasion of the hydrocarbon phases by particulate matter (proteins?) as shown in Figure 25. Even in the Schwann cell plasma membranes, which are continuous with the compound myelin membranes, fractures uncover rough faces (Branton, 1967, 1971).

Although the myelin-type proteolipids can be found in other membranes, the basic proteins are unique to myelin. That the non-myelin membranes are, nevertheless, somewhat similar to myelin appears from electron-density plots, obtained from X-ray diffraction studies by Blaurock and Wilkins (1969), of the discs stacked in retinal rod outer segments. The density distribution resembles that of myelin in that characteristic lipid troughs are observed, on both sides of which occur high levels corresponding to polar phases. The same system is reported to yield electron micrographic and X-ray diffraction evidence for particles 40 A in diameter arranged on a square net of 70 A sides, although the location of these particles within or at the membranous surfaces of the disc was not determined (Blasie et al., 1965). X-ray diffraction studies by Wilkins, Blaurock, and Engelman (1971) of a broad range of membranes strongly suggest that the electron-density profiles also are similar to the myelin profile. Although chloroplasts yield pertinent X-ray diffraction data (Finean et al., 1953) and show in the electron microscope lamella structure (Hodge, 1959) as well as particulate elements related to the grana layers (Park, 1965) and implicated in the photosynthetic process (Ogawa and Vernon, 1969), detailed structural analyses have not yet been achieved.

Finean and his collaborators (1966) were able to prepare pellets of membranes from erythrocytes, intestinal epithelium, and liver cells that, when properly hydrated, gave small-angle diffraction spacings of 110 A to 140 A and resembled the behavior of myelin upon drying in the separation out of typical lipid phases. No evidence for subunit structure within membrane layers was achieved.

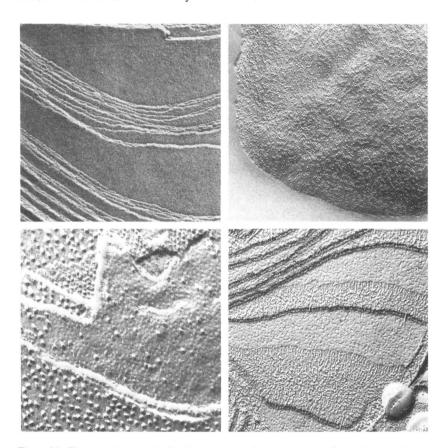

Figure 25. Electron micrographs showing membrane faces exposed by freeze-fracture of nerve myelin, upper left x 100 000 [Branton, 1967]; erythrocyte, upper right x 40,000 [Engstrom, 1970]; chloroplast, lower left x 120,000 [Branton and Park, 1967]; and retinal rod outer segment, lower right x 63,000.[Clark and Branton, 1968]

Chemical Evidence: J. S. O'Brien

The lipids of myelin are also significantly different in kind from those of other membranes (O'Brien, 1967): cholesterol is about equally present in myelin and erythrocyte membranes but is much diminished in mitochondria and absent from chloroplasts; glycerophosphatides are somewhat more prevalent in erythrocytes than in myelin but are more than doubled in mitochondria and almost doubled in chloroplasts; sphingolipids diminish and disappear as one passes from myelin to erythrocytes and mitochondria and are absent from chloroplasts; chloroplasts make up the bulk of their lipids and lipid-miscible constituents by way of some unique constituents, such as galactosyl diglycerides and, of course, chlorophyll and carotenoids.

O'Brien (1965) has also pointed out that myelin contains lipids that, in contrast to those of other membranes, render it more stable and metabolically inert; there are fewer polyunsaturated paraffin chains, more of the long sphingolipid tails, and a higher percentage of lipids stabilizing the structure by hydrophobic tail interactions. Demyelinating diseases tend to accompany deficiencies in the sphingolipids.

Many of the remarkable special properties of cell membranes that have been under considerable investigation recently have not been observed in myelin, or perhaps they are obscured there by the compound nature of the myelin structure. The extraordinary effectiveness and specificity of certain cyclic antibiotic peptides (enniatins, valinomycin, gramicidins, the actin macrotetralides; Lardy et al., 1967; Eigen and De Maeyer, 1971) in inducing or inhibiting ion transport in mitochondria have been ascribed to particular flexible cagelike structures capable of grasping and discharging specific ions before and after transport through a membrane, or even to the possibility that stacks of the ring molecules provide specific channels through membranes. The structure of the ionophore nonactin, as a complex with KSCN, has been worked out crystallographically by Kilbourn and his co-workers (1967).

Certain antifungal polyene antibiotics (nystatin, amphotericin B, filipin, pimaricin) readily penetrate cell membranes or lipid monolayers in which sterols (cholesterol, ergosterol) are present (Demel et al., 1965). Resistant bacteria owe their resistance to a lack of cholesterol in their plasma membranes. Certain fish poisons, such as tetrodotoxin (Moore and Narahashi, 1967), in very small amounts (10 to 15 nM) act to block infrequent $Na^+$ channels at the outside of lobster or squid axonal membranes and prevent excitation. Comparable evidence of gatelike ion-controlling structures in membranes comes from the work of Watanabe and collaborators (1967), who found that squid axon sheaths emptied of axoplasm by extrusion and refilled with electrolyte solutions retain excitability, provided that a suitable ion gradient is maintained across the membrane. Gradients can be maintained with a number of univalent cations, provided that a few divalent cations, such as $Ca^{2+}$, are present on the exterior surface; this presumably prepares certain membrane conformations for exchange with and transport of the univalent ions upon stimulation. Suggestions have been made that membrane changes of state, which are influenced by very few triggering molecules, are of a cooperative sort; thus, the few triggering molecules provide a considerable change in the membrane (Changeux et al., 1967).

The possibility of finding particles or channels that interrupt the layer continuities of biological membranes in such a way as to explain phenomena such as those that have just been described is an elusive one in terms of arriving at definitive structures. It is difficult to apply to a single membrane any of the physical "image-forming" methods (electron microscopy and X-ray diffraction) in sufficient detail. Nevertheless, electron-microscopic evidence for mosaic structure of membrane surfaces has occasionally been observed. For example, Robertson (1963) has described hexagonal patterns of subunits in frontal views of synaptic discs at the club endings of Mauthner cell synapses in goldfish brains. A most extraordinary hexagonal pattern of material applied to the surface membranes of mouse urinary bladder cells has been reported by Vergara and his co-workers (1969). The question that arises in such cases is whether these patterns reflect features of the structure of the underlying membrane or are consequences of the structure of an added surface layer. A bewildering variety of synaptic or close-contact intercellular membrane sub- and superstructures has been reviewed by Robertson (1965). Some of these result in extraordinary reduction of permeability barriers between cells (Loewenstein, 1966; Furshpan, 1964).

Although it seems clear that the relatively simple and stable structure of myelin cannot completely provide a model for other membranes, one gains the general impression that all membranes contain interleaved proteinaceous and lipid layers much like the single membrane half of the myelin radial period, but that active particulate ingredients may be added to the proteinaceous polar layers or even penetrate into the myelin when specific functional requirements develop.

## Future Prospects

It would seem that the general field of myelin structure and composition has achieved a considerable degree of sophistication. The ultrastructural images provided by the electron-density plots of the X-ray diffractionists, assisted by their electron optical and biochemical colleagues, furnish base lines upon which rapid progress toward reliable additional information will develop.

The physical methods should be widened in application to a broader range of nerve tissue. Specifications for the myelin periods will thus be developed to permit examination of the differences between myelins. Past studies have been limited to relatively few sources, chiefly representative CNS and PNS nerves of avian, amphibian, and mammalian animals (vertebrates). Although invertebrate nerves are not usually heavily myelinated, there are a few cases where sufficient myelination occurs for the application of the physical methods. Histological investigations indicated long ago that giant fibers of certain annelids and arthropods (Friedlaender, 1889) and many fibers of prawns and shrimps (Retzius, 1890) are well myelinated. The phylogenetic separation between these fibers and those of the vertebrates suggests that illuminating variations on the myelin theme might be present. Among the vertebrates, a broader sampling of fish specimens seems particularly desirable because of the demonstration of two periods within individual nerves.

It is hoped that the analytical applications will be sufficiently improved to permit gathering of data from smaller quantities of nerve so that information about both physical and chemical aspects from a broader range of sources for the same specific myelins becomes available. Analyses should reveal all details: kinds and amounts of protein and lipid, details of amino acid sequence or acyl chain characteristics, and the possible presence and properties of relatively rare constituents, all done with sufficient precision so that questions of the kind raised in this discussion can be examined with greater confidence.

It is probable that the electron-density plots of myelin will not be improved greatly beyond the resolution already achieved because myelin is a mixed paracrystalline substance that inherently does not have the perfection of atomic arrangement characteristic of crystals. Some alleviation of this difficulty may be found in studies that will modify myelin enough to vary the electron-density plots meaningfully.

These modifications may be obtained by in vivo dietary alterations that could change some lipid components or introduce deuterated acyl groups for neutron diffraction; or lipid or protein variations may be possible in vitro by way of the externally connected appositional channels of swollen myelin.

In this connection, considerable theoretical interest attaches to the phenomenon of "action-at-a-distance" that permits the separated membrane pairs to space themselves regularly and far apart in swollen myelin. This matter deserves careful attention by study in which systematic variations of osmotic pressure, ionic strength, and types of ion are achieved by adding simple anions and cations, or by coating or introducing proteins and lipids with differing net charges or amphoteric groups to the externally connected surface.

To date, studies of myelin swelling have found this phenomenon to be erratic and nonreproducible. The cause is probably the presence of variable amounts of connective tissue that restrict free swelling to equilibrium. Robertson* found that pretreatment of nerves with collagenase and trypsin improves the reproducibility as judged by polarized light observations. Consequently, it seems desirable in investigations of swelling to employ this pretreatment, or to use nerves relatively free of connective tissue, such as spinal roots. It might be possible to induce the myelin of still purer CNS nerves to swell by artificial introduction of anionic lipids.

The general method of varying myelin composition by way of the externally connected channels opened during swelling may have even wider usefulness. One might seek to modify myelin structure in the direction of imitation of other membranes. It is known that artificial bilayers of lipid can be prepared, and that these assume properties of excitability and action potential development much like natural membranes when treated with certain polypeptides and macrocyclic peptides (Mueller and Rudin, 1968). This suggests the possibility that the natural bilayers of myelin might also assimilate the various surface-active substances that affect membrane properties so remarkably. Although relatively minor amounts are required, if the effects are cooperative and change appreciable volumes of structure, these may be detectable by X-ray diffraction. The advantage of using myelin for such studies, rather than making a direct attack with other more active membranes, derives from the well-organized multilamellar nature of

*J.D. Robertson, unpublished observations.

myelin that permits the gathering of better diffraction information regarding structure and structural change.

Now that it is possible to phase the small-angle diffractions of myelin, repeating the different ways in which the structure may be varied from the normal condition (dehydration, temperature change, solvent application, staining) is worth doing again; however, this time the more significant objectives of analysis in terms of specific ways in which the electron density is redistributed should be kept in mind. It would be particularly interesting if lipid-free or nearly lipid-free myelin could be made to diffract X-rays by devising means to remove or displace the lipids less destructively than in the past. The result would be electron-density information regarding the protein alone, and by noting the difference from normal myelin the same knowledge regarding lipid distribution would be achieved.

Fortunately, the lipids and proteins readily self-assemble into lamellar structures similar to that of myelin. The success of Levine and his co-workers (1968) in preparing built-up multilayers of dipalmitoyl lecithin suitable for diffraction study suggests that this kind of effort may be particularly useful in attempts to synthesize model systems having the constituents and structure of myelin. Preliminary work with pure lipids of several kinds, followed by systematic investigation of mixtures, would do much to overcome the difficulty of deriving the details of lipid organization from results on natural myelin alone. Protein-lipid models in which the actual myelin proteins are used would help to reveal the role each plays in the structure.

Clearly it will be some time before the possibilities for improved knowledge regarding myelin structure are exhausted.

## Summary of Ideas About Myelin
## Structure and Constituents

The following are the major conclusions and speculations that have been developed from the extensive information now available about the constitution and structure of nerve myelin:

1. Early optical and chemical investigations indicated that myelin is a compound lamellar structure with interleaved cylindrical layers of protein and lipids concentric about the nerve fiber axis.

2. X-ray diffraction evidence then revealed that the unit period of the pattern along radii in the myelin sheath consists of two bilayers of lipid with alternating polar-head surfaces covered by different protein-aqueous phases. The periods observed are more than twice that of individual bilayers, apparently clustering about the values 156 A (typically in vertebrate CNS myelin) and 173 A and 183 A (typically in PNS myelin). The lipid acyl groups provide fluid hydrocarbon layers (short-range disorder) despite the good long-range order of the total system.

3. Electron micrographs of fixed stained material disclose the polar layers as alternating major dense and minor intraperiod lines that are traceable to continuity with the cytoplasmic and external surfaces, respectively, of the axon's enveloping Schwann cells (PNS) or oligodendrocytes (CNS). Centers of symmetry thus occur at the cytoplasmic and external appositions (the stained lines). Hypotonic solutions cause a remarkably regular separation along external appositions with the insertion of broad, equally thick lakes of aqueous phase between membrane pairs enlarging the radial period to several hundred Angstroms.

4. The center of symmetry plus the sampling of the membrane pair's diffraction transform at several swelling states have simplified the problem of diffraction phase determination, enabling crystallographers to obtain radial electron-density distribution plots for several myelins from frog and rabbit sciatic and rabbit optic nerves. These show two troughs of low density per period, corresponding to the paraffin chains of the two bilayers. The troughs are asymmetric with unequal shoulders on both sides of each trough that are believed to indicate unequal distribution of the cholesterol that is more prevalent toward the external appositions. Each trough is bounded by a high peak of density, related to the location of P atoms of the phospholipids; finally, plateaus of moderately high density, corresponding to the protein-aqueous layers, are located at the cytoplasmic and externally connected centers of symmetry. Examination of deuterated specimens by neutron diffraction generally confirms this analysis.

5. Results obtained from diffraction studies of modified myelin and of model systems consisting of lipids and proteins generally support and illuminate structures of the type derived for myelin.

6. The proteins of myelin are of three main kinds: (a) the very basic ones exhibit allergenic properties when misplaced into the circulatory system; the basic proteins may be located at the cyto-

plasmic appositions and are relatively constant in amount in both CNS and PNS myelins (20% to 30% of total protein); (b) the Folch-Pi-Lees proteolipid protein seems to be moderately basic and probably is distributed at both cytoplasmic and external appositions, particularly in CNS myelin where it is the major protein component (54%); (c) the Wolfgram proteolipid protein is relatively neutral or even anionic and largely replaces its Folch-Pi-Lees counterpart in PNS myelin; it may be chiefly at external appositions and contribute to the swelling properties of myelin. The heavy osmophilia of the cytoplasmic appositions may be the result of the (hypothesized) distribution of these proteins; also, because of their net positive charge, the proteins may draw anionic lipid groups toward the cytoplasmic appositions, excluding neutral lipids such as cholesterol toward the external appositions.

7. The protein content of myelin is relatively low (about 20% of dry weight) compared to other cellular membranes. It is barely sufficient to cover the lipid layers in the most expanded conformation possible, namely, the pleated $\beta$ state with side chains extending flat in the polar layers. Patches of parallel $\beta$-chains oriented randomly in the aqueous layers may account for a diffuse 12 A diffraction recently reported for frog sciatic myelin.

8. The lipids of myelin seem to be involved in providing stability and biological inertness. Lipids constitute about 75% of myelin, thus providing considerable capability for establishment of stable hydrophobic layers. The prevailing and characteristic kinds of myelin lipid are cholesterol and the sphingolipids; the former interferes with general mobility of other lipids in the bilayers, and the latter bear unusually long acyl paraffin chains. Although the average paraffin chain lengths in both CNS and PNS myelin representatives are similar (about 18 carbon atoms), there may be a significant difference between the acyl chain unsaturation that seems to be greater in CNS than in PNS.

9. Acidic lipids in both CNS and PNS approximately balance the net positive charges at neutral pH of the myelin proteins. PNS myelin may be more distinctly negatively charged at external appositions because it swells via these appositions, whereas CNS myelin resists swelling in water. Variations in the myelin radial period between CNS and PNS may be due to this fact, i.e., CNS myelin seems normally to approach contact between membrane pairs, whereas PNS myelin may remain somewhat swollen.

10. Myelin structure is not entirely that of a pair of typical biological membranes; it seems especially designed for axon protection.

Proteins of other membranes exhibit metabolic and transport phenomena that are not discernible in myelin. Often the more active membrane phenomena involve cooperative events evoked by a few ions or surface substances. Membranes generally, however, would appear to resemble myelin structure in the interleaving of polar and lipid layers with possibly particulate structures added when specific functional requirements are present.

11. The recent achievement of greater sophistication regarding the constituents and structure of myelin provides base lines for further investigation. Both physical and chemical studies can now be broadened so that both kinds of information are available for the same substances over a wider range of sources. The phenomenon of swelling will be particularly useful in providing means for introducing variations in the normal structure to imitate that of other membranes. Self-assembling artificial models, in which the natural lipids and proteins are used, will be helpful in illuminating the functions and structures of myelin and other membranes.

## List of Abbreviations

| | |
|---|---|
| AMP | adenosine monophosphate |
| ATP | adenosine triphosphate |
| BP | basic protein |
| CD | circular dichroism |
| CM | chloroform-methanol |
| CNS | central nervous system |
| DEAE | diethylaminoethyl |
| DNA | deoxyribonucleic acid |
| EAE | experimental allergic encephalomyelitis |
| EAN | experimental allergic neuritis |
| EBP | encephalitogenic basic protein |
| EM | electron microscope |
| FLPL | Folch-Pi-Lees proteolipid |
| GM-1 | monosialoganglioside |
| GP | glycerophosphatides |
| NADP | nicotinamide adenine dinucleotide phosphate |
| NANA | N-acetylneuraminic acid |
| ORD | optical rotatory dispersion |
| PNS | peripheral nervous system |
| PTA | phosphotungstic acid |
| RNA | ribonucleic acid |
| TPI | triphosphoinositide |
| UDP | uridine diphosphate |
| WPRL | Wolfgram proteolipid |

# BIBLIOGRAPHY

This bibliography contains two types of entries: (1) citations given or work alluded to in the report, and (2) additional references to pertinent literature by conference participants and others. Citations in group (1) may be found in the text on the pages listed in the right-hand column.

Page

Adams, C.W., Davison, A.N., and Gregson, N.A. (1963): Enzyme inactivity of myelin: histochemical and biochemical evidence. *J. Neurochem.* 10:383-395.  — 480,485, 487,561

Agrawal, H.C., Banik, N.L., Bone, A.H., Davison, A.N., Mitchell, R.F., and Spohn, M. (1970a): The identity of a myelin-like fraction isolated from developing brain. *Biochem. J.* 120:635-642.  — 466,467, 468,486

Agrawal, H.C., Bone, A.H., and Davison, A.N. (1970b): Effect of phenylalanine on protein synthesis in the developing rat brain. *Biochem. J.* 117:325-331.

Akers, C.K. and Parsons, D.F. (1970): X-ray diffraction of myelin membrane. II. Determination of the phase angles of the frog sciatic nerve by heavy atom labeling and calculation of the electron density distribution of the membrane. *Biophys. J.* 10:116-136.  — 541

Aleu, F.P., Katzman, R., and Terry, R.D. (1963): Fine structure and electrolyte analyses of cerebral edema induced by alkyl tin intoxication. *J. Neuropathol. Exp. Neurol.* 22:403-413.  — 494

Alvord, E.C. (1970): Acute disseminated encephalomyelitis and allergic neuro-encephalopathies. *In: Handbook of Clinical Neurology, Vol. 9. Multiple Sclerosis and Other Demyelinating Diseases*, Vinken, P.J. and Bruyn, G.W., eds. Amsterdam: North-Holland Publishing Co., pp. 500-571.

Ambronn, H. (1890): Das optische Verhalten markhaltiger und markloser Nervenfasern. *Ber. Verh. königl. sächs. Ges. Wissensch. Leipzig* 42:419-429.  — 446,508

Apáthy, S. (1889): Nach welcher Richtung hin soll die Nervenlehre reformiert werden? *Biol. Zbl.* 9:625-648.  — 446

Arnason, B. (1971): Idiopathic polyneuritis. *In: Immunologic Disorders of the Nervous System,* Rowland, L., ed. New York: Academic Press. (In press)  — 501

Arnason, B.G., Winkler, G.F., and Hadler, N.M. (1969): Cell-mediated demyelination of peripheral nerve in tissue culture. *Lab. Invest.* 21:1-10.

Asbury, A.K., Arnason, B.G., and Adams, R.D. (1969): The inflammatory lesion in idiopathic polyneuritis: Its role in pathogenesis. *Medicine (Baltimore)* 48:173-215.

Page

August, C., Davison, A.N., and Maurice-Williams, F. (1961): Phospholipid metab-    451
olism in nervous tissue. 4. Incorporation of $^{32}$P into the lipids of subcellular
fractions of the brain. *Biochem. J.* 81:8-12.

Autilio, L. (1966): Fractionation of myelin proteins, *Fed. Proc.* 25:764 (Abstr.)    457

Autilio, L.A., Norton, W.T., and Terry, R.D. (1964): The preparation and some    451,455,
properties of purified myelin from the central nervous system. *J. Neurochem.*    502,543
11:17-27.

Banik, N.L., Blunt, M.J., and Davison, A.N. (1968): Changes in the osmiophilia of    465
myelin and lipid content in the kitten optic nerve. *J. Neurochem.* 15:471-475.

Banik, N.L. and Davison, A.N. (1969): Enzyme activity and composition of myelin    466,467,
and subcellular fractions in the developing rat brain. *Biochem. J.*    468,483
115:1051-1062.

Banik, N.L. and Davison, A.N. (1971): Exchange of sterols between myelin and    482
other membranes of developing rat brain. *Biochem. J.* 122:751-758.

Barman, T.E. and Koshland, D.E. (1967): A colorimetric procedure for the    501
quantitative determination of tryptophan residues in proteins. *J. Biol. Chem.*
242:5771-5776.

Bear, R.S., Palmer, K.J., and Schmitt, F.O. (1941): X-ray diffraction studies of    514
nerve lipides. *J. Cell. Comp. Physiol.* 17:355-367.

Bear, R.S. and Schmitt, F.O. (1936): The optics of nerve myelin. *J. Opt. Soc. Am.*
26:206-212.

Bear, R.S. and Schmitt, F.O. (1937): Optical properties of the axon sheaths of
crustacean nerves. *J. Cell. Comp. Physiol.* 9:275-287.

Bear, R.S., Schmitt, F.O., and Young, J.Z. (1937): Sheath components of the giant
nerve fibres of the squid. *Proc. Roy. Soc. B.* 123:496-505.

Beck, C.S., Hasinoff, C.M., and Smith, M.E. (1968): L-alanyl-β-napthylamidase in    480,487
rat spinal cord myelin. *J. Neurochem.* 15:1297-1301.

Blasie, J.K., Dewey, M.M., Blaurock, A.E., and Worthington, C.R. (1965): Electron    561
microscope and low-angle x-ray diffraction studies on outer segment membranes
from the retina of the frog. *J. Mol. Biol.* 14:143-152.

Blaurock, A.E. (1967): X-ray diffraction studies of the myelin sheath of nerve.    531
Dissertation, University of Michigan.

Blaurock, A.E. (1971): Structure of the nerve myelin membrane: Proof of the    528,530,
low-resolution profile. *J. Mol. Biol.* 56:35-52.    531,538,540,560

Blaurock, A.E. and Wilkins, M.H.F. (1969): Structure of frog photoreceptor    561
membranes. *Nature* 223:906-909.

Page

Blaurock, A.E. and Worthington, C.R. (1966): Treatment of low angle X-ray data            530
from planar and concentric multilayered structures. *Biophys. J.* 6:305-312.

Blaurock, A.E. and Worthington, C.R. (1969): Low-angle X-ray diffraction patterns     512,513
from a variety of myelinated nerves. *Biochim. Biophys. Acta* 173:419-426.

Boehm, G. von (1933): Das Röntgendiagramm der Nerven. *Kolloidzschr.* 62:22-26.         510

Bourne, G.H. (1969): *The Structure and Function of Nervous Tissue, Vol. 3.*
*Biochemistry and Disease.* New York: Academic Press.

Bowen, D.M. and Radin, N.S. (1969): Hydrolase activities in brain of neurological        493
mutants: cerebroside galactosidase, nitrophenyl galactoside hydrolase, nitrophenyl glucoside hydrolase and sulphatase. *J. Neurochem.* 16:457-460.

Brady, R.O. (1970): Prenatal diagnosis of lipid storage diseases. *Clin. Chem.* 16:811-   489
815.

Bragg, W.L. (1914a): The analysis of crystals by the X-ray spectrometer. *Proc. Roy.*    447
*Soc.* 89A:468-489.

Bragg, W.L. (1914b): The structure of some crystals as indicated by their                447
diffraction of X-rays. *Proc. Roy. Soc.* 89A:248-277.

Brante, G. (1949): Studies on lipids in the nervous system, with special reference to     450
quantitative chemical determination and topical distribution. *Acta Physiol.*
*Scand. 18 Suppl.* 63:1-284.

Branton, D. (1967): Fracture faces of frozen myelin. *Exp. Cell Res.* 45:203-207.     561,562

Branton, D. (1971): Membrane structure. *Protoplasmatologia.* (In press)                 561

Branton, D. and Park, R.B. (1967): Subunits in chloroplast lamellae. *J. Ultrastruct.*    562
*Res.* 19:283-303.

Bunge, M.B., Bunge, R.P., and Ris, H. (1961): Ultrastructural study of re-               523
myelination in an experimental lesion in adult cat spinal cord. *J. Biophys.*
*Biochem. Cytol.* 10:67-94.

Bunge, R.P. (1968): Glial cells and the central myelin sheath. *Physiol. Rev.*          522
48:197-251.

Bunge, R.P. (1970a): Observations on the repair of myelin sheaths after experi-          497
mental damage. *J. Cell Biol.* 47:27a. (Abstr.)

Bunge, R.P. (1970b): Structure and function of neuroglia: some recent observations. *In: The Neurosciences: Second Study Program.* Schmitt, F.O., editor-
in-chief. New York: Rockefeller University Press, pp. 782-797.

Caley, D.W. and Maxwell, D.S. (1968): An electron microscopic study of the
neuroglia during postnatal development of the rat cerebrum. *J. Comp. Neurol.*
133:45-69.

Page

Caspar, D.L.D. and Cohen, C. (1969): Polymorphism of tropomyosin and a view of
protein function. *In: Nobel Symposium 11. Symmetry and Function of
Biological Systems at the Macromolecular Level.* Engstrom, A. and Strandberg,
B., eds. Stockholm: Almquist and Wiksell, pp. 393-414.

Caspar, D.L.D. and Kirschner, D.A. (1971): Myelin membrane structure at 10 A    513,536,
resolution. *Nature (New Biology)* 231:46-52.                                   537,538,539,
                                                                                540,559

Caspar, D.L.D. and Klug, A. (1962): Physical principles in the construction of
regular viruses. *Cold Spring Harbor Symp. Quant. Biol.* 27:1-24.

Cecil, R. and Louis, C.F. (1970): Protein-hydrocarbon interactions: Interactions of
various proteins with pure decane: Interactions of various proteins with decane
in the presence of alcohols. *Biochem. J.* 117:139-156.

Changeux, J.P., Thiery, J., Tung, Y., and Kittel, C. (1967): On the cooperativity of    564
biological membranes. *Proc. Nat. Acad. Sci.* 57:335-341.

Chinn, P. and Schmitt, F.O. (1937): On the birefringence of nerve sheaths as    447,509
studied in cross sections. *J. Cell. Comp. Physiol.* 9:289-296.

Clark, A.W. and Branton, D. (1968): Fracture faces in frozen outer segments from    562
the guinea pig retina. *Z. Zellforsch. Mikrosk. Anat.* 91:586-603.

Costantino-Ceccarini, E. and Morell, P. (1971): Quaking mouse: *in vitro* studies
of brain sphingolipid biosynthesis. *Brain Res.* 29:75-84.

Cotman, C.W. and Mahler, H.R. (1967): Resolution of insoluble proteins in rat    457
subcellular fractions. *Arch. Biochem. Biophys.* 120:384-396.

Cuzner, M.L. and Davison, A.N. (1968): The lipid composition of rat brain myelin    455
and subcellular fractions during development. *Biochem. J.* 106:29-34.

Cuzner, M.L., Davison, A.N., and Gregson, N.A. (1965): The chemical composition    466,477
of vertebrate myelin and microsomes. *J. Neurochem.* 12:469-481.

Danielli, J.F. and Davson, H. (1935): A contribution to the theory of permeability    447,448,
of thin films. *J. Cell. Comp. Physiol.* 5:495-508.                               510,522

Davison, A.N. (1961): Metabolically inert proteins of the central and peripheral    480
nervous system, muscle and tendon. *Biochem. J.* 78:272-282.

Davison, A.N. (1968): Progress in pedology. *In: Fortschritte der Pädologie, Vol. 2,*
Linneweh, F., ed. Heidelberg: Springer-Verlag, p. 65.

Davison, A.N., Dobbing, J., Morgan, R.S., and Payling Wright, G. (1958): The    477
deposition and disposal of $(4-{}^{14}C)$ cholesterol in the brain of growing chickens.
*J. Neurochem.* 3:89-94.

Davison, A.N., Dobbing, J., Morgan, R.S., and Payling Wright, G. (1959): Metab-    477
olism of myelin: the presence of $(4-{}^{14}C)$ cholesterol in the mammalian central
nervous system. *Lancet* 1:658.

Page

Davison, A.N. and Peters, A. (1970): *Myelination*. Springfield, Ill.: Charles C        452
Thomas.

Davison, A.N. and Wajda, M. (1959): Persistence of cholesterol-4-$^{14}$C in the central
nervous system. *Nature* 183:1606-1607.

Davson, H. and Danielli, J.F. (1952): *The Permeability of Natural Membranes. 2nd*      448
*Ed.* Cambridge: Cambridge University Press.

Deamer, D.W. and Branton, D. (1967): Fracture planes in an ice-bilayer model
membrane system. *Science* 158:655-657.

Deamer, D.W., Leonard, R., Tardieu, A., and Branton, D. (1970): Lamellar and
hexagonal lipid phases visualised by freeze-etching. *Biochim. Biophys. Acta*
219:47-60.

Demel, R.A., van Deenen, L.L.M., and Kinsky, S.C. (1965): Penetration of lipid          563
monolayers by polyene antibiotics. Correlation with selective toxicity and mode
of action. *J. Biol. Chem.* 240:2749-2753.

De Robertis, E., Pellegrino de Iraldi, A., Rodriguez de Lores Arnaiz, G., and            451
Salganicoff, L. (1962): Cholinergic and non-cholinergic nerve endings in rat
brain. I. Isolation and subcellular distribution of acetylcholine and acetylcholinesterase, *J. Neurochem.* 9:23-35.

Deshmukh, D.S., Inoue, T., and Pieringer, R.A. (1970): Biosynthesis and degrada-    489,493
tion of galactolipids in the brain of myelin deficient mutant (Jimpy) mouse.
*Fed. Proc.* 29:409. (Abstr.)

DeVries, G.H. and Norton, W.T. (1971): Evidence for the absence of myelin and           503
the presence of galactolipid in an axon-enriched fraction from bovine CNS. *Fed.*
*Proc.* 30:1248 (Abstr.)

Dickinson, J.P., Jones, K.M., Aparicio, S.R., and Lumsden, C.E. (1970): Localization of encephalitogenic basic protein in the intraperiod line of lamellar
myelin. *Nature* 227:1133-1134.

Dobbing, J. (1963): The entry of cholesterol into rat brain during development. *J.*    477
*Neurochem.* 10:739-742.

Drummond, G.I., Iyer, N.T., and Keith, J. (1962): Hydrolysis of ribonucleoside          486
2'3'-cyclic phosphates by a diesterase from brain. *J. Biol. Chem.*
237:3535-3539.

Edström, A. (1964): The ribonucleic acid in the Mauthner neuron of the goldfish.        458
*J. Neurochem.* 11:309-314.

Eigen, M. and De Maeyer, L. (1971): Carriers and specificity in membranes.          452,563
*Neurosciences Res. Prog. Bull.* 9:299-437.

Page

Elkes, J. and Finean, J.B. (1949): The effect of drying upon the structure of myelin          514
in the sciatic nerve of the frog. *Discuss. Faraday Soc.* 6:134-141.

Elkes, J. and Finean, J.B. (1953a): Effects of solvents on the structure of myelin in          515
the sciatic nerve of the frog. *Exp. Cell Res.* 4:82-95.

Elkes, J. and Finean, J.B. (1953b) X-ray diffraction studies on the effect of          515
temperature on the structure of myelin in the sciatic nerve of the frog. *Exp. Cell
Res.* 4:69-81.

Eng, L.F., Chao, F.C., Gerstl, B., Pratt, D., and Tavaststjerna, M.G. (1968): The          545,546,
maturation of human white matter myelin. Fractionation of the myelin          547,548,
membrane proteins. *Biochemistry* 7:4455-4465.          555,558

Eng, L.F. and Noble, E.P. (1968): The maturation of rat brain myelin. *Lipids*          459
3:157-162.

Eng, L.F. and Smith, M.E. (1966): The cholesterol complex in the myelin          481
membrane. *Lipids* 1:296.

Engstrom, L.H. (1970): Structure in the erythrocyte membrane. Dissertation,
University of California, Berkeley.

Evans, M.J. and Finean, J.B. (1965): The lipid composition of myelin from brain          456
and peripheral nerve. *J. Neurochem.* 12:729-734.

Ewald, A. and Kühne, W. (1877): Uber einen neuen Bestandteil des Nervensystems.          449
*Verhandl. naturhist.-med. Vereins, Heidelberg* 1:457-464.

Eylar, E.H. (1970): Amino acid sequence of the basic protein of the myelin          462,546
membrane. *Proc. Nat. Acad. Sci.* 67:1425-1431.

Eylar, E.H. (1971): *In: Immunologic Disorders of the Nervous System.* Rowland,          457,499,
L., ed. New York: Academic Press. (In press)          501

Eylar, E.H., Caccam, J., Jackson, J.J., Westall, F.C., and Robinson, A.B. (1970):          546
Experimental allergic encephalomyelitis: Synthesis of the disease-inducing site
of the basic protein. *Science* 168:1220-1223.

Eylar, E.H. and Hashim, G.A. (1968): Allergic encephalomyelitis: the structure of
the encephalitogenic determinant. *Proc. Nat. Acad. Sci.* 61:644-650.

Eylar, E.H., Salk, J., Beveridge, G.C., and Brown, L.V. (1969): Experimental allergic          461,499,
encephalomyelitis: An encephalitogenic basic protein from bovine myelin. *Arch.*          545
*Biochem. Biophys.* 132:34-48.

Eylar, E.H. and Thompson, M. (1969): Allergic encephalomyelitis: the physico-          463,464,
chemical properties of the basic protein encephalitogen from bovine spinal cord.          499,545,
*Arch. Biochem. Biophys.* 129:468-479.          546,547

Page

Farkas, E., Zahnd, J.P., Nussbaum, J.L. and Mandel, P. (1970): Etude histologique,          493
histochimique et ultrastructurale des souris de souche Jimpy. *In: Les Mutants
Pathologiques chez l'Animal. Leur Intérêt dans la Recherche Bio-Médicale. No.
924, Coll. Internat. CNRS.* Sabourdy, M., ed. Paris:CNRS.

Fernández-Morán, H. (1950a): Electron microscope observations on the structure of          448
the myelinated nerve fiber sheath. *Exp. Cell Res.* 1:143-149.

Fernández-Morán, H. (1950b): Sheath and axon structures in the internode portion          448
of vertebrate myelinated nerve fibers, an electron microscope study of rat and
frog sciatic nerves. *Exp. Cell Res.* 1:309-340.

Fernández-Morán, H. (1952): The submicroscopic organization of vertebrate nerve          448
fibres. An electron microscope study of vertebrate nerve fibres. An electron
microscopic study of myelinated and unmyelinated nerve fibres. *Exp. Cell Res.*
3:282-350.

Fernández-Morán, H. and Finean, J.B. (1957): Electron microscope and low-angle          519
X-ray diffraction studies of the nerve myelin sheath. *J. Biophys. Biochem.
Cytol.* 3:725-748.

Fiil, A. and Branton, D. (1969): Changes in the plasma membrane of *Escherichia
coli* during magnesium starvation. *J. Bacteriol.* 98:1320-1327.

Finean, J.B. (1953): Phospholipid-cholesterol complex in the structure of myelin.          554
*Experientia* 9:17-19.

Finean, J.B. (1957): The role of water in the structure of peripheral nerve myelin.          559
*J. Biophys. Biochem. Cytol.* 3:95-102.

Finean, J.B. (1960): Electron microscope and x-ray diffraction studies of the      559,560
effects of dehydration on the structure of nerve myelin. I. Peripheral nerve. *J.
Biophys. Biochem. Cytol.* 8:13-29.

Finean, J.B. (1965): Molecular parameters in the nerve myelin sheath. *Ann. N.Y.*          543
*Acad. Sci.* 122:51-56.

Finean, J.B. and Burge, R.E. (1963): The determination of the Fourier transform      529,530,
of the myelin layer from a study of swelling phenomena. *J. Mol. Biol.*      539,556
7:672-682.

Finean, J.B., Coleman, R., and Green, W.A. (1966): Studies of isolated plasma          562
membrane fractions. *Ann. N.Y. Acad. Sci.* 137:414-420.

Finean, J.B., Sjöstrand, F.S., and Steinmann, E. (1953): Submicroscopic orga-          561
nisation of some layered lipoprotein structures. (Nerve myelin, retinal rods, and
chloroplasts). *Exp. Cell Res.* 5:557-559.

Page

Folch, J. (1951): The chemistry of phosphatides. *In: Phosphorus Metabolism, Vol.*          450
*2.* McElroy, W.D. and Glass, B., eds. Baltimore: Johns Hopkins Press, pp.
186-200.

Folch-Pi, J. (1955): Composition of the brain in relation to maturation. *In:*          450,469
*Biochemistry of the Developing Nervous System.* Waelsch, H., ed. New York:
Academic Press, pp. 121-136.

Folch-Pi, J. (1959): Recent studies on the chemistry of the brain and their relation          457,461
to the structure of the myelin sheath. *Expos. Annu. Biochim. Med.* 21:81-95.

Folch-Pi, J. (1964): Some considerations on the structure of proteolipids. *Fed.*
*Proc.* 23:630-633.

Folch-Pi, J. (1966): Proteolipids, neurokeratin, neurosclerin, and copper proteins.          449,544,
*In: Protides of the Biological Fluids,* Peeters, H., ed. Amsterdam: Elsevier, pp.          548
21-34.

Folch-Pi, J. (1967): Interrelationship between lipids and proteins in membranes.
*Protoplasma* 63:160-164.

Folch-Pi, J. (1968): The composition of nervous membranes. *Progr. Brain Res.*
29:1-17.

Folch, J., Ascoli, M., Lees, M., Meath, J.A.,and Le Baron, F.N. (1951): Preparation          449
of lipide extracts from brain tissue. *J. Biol. Chem.* 191:833-841.

Folch, J., Casals, J., Pope, A., Meath, J.A., Le Baron, F.N.,and Lees, M. (1959):
Chemistry of myelin development. *In: The Biology of Myelin,* Korey, S.R., ed.
New York: Hoeber-Harper, pp. 122-137.

Folch, J. and Lees, M. (1951): Proteolipides, a new type of tissue lipoproteins.          458,504,
Their isolation from brain. *J. Biol. Chem.* 191:807-817.          544

Friedel, G. (1922): Les états mesomorphes de la matière. *Ann. Physique*          445,446
18:273-474.

Friedlaender, B. (1889): Uber die markhaltigen Nervenfasern und Neurochorde der          446,565
Crustaceen und Anneliden. *Mitt. Zool. Station Neapel.* 9:205-265.

Friedrich, W., Knipping, P.,und Laue, M. (1913): Interferenzerscheinungen bei          447
Röntgenstrahlen. *Ann. Phys.* 41:971-988.

Fumagalli, R., Smith, M.E., Urna, G., and Paoletti, R. (1969): The effect of          482
hypocholesteremic agents on myelinogenesis. *J. Neurochem.* 16:1329-1339.

Furshpan, E.J. (1964): "Electrical transmission" at an excitatory synapse in a          564
vertebrate brain. *Science* 144:878-880.

Page

Geren, B.B. (1954): The formation from the Schwann cell surface of the myelin in       448,520
the peripheral nerves of chick embryos. *Exp. Cell Res.* 7:558-562.

Geren, B.B. and Schmitt, F.O. (1954): The structure of the Schwann cell and its            448
relation to the axon in certain invertebrate nerve fibers. *Proc. Nat. Acad. Sci.*
40:863-870.

Geren, B.B. and Schmitt, F.O. (1955): Electron microscope studies of the Schwann
cell and its constituents with particular reference to their relation to the axon.
*In: Symposium on Fine Structure of Cells. (VIII Congress of Cell Biology.)*
Groningen, The Netherlands: D. Noordhoff, Ltd., pp. 251-260.

Gerstl, B., Eng, L.F., Hayman, R.B., Tavaststjerna, M.G.,and Bond, P.R. (1967): On        457
the composition of human myelin. *J. Neurochem.* 14:661-670.

Göthlin, G.F. (1913): Die doppelbrechenden Eigenschaften des Nervengewebes.               446
Ihre Ursachen und ihre biologischen Konsequenzen. *Kungl. Svensk. Vetenskap.*
*Handl.* 51:1-92.

Greenfield, S., Norton, W.T., and Morell, P. (1971): Myelin proteins of the Quaking       503
mouse. *Trans. Am. Soc. Neurochem.* 2:76. (Abstr.)

Grove, S.N., Bracker, C.E., and Morré, D.J. (1968): Cytomembrane differentiation          560
in the endoplasmic reticulum-Golgi apparatus-vesicle complex. *Science* 161:
171-173.

Gulik-Krzywicki, T., Rivas, E., and Luzzati, V. (1967): Structure et polymorphisme        517
des lipides: étude par diffraction des rayons X du système formé de lipides de
mitochondries de coeur de boeuf et d'eau. *J. Mol. Biol.* 27:303-322.

Gulik-Krzywicki, T., Shechter, E., Iwatsubo, M., Ranck, J.L., and Luzzati, V.
(1970): Correlations between structure and spectroscopic properties in membrane model systems. Tryptophan and 1-anilino-8-naphthalene sulfonate fluorescence in protein-lipid-water phases. *Biochim. Biophys. Acta.* 219:1-10.

Gulik-Krzywicki, T., Shechter, E., Luzzati, V.,and Faure, M. (1969a): Interactions        518
of proteins and lipids: structure and polymorphism of protein-lipid-water
phases. *Nature* 223:1116-1121.

Gulik-Krzywicki, T., Tardieu, A.,and Luzzati, V. (1969b): The smectic phase of            516
lipid-water systems: properties related to the nature of the lipid and to the
presence of net electrical charges. *Molecular Crystals and Liquid Crystals*
8:285-291.

Hall, S.M. and Williams, P.L. (1970): Studies on the 'incisures' of Schmidt and           525
Lanterman. *J. Cell Sci.* 6:767-791.

Halliburton, W.D. (1894): The proteids of nervous tissue. *J. Physiol.* 15:90-107.        449

Page

Hallpike, J.F. and Adams, C.W. (1969): Proteolysis and myelin breakdown: review of recent histochemical and biochemical studies. *Histochem. J.* 1:559-578.

Hardesty, I. (1905): On the occurrence of sheath cells and the nature of the axone sheaths in the central nervous system. *Am. J. Anat.* 4:329-354.　　484

Hashim, G.A. and Eylar, E.H. (1969a): Allergic encephalomyelitis: enzymatic degradation of the encephalitogenic basic protein from bovine spinal cord. *Arch. Biochem. Biophys.* 129:635-644.　　500

Hashim, G.A. and Eylar, E.H. (1969b): Allergic encephalomyelitis: isolation and characterization of encephalitogenic peptides from the basic protein of bovine spinal cord. *Arch. Biochem. Biophys.* 129:645-654.　　500

Hashim, G.A. and Eylar, E.H. (1969c): The structure of the terminal regions of the encephalitogenic basic protein from bovine myelin. *Arch. Biochem. Biophys.* 135:324-333.

Hedley-Whyte, E.T., Darrah, H.K., Stendler, F., and Uzman, B.G. (1968): The value of cholesterol-1,2-$H^3$ as a long term tracer for autoradiographic study of the nervous system of mice. *Lab. Invest.* 19:526-529.　　477

Hedley-Whyte, E.T., Rawlins, F.A., Salpeter, M.M. and Uzman, B.G. (1969): Distribution of cholesterol-1,2-$H^3$ during maturation of mouse peripheral nerve. *Lab. Invest.* 21:536-547.　　478,479

Hedley-Whyte, E.T., and Uzman, B.G. (1968): Comparison of cholesterol extraction from tissues during processing for electron microscopic radioautography. *In: Proc. Electron Microscopy Society of America, 26th Annual Meeting,* Arceneaux, C.J., ed. Baton Rouge: Claitor's Publ. Div., pp. 92-93.　　477

Hirano, A. (1968): A confirmation of the oligodendroglial origin of myelin in the adult rat. *J. Cell Biol.* 38:637-640.

Hirano, A. (1969): The fine structure of brain edema. *In: The Structure and Function of Nervous Tissue. Vol. 2.* Bourne, G.H., ed. New York: Academic Press, pp. 69-135.　　493,494

Hirano, A. (1971): The pathology of the central myelinated axon. *In: The Structure and Function of Nervous Tissue. Vol. 4.* Bourne, G.H., ed. New York: Academic Press. (In press)

Hirano, A., Becker, N.H. and Zimmerman, H.M. (1969): Isolation of the periaxonal space of the central myelinated nerve fiber with regard to the diffusion of peroxidase. *J. Histochem. Cytochem.* 17:512-516.

Hirano, A., Cook, S.D., Whitaker, J.N., Dowling, P.C. and Murray, M.R. (1971): Fine structural aspects of demyelination *in vitro*. The effects of Guillain-Barré serum. *J. Neuropathol. Exp. Neurol.* (In press)　　494

Page

Hirano, A. and Dembitzer, H.M. (1967): A structural analysis of the myelin sheath          493,494,
in the central nervous system. *J. Cell Biol.* 34:555-567.                          521,522,523,525

Hirano, A. and Dembitzer, H.M. (1969): The transverse bands as a means of access          483,484,
to the periaxonal space of the central myelinated nerve fiber. *J. Ultrastruct. Res.*          524
28:141-149.

Hirano, A., Dembitzer, H.M., Becker, N.H., and Zimmerman, H.M. (1969): The
distribution of peroxidase in the triethyltin-intoxicated rat brain. *J.
Neuropathol. Exp. Neurol.* 28:507-511.

Hirano, A., Levine, S., and Zimmerman, H.M. (1967): Experimental cyanide                493
encephalopathy: electron microscopic observations of early lesions in white
matter. *J. Neuropathol. Exp. Neurol.* 26:200-213.

Hirano, A., Levine, S.,and Zimmerman, H.M. (1968a): Remyelination in the central          495
nervous system after cyanide intoxication. *J. Neuropathol. Exp. Neurol.*
27:234-245.

Hirano, A., Sax, D.S., and Zimmerman, H.M. (1969): The fine structure of the
cerebella of jimpy mice and their "normal" litter mates. *J. Neuropathol. Exp.
Neurol.* 28:388-400.

Hirano, A. and Zimmerman, H.M. (1971): Glial filaments in the myelin sheath after
vinblastine implantation. *J. Neuropathol. Exp. Neurol.* 30:63-67.

Hirano, A., Zimmerman, H.M.,and Levine, S. (1964): The fine structure of cerebral          493
fluid accumulation. III. Extracellular spread of cryptococcal polysaccharides in
the acute stage. *Am. J. Pathol.* 45:1-11.

Hirano, A., Zimmerman, H.M.,and Levine, S. (1966): Myelin in the central nervous          493
system as observed in experimentally induced edema in the rat. *J. Cell Biol.*
31:397-411.

Hirano, A., Zimmerman, H.M.,and Levine, S. (1968b): Intramyelinic and extra-          494
cellular spaces in triethyltin intoxication. *J. Neuropathol. Exp. Neurol.*
27:571-580.

Hirano, A., Zimmerman, H.M., and Levine, S. (1969): Electron microscopic
observations of peripheral myelin in a central nervous system lesion. *Acta
Neuropathol.* 12:348-365.

Hodge, A.J. (1959): Fine structure of lamellar systems as illustrated by chloro-          562
plasts. *Rev. Mod. Phys.* 31:331-341.

Höglund, G. and Ringertz, H. (1961): X-ray diffraction studies on peripheral nerve          512
myelin. *Acta Physiol. Scand.* 51:290-295.

Hopfer, U., Lehninger, A.L., and Lennarz, W.J. (1970): The effect of the polar
moiety of lipids on bilayer conductance induced by uncouplers of oxidative
phosphorylation. *J. Membrane Biol.* 3:142-155.

Page

Horrocks, L.A. (1967): Composition of myelin from peripheral and central nervous systems of the squirrel monkey. *J. Lipid Res.* 8:569-576.　　456

Jakoubek, B. and Edström, J.E. (1965): RNA changes in the Mauthner axon and myelin sheath after increased functional activity. *J. Neurochem.* 12:845-849.　　458

Johnson, A.C., McNabb, A.R., and Rossiter, R.J. (1950): Chemistry of Wallerian degeneration. *Arch. Neurol. Psychiat.* 64:105-121.　　450

Joos, P. (1970): Cholesterol as liquifier in phospholipid membranes studied by surface viscosity measurements of mixed monolayers. *Chem. Phys. Lipids* 4:162-168.

Kandutsch, A.A. and Saucier, S.E. (1969): Regulation of sterol synthesis in developing brains of normal and jimpy mice. *Arch. Biochem. Biophys.* 135:201-208.　　492

Kies, M.W. (1971): Myelin basic proteins. *In: Abstracts, 3rd Intl. Meeting, Intl. Soc. Neurochem.* Budapest: Akadémiai Kiadó, p. 419.

Kies, M.W., Thompson, E.B., and Alvord, E.C., Jr. (1965): The relationship of myelin proteins to experimental allergic encephalomyelitis. *Ann. N.Y. Acad. Sci.* 122:148-160.　　457

Kilbourn, B.T., Dunitz, J.D., Pioda, L.A.R., and Simon, W. (1967): Structure of the $K^+$ complex with nonactin, a macrotetralide antibiotic possessing highly specific $K^+$ transport properties. *J. Mol. Biol.* 30:559-563.　　563

King, G. and Worthington, C.R. (1971): Refinement of myelin structure: Part II. *Biophysical Society Abstracts, Fifteenth Annual Meeting*, p. 292a. (Abstr.)　　540

Koch, W. (1904): Methods for the quantitative chemical analysis of the brain and spinal cord. *Am. J. Physiol.* 11:303-329.　　450

Koch, W. (1905): On the presence of a sulphur compound in nerve tissues. *Science* 21:884-885.　　450

Koch, W. (1907): Zur Kenntnis der Schwefelverbindungen des Nervensystems. *Z. physiol. Chem.* 53:496-507.　　450

Koch, W. and Upson, F.W. (1909): Quantitative chemical analysis of animal tissues. IV. Estimation of the elements, with special reference to sulphur. *J. Am. Chem. Soc.* 31:1355-1364.　　450

Korey, S.R., Orchen, M., and Brotz, M. (1958): Studies of white matter. I. Chemical constitution and respiration of neuroglial and myelin enriched fractions of white matter. *J. Neuropathol. Exp. Neurol.* 17:430-438.　　455

Korn, E.D. (1969a): Cell membranes: structure and synthesis. *Ann. Rev. Biochem.* 38:263-288.　　452

Korn, E.D. (1969b): Current concepts of membrane structure and function. *Fed. Proc.* 28:6-11.　　452

Page

Kostic, D., Nussbaum, J.L.,and Mandel, P. (1969): A study of brain gangliosides in
    "Jimpy" mutant mice. *Life Sci.* 8:1135-1143.

Kurihara, T., Nussbaum, J.L., and Mandel, P. (1969): 2',3'-Cyclic nucleotide          485,486
    3'-phosphohydrolase in the brain of the "Jimpy" mouse, a mutant with
    deficient myelination. *Brain Res.* 13:401-403.

Kurihara, T., Nussbaum, J.L., and Mandel, P. (1970): 2',3'-Cyclic nucleotide          485,486,
    3'-phosphohydrolase in brains of mutant mice with deficient myelination. *J.*         492
    *Neurochem.* 17:993-997.

Kurihara, T., Nussbaum, J.L., and Mandel, P. (1971): 2',3'-Cyclic nucleotide          485,486,
    3'-phosphohydrolase in purified myelin from brain of Jimpy and normal young           492
    mice. *Life Sci.* 10:421-429.

Kurihara, T. and Tsukada, Y. (1967): The regional and subcellular distribution of     485,486,
    2',3'-cyclic nucleotide 3'-phosphohydrolase in the central nervous system. *J.*        487
    *Neurochem.* 14:1167-1174.

Kurihara, T. and Tsukada, Y. (1968): 2',3'-Cyclic nucleotide 3'-phosphohydrolase in   485,486,
    the developing chick brain and spinal cord. *J. Neurochem.* 15:827-832.               487

Kurtz, D.J. and Kanfer, J.N. (1970): Cerebral acid hydrolase activities: comparison      493
    in "quaking" and normal mice. *Science* 168:259-260.

Lampert, P. and Carpenter, S. (1965): Electron microscopic studies on the vascular       494
    permeability and the mechanism of demyelination in experimental allergic
    encephalomyelitis. *J. Neuropathol. Exp. Neurol.* 24:11-24.

Lardy, H.A., Graven, S.N., and Estrado-O, S. (1967): Specific induction and              563
    inhibition of cation and anion transport in mitochondria. *Fed. Proc.*
    26:1355-1360.

Le Baron, F.N. and Folch-Pi, J. (1956): The isolation from brain tissue of a             449
    trypsin-resistant protein fraction containing combined inositol, and its relation
    to neurokeratin. *J. Neurochem.* 1:101-108.

Lees, M.B., Messinger, B.F., and Burnham, J.D. (1967): Tryptic hydrolysis of brain       461
    proteolipid. *Biochem. Biophys. Res. Commun.* 28:185-190.

Lehman, O. (1911): *Die neue Welt der flüssigen Kristalle.* Leipzig: Akademische      445,446
    Gesellschaft.

Lehman, O. (1918): Die Lehre von den flüssigen Kristallen und ihre Beziehung zu       445,446
    den Problemen der Biologie. *Ergeb. Physiol.* 16:255-509.

Lehninger, A.L. (1965): *Bioenergetics.* New York: W.A. Benjamin, Inc.

Lehninger, A.L. (1966): The supramolecular organization of enzyme and mem-
    brane systems. *Naturwissenschaften* 53:57-63.

Lehninger, A.L. (1968): The neuronal membrane. *Proc. Nat. Acad. Sci.*                   554
    60:1069-1080.

Page

Levine, Y.K. (1970): X-ray diffraction studies of oriented bimolecular layers of    538
phospholipids. Dissertation, University of London.

Levine, Y.K., Bailey, A.I., and Wilkins, M.H. (1968): Multilayers of phospholipid    567
bimolecular leaflets. *Nature* 220:577-578.

Levine, Y.K. and Wilkins, M.H.F. (1971): Structure of oriented lipid bilayers.    538
*Nature (New Biology)* 230:69-72.

Loewenstein, W.R. (1966): Permeability of membrane junctions. *Ann. N.Y. Acad.*    564
*Sci.* 137:441-472.

Lowden, J.A. and Wolfe, L.S. (1964): Studies on brain gangliosides. III. Evidence    553
for the location of the gangliosides specifically in neurones. *Can. J. Biochem.*
42:1587-1594.

Luzzati, V. (1968): X-ray diffraction studies of lipid-water systems. *In: Biological*    515,516
*Membranes,* Chapman, D., ed. New York: Academic Press, pp. 71-123.

Luzzati, V., Gulik-Krzywicki, T., Rivas, E., Reiss-Husson, F., and Rand, R.P.
(1968a): X-ray study of model systems: structure of the lipid-water phases in
correlation with the chemical composition of the lipids. *J. Gen. Physiol.*
51:37s-43s.

Luzzati, V., Gulik-Krzywicki, T., and Tardieu, A. (1968b): Polymorphism of    516
lecithins. *Nature* 218:1031-1034.

Luzzati, V., Gulik-Krzywicki, T., Tardieu, A., Rivas, E., and Reiss-Husson, F.    516
(1969): Lipids and membranes. *In: The Molecular Basis of Membrane Function.*
Tosteson, D.C., ed. Englewood Cliffs, N.J.: Prentice-Hall, pp. 79-108.

Luzzati, V., Tardieu, A., and Gulik-Krzywicki, T. (1968c): Polymorphism of lipids.    516
*Nature* 217:1028-1030.

Luzzati, V., Tardieu, A., Gulik-Krzywicki, T., Rivas, E., and Reiss-Husson, F.    516
(1968d): Structure of the cubic phases of lipid-water systems. *Nature*
220:485-488.

MacArthur, C.G. and Doisy, E.A. (1919): Quantitative chemical changes in the    450
human brain during growth. *J. Comp. Neurol.* 30:445-486.

Mandel, P., Borkowski, T., Harth, S., and Mardell, R. (1961): Incorporation of $^{32}$P    451
in ribonucleic acid of subcellular fractions of various regions of the rat central
nervous system. *J. Neurochem.* 8:126-138.

Mandel, P. and Nussbaum, J.L. (1966): Incorporation of $^{32}$P into the phosphatides    485
of myelin sheaths and of intracellular membranes. *J. Neurochem.* 13:629-642.

Mandel, P. and Nussbaum, J.L. (1967): Distribution and turnover of the phosphatides in the intracellular membranes and in the myelin sheaths of the rat brain.
*Protoplasma* 63:110-111.

Page

Marks, N. (1971): Myelin enzymes and protein metabolism. *In: Abstracts: 3rd Intl. Meeting, Intl. Soc. Neurochem.* Budapest: Akadémiai Kiádo, p. 415.

Martenson, R.E., Deibler, G.E., and Kies, M.W. (1970): Myelin basic proteins of the rat central nervous system. Purification, encephalitogenic properties, and amino acid compositions. *Biochim. Biophys. Acta* 200:353-362.                               461

Martenson, R.E. and Le Baron, F.N. (1966): Studies on the acid-extractable proteins of bovine brain white matter. *J. Neurochem.* 13:1469-1479.               457,460

Masurovsky, E. and Bunge, R. (1971): Patterns of myelin degeneration following the rapid death of cells in cultures of peripheral nervous tissue. *J. Neuropathol. Exp. Neurol.* (In press)                                                            496

Maturana, H.R. (1960): The fine anatomy of the optic nerve of Anurans—An electron microscope study. *J. Biophys. Biochem. Cytol.* 7:107-119.               522

Mc Gregor, H.H. (1917): Proteins of the central nervous system. *J. Biol. Chem.* 28:403-427.                                                                   449

Mehl, E. and Wolfgram, F. (1969): Myelin types with different protein components in the same species. *J. Neurochem.* 16:1091-1097.                           465,504

Merrifield, R.B. (1963): Solid phase peptide synthesis. I. The synthesis of a tetrapeptide. *J. Am. Chem. Soc.* 85:2149-2154.                                 501

Miani, N., Cavallotti, C., and Caniglia, A. (1969): Synthesis of adenosine triphosphate by myelin of spinal nerves of rabbit. *J. Neurochem.* 16:249-260.     486

Mokrasch, L.C. (1966): Incorporation of [$^{14}$C] amino acids into the proteolipid of subcellular preparations of rat brain in vitro. *J. Neurochem.* 13:49-58.    485

Mokrasch, L.C. (1967): A rapid purification of proteolipid protein adaptable to large quantities. *Life Sci.* 6:1905-1909.                                    459,461

Mokrasch, L.C. (1969): Myelin. *In: Handbook of Neurochemistry, Vol. I. Chemical Architecture of the Nervous System.* Lajtha, A., ed. New York: Plenum Press, pp. 171-193.                                                                  452

Mokrasch, L.C. (1971): Purification and properties of isolated myelin. *In: Methods of Neurochemistry, Vol. 1.* Fried, R., ed. New York: Marcel Dekker, Inc., pp. 1-29.                                                                       462

Mokrasch, L.C. (1972): Preparation and properties of animal proteins soluble in organic solvents. *Preparative Biochemistry.* (In press)                      459,461

Mokrasch, L.C. and Andelman, R. (1968): Non-enzymic incorporation of amines into proteolipid protein. *J. Neurochem.* 15:1207-1216.

Mokrasch, L.C. and Manner, P. (1963): Incorporation of $^{14}$C-amino acids and [$^{14}$C] palmitate into proteolipids of rat brains in vitro. *J. Neurochem.* 10:541-547.                                                                     484

Moody, M.F. (1963): X-ray diffraction pattern of nerve myelin: a method for determining the phases. *Science* 142:1173-1174.                              528,529, 539

Page

Moore, J.W. and Narahashi, T. (1967): Tetrodotoxin's highly selective blockage of    563
an ionic channel. *Fed. Proc.* 26:1655-1663.

Moscona, A.A. (1962): Cellular interactions in experimental histogenesis. *Int. Rev.*    560
*Exp. Pathol.* 1:371-428.

Mueller, P. and Rudin, D.O. (1968): Action potentials induced in bimolecular lipid    566
membranes. *Nature* 217:713-719.

Murray, M.R. (1965): Nervous tissue in vitro. *In: Cells and Tissues in Culture, Vol.*    482
2. Willmer, E.N., ed., New York: Academic Press, pp. 373-455.

Nakao, A., Davis, W., and Einstein, E. (1966a): Basic proteins from the acidic    461,546,
extract of bovine spinal cord. II. Encephalitogenic, immunologic and structural    547
interrelationships. *Biochim. Biophys. Acta* 130:171-179.

Nakao, A., Davis, W.J., and Einstein, E.R. (1966b): Basic proteins from the acidic    461,546,
extract of bovine spinal cord. I. Isolation and characterization. *Biochim.*    547
*Biophys. Acta* 130:163-170.

Napolitano, L., Le Baron, F., and Scaletti, J. (1967): Preservation of myelin lamellar    525
structure in the absence of lipid. A correlated chemical and morphological
study. *J. Cell Biol.* 34:817-826.

Napolitano, L.M. and Scallen, T.J. (1969): Observations on the fine structure of    524,559
peripheral nerve myelin. *Anat. Rec.* 163:1-6.

Neskovic, N.M., Nussbaum, J.L., and Mandel, P.(1969): Enzymatic synthesis of    490,491
psychosine in "Jimpy" mice brain. *FEBS Letters* 3:199-201.

Neskovic, N., Nussbaum, J.L., and Mandel, P. (1969): Etude de la galactosyl-    490,491
sphingosine transférase du cerveau de souris mutante "Quaking." *C.R. Soc. Biol.*
269:1125-1128.

Neskovic, N., Nussbaum, J.L., and Mandel, P. (1970a): A study of glycolipid    491,492
metabolism in myelination disorder of Jimpy and Quaking mice. *Brain Res.*
21:39-53.

Neskovic, N.M., Nussbaum, J.L., and Mandel, P. (1970b): Enzymatic deficiency in    490,491
neurological mutants. Brain uridine diphosphate galactose: ceramide galactosyl
transferase in Jimpy mouse. *FEBS Letters* 8:213-216.

Norton, W.T. (1971a): The myelin sheath. *In: The Cellular and Molecular Basis of*    452,504,
*Neurologic Disease.* Shy, G.M., Goldensohn, E.S., and Appel, S.M., eds. Phila-    505
delphia: Lea and Febiger. (In press)

Norton, W.T. (1971b): Recent developments in the investigation of purified    502,503
myelin. *In: Advances in Experimental Medicine and Biology, Vol. 13. Chemistry*
*of Brain Development.* Paoletti, R. and Davison, A.N., eds. New York: Plenum
Press, pp. 327-337.

Norton, W.T. (1972): The chemistry of myelin. *In: Basic Neurochemistry.* Albers,
R.W., Agranoff, B., Katzman, R. and Siegel, G.S., eds. Boston: Little, Brown &
Co. (In press)

Page

Norton, W.T. and Autilio, L.A. (1965): The chemical composition of bovine CNS myelin. *Ann. N.Y. Acad. Sci.* 122:77-85.

Norton, W.T. and Autilio, L.A. (1966): The lipid composition of purified bovine brain myelin. *J. Neurochem.* 13:213-222.                                                  456,457,
                                                                                                                                                          550,554

Norton, W.T. and Poduslo, S. (1966): Metachromatic leucodystrophy: chemically abnormal myelin and cerebral biopsy studies of three siblings. *In: Variation in Chemical Composition of the Nervous System as Determined by Developmental and Genetic Factors.* Ansell, G.B., ed. New York: Pergamon, p. 82. (Abstr.)

Norton, W.T., Poduslo, S.E.,and Raine, C.S. (1971): The isolation, composition and      503,504
ultrastructure of bovine oligodendroglia. *Trans. Am. Soc. Neurochem.* 2:98.
(Abstr.)

Norton, W.T., Poduslo, S.E., and Suzuki, K. (1966): Subacute sclerosing leuko-encephalitis. II. Chemical studies including abnormal myelin and an abnormal ganglioside pattern. *J. Neuropathol. Exp. Neurol.* 25:582-597.

Norton, W.T. and Turnbull, J.M. (1970): The isolation and lipid composition of a         503,506
myelin-free, axon-enriched fraction from the CNS. *Fed. Proc.* 29:472. (Abstr.)

Nussbaum, J.L., Bieth, R.,and Mandel, P. (1963): Phosphatides in myelin sheaths              456
and repartition of sphingomyelin in the brain. *Nature* 198:586-587.

Nussbaum, J.L., Neskovic, N.,and Kostic, D. (1970): Modifications portant sur les        489,490
lipides du cerveau chez la souris de souche Jimpy. *In: Les Mutants Pathologiques chez l'Animal. Leur Intérêt dans la Recherche Bio-Médicale. No. 924, Coll. Internat. CNRS.* Sabourdy, M., ed. Paris: CNRS.

Nussbaum, J.L., Neskovic, N.M., Kostic, D.M.,and Mandel, P. (1968): Etude des                469
lipides du cerveau d'un mutant de souris (Jimpy) atteint d'un trouble de la myélinisation. *Bull. Soc. Chim. Biol.* 50:2194-2196.

Nussbaum, J.L., Neskovic, N.,and Mandel, P. (1969): A study of lipid components          469,489
in brain of the "Jimpy" mouse, a mutant with myelin deficiency. *J. Neurochem.* 16:927-934.

Nussbaum, J.L., Neskovic, N.M.,and Mandel, P. (1971): Fatty acid composition of              492
phosphatides and glycolipids in Jimpy mouse brain. *J. Neurochem.* (In press)

O'Brien, J.S. (1965): Stability of the myelin membrane. *Science* 147:1099-1107.         551,563

O'Brien, J.S. (1967): Cell membranes—composition: structure: function. *J. Theor.*      452,480,
*Biol.* 15:307-324.                                                                      549,562

O'Brien, J.S. (1969): Five gangliosidoses. *Lancet* 2:805.

O'Brien, J.S., Fillerup, D.L.,and Mead, J.F. (1964): Brain lipids: I. Quantification     550,558
and fatty-acid composition of cerebroside sulfate in human cerebral gray and white matter. *J. Lipid Res.* 5:109-116.

Page

O'Brien, J.S. and Sampson, E.L. (1965a): Fatty acid and fatty aldehyde compo- 454,456,
sition of the major brain lipids in normal human gray matter, white matter, and 457,550,
myelin. *J. Lipid Res.* 6:545-551. 552

O'Brien, J.S. and Sampson, E.L. (1965b): Lipid composition of the normal human 550,557
brain: gray matter, white matter, and myelin. *J. Lipid Res.* 6:537-544.

O'Brien, J.S. and Sampson, E.L. (1965c): Myelin membrane: a molecular abnormal- 455
ity. *Science* 150:1613-1614.

O'Brien, J.S., Sampson, E.L.,and Stern, M.B. (1967): Lipid composition of myelin 454,456,
from the peripheral nervous system. Intradural spinal roots. *J. Neurochem.* 549,550,
14:357-365. 551,552,561

Ogawa, T. and Vernon, L.P. (1969): A fraction from *Anabaena variabilis* enriched 562
in the reaction center chlorophyll P700. *Biochim. Biophys. Acta* 180:334-346.

Okada, S. and O'Brien, J.S. (1968): Generalized gangliosidosis: beta-galactosidase
deficiency. *Science* 160:1002-1004.

Okada, S. and O'Brien, J.S. (1969): Tay-Sachs disease: Generalized absence of a 489
beta-D-N-acetylhexosaminidase component. *Science* 165:698-700.

Oshiro, Y. and Eylar, E.H. (1970): Allergic encephalomyelitis: A comparison of the 463,464,
encephalitogenic A1 protein from human and bovine brain. *Arch. Biochem.* 546
*Biophys.* 138:606-613.

Palmer, K.J. and Schmitt, F.O. (1941): X-ray diffraction studies of lipide emul- 517
sions. *J. Cell. Comp. Physiol.* 17:385-394.

Palmer, K.J., Schmitt, F.O., and Chargaff, E. (1941): X-ray diffraction studies of 448,517
certain lipide-protein complexes. *J. Cell. Comp. Physiol.* 18:43-47.

Park, R.B. (1965): Substructure of chloroplast lamellae. *J. Cell Biol.* 27:151-161. 562

Patterson, J.D. and Finean, J.B. (1961): Ultracentrifugal fractionation of nerve 451
tissue. *J. Neurochem.* 7:251-258.

Pauling, L. and Corey, R.B. (1951): Configurations of polypeptide chains with fa- 555
vored orientations around single bonds: two new pleated sheets. *Proc. Nat.*
*Acad. Sci.* 37:729-740.

Peters, A. (1960): The formation and structure of myelin sheaths in the central ner- 522
vous system. *J. Biophys. Biochem. Cytol.* 8:431-446.

Peters, A., Palay, S.L.,and Webster, H.deF. (1970): *The Fine Structure of the Ner-*
*vous System: The Cells and Their Processes.* New York: Hoeber Medical Division, Harper and Row.

Page

Peters, A. and Vaughn, J.E. (1970): Morphology and development of the myelin        484
sheath. *In: Myelination*. Davison, A.N. and Peters, A., eds. Springfield, Ill.:
Charles C Thomas, pp. 3-79.

Peterson, E.R. and Murray, M.R. (1965): Patterns of peripheral demyelination in     496
vitro. *Ann. N.Y. Acad. Sci.* 122:39-50.

Pilz, H. and Mehl, E. (1966): Untersuchungen zur Lipoidzusammensetzung des          457
menschlichen Myelins. *Hoppe Seylers Z. Physiol. Chem.* 346:306-309.

Pinto da Silva, P. and Branton, D. (1970): Membrane splitting in freeze-etching. Covalently bound ferritin as a membrane marker. *J. Cell Biol.* 45:598-605.

Poduslo, S.E. and Norton, W.T. (1971): The bulk separation of neuroglia and          504
neuron perikarya. *In: Methods in Neurochemistry. Vol. I.* Marks, N. and
Rodnight, R., eds. New York: Plenum Press. (In press)

Rambourg, A., Hernandez, W., and Leblond, C.P. (1969): Detection of complex          560
carbohydrates in the Golgi apparatus of rat cells. *J. Cell Biol.* 40:395-414.

Rambourg, A., Neutra, M., and Leblond, C.P. (1966): Presence of a "cell coat" rich    560
in carbohydrate at the surface of cells in the rat. *Anat. Rec.* 154:41-71.

Rand, R.P. and Luzzati, V. (1968): X-ray diffraction study in water of lipids    538,555
extracted from human erythrocytes: The position of cholesterol in the lipid
lamellae. *Biophys. J.* 8:125-137.

Ranvier, L. (1889): *Traité Technique d'Histologie.* Paris: Savy.                    445

Rawlins, F.A., Hedley-Whyte, E.T., Villegas, G., and Uzman, B.G. (1970): Reutilization of cholesterol-1,2-H$^3$ in the regeneration of peripheral nerve. An     479
autoradiographic study. *Lab. Invest.* 22:237-240.

Rawlins, F.A. and Uzman, B.G. (1970): Effect of AY-9944, a cholesterol biosynthesis inhibitor, on peripheral nerve myelination. *Lab. Invest.* 23:184-189.

Rawlins, F.A. and Uzman, B.G. (1970): Retardation of peripheral nerve myelination in mice treated with inhibitors of cholesterol biosynthesis. A quantitative electron microscopic study. *J. Cell Biol.* 46:505-517.

Retzius, G. (1890): Zur Kenntniss des peripherischen Nervensystems bei               565
Crustaceen. *In: Biologische Untersuchungen, Vol. 1.* Stockholm: Samson and
Wallin, pp. 38-50.

Revel, J.-P. and Hamilton, D.W. (1969): The double nature of the intermediate        524
dense line in peripheral nerve myelin. *Anat. Rec.* 163:7-16.

Rivas, E. and Luzzati, V. (1969): Polymorphisme des lipides polaires et des
galacto-lipides de chloroplastes de maïs, en présence d'eau. *J. Mol. Biol.*
41:261-275.

Page

Robertson, J.D. (1955): The ultrastructure of adult vertebrate peripheral mye-     521
linated nerve fibers in relation to myelinogenesis. *J. Biophys. Biochem. Cytol.*
1:271-278.

Robertson, J.D. (1957): New observations on the ultrastructure of the membranes     484
of frog peripheral nerve fibers. *J. Biophys. Biochem. Cytol.* 3:1043-1048.

Robertson, J.D. (1958a): Structural alterations in nerve fibers produced by     522,524,
hypotonic and hypertonic solutions. *J. Biophys. Biochem. Cytol.* 4:349-364.     525

Robertson, J.D. (1958b): The ultrastructure of Schmidt-Lanterman clefts and     525
related shearing defects of the myelin sheath. *J. Biophys. Biochem. Cytol.*
4:39-46.

Robertson, J.D. (1960a): A molecular theory of cell membrane structure. *In:*     522,560
*Fourth International Conference on Electron Microscopy, Vol. 2. (Biol. Med.-
Teil. Berlin, 10-17 Sept., 1958.)* Bargmann, W., Möllenstedt, G., Niers, H.,
Peters, D., Ruska, E., and Wolpers, C., eds. Berlin: Springer-Verlag. pp. 159-171.

Robertson, J.D. (1960b): The molecular biology of cell membranes. *In: Molecular*     560
*Biology.* Nachmansohn, D., ed. New York: Academic Press, pp. 87-151.

Robertson, J.D. (1960c): The molecular structure and contact relationships of cell     477,522,
membranes. *Progr. Biophys.* 10:344-418.     525,560

Robertson, J.D. (1961a): Cell membranes and the origin of mitochondria. *In:*     524
*Regional Neurochemistry.* Kety, S.S. and Elkes, J., eds., Oxford: Pergamon
Press, pp. 497-534.

Robertson, J.D. (1961b): The unit membrane. *In: Electron Microscopy in Anatomy.* Boyd, J.D., Johnson, F.R. and Lever, J.D., eds. London: Edward Arnold,
pp. 74-99.

Robertson, J.D. (1961c): Ultrastructure of excitable membranes and the crayfish     522
median-giant synapse. *Ann. N.Y. Acad. Sci.* 94:339-389.

Robertson, J.D. (1962): The unit membrane of cells and mechanisms of myelin     470
formation. *Res. Publ. Assoc. Res. Nerv. Ment. Dis.* 40:94-158.

Robertson, J.D. (1963): The occurrence of a subunit pattern in the unit membranes     564
of club endings in Mauthner cell synapses in goldfish brains. *J. Cell Biol.*
19:201-221.

Robertson, J.D. (1964): Unit membranes: a review with recent new studies of
experimental alterations and a new subunit structure in synaptic membranes. *In:*
*Cellular Membranes in Development.* Locke, M., ed. New York: Academic Press,
pp. 1-79.

Robertson, J.D. (1965): The synapse: morphological and chemical correlates of     564
function. *Neurosciences Res. Prog. Bull.* 3(4):1-79. Also *In: Neurosciences
Research Symposium Summaries, Vol. 1.* Schmitt, F.O. and Melnechuk, T., eds.
Cambridge, Mass.: M.I.T. Press, pp. 463-541.

Page

Robertson, J.D. (1966a): Granulo-fibrillar and globular substructure in unit       483
membranes. *Ann. N.Y. Acad. Sci.* 137:421-440.

Robertson, J.D. (1966b): The unit membrane and the Danielli-Davson model. *In:*    520
*Intracellular Transport, Vol. 5, Symposium of the International Society for Cell*
*Biology.* Warren, K.B., ed. New York: Academic Press, pp. 1-31.

Robertson, J.D. (1967): Origin of the unit membrane concept. *Protoplasma*
63:218-245.

Robertson, J.D. (1969): Molecular structure of biological membranes. *In: Hand-*   525
*book of Molecular Cytology.* Lima-de-Faria, A., ed. Amsterdam: North-Holland
Publ. Co., pp. 1403-1443.

Robertson, J.D. (1970): The ultrastructure of synapses. *In: The Neurosciences:*   525
*Second Study Program.* Schmitt, F.O., editor-in-chief. New York: Rockefeller
University Press, pp. 715-728.

Rosenbluth, J. (1966): Redundant myelin sheaths and other ultrastructural features  525
of the toad cerebellum. *J. Cell Biol.* 28:73-93.

Rossiter, R.J. (1962): Chemical constituents of brain and nerve. *In: Neuro-*
*chemistry, 2nd Ed..* Elliott, K.A.C., Page, I.H., and Quastel, J.H., eds.
Springfield, Ill.: Charles C Thomas, pp. 10-54.

Rumsby, M.G. and Finean, J.B. (1966a): The action of organic solvents on the       504
myelin sheath of peripheral nerve tissue. I. Methanol, ethanol, chloroform and
chloroform-methanol (2:1, v/v). *J. Neurochem.* 13:1501-1507.

Rumsby, M.G. and Finean, J.B. (1966b): The action of organic solvents on the       504
myelin sheath of peripheral nervous tissue. II. Short-chain aliphatic alcohols. *J.*
*Neurochem.* 13:1509-1511.

Rumsby, M.G. and Finean, J.B. (1966c): The action of organic solvents on the       504
myelin sheath of peripheral nervous tissue. III. Chlorinated hydrocarbons. *J.*
*Neurochem.* 13:1513-1515.

Rumsby, M.G., Riekkinen, P.J., and Arstila, A.V. (1970): A critical evaluation of
myelin purification. Non-specific esterase activity associated with central nerve
myelin preparations. *Brain Res.* 24:495-516.

Salway, J.G., Kai, M., and Hawthorne, J.N. (1967): Triphosphoinositide phospho-    487
monoesterase activity in nerve cell bodies, neuroglia and subcellular fractions
from whole rat brain. *J. Neurochem.* 14:1013-1024.

Schmidt, W.J. (1924): *Die Bausteine des Tierkörpers in polarisiertem Lichte.* Bonn:  445
Friedrich Cohen.

Schmidt, W.J. (1936): Doppelbrechung und Feinbau der Markscheide der               445,447,
Nervenfasern. *Z. Zellforsch. Mikrosk. Anat.* 23:657-676.                          507,509

Page

Schmitt, F.O. (1936): Nerve ultrastructure as revealed by X-ray diffraction and          448
polarized light studies. *Cold Spring Harbor Symp. Quant. Biol.* 4:7-12.

Schmitt, F.O. (1941): Some protein patterns in cells. *Growth, Third Growth
Symposium* 5:1-20.

Schmitt, F.O. (1944): Structural proteins of cells and tissues. *Adv. Protein Chem.*       448
1:25-68.

Schmitt, F.O. (1950a): The colloidal organization of the nerve fiber. *In: Genetic
Neurology.* Weiss, P., ed. Chicago: University of Chicago Press, pp. 40-52.

Schmitt, F.O. (1950b): The ultrastructure of the nerve myelin sheath. *Res. Publ.*         448
*Assoc. Res. Nerv. Ment. Dis.* 28:247-254.

Schmitt, F.O. (1957): Structure and properties of nerve membranes. *In: The
Metabolism of the Nervous System.* Richter, D., ed. New York: Pergamon Press,
pp. 35-47.

Schmitt, F.O. (1958): Axon-satellite cell relationships in peripheral nerve fibers.
*Exp. Cell Res. Suppl.* 5:33-57.

Schmitt, F.O. (1959): Molecular organization of the nerve fiber. *Rev. Mod. Phys.*
31:455-465.

Schmitt, F.O. (1959): Ultrastructure of nerve myelin and its bearing on fundamental concepts of the structure and function of nerve fibers. *In: The Biology
of Myelin.* Korey, S.R., ed. New York: Hoeber-Harper, pp. 1-36.

Schmitt, F.O. and Bear, R.S. (1937): The optical properties of vertebrate nerve           507
axons as related to fiber size. *J. Cell. Comp. Physiol.* 9:261-273.

Schmitt, F.O. and Bear, R.S. (1939): The ultrastructure of the nerve axon sheath.       447,510
*Biol. Rev.* 14:27-50.

Schmitt, F.O., Bear, R.S., and Clark, G.L. (1935a): The role of lipoids in the X-ray     447,511
diffraction patterns of nerve. *Science* 82:44-45.

Schmitt, F.O., Bear, R.S., and Clark, G.L. (1935b): X-ray diffraction studies on       447,510,
nerve. *Radiology* 25:131-151.                                                          512,514

Schmitt, F.O., Bear, R.S., and Palmer, K.J. (1941): X-ray diffraction studies on the    448,513,
structure of the nerve myelin sheath. *J. Cell. Comp. Physiol.* 18:31-42.               517,559

Schmitt, F.O. and Geren, B.B. (1956): On the significance of the Schwann cell in
the structure and function of peripheral nerve. *Protoplasma* 46:659-662.

Schmitt, F.O. and Geschwind, N. (1957): The axon surface. *In: Progress in
Biophysics and Biophysical Chemistry, Vol. 8.* Butler, J.A., ed. New York:
Pergamon Press, pp. 165-215.

Page

Schmitt, F.O. and Palmer, K.J. (1940): X-ray diffraction studies of lipide and lipide-protein systems. *Cold Spring Harbor Symp. Quant. Biol.* 8:94-101.

Schmitt, F.O. and Samson, F.E., Jr. (1969): Brain cell microenvironment. *Neuro-sciences Res. Prog. Bull.* 7(4):277-417. Also *In: Neurosciences Research Symposium Summaries, Vol. 4.* Schmitt, F.O. et al., eds. Cambridge, Mass.: M.I.T. Press 1970. pp. 191-325.          452,560

Schneider, W.C. and Hogeboom, G.H. (1952): Intracellular distribution of enzymes. IX. Certain purine-metabolizing enzymes. *J. Biol. Chem.* 195:161-166.          451

Schwarz, H.P., Kostyk, I., Marmolejo, A.,and Sarappa, C. (1967): Long-chain bases of brain and spinal cord of rabbits. *J. Neurochem.* 14:91-97.          457

Seminario, L.M., Hren, N., and Gómez, C.J. (1964): Lipid distribution in subcellular fractions of the rat brain. *J. Neurochem.* 11:197-209.          454,455

Sherman, G. and Folch-Pi, J. (1970): Rotatory dispersion and circular dichroism of brain "proteolipid" protein. *J. Neurochem.* 17:597-605.          461

Shiraki, H. and Otani, S. (1959): Clinical and pathological features of rabies post-vaccinal encephalomyelitis in man. *In: "Allergic" Encephalomyelitis.* Kies, M.W. and Alvord, E.C., eds., Springfield, Ill.: Charles C Thomas. pp. 58-129.          502

Siakotos, A.N., Rauser, G., and Fleischer, S. (1969): Phospholipid composition of human, bovine and frog myelin isolated on a large scale from brain and spinal cord. *Lipids* 4:239-242.          452

Sidman, R.L., Dickie, M.M., and Appel, S.H. (1964): Mutant mice (quaking and jimpy) with deficient myelination in the central nervous system. *Science* 144:309-311.          489

Singer, M. and Bryant, S.V. (1969): Movements in the myelin Schwann sheath of the vertebrate axon. *Nature* 221:1148-1150.          482

Sjöstrand, F.S. (1949): An electron microscope study of the retinal rods of the guinea pig eye. *J. Cell. Comp. Physiol.* 33:383-403.          447

Sjöstrand, F.S. (1950): Electron-microscopic demonstration of a membrane structure isolated from nerve tissue. *Nature* 165:482-483.          448

Sjöstrand, F.S. (1953): The lamellated structure of the nerve myelin sheath as revealed by high resolution electron microscopy. *Experientia* 9:68-69.          448

Smith, M.E. (1967): The metabolism of myelin lipids. *Adv. Lipid Res.* 5:241-278.          479,485

Smith, M.E. (1968): The turnover of myelin in the adult rat. *Biochim. Biophys. Acta* 164:285-293.          480

Smith, M.E. (1969a): An in vitro system for the study of myelin synthesis. *J. Neurochem.* 16:83-92.          480

Page

Smith, M.E. (1969b): Myelin metabolism in vitro in experimental allergic encephalomyelitis. *J. Neurochem.* 16:1099-1104.

Smith, M.E. and Eng, L.F. (1965): The turnover of the lipid components of myelin. *J. Am. Oil Chem. Soc.* 42:1013-1018.

Smith, M.E., Fumagalli, R.,and Paoletti, R. (1967): The occurrence of desmosterol      457
in myelin of developing rats. *Life Sci.* 6:1085-1091.

Smith, M.E., Hasinoff, C.,and Fumagalli, R. (1970): Inhibitors of cholesterol      482
synthesis and myelin formation. *Lipids* 5:665-671.

Smith, M.E. and Hasinoff, C. (1971): Biosynthesis of myelin proteins in vitro. *J.*      480
*Neurochem.* 18:739-747.

Stoeckenius, W. (1962): Some electron microscopical observations on liquid-      522
crystalline phases in lipid-water systems. *J. Cell Biol.* 12:221-229.

Suzuki, K., Poduslo, S.E., and Norton, W.T. (1967): Gangliosides in the myelin      468
fraction of developing rats. *Biochim. Biophys. Acta* 144:375-381.

Suzuki, K., Poduslo, J.F.,and Poduslo, S.E. (1968): Further evidence for a specific      468,553,
ganglioside fraction closely associated with myelin. *Biochim. Biophys. Acta*      560
152:576-586.

Tardieu, A. and Luzzati, V. (1970): Polymorphism of lipids. A novel cubic phase—a
cage-like network of rods with enclosed spherical micelles. *Biochim. Biophys.*
*Acta* 219:11-17.

Tenenbaum, D. and Folch-Pi, J. (1966): The preparation and characterization of      459,460,
water-soluble proteolipid protein from bovine brain white matter. *Biochim.*      544,546,
*Biophys. Acta* 115:141-147.      547

Thorun, W. and Mehl, E. (1968): Determination of molecular weights of microgram      545
quantities of protein components from biological membranes and other complex mixtures: gel electrophoresis across linear gradients of acrylamide.
*Biochim. Biophys. Acta* 160:132-134.

Thudicum, J.W. (1884): *A Treatise on the Chemical Composition of the Brain.*      449
London: Ballière, Tindall and Cox.

Tolani, A.J. and Mokrasch, L.C. (1967): Incorporation of $^{14}$C-amino acids into      485
proteolipid protein of subcellular fractions from rat brain, heart, and liver. *Life*
*Sci.* 6:1771-1774.

Tourtellotte, M.E., Branton, D.,and Keith, A. (1970): Membrane structure: spin
labeling and freeze etching of *Mycoplasma laidlawii. Proc. Nat. Acad. Sci.*
66:909-916.

Uzman, B.G. and Nogueira-Graf, G. (1957): Electron microscope studies of the      470
formation of nodes of Ranvier in mouse sciatic nerves. *J. Biophys. Biochem.*
*Cytol.* 3:589-598.

Page

Uzman, L.L. (1958): Lipophilic peptides and proteins of brain. I. Their relation to           461
development of the brain and myelin formation. *Arch. Biochem. Biophys.*
76:474-489.

Uzman, L.L. and Rosen, H. (1958): Lipophilic peptides and proteins of brain. II.            461
Composition of the neurosclerin fraction. *Arch. Biochem. Biophys.* 76:490-495.

Valentin, G. (1862): Histologische und Physiologische Studien. *Z. rat. Med.*              445
14:122-181.

van Deenen, L.L.M. (1966): Some structural and dynamic aspects of lipids in                554
biological membranes. *Ann. N.Y. Acad. Sci.* 137:717-730.

Vandenheuvel, F.A. (1965): Structural studies of biological membranes: the             534,535,
structure of myelin. *Ann. N.Y. Acad. Sci.* 122:57-76.                                 554,558,
                                                                                            559
Vergara, J., Longley, W.,and Robertson, J.D. (1969): A hexagonal arrangement of            564
subunits in membrane of mouse urinary bladder. *J. Mol. Biol.* 46:593-596.

Virchow, R. (1854): Ueber das ausgebreitete Vorkommen einer dem Nervenmark                 445
analogen Substanz in den tierischen Geweben. *Arch. path. Anat. Physiol. klin.*
*Med.* 6:562-572.

von Bibra, E. (1854): *Vergleichende Untersuchungen über das Gehirn des Menschen*          449
*und der Wirbelthiere.* Mannheim, Germany: Basserman & Mathy.

Waelsch, H., Sperry, W.M.,and Stoyanoff, V.A. (1940): A study of the synthesis             450
and deposition of lipids in brain and other tissues with deuterium as an
indicator. *J. Biol. Chem.* 135:291-296.

Waelsch, H., Sperry, W.M.,and Stoyanoff, V.A. (1941): The influence of growth              450
and myelination on the deposition and metabolism of lipids in the brain. *J. Biol.*
*Chem.* 140:885-897.

Waksman, A., Rubinstein, M.K., Kuriyama, K.,and Roberts, E. (1968): Localization            487
of $\gamma$-aminobutyric-$\alpha$-oxoglutaric acid transaminase in mouse brain. *J.*
*Neurochem.* 15:351-357.

Wallach, D.F.H., Ferber, E., Selin, D., Weidekamm, E., and Fisher, H. (1970): The
study of lipid-protein interactions in membranes by fluorescent probes.
*Biochim. Biophys. Acta* 203:67-76.

Warren, B.E. (1933): X-ray diffraction in long chain liquids. *Phys. Rev.* 44:969-973.      512

Watanabe, A., Tasaki, I., and Lerman, L. (1967): Bi-ionic action potentials in squid        563
giant axons internally perfused with sodium salts. *Proc. Nat. Acad. Sci.*
58:2246-2252.

Webster, H. deF. (1964): Some ultrastructural features of segmental demyelination
and myelin regeneration in peripheral nerve. *Progr. Brain Res.* 13:151-174.

Page

Webster, H. deF. (1965): The relationship between Schmidt-Lantermann incisures
and myelin segmentation during Wallerian degeneration. *Ann. N.Y. Acad. Sci.*
122:29-38.

Webster, H. deF. (1971): The geometry of peripheral myelin sheaths during their          470,471,
formation and growth in rat sciatic nerves. *J. Cell Biol.* 48:348-367.                  472,473,
                                                                                         474,475,525

Webster, H. deF. and O'Connell, M.F. (1970): Myelin formation in peripheral
nerves. A morphological reappraisal and its neuropathological significance.
*In: Proceedings VI International Congress Neuropathology*. Paris: Masson & Cie.,
579-588.

Webster, H. deF. and Spiro, D. (1960): Phase and electron microscopic studies of             474
experimental demyelination. I. Variations in myelin sheath contour in normal
guinea pig sciatic nerve. *J. Neuropathol. Exp. Neurol.* 19:42-69.

Westall, F.C., Robinson, A.B., Caccam, J., Jackson, J.,and Eylar, E.H. (1971):           500,501
Essential chemical requirements for induction of allergic encephalomyelitis.
*Nature* 229:22-24.

Wiener, O. (1912): Die Theorie des Mischkörpers für das Feld der stationären                 447
Strömung, *Abh. math.-phys. Klasse königl. sächs. Ges. Wissensch. Leipzig*
32:507-604.

Wilkins, M.H.F., Blaurock, A.E., and Engelman, D.M. (1971): Bilayer structure in         533,543
membranes. *Nature (New Biology)* 230:72-76.                                                 561

Wiśniewski, H. and Morell, P. (1971): Quaking mouse: ultrastructural evidence
for arrest of myelinogenesis. *Brain Res.* 29:63-73.

Wolfe, L.S., Mossard, J.M., and Jossot, G. (1970): La maladie de Fabry. *La Presse*          489
*Medicale* 78:2051-2052.

Wolfgram, F. (1966): A new proteolipid fraction of the nervous system. I. Isolation      460,465,
and amino acid analyses. *J. Neurochem.* 13:461-470.                                     545,546,547

Wolfgram, F. and Kotorii, K. (1968a): The composition of the myelin proteins of          459,546,
the central nervous system. *J. Neurochem.* 15:1281-1290.                                    547

Wolfgram, F. and Kotorii, K. (1968b): The composition of the myelin proteins of          459,546,
the peripheral nervous system. *J. Neurochem.* 15:1291-1295.                             547,560

Wolfgram, F. and Rose, A.S. (1961): A study of some component proteins of                504,546,
central and peripheral nerve myelin. *J. Neurochem.* 8:161-168.                              547

Wolfgram, F. and Rose, A.S. (1962): The amino acid composition of central and
peripheral nerve neurokeratin. *J. Neurochem.* 9:623-627.

Wolman, M. (1957): Histochemical study of myelinization in the rat. *Bull. Res.*             466
*Counc. Israel* 6E:163-167.

Page

Wolman, M. (1965): Structure of biological membranes: difference between          504
membrane layers. *Biol. Conf. "Ohola"* 10:33-35.

Wolman, M. and Hestrin-Lerner, S. (1960): A histochemical contribution to the      560
study of the molecular morphology of myelin sheath. *J. Neurochem.* 5:114-120.

Wolman, M. and Weiner, H. (1965): Role of calcium in stabilizing the structure of   504
biological membranes. *Biol. Conf. "Ohola"* 10:36-39.

Wood, J.C. and King, N. (1971): Turnover of basic protein of rat brain. *Nature*    480
229:56-57.

Worthington, C.R. (1969): The interpretation of low-angle X-ray data from planar   531,533
and concentric multilayered structures. The use of one-dimensional electron
density strip models. *Biophys. J.* 9:222-234.

Worthington, C.R. (1970): On the interpretation of X-ray diffraction intensities     542
from chemically treated frog sciatic nerve. *Biophys. J.* 10:675-677.

Worthington, C.R. and Blaurock, A.E. (1968): Electron density model for nerve
myelin. *Nature* 218:87-88.

Worthington, C.R. and Blaurock, A.E. (1969a): A low-angle X-ray diffraction study   529,539
of the swelling behavior of peripheral nerve myelin. *Biochim. Biophys. Acta*
173:427-435.

Worthington, C.R. and Blaurock, A.E. (1969b): A structural analysis of nerve        528,531,
myelin. *Biophys. J.* 9:970-990.                                                     533,539

Yabuuchi, H. and O'Brien, J.S. (1968): Positional distribution of fatty acids in
glycerophosphotides of bovine gray matter. *J. Lipid Res.* 9:65-67.

Zahnd, J.P. and Bonaventure, N. (1969): Données ultrastructurales et électro-
physiologiques obtenues au niveau du système nerveux central chez la souris
"Jimpy." *C.R. Soc. Biol.* 163(7):1631-1635.

# INDEX

# Are Apes Capable of Language?

A report based on an NRP Conference
held April 19-20, 1970

by

**Detlev Ploog**
(Conference Chairman)
Max Planck Institute for Psychiatry
Munich, West Germany

and

**Theodore Melnechuk**
Neurosciences Research Program
Brookline, Massachusetts

Dorothy W. Bishop
NRP Writer-Editor

CONTENTS

## LIST OF PARTICIPANTS

Mr. Julian H. Bigelow*
Permanent Member
Institute for Advanced Study
Princeton, New Jersey 08540

Dr. Roger W. Brown
Department of Psychology
William James Hall
Harvard University
Cambridge, Massachusetts 02138

Dr. Jerome S. Bruner
Department of Psychology
and Center for Cognitive
Development
William James Hall
Harvard University
Cambridge, Massachusetts 02138

Dr. Irven DeVore
Department of Anthropology and
Social Relations
William James Hall
Harvard University
Cambridge, Massachusetts 02138

Dr. Merrill F. Garrett
Department of Psychology
Massachusetts Institute of Technology
Cambridge, Massachusetts 02139

Dr. Richard Hirsh*
Division of Biology
California Institute of Technology
Pasadena, California 91109

Dr. Eric H. Lenneberg†
Departments of Psychology and
Neurobiology
Cornell University
Ithaca, New York 14850

Mr. Theodore Melnechuk
Neurosciences Research Program
280 Newton Street
Brookline, Massachusetts 02146

Dr. Detlev Ploog*
Max Planck Institute for
Psychiatry
Kraepelinstrasse 2/10
8 Munich 23, West Germany

Dr. David Premack
Department of Psychology
University of California
Santa Barbara, California 93106

Dr. Francis O. Schmitt
Neurosciences Research Program
280 Newton Street
Brookline, Massachusetts 02146

Dr. Howard H. Wang*
Natural Sciences Building I
University of California
Santa Cruz, California 95060

Dr. Frederic G. Worden
Neurosciences Research Program
280 Newton Street
Brookline, Massachusetts 02146

---

*Current address. Bigelow, Hirsh, Ploog, and Wang were Staff Scientists at the Neurosciences Research Program Center in Brookline at the time of the Conference in April 1970.
†Currently a Staff Scientist at the NRP Center.
Note: NRP Work Session summaries are reviewed and revised by participants prior to publication.

A. Chimpanzee Sarah, with three plastic "words" on magnetic board. Estimated age, 6 years. [Premack]

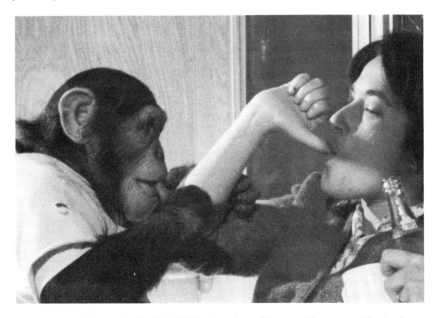

B. Chimpanzee Washoe, signing "drink." Estimated age, 2½ years. [Gardner and Gardner]

Figure 1.

## ARE APES CAPABLE OF LANGUAGE?

He who understands baboon would do more
toward metaphysics than Locke.

Charles Darwin

## I. INTRODUCTION

This *Bulletin* reports a one-day Conference, including background material and second thoughts of the participants, that grew out of our earlier longer Work Session on "Primate Communication" (Ploog and Melnechuk, 1969) where the work of Gardner and Gardner (1969) and Premack and Schwartz (1966) was mentioned; in these experiments humans taught themselves linguistic systems in which a chimpanzee could participate.

The subject was not pursued in further detail at that Work Session, but the Work Session participants insisted it would be of the utmost interest to follow up this remarkable research.

The original plan of the Conference proposed four main components: (1) a review of results of R. Allen Gardner and Beatrice T. Gardner, (2) a review of David Premack's recent results, (3) a comparison of results of these two experimental programs, and (4) a discussion of that comparison relative to our title question, "Are Apes Capable of Language?"

Unfortunately, the Gardners were unable to attend because of their commitment to their laboratory program. The meeting therefore centered primarily on Premack's results; however, our participant panel included Roger Brown, a recipient of many personal communications from the Gardners, who was able to speak about their work in some detail and bring into the discussion some of the Gardners' available data and complementary evidence on questions related to the topic. The Gardners, like other invited participants, were given and accepted the opportunity to review, revise, and augment this report, and their additional comments appear in Appendix I.

Marler, who was instrumental in refocusing our attention on the topic, was also unable to attend the meeting at the only time at which it could be scheduled.* Other participants represented a wide spectrum

*Peter Marler's suggestion that we hold the Conference and his willingness to read the draft of this report are appreciated. The authors also gratefully acknowledge the usefulness of the meeting notes and criticisms of the draft made by the participants and the NRP staff, especially those by Julian H. Bigelow.

of interests in the field of communication, ranging from the study of primates, through examination of the evolution and development of human verbal and nonverbal capabilities, to modern linguistics.

## Reasons for the Conference

### Cognitive Fundamentals of Communication

In opening the Conference, Ploog referred to one of the reasons for a neuroscientific interest in the linguistic capabilities of nonhuman primates: that is, the desire to learn whether language and nonverbal communication, and perhaps even noncommunicative behavior, have in common any rule-generating brain mechanisms. He quoted the end of his concluding essay in "Primate Communication" (Ploog and Melnechuk, 1969):

> ... the general problem of cognition ... is a central problem for research into the biological foundation of human language. *What are the common cognitive processes of nonverbal primate communication and of human language?* There are so far no grounds on which this question can be answered. ... Are there common [brain] mechanisms that contribute to the generation of nonverbal signals and of language as well? ...
>
> If there is a neuronal mechanism that serves as a common denominator for both nonverbal and verbal communication, one could speculate about this mechanism as a rule-generating system that sets the stage for communicative behavior of primates in general. ... [However,] nonverbal communication and language may be based on totally separate mechanisms, and [any] rule-generating systems of the two forms of communication may operate on different principles.

### Antecedents of Language

Brown mentioned another motivating scientific question that asks how animals that have language—i.e., men—evolved from animals that are assumed to have lacked it—e.g., hominoid primates. The frustrating absence of historical and fossil evidence on the question makes welcome the suggestion that man's nonhuman ancestors did not lack at least some competence in language, even though this suggestion seems merely to push the problem back in time a few million years, for it turns the difficult task of explaining a supposed quantum jump into the apparently less difficult task of explaining a developmental process.

## Rank-Ordering of Languages

Brown next pointed out that linguists, anthropologists, and neurologists wish it were possible to rank-order natural languages along some dimension of complexity. This does not appear to be feasible with contemporaneous languages, all of which seem to be more or less equivalent in complexity. For example, Garrett said that all known languages had only three ways to combine two sentences: (1) relatively, as in "I like a drink which is weaker than whiskey"; (2) conjunctively, as in "I like wine, but I dislike whiskey"; and (3) complementarily, as in "John knows that I like wine."

Natural languages differ in vocabulary size, but this is merely an index of environmental and technical variety. However, a chimpanzee-human language could conceivably throw light on the theoretical as well as evolutionary organization of language.

## Therapeutic Possibilities

The possibility of improving the linguistic performance of aphasic children and adults gave added impetus to an inquiry into methods and results of language training for juvenile nonhuman primates.

## Philosophical Import

Brown (1970) wrote half-humorously, "It is lonely being the only language-using species in the Universe." And Lenneberg (1967, p. viii) wrote, "[O.] Marx's historical survey ... makes it clear that there has been an implicit assumption since ancient times that language is somehow dependent upon man's nature."* An unequivocal demonstration that at least one other species besides man is capable of language would add to the series of man's great reconceptions of himself, including the ideas of heliocentrism, evolution, economic determinism, the unconscious, and conditionability—concepts that in our time have already been augmented by evolving ecological and ethological attitudes.

*Lenneberg later commented that neither the dependence of language "upon man's nature" nor its species-specific quality would exclude homologies in other primates.

These reorientations, while they were humbling from one point of view, insofar as they seemed to demote man from his assumed status as the uniquely rational, articulate, and free-willing heir of Creation, have from another point of view enriched him, not only with the benefits of greater realism but also with the revelation of his unsuspected nearness to his fellow creatures. DeVore (Ploog and Melnechuk, 1969, p. 459) has pointed out that of the three behavioral attributes once thought to be specific to the human species—namely, tool-making, incest avoidance, and language—forms of the first two have been found in primate species, and the supposed exclusiveness of the last is now under experimental study.

### Artificial Languages

Investigations of candidate languages have gone so far that scientists have seriously prepared to interpret and attempted to detect messages from across interplanetary space, first in the 1960's as in Project Ozma (Horowitz, 1966) and currently in proposals for projects that would be many times more sophisticated than Ozma, such as Project Cyclops (at Stanford University), or those discussed at the international conference on CETI (communication with extraterrestrial intelligence). Also, computers have been programmed to model aspects of human competence in language (Minsky, 1968; Winograd, 1971).

These two kinds of artificial communication systems were excluded from the scope of the Conference, which, at Ploog's suggestion, was concerned only with the structure and function of natural human language and the forms of it that seem to be manageable by chimpanzees.

## II. RESULTS IN BRIEF

### Earlier Experiments

Before summarizing the results of Gardner and Gardner and of Premack, it seems appropriate to give a brief account of the prior state of the art. This is drawn from the review by Kellogg (1968), to which the reader is referred for further details.

Kellogg's article in *Science* was entitled "Communication and Language in the Home-Raised Chimpanzee." It dealt with the communication and language aspects of six experiments in which psychobiologists observed and measured the reactions of chimpanzees raised in controlled experimental surroundings.

Table 1 gives some of the characteristics of the different experiments, including (1) the approximate duration of each, (2) the number of human child controls, if any, (3) the ages of the chimpanzees, (4) the names of the investigators, and (5) references to the investigators' relevant publications, if any.

Kellogg's own summary of his review follows:

> Although often misunderstood, the scientific rationale for rearing an anthropoid ape in a human household is to find out just how far the ape can go in absorbing the civilizing influences of the environment. To what degree is it capable of responding like a child and to what degree will genetic factors limit its development? At least six comprehensive studies by qualified investigators have been directed wholly or partly to this problem. All of these studies employed young chimpanzees as subjects and some also had in-house child controls whose day-to-day development could be compared directly with that of the experimental animal. In general, the results of this sort of research show that the home-raised chimp adapts rapidly to the physical features of the household. It does many things as well as a human child and some of them better (for example, those involving strength and climbing).
>
> By far the greatest deficiency shown by the ape in the human environment is its lack of language ability. This eliminates the verbal communication which humans enjoy, and with it the vast amount of social intercourse and learning which are dependent on language. Even amid human surroundings a chimp never prattles or babbles as a young

TABLE 1

Principal Chimpanzee-Raising Experiments
[Adapted from Kellogg, 1968]

| Investigator and Publication Dates | Approximate Duration of Experiment | Approximate Age of Chimpanzee | Sex and Name | | Child Controls Number |
|---|---|---|---|---|---|
| C. F. Jacobsen, M. M. Jacobsen, and J. G. Yashioka (1932) | 1 year | A few days | F. | Alpha | 0 |
| W. N. Kellogg and L. A. Kellogg (1933; reprinted 1967; 1968) | 9 months | 7 ½ months | F. | Gua | 1 |
| Nadezhna N. Ladygina-Kohts* (1935) | 2 ½ years | 1 ½ years | M. | Joni | 1 |
| Finch† | 3 years | 3 days | M. | Fin | 2 |
| K. J. Hayes and C. Hayes (1950, 1951, 1952, 1953, 1954a, 1954b); C. Hayes (1951, 1970) | 6 ½ years | 3 days | F. | Viki | 0 |
| R. A. Gardner and B. T. Gardner (1968, 1969; 1971a, 1971b) | In progress | 8 to 14 months | F. | Washoe | 0‡ |
| D. Premack and A. Schwartz (1966) | 1 ½ years | 2 to 2 ½ years | F. | Sarah§ | 0 |
| D. Premack (1970a, 1970b, 1971) | In progress | 5 to 6 years | F. | Sarah§ | 0 |

*This is the full version of the name given by R. M. Yerkes as "N. Kohts" and since cited as such by authors referring to her book, which is not generally obtainable in the West.
†Did not publish, according to Kellogg.
‡See Appendix I for discussion of Washoe's child controls.
§This is the same animal.

child does when beginning to talk. Although it imitates the behavior of others readily, it seems to lack the ability for vocal imitation. The neural speech centers of the brain are no doubt deficient in this respect and it is possible also that the larynx and speech organs are incapable of producing the complex sound patterns of human language.* One long-time attempt to teach a home-raised chimp to pronounce human words succeeded only in getting the animal to mouth unvoiced whispers of the words "mama," "papa," "cup," and "up."

At the same time, a chimpanzee in the home, as in the wild state, uses gestures or movements as communicating signals. This suggests the possibility of training a home-raised ape to employ a standardized system of gestures as a means of two-way communication. Such an investigation is now under way, using a gesture language devised for the deaf. Considerable progress has already been made in both the receiving and sending of gesture signals by this method. The technique seems to offer a much greater likelihood of success than other methods of intercommunication between chimpanzees and humans.

Note Kellogg's emphasis on the human homelike environment in which his own and the other earlier chimp-raising experiments were conducted, and the human child controls employed in some of them. Neither the Gardners nor Premack have made any special attempt to simulate a homelike environment, although Washoe's is much more homelike than is Sarah's (see Figure 1).

The Gardners and Premack also differ from their predecessors in having given up any attempt to teach apes to understand and produce spoken language in favor of trying to teach them some alternative form of communication based on aspects of language.

### Results of R. Allen Gardner and Beatrice T. Gardner with Washoe

In addition to the publications listed in Table 1, the Gardners have been writing "Summaries of Washoe's Diary" and from time to time circulating installments to psychologists who might be expected to

---

*See Lieberman (1968) and Lieberman et al. (1969) for an argument that man alone among the primates possesses the capacity for continuous modification of the cross-section of pharyngeal portions of the vocal tract while vocalizing, which permits the finely regulated sounds of human speech.

take an interest. A brief account of their results as of late summer 1967 concludes the Kellogg review already cited. Still more information about Washoe's performance, and a comparison of it to the language behavior of two-year-old children, is given in "The First Sentences of Child and Chimpanzee" (Brown, 1970). What follows is the Gardners' own summary of their August 1969 paper, "Teaching Sign Language to a Chimpanzee," together with a table given in it:

> We set ourselves the task of teaching an animal to use a form of human language. Highly intelligent and highly social, the chimpanzee is an obvious choice for such a study, yet it has not been possible to teach a member of this species more than a few spoken words. We reasoned that a spoken language, such as English, might be an inappropriate medium of communication for a chimpanzee. This led us to choose American Sign Language (ASL), the gestural system of communication used by the deaf in North America, for the project.
>
> The youngest infant that we could obtain was a wild-born female, whom we named Washoe, and who was estimated to be between 8 and 14 months old when we began our program of training. The laboratory conditions, while not patterned after those of a human family (as in the studies of Kellogg and Kellogg [1967] and of Hayes and Hayes [1951, 1952]), involved a minimum of confinement and a maximum of social interaction with human companions. For all practical purposes, the only verbal communication was in ASL, and the chimpanzee was maximally exposed to the use of this language by human beings.
>
> It was necessary to develop a rough-and-ready mixture of training methods. There was evidence that some of Washoe's early signs were acquired by delayed imitation of the signing behavior of her human companions, but very few, if any, of her early signs were introduced by immediate imitation. Manual babbling was directly fostered and did increase in the course of the project. A number of signs were introduced by shaping and instrumental conditioning. A particularly effective and convenient method of shaping consisted of holding Washoe's hands, forming them into a configuration, and putting them through the movements of a sign.
>
> We have listed more than 30 signs that Washoe acquired and could use spontaneously and appropriately by the end of the 22nd month of the project. [See Table 2.] The signs acquired earliest were simple demands. Most of the later signs have been names for objects,

TABLE 2

Signs Used Reliably by Chimpanzee Washoe Within 22 Months of the Beginning of Training.
Signs Listed in Order of Their Original Appearance in Her Repertoire.
[Gardner and Gardner, 1969]*

| Signs | Description | Context |
|---|---|---|
| Come-gimme | Beckoning motion, with wrist or knuckles as pivot. | Sign made to persons or animals, also for objects out of reach. Often combined: "come tickle," "gimme sweet," etc. |
| More | Fingertips are brought together, usually overhead. (Correct ASL form: tips of the tapered hand touch repeatedly.) | When asking for continuation or repetition of activities such as swinging or tickling, for second helpings of food, etc. Also used to ask for repetition of some performance, such as a somersault. |
| Up | Arm extends upward, and index finger may also point up. | Wants a lift to reach objects such as grapes on vine, or leaves; or wants to be placed on someone's shoulders; or wants to leave potty-chair. |
| Sweet | Index or index and second fingers touch tip of wagging tongue. (Correct ASL form: index and second fingers extended side by side.) | For dessert; used spontaneously at end of meal. Also, when asking for candy. |
| Open | Flat hands are placed side by side, palms down, then drawn apart while rotated to palms up. | At door of house, room, car, refrigerator, or cupboard; on containers such as jars; and on faucets. |
| Tickle | The index finger of one hand is drawn across the back of the other hand. (Related to ASL "touch.") | For tickling or for chasing games. |
| Go | Opposite of "come-gimme." | While walking hand-in-hand or riding on someone's shoulders. Washoe usually indicates the direction desired. |
| Out | Curved hand grasps tapered hand; then tapered hand is withdrawn upward. | When passing through doorways; until recently, used for both "in" and "out." Also, when asking to be taken outdoors. |
| Hurry | Open hand is shaken at the wrist. (Correct ASL form: index and second fingers extended side by side.) | Often follows signs such as "come-gimme," "out," "open," and "go," particularly if there is a delay before Washoe is obeyed. Also, used while watching her meal being prepared. |
| Hear-listen | Index finger touches ear. | For loud or strange sounds: bells, car horns, sonic booms, etc. Also, for asking someone to hold a watch to her ear. |

*See text and this reference for the criterion of reliability and for the method of assigning the date of original appearance.

613

TABLE 2 (Continued)

| Signs | Description | Context |
|---|---|---|
| Toothbrush | Index finger is used as brush, to rub front teeth. | When Washoe has finished her meal, or at other times when shown a toothbrush. |
| Drink | Thumb is extended from fisted hand and touches mouth. | For water, formula, soda pop, etc. For soda pop, often combined with "sweet." |
| Hurt | Extended index fingers are jabbed toward each other. Can be used to indicate location of pain. | To indicate cuts and bruises on herself or on others. Can be elicited by red stains on a person's skin or by tears in clothing. |
| Sorry | Fisted hand clasps and unclasps at shoulder. (Correct ASL form: fisted hand is rubbed over heart with circular motion.) | After biting someone, or when someone has been hurt in another way (not necessarily by Washoe). When told to apologize for mischief. |
| Funny | Tip of index finger presses nose, and Washoe snorts. (Correct ASL form: index and second fingers used; no snort.) | When soliciting interaction play, and during games. Occasionally, when being pursued after mischief. |
| Please | Open hand is drawn across chest. (Correct ASL form: fingertips used, and circular motion.) | When asking for objects and activities. Frequently combined: "Please go," "Out, please," "Please drink." |
| Food-eat | Several fingers of one hand are placed in mouth. (Correct ASL form: fingertips of tapered hand touch mouth repeatedly.) | During meals and preparation of meals. |
| Flower | Tip of index finger touches one or both nostrils. (Correct ASL form: tips of tapered hand touch first one nostril, then the other.) | For flowers. |
| Cover-blanket | Draws one hand toward self over the back of the other. | At bedtime or naptime, and, on cold days, when Washoe wants to be taken out. |
| Dog | Repeated slapping on thigh. | For dogs and for barking. |
| You | Index finger points at a person's chest. | Indicates successive turns in games. Also used in response to questions such as "Who tickle?" "Who brush?" |
| Napkin-bib | Fingertips wipe the mouth region. | For bib, for washcloth, and for Kleenex. |

614

TABLE 2 (Continued)

| Signs | Description | Context |
|---|---|---|
| In | Opposite of "out." | Wants to go indoors, or wants someone to join her indoors. |
| Brush | The fisted hand rubs the back of the open hand several times. (Adapted from ASL "polish.") | For hairbrush, and when asking for brushing. |
| Hat | Palm pats top of head. | For hats and caps. |
| I-me | Index finger points at, or touches, chest. | Indicates Washoe's turn, when she and a companion share food, drink, etc. Also used in phrases, such as "I drink," and in reply to questions such as "Who tickle?" (Washoe: "you"); "Who I tickle?" (Washoe: "Me.") |
| Shoes | The fisted hands are held side by side and strike down on shoes or floor. (Correct ASL form: the sides of the fisted hands strike against each other.) | For shoes and boots. |
| Smell | Palm is held before nose and moved slightly upward several times. | For scented objects: tobacco, perfume, sage, etc. |
| Pants | Palms of the flat hands are drawn up against the body toward waist. | For diapers, rubber pants, trousers. |
| Clothes | Fingertips brush down the chest. | For Washoe's jacket, nightgown, and shirts; also for our clothing. |
| Cat | Thumb and index finger grasp cheek hair near side of mouth and are drawn outward (representing cat's whiskers). | For cats. |
| Key | Palm of one hand is repeatedly touched with the index finger of the other. (Correct ASL form: crooked index finger is rotated against palm.) | Used for keys and locks and to ask us to unlock a door. |
| Baby | One forearm is placed in the crook of the other, as if cradling a baby. | For dolls, including animal dolls such as a toy horse and duck. |
| Clean | The open palm of one hand is passed over the open palm of the other. | Used when Washoe is washing, or being washed, or when a companion is washing hands or some other object. Also used for "soap." |

615

which Washoe has used both as demands and as answers to questions. Washoe readily used noun signs to name pictures of objects as well as actual objects and has frequently called the attention of her companions to pictures and objects by naming them. Once acquired, the signs have not remained specific to the original referents but have been transferred spontaneously to a wide class of appropriate referents. At this writing, Washoe's rate of acquisition of new signs is still accelerating.

From the time she had eight or ten signs in her repertoire, Washoe began to use them in strings of two or more. During the period covered by this article we made no deliberate effort to elicit combinations other than by our own habitual use of strings of signs. Some of the combined forms that Washoe has used may have been imitative, but many have been inventions of her own. Only a small proportion of the possible combinations have, in fact, been observed. This is because most of Washoe's combinations include one of a limited group of signs that act as combiners. Among the signs that Washoe has recently acquired are the pronouns "I-me" and "you." When these occur in combinations the result resembles a short sentence. In terms of the eventual level of communication that a chimpanzee might be able to attain, the most promising results have been spontaneous naming, spontaneous transfer to new referents, and spontaneous combinations and recombinations of signs. (Gardner and Gardner, 1969)

## David Premack's Results with Sarah

At the time of the NRP Conference, Premack's results had not been published except in the form of an invited address presented at the American Psychological Association in Washington, D.C., September 1969, and at the Symposium on Cognitive Processes of Nonhuman Primates in Pittsburgh in March 1970. That paper has since been published (Premack, 1970a), as has a popular article (Premack, 1970b). A fuller, later account—partly in response to issues raised at the NRP Conference—was published in May 1971 (Premack, 1971a), and Premack is working on the manuscript of a full-length book on the topic, which he hopes will enable other investigators to attempt replication of his work.*

*Premack has asked us especially to acknowledge the patience and ingenuity of the six research assistants who have participated in the study: Mary Morgan, the main trainer, who has been assisted at various times by James Olson, Randolph Funk, Deborah Peterson, J. Scott, and Ann Premack.

We base the following account of his work on the two scientific articles and on additional data brought out at the NRP Conference and in later conversations. It is brief, giving only the essence of his method and highlights of his results. Additional details are brought out in later sections.

## Purpose of the Experiment

Premack believes that a functional definition of language ought to supplement the predominantly structural definition given it by many modern linguists. Without attempting an exhaustive list of what an organism would have to do in order to give evidence of having language, Premack suggests that such a list would almost certainly include at least three capacities: (1) to recognize and use new words, i.e., the names of concepts; (2) to understand and generate sentences, i.e., statements, demands, and questions; and (3) by means of language, to learn more language.

In man, these three functions are expressed in certain complex ways that may not be the only possible ways. For example, each of the approximately 100,000 words* of a given natural language has a spoken form that is some combinatorial sequence of several of the 40 to 50 different phonemes† of that language. Such a vocal system is unique to man, and if such vocalization is considered essential to language, we may foreclose the title question, for no ape appears to be capable of it.‡ However, semantics and logic may be more widely distributed

---

*With regard to this estimate, two facts may be cited: First, a computer analysis of the *Standard Corpus of Present-Day Edited American English,* a collection of 500 randomly chosen 2,000-word samples of 15 types of published prose, showed that in its 1,014,232 running words of text there were 50,406 different words. The *Corpus* is described in Kučera and Francis (1967). Second, *The Random House Dictionary of the English Language, College Edition* (Urdang and Flexner, 1968) is estimated to contain approximately 100,000 entries, and *The American Heritage Dictionary of the English Language* (Morris, 1969) has approximately 155,000 entries, as reported in a review of both by Quine (1969).

†Phonemes are the smallest units of speech that distinguish one utterance or word from another in a given language. The *m* of *mat* and the *b* of *bat* are two English phonemes. According to Smith (1969), English has an inventory of 21 consonant, 9 vowel, and 3 glide semivowel phonemes, and 12 phonemes of pitch, stress, and juncture, for a total of 45.

‡Premack, reviewing this report, has added the following comment:

It is clear that the chimp cannot reproduce human language sounds and also that man enjoys apparently innate perceptual advantages with regard to these sounds (see Eimas et al., 1971). But it does not follow that the chimp is incapable of managing a system of primitive meaningless elements–based upon a medium suitable to the chimp–which combine to produce words in the same way that phonemes combine to produce words. Indeed, we have not yet tested the perceptual basis on which Sarah discriminates among the plastic words. For all we know, she may do so on the basis of a small set of minimal differences which she invented or discovered, despite the fact that we did not construct the words on the basis of any such set. We think this unlikely, but it is not a possibility that we can rule out at this time.

across the species, and they may be a sufficient basis for the three functions of language cited by Premack as fundamental. Therefore, to approach a more general possible definition of language, he thinks it desirable to demonstrate the capacity for the three functions in at least one species other than man.

### Reason for Choosing a Chimpanzee

The most illuminating demonstration of a capacity for the fundamental functions of language would be in a nonprimate species. However, Premack asserts that teaching an organism language amounts in large part to mapping concepts that it already possesses—in the human case, concepts such as color, shape, size, identity, numerosity, etc. He believes that it is difficult, perhaps impossible, to disinter the corresponding concepts in species far removed from man. He gives the greater accessibility of this built-in knowledge in the chimpanzee as the main reason he began with it, rather than the facts of its playfulness and brightness.

In 1964, Premack obtained a young female chimp named Sarah, then apparently 2 years old. For the first 1½ years, he collaborated with the linguist A. Schwartz in designing and proposing an experiment that would have involved an artificial manual language expressed by movements of a joystick (Premack and Schwartz, 1966). In 1966, after the end of the collaboration, Premack designed a different method, which, with some later developments, he has used ever since.

### Physical Arrangement and Teaching Procedure

The elements of the language that Premack has been teaching Sarah are "words," expressed physically in the form of pieces of plastic that differ in shape, size, texture, and color. Metal-backed, they will adhere to a magnetized wallboard (see Figure 1). The words for a given lesson are laid out on a table top, or occasionally supplied in a bucket, and generally both teacher and pupil handle them. When real objects have to be put with the plastic words, a table top or even the floor is used; this makes no difference to Sarah. However, at the start of her training, she seemed to prefer that, on whatever plane, the words be arrayed in a column rather than in a row, and this "Chinese" convention has since been adhered to.

When Sarah was younger, she was unconfined, and Premack and his assistants worked with her across a table. Now, being more dangerous, she is kept in a mesh cage 3 x 6 m, which has an opening through which a trainer reaches in and "writes" on the nearer half of the wallboard. Sarah responds on the other half.

Sessions of 20 trials have an average duration of 12 minutes; an exceptionally good session lasts 40 minutes. Sarah is emotional, and there are days on which she will not learn.

Premack points out that several consequences follow from the use of plastic words:

1. Being "written" rather than spoken or gestured, an expression is relatively permanent, eliminating the short-term memory problem with respect to the given task.

2. Production can occur as early as comprehension because sentence production requires only that the chimp cause the pieces of plastic to adhere to the board; it does not depend upon complex response differentiation, as is the case when words are spoken or gestured.

3. Most important, perhaps: Because the experimenter produces the words, while the subject merely uses them, the experimenter can make as few or as many words available at any moment in time as he chooses. This makes possible a one-to-one substitution training method, which may be the simplest of all training methods.

### Training Strategy

For each of the language functions Premack wanted to try to demonstrate in the animal, he had first to decide what exemplars would be convincing evidence of that function—for example, at least partial competence in the interrogative mode would be shown by the ability to answer questions. He had then to devise at least one training procedure that would produce the selected competence.

Each training procedure consisted of an ordered series of small steps manageable by a docile organism. At each step, the desired behavior was brought about in part by limiting the possibility of any other behavior. There was no algorithm for generating such behavior recipes; each one was based on judgment, and constituted only a method, not a theory.

As will be seen shortly, Premack succeeded in devising training procedures that have led Sarah to possess a vocabulary of more than 125 "words" with which—at a correctness level generally between 75% and 80%—she can complete or correct imperative, declarative, and interrogative "sentences," some of which are compound, using correct word order.

However, preceding all of the training there was a stage in which Sarah became at ease with her trainers. Friendly and intimate relationships were established. These were very important, even essential, for Sarah's willingness to interreact with human beings in an artificial mode of communication. (Washoe, too, required warm familiarity in order to communicate in her corresponding sign language.) The emotional bond with Sarah was forged through such simple social transactions as feeding her her favorite fruit, such as bananas, and by gift exchanges in which she got the better of the deal.

Among the first words taught to Sarah—such as a blue plastic triangle for "apple"—were the names of fruits that she was allowed to eat in the process. Other contingent rewards have included patting Sarah, saying "good girl" to her in an approving manner, and giving her the coveted cherry in a fruit cocktail.

These emotional aspects of Sarah's training will be brought up again in later sections of this report on the possibility of unwitting signals and on questions concerning motivation.

### Toward Competence in the Imperative Mode

Premack gives the following account of one major training sequence:

> Sarah's original lexicon, words like "apple," "banana," "give," "take," her own name "Sarah," and the names of her trainers, were taught her in the following manner: A social transaction was first established in which the trainer, sitting across the table, offered her pieces of fruit which she became accustomed to accepting and eating. After she had come to believe that fruit would always be given her, a change was made in the procedure and rather than receive fruit she had henceforth to ask for it. A sentence she might use for this purpose would be, "Mary give apple Sarah," where "Mary" is the name of one of her trainers, and

"Sarah" is her own name. Such sentences were the target of the training. Notice that this situation or transaction between trainer and Sarah is subject to variation at four points, and that the four slots in the sentence may be seen as attesting to that fact.

First, it is possible to change *donors* so that on some trials it would be necessary to say, "Mary do so and so" but on other trials necessary to address the same request to Randy, or Jim, or Debby, etc. Similarly, the *object* might vary so that on some trials one would request apple but on other trials banana, etc. The *recipient,* too, is subject to variation, and even though Sarah may be reluctant to write sentences such as "Mary give apple Jim," directing that the apple be given to a recipient other than herself, the situation contains the possibility of this kind of variation. In other words, the sentence contains a slot for the indirect oject, in which such changes would be registered. Finally, giving is not the only *operation* that can be applied to the objects in question. It may be Sarah's preferred operation, and thus one from which it will be difficult to distract her, but apples, oranges, and the like can also be washed, cut, inserted into containers, etc.

Sarah was taught to map this situation one word class (or slot) at a time. Variation was made in the membership of each of the four classes, accompanied by variation in corresponding language elements. When apple was the object present, she was given one piece of plastic and required to place it on the writing board, a different word when banana was present, and still a different word when orange was the fruit offered. Not until she placed the word on the board did she receive the fruit. Choice trials were used to determine whether she had formed the appropriate association between word and fruit. That is, she was given several words, but only one piece of fruit, and required to put the correct word on the board. Her associations were also tested by obtaining independent preference orderings for the fruits on the one hand and their would-be names on the other. This step was necessary, because errors on choice trials might not mean a failure in learning, but rather constitute a request for a preferred fruit not offered on the trial in question. A

concordance of better than 80% between preference orderings for fruits and words, which was obtained on four
occasions, established that her associations were better than
the choice trials indicated; more important, it indicated that
she was using the individual names essentially as requests.

Members of the other classes were mapped in the
same fashion. Donors were changed, and at the same time a
change was made in the language element. In addition, as
each new class was mapped, the language requirement was
increased. In the beginning, Sarah merely took the fruit,
observed benevolently by the trainer. Then one word was
required, then two, then three, until finally a large set of
words could be placed before her, from which she produced
such target sentences as "Mary give apple Sarah," with the
sentence correct at all four points.

## Word Order

Correct word order is always used by the experimenter, and it is
used reliably—approximately 80% of trials—by Sarah. However, as
Premack says, "The subject may produce properly ordered strings of
words and yet give no essential evidence that the string is a sentence. A
sentence differs from a string of words in that it has an internal
organization, a knowledge of which is a necessary condition for
responding to it." Therefore, to test Sarah's knowledge of a hierarchical
sentence organization, he taught her to comprehend the target sentence
"Sarah insert banana pail apple dish."

The training proceeded by three steps. First, Sarah learned four
simple sentences from which this target sentence could be com-
pounded. These were,

> Sarah insert banana pail.
> Sarah insert apple pail.
> Sarah insert banana dish.
> Sarah insert apple dish.

Next, she was given, side by side, all possible pairs of these sentences;
for example,

> Sarah insert banana dish.
> Sarah insert apple pail.

This was done to accustom her to carry out two acts of insertion in tandem. Finally, all possible pairs were again combined but this time one after the other, and then each conjunction of two simple sentences was gradually converted into one compound sentence, as follows:

| | |
|---|---|
| Sarah insert banana pail Sarah insert apple dish. | (1) |
| Sarah insert banana pail insert apple dish. | (2) |
| Sarah insert banana pail apple dish. | (3) |

The deletions in (2) and (3) did not disrupt her usual 75% to 80% performance; neither did subsequent generalization tests in which the fruit names were changed. Premack has since changed the verb, e.g., "take" instead of "insert," without disrupting her performance. As usual, her comprehension was examined both by having her perform the described action correctly and by having her fill in each of the possible blank slots in a test sentence.

**Declarative Mode**

Up to this point, the ability described has been to respond appropriately to sentences in the imperative mode. However, Sarah also learned to understand and to complete formulations that can be interpreted as declarative sentences, as in descriptions of the placement of colored cards, such as "Red on green," etc.

**Toward Competence in the Interrogative Mode**

Here is Premack's own account of another major training sequence:

*Same-Different*

Two of the first words taught Sarah were "same" and "different." These simple predicates represent a good starting point for several reasons: (1) They can be introduced as the relation between unnamed objects and thus have no linguistic prerequisites. (2) With the use of language-free procedures, such as "match to sample," it is possible to determine whether or not the subject is capable of the perceptual judgments that underlie the linguistic distinction. (3) An unsuspected bonus is associated with "same-different"; this comparison provides the simplest context in which to introduce the interrogative.

After match-to-sample procedures had shown that Sarah was indeed capable of matching like objects, words to label these relations were introduced in the following manner: Two like objects (cups or spoons) were placed before her, she was given a plastic word meaning "same," and was induced to place the word between the objects. On other trials she was given two unlike objects (cup and spoon), a second word intended to mean "different," and induced to place it between them. Because she was given only one word on trials of this kind, no errors were possible. On the next step she was given both words in the presence of the same materials, e.g., cup-cup or spoon-cup, and required to choose between the two words. Typically, she was given about five trials on each of the two cases, and typically she performed at about the 80% level. Once she achieved this criterion, she was advanced to the transfer test in which the same kind of choice trial was repeated, now, however, with items not used in training. So, for example, she had to decide whether to call two balls "same" or "different," and likewise a ball and a nail. She was given at least twenty transfer trials and typically she performed at about the 80% level, on this and on other words and objects. Errors were far more frequent on the last ten than on the first ten trials, suggesting inattention or boredom rather than a learning failure. Errors on the first trial of a transfer test were practically nonexistent, or about two in over forty such tests.

*Interrogative Mode*

The interrogative was introduced by placing what amounted to a question mark in the formerly blank space between the two objects and requiring Sarah to displace this marker with the correct answer, either "same" or "different." This method of introducing the question is based on the argument that a question, despite the diversity of mechanisms that natural languages employ to mark the form, is essentially a "missing element in a potentially complete construction." "X same X" is a complete construction, as is "X different Y"; whereas both "X ? X" and "X ? Y" are incomplete. Both forms raise the question of the word through which they can find completion.

Three kinds of questions were introduced in this simple context. Because questions are based on missing elements, with a two-term relation such as *same-different,* two question forms were generated directly, one by removing the predicate ("same" or "different"), another by removing one or even both of the arguments instancing the predicate (X or Y). A third form was generated indirectly by appending the interrogative marker, which itself stands for missing element(s), to the head of the construction and then requiring that it be replaced by a further element, specifically "yes" or "no." "No" (the negative particle) was taught the subject earlier in another context before training her on the yes-no form of the question. "Yes" (the positive particle) was introduced through training on the yes-no question. Examples and paraphrases of all three question forms are shown in Figures 2 and 3.

An example of two versions of a *wh*-type is shown in the upper panel of Figure 2. These questions can be paraphrased as "X is what to X?" and "X is what to Y?" The alternatives are "same" and "different," and Sarah's task was to replace the interrogative marker with the appropriate predicate.

Two versions of a second type of *wh*-question are shown in the lower panel of Figure 2; they can be paraphrased as "What is X the same as?" and "What is X different from?" Now the alternatives are no longer the predicates "same" and "different" but the objects themselves. Sarah's task remained the same, however: to replace the interrogative marker with the proper object and thereby complete the construction.

The yes-no question, the third form that can be generated in this context, is shown in four versions in Figure 3. They can be paraphrased as (1) "Is X the same as X?", (2) "Is X different from X?", (3) "Is X different from Y?", and (4) "Is X the same as Y?" These questions were formed not by removing any item from the string, but rather by appending the interrogative marker to the head of the string. Her alternatives were now "yes" and "no" rather than either the predicates or the objects.

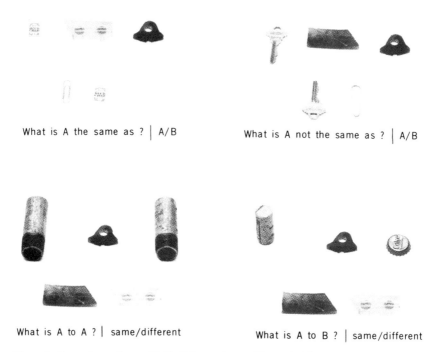

What is A the same as ?  |  A/B                  What is A not the same as ?  |  A/B

What is A to A ?  |  same/different              What is A to B ?  |  same/different

Figure 2.  Four *wh* questions, with English paraphrases. [Premack, 1971a]

Sarah was taught all of the question forms and was ultimately able to apply them widely. The purpose of this procedure is to accustom the subject to the question form at the earliest possible stage in language training, because the question provides a means both for collecting evidence about what the subject already knows on the one hand and for teaching new material on the other. The success of the device depends in part on the subject's ability to comprehend the interrogative marker abstractly. Does the interrogative marker, although introduced in the context of the same-different construction, nevertheless mean "missing element(s)" in general, or does it mean merely "missing element(s) in the same-different construction"? If the latter, then a new interrogative marker would have to be introduced for each of the (in principle, infinitely many) recognizably different completable constructions that are to be taught to the subject in the course of language learning. However, if the subject defines the marker more abstractly, then it will stand

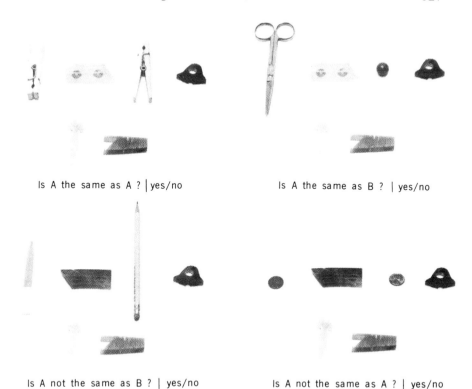

Is A the same as A ? | yes/no            Is A the same as B ? | yes/no

Is A not the same as B ? | yes/no        Is A not the same as A ? | yes/no

Figure 3. Four yes-no questions, with English paraphrases. "Different" would be a more suitable paraphrase than "not same," because a negative particle was not used. [Premack, 1971a]

for "missing element(s)" in any construction; in other words, the subject recognizes the sentence as being completable, as having the possibility of being well formed. The evidence strongly suggested that Sarah did interpret the particle abstractly, making it possible to use the same interrogative marker in all constructions.

*Name Of*

Other predicates, including some of the logical connectives, all of the quantifiers, the names of property classes (color, shape, size), and so forth, were taught to Sarah in a manner analogous to the one in which she was taught the simple predicates "same" and "different." For example, the important concept "name of" was taught her by presenting a string consisting of the name for apple, the interrogative marker, and an actual apple. The string, "'apple' ? apple," can be paraphrased as asking the question, "What is the relation between the word 'apple' and the object apple?" The answer supplied here was a piece of plastic meaning "name of," with which she displaced the interrogative marker forming the sentence, "'apple' name of apple." She was given the same training on the relation between the word "banana," which had been taught to her in an earlier lesson, and the object banana. Following the customary five trials on each of the two positive instances of the concept, she was given five trials on each of two negative instances. In effect, she was asked the question, "What is the relation between the word 'apple' and the object banana?" as well as "What is the relation between the word 'banana' and the object apple?" On both occasions the answer given her was "not-name-of," which had been previously taught, i.e., the negative particle appended to the word "name of." In this way she was induced to write both "'apple' not name of banana," and "'banana' not name of apple." These trials were then followed by choice trials where she was presented the same questions and both the words "name of" and "not-name-of" and was required to choose between them. Once she attained the 80% criterion, she was advanced to the transfer test, where she was required to answer the same questions with respect to items not used in training. For example, she was asked such questions as, "'raisin' ? apricot," i.e., "What is the relation between the word 'raisin' and the object apricot?" Her performance on transfer was, in this test and others, comparable to her terminal performance on the choice trials.

"Name of" has therefore been used productively, i.e., to generate new instances of itself. Thus, to teach a noun to Sarah now, Premack and his assistants no longer go through the laborious routine of rotating a class through a range of values while simultaneously changing the corresponding language element. Instead, they show Sarah the new word to be learned, the word "name of," and the object being named as a positive instance; then they give a negative instance to rule out the possibility that the name is conferred by mere physical contiguity to the object—and Sarah learns the new name.

**Vocabulary**

Sarah now has a vocabulary of more than 125 words that she can use with a reliability of 75% to 80%. About 45 of these words are listed in Table 3. Premack believes that her vocabulary could be expanded indefinitely.

Premack has not yet allowed Sarah the opportunity to make a word—that is, to assign a meaning of her own to an arbitrary piece of plastic previously not a word in the language.

TABLE 3

Some of the More Than 125 Terms Associated with the
Plastic "Words" Reliably Used by Chimpanzee Sarah.
[Derived from Premack, 1970a,b; 1971a]

| Term | Part of Speech | Term | Part of Speech |
|---|---|---|---|
| apple | "Nouns" | Jim | "Proper nouns" |
| apricot | | Mary | |
| banana | | Randy | |
| cherry | | Sarah | |
| fig | | cut | "Verbs" |
| grape | | eat | |
| orange | | give | |
| raisin | | insert | |
| caramel | | take | |
| chocolate | | same as | "Adjectives" |
| crackerjack | | different from | |
| gumdrop | | blue | |
| honey-bread | | green | |
| jam-cracker | | red | |
| cup | | yellow | |
| dish | | brown | |
| pail | | square | |
| the color of | | round | |
| not the color of | | in | "Prepositions" |
| the shape of | | on | |
| not the shape of | | in front of | |
| the name of | | to the side of | |
| not the name of | | no | "Particles" |
| | | yes | |
| | | if/then | "Conjunction" |

**If/Then**

Lately, Premack has taught Sarah the conditional "if/then" by presenting two fruits, rewarding the choice of one but not the other, and showing a changed preference for the rewarded fruit. Premack chose to symbolize the term "if/then" by a single piece of plastic, placed between the two expressions to be conjoined, in the way logicians place their corresponding symbol for material implication, as in "A ⊃ B," read "if A, then B." Sarah can now handle a sentence such as "If Sarah give apple Jim, then Mary give cherry Sarah."

**Conclusion**

As is the case with some children in primary school, grammar is not Sarah's favorite subject; there is a limit to the number of tests she has been willing to accept on a grammatical topic, and that limit is not always sufficient to include all of the control sentences that might be desired. However, on the basis of many tests, Premack feels that he has not yet met any ceiling to Sarah's linguistic capacity, for she has ultimately learned every word and usage he has tried to teach her, and she has responded to every test of variant examples at a level of 70% to 80% correct.

## III. ARE APES ALREADY USING LANGUAGE?

The participants were impressed by the reported achievements of Washoe and Sarah—so much so that, as Brown later commented, they postponed discussing the question "Are Apes Capable of Language?" in favor of "Are Apes Already Using Language?"

### Possibility of Unwitting Signals ("Clever Hans")

Before reviewing the profiles of Sarah's and Washoe's achievements and considering the capacities that underlay them, the conferees attempted to rule out the possibility that what is exhibited in the chimps' performance is a phenomenon of the sort associated with "Der Kluge Hans," the horse that astonished the world with apparent feats of computation signaled by hoof-taps in response to spoken questions, until the discovery that its trainer gave "Clever Hans" cues more easily perceived by the horse than by the observers.

With regard to Washoe, Brown and others present vouched for her ability to do what was claimed for her. A motion-picture record of Washoe using her manual signs has been kept since 1966 (Gardner and Gardner, 1966). A small part of this record was seen by Ploog subsequent to the Conference. There are no comparable films of Sarah at work with her plastic pieces, and Premack decries as too brief to be meaningful the sequences of a recent film made for educational television that show glimpses of Sarah, who was disturbed by the experience, perhaps by the unfamiliarity of being filmed in her narrow quarters under crowded conditions, and certainly by the filmers, who were strangers.*

Granting the animals' manual performance, questions remain as to whether either was a "Clever Hans," or, as Premack put it (1971a), both chimpanzees being female, a "Clever Gretel." (Perhaps it should be noted that the majority of the other chimpanzees listed in Table 1 were also females.)

Lenneberg mentioned that since the days of "Clever Hans," dogs, too, have been conclusively shown to be sensitive to fine changes

---

*This film, "The Mind of Man," produced by the British Broadcasting Company, has been shown on television throughout the world; it is not yet available as a classroom film. A subsequent book, *The Mind of Man* (1970), by Nigel Calder, author of the film, is a popular introduction to research on brain and behavior.

in the behavior of their trainers. He commented that the best test is always double-blind, or one in which no human observer knows the correct answer.

For a discussion of Washoe's double-blind tests, see Appendix I by Gardner and Gardner. As for Sarah, Premack up to the time of the Conference had not yet used the double-blind methodology, but had used several other tests, described below.

In addition to a number of informal steps, such as putting words in opaque containers so that the trainer could not influence Sarah's choice of words and concealing the trainer behind a blind when Sarah was required to choose among both verbal and nonverbal alternatives, Sarah was given more formal tests on the possible influence of nonlinguistic cues. The most obvious approach would be to expand this procedure. In an ideal case, the trainer might leave ten yes-no questions on her board, stroll out of the room in the manner of a lazy proctor, and return later to score her test. But this approach failed. She refused to work in the absence of the trainer, suggesting that her primary motivation may be social.

It was necessary to take the opposite approach. Rather than have a trainer who knew the language divorce himself from normal social behavior, Premack introduced a trainer who did not know the language but who allowed normal social behavior. Sarah was first adapted to the uninitiated, "dumb" trainer by a series of feeding experiences, so that she would work for him, and then was tested on both production and comprehension. At each lesson the dumb trainer was given an instruction sheet consisting of numbered pictures of the words, and all instructions were given to him in this number code. Instructions told him what words to set out, and Sarah was expected to produce sentences appropriate to the object before her. Other instructions told him what sequence of words (i.e., what sentence) to write on Sarah's board and served to test her comprehension. When Sarah had completed a sentence, this trainer, equipped with earphones and microphone, read the corresponding sequence of numbers to a schooled trainer in the hall. The unschooled trainer then heard either "yes" or "no," depending of course on whether her sentence was correct or not, after which he either did or did not praise Sarah and either did or did not give her the object in question.

Under these conditions, Sarah's performance was poorer than usual, both in substance and in form. Her percentage of successful response fell from about 80% to about 70%; this was still above chance performance. If word order was not considered, chance on production tests ranged from 13% to 20%; if word order was considered, chance was substantially less. However, final word order was less often correct than usual; and even when the final word order was correct, she did not produce entire sentences in their final order; instead, she reverted to an earlier style in which she placed correct words on the board in incorrect order, and then made one or two rearrangements before settling on a final order. The verticality of her sentences decayed to a sprawl, also characteristic of an earlier stage. These changes notwithstanding, Sarah's performance was substantially above chance. Premack concludes that Sarah did give proper responses for a trainer who did not know the language.* One can only speculate about the sources of the decrements, emotional factors seeming the most likely. Premack considers the most interesting among the possibilities the inconsistency of reinforcement and wonders how well the human being would do if tested in the same situation, i.e., if led to "talk" to someone who was actually dumb but who gave at least a partial impersonation of someone who knew the language.

Lenneberg suggested that, since several people know the "language" that has been developed with the chimp, Premack should try having one of them ask a question and another one observe its solution. Bruner added that the first experimenter might also take a handful of plastic words from the bucket without seeing what they were.

Hirsh pointed out that "Clever Hans" answered questions by interrupting an otherwise continuous repetitive activity when signaled; he asked if, in contrast, Sarah answered a question by initiating a response not otherwise given. Premack said yes, but what seems more conclusive to him is that, in preparing to solve a problem, Sarah will look over a large number of plastic words, pull out the relevant ones, push the irrelevant ones aside, and produce her answer only from the relevant set.

Lenneberg suggested that a check on "leakage" is not the only purpose of a double-blind test; it will also check the possibility that the animal's responses are being "overinterpreted" by the observer. He further asked if one experimenter could give her a command while an

---

*After the test series, the "dumb" trainer was given tests for language comprehension. They showed that during the series, he had learned some things correctly, others incorrectly, in a pattern that did not conform to that of Sarah's errors. Moreover, Sarah's performance data reported above were taken only from first sessions of each kind, when the trainer was least likely to have learned anything. (For more details see Premack, 1971a.)

observer who does not know what was asked of the animal records the answer, which would then, after the session, be compared with the input. Perhaps there will be an answer to his question as testing continues.

Shortly after the Conference, Ploog visited both laboratories, and he is personally convinced that no answer "leakage" à la "Clever Hans" accounts for the results with either Washoe or Sarah.

## Definitions of Language

### Need for an Explicit Theory

It is sometimes assumed that a good strategy for throwing light on the question of whether apes are capable of language is a straightforward comparison of the chimpanzees' achievements with a list of the characteristics that presumably can be attributed to linguistic capacity in some acknowledged prior definition.

However, some linguists and logicians of science deny that this strategy is satisfactory (for example, see Black, 1968), and Ploog contrasted Premack's question, "What must an organism do in order to give evidence that it has language?" with the Gardners' feeling that such questions are foreign to the spirit of their research. For, as they wrote in 1969, such questions "imply a distinction between one class of communicative behavior that can be called language and another class that cannot. This in turn implies a well-established theory that could provide the distinction." If their objectives had required such a theory, they would not have been able to begin their project as early as they did, for no theory of language is "well established."

### Theoretical Disputes

Many controversies over theories animate the field. In large part, the disagreements stem from different purposes and interests, dictating different emphases, and in many cases both of two positions may be quite "correct":

1. *Some investigators hold that all observable behavior is communicative; others, that communicative behavior can be distinguished from noncommunicative behavior.*

To quote Lenneberg (1969b), "Communication can only be defined in somewhat arbitrary terms and it never represents a unified or even clearly related behavioral unity." At the NRP Work Session on "Primate Communication" (Ploog and Melnechuk, 1969), Kaufman, for one, saw no difference between social behavior and communicative behavior, and Ploog agreed that it is not easy to draw a line between the two at this stage of the analysis of signal function. Some of the Work Session participants thought that a distinction ought to be possible between the communicative and other aspects of social behavior.

2. *Some emphasize the vocal nature of language (i.e., its similarities to nonverbal vocalization systems); others emphasize its denotational and propositional capacity (i.e., its differences from nonverbal vocalization systems).*

In this connection Ploog cited an article on "Primate Communication Systems and the Emergence of Human Language," in which Jane Lancaster (1968) stated:

> The communication systems of nonhuman primates are important to understand if human language is to be understood, not because they are similar but because they are different; understanding here comes from analyzing contrast, and not similarities.

Previously, Ploog had commented that he was negative about efforts to relate primate vocalizations to human language. He saw primate vocalizations as not much different from gestures, whereas, as Altmann (1967) had pointed out, language has sounds that may be detached from action and from the immediate and contemporaneous environment, that give command over objects and partners without moving, except for the mouth. Primate vocalizations express mood, motive states, and certain interactions between partners; but this is not language in the human sense.

3. *Some emphasize the use of language for communication (i.e., its similarity to nonverbal communication systems); others emphasize its abstract nature.*

Lenneberg (1970) distinguishes the essence of language from its most familiar use (which is not its only use, as poets know):

> Just as arithmetic is not "solving equations" or "figuring out percentages," but a system of relationships (the byproducts or benefits of which are the solutions of practical problems), so language is not essentially defined as communication or exchange of information.

These are its byproducts, benefits, or applications, but the nature of language is a particular system of relationships, and language knowledge is the capacity to relate in specified ways.

4. *Some emphasize the supposed vulnerability of language, as learned behavior, to analysis in terms of the familiar Skinnerian three-term contingency "Discriminative Stimulus → Response → Reinforcer"; others emphasize its supposed requirement, because of hierarchical structuring and feed-forward in utterance, for analysis in terms of rules.*

The conflict between these two positions was discussed at the earlier Work Session on "Primate Communication," where George A. Miller said that linguists find it natural, and indeed almost unavoidable, to discuss linguistic behavior in terms of rules. They do so because it is inconceivable that the vast variety of human utterances could all have been memorized. Rather, sentences must be produced as needed according to some underlying set of rules that a language user knows, much as a large number of mathematical theorems are produced from a small set of axioms and logical rules.

Miller also distinguished the actual use of language, which Noam Chomsky calls "performance," from the inferred underlying knowledge that a language user must have in order to produce and interpret language, knowledge that Chomsky calls "competence" (Chomsky, 1965). The reason for making the distinction is that it is in the description of linguistic competence that rules seem unavoidable as explanatory concepts. Rules are not causes; they can be violated for a number of reasons, but then the exceptions can be regularized, as in the familiar English spelling rule, "I before E, except after C, etc." (See also Ploog and Melnechuk, 1969.)

5. *Some emphasize the progress that has been made toward the goal of listing and ranking linguistic rules; others emphasize the doubts about the reality—as distinct from the utility—of the rules model.*

In the last two decades, Chomsky and others have made a noteworthy advance in their attempt to formulate grammatical rules that can account for—"generate," as they say—every conceivable sentence that native speakers of a given language (English, in this case) feel to be "well formed." They posit two sorts of such rules: *phrase structure* rules, such as "Sentence → [read the arrow as "may consist of"] Noun Phrase + Predicate" and "Predicate → Verb + Noun Phrase," and economical *transformation rules,* such as the rule for the passive that changes the underlying structure of an active form into its corresponding passive form—for example, that of "A psychologist studied a chimpanzee," which is (ignoring articles for simplicity) "Noun Phrase$_1$ + Past Tense + Verb + Noun Phrase$_2$," to that of "The chimpanzee was studied by a psychologist," which is "Noun Phrase$_2$ + Past Tense + *be* + *by* + Noun Phrase$_1$." Their generative grammar is also intended to assign to each fully grammatical sentence a correct structure in which, for example, the native speaker's feeling for a hierarchy of units (from subject and predicate down to words and morphemes*) is formally represented. Finally, this generative grammar also attempts to link a description of the facts of a language to what is known about human cognitive capacities, so that it is a theory of the knowledge, conscious or not, had by a fluent speaker; as such, it is a part of human psychology (Chomsky, 1967; Lyons, 1970).

Lenneberg commented that the logic of generative syntax is also applicable to certain aspects of perception—for instance, of pictures. However, he continued, this may be no more surprising than that the number system can be used to count eggs as well as stars.

The specific assumptions of the school of generative grammarians have been, and continue to be, fruitful, but to quote Ohmann (1969), on whose brief account these paragraphs are based, "There remain many unsolved problems in grammar—indeed, many areas of confusion and controversy—[for generative grammar] provides a model of *part* of what every native speaker knows."

In a description of a computer program for understanding natural language, Winograd (1971) writes,

> What is needed is an approach which can deal meaningfully with the question "How is language organized to convey meaning?" rather than "How are syntactic structures organized when viewed in isolation?" ... We use a type of syntactic analysis which is designed to deal with questions of semantics. Rather than concentrating on the exact form of rules for shuffling around linguistic symbols, it studies the way language is structured around choices for conveying meaning.

6. *Some emphasize the generation of language; others, the interpretation of language.*

---

*A morpheme is a linguistic unit of relatively stable meaning that cannot be divided into smaller meaningful parts; e.g., words such as *man* or *most,* or word elements such as *-ly* or *al-* as found in *manly* or *almost.*

Winograd (1971) states,

> Most current theories are "generative," but it seems more
> interesting to look at the interpretive side [see Winograd, 1969, for a
> discussion of the issues involved]. The first task a child faces is
> understanding rather than producing language, and he understands
> many utterances before he can speak any. At every stage of develop-
> ment, a person can understand a much wider range of patterns than he
> produces. ... Syntax, semantics, and inference must be integrated in a
> close way, so they can share in the responsibility for interpretation.

7. *Some emphasize the distinction between semantics and
syntax; others, the difficulty of distinguishing between them.*
   "Semantics" is defined as the branch of the science of grammar
that studies the meanings of words and sentences, as distinguished from
"syntax," which studies the arrangement of words in sentences. It does
not always seem possible to separate semantic from syntactical considerations, as for example in attempting to resolve an ambiguous sentence
such as "I had three books stolen," which has at least three meanings,
according to which of three ways its words are syntactically related to
each other, as evident in these three explanations:

> I had three books stolen ... from me.
> I had three books stolen ... for me.
> I had three books stolen ... when I was caught.

Lenneberg (1971) gives another reason why a sharp dichotomy
between semantics and syntax appears to be impossible:

> In a highly inflected language, practically all relationships are or
> at least can be expressed by the words themselves (or by some
> morpheme). And even in the so-called synthetic languages, such as
> English, some relationships are signalled by words and some by such
> syntactic devices as word order. If we look at historical changes in
> language, it is clear that no relationship whatever is in principle exempt
> from being taken out of the realm of pure syntax and being incorpo-
> rated into the lexicon.

8. *Some emphasize necessary characteristics of language;
others, the systemic nature of language.*
   In the absence of a complete theory of language, Ploog asked
about the wisdom of making an attempt to select at least some of its
necessary conditions.

One approach to a set of necessary conditions is the search for those attributes true of all known natural human languages.

Ploog mentioned various attributes that had been suggested as essential by several authors; for example, the features listed by Otto Koehler (1954a,b), or the "universals of language" found by Hockett (1960, 1963; see also Greenberg, 1966, 1969), or the somewhat different formulation of what Hockett is sure is shared by all human languages of today and of a long time in the past (Hockett, 1968). Lenneberg (1969a) has also listed six characteristics of language. Altmann, starting from Hockett, has made a more extensive list, with critical discussion of criteria in primates (Altmann, 1967).

A broader approach was outlined by Hockett and Altmann in 1968, when they listed the "design features" of any communication system; and Marler (1969) pointed out that all of the design features proposed by Hockett and Altmann are exemplified somewhere in the animal kingdom but appear together only in human language. Brown commented that it may be worth noting that if such a list were to be taken as defining the notion of language, then neither Sarah nor Washoe is yet using language, since neither has been shown to manifest all the capabilities included.

A more selective approach is typified by Brown, who feels that some of these universals are more consequential than others; he has selected two of them, "semanticity" (or meaningfulness) and "openness" (or productivity) for special emphasis in his paper on Washoe (Brown, 1970).

Premack, in this report and elsewhere, has already listed three capacities that he feels an organism would have to show in order to give evidence of having language.

But Ploog also cited a paper in which Lenneberg (1971) listed four criteria that are commonly, but in Lenneberg's opinion, erroneously, used to answer the question of whether a given man or other creature has language knowledge. These are

1. The number of different words the subject produces.

2. The capacity to extend the meaning of the word to more than a single situation or object.

3. The capacity to emit an utterance in an "appropriate" situation.

4. The propensity for combining words.

In his paper, Lenneberg (1971) examines these popular criteria in detail and comes to the conclusion that they are insufficient for providing definitive evidence of language knowledge. At the Conference, he repeated his conclusion, emphasizing the irrelevance of the first criterion, for a communication system does not become more language-like as the vocabulary increases; a language can have few words.

Lenneberg's objection was not only to the particular set of criteria but also to the general approach of establishing touchstones, which he did not think was too useful, on the grounds that language is not simply the sum of its features, instead, it is a system. Thus, while he agrees to the importance of at least some of what he called Koehler's "19 biological prerequisites of language" and of Hockett's "13 logical design features," he warned against the assumption, sometimes made implicitly by followers of this approach, that language is a loose association of relatively independent abilities.

Lenneberg further commented that if language is instead a system, could another communication system be acceptably called language if there were merely an ill-defined similarity? Similarity does not necessarily mean homology in an evolutionary sense. A phylogenetically related system would not necessarily have to be formally isomorphic to human language, though some homomorphism would have to exist.

*Vocal Incapacity*

On at least one theoretical point, there was no dispute. The participants agreed that a certain one of the "universals of language" was *not* a necessary condition. The consensus was that human vocalization did not have to be produced by an organism in order for it to qualify as having language. As Brown (1970) says,

> The essential properties of language can be divorced from articulation. Meaning or "semanticity" and grammatical productivity appear not only in speech but in writing and print and in sign language. ... The chimpanzee articulatory apparatus is quite different from the human.* ... It is possible, therefore, that Viki and Gua failed not because of an incapacity that is essentially linguistic but because of a motoric ineptitude that is only incidentally linguistic.

In discarding vocalization as a criterion of language capacity, the conferees tacitly agreed not to consider a question raised by Ploog.

---

*See Lieberman, 1968, and Lieberman et al., 1969.

Pointing out that in human speech, input is normally audiovisual, while in the case of Washoe* and Sarah, the input was only visual, he had asked whether such one-channel communication was different in principle from two-channel social signaling, lacking for example its sensorimotor feedback capacity for motor matching.

## Semantic Capacities

Discussion of the empirical results started with the two animals' semantic achievements, using the admittedly vague term "semantic" for phenomena in which naming appears to be emphasized more than the organization of strings of symbols according to set principles.

### Comprehension and Production of Signs

*Perceptual Stability*

Premack made an allusion, the first of many that followed, to the partial similarity of chimpanzee and human cognition when he said that, at the very least, Sarah shared with man the cognitive process of dividing visual perceptive experience into stable elements, "things." His point was that an organism like the sea slug might perceive the world not as an aggregate of objects but in terms of some other paradigm, perhaps of gradients, as different from that of objects as the physicist's elementary-particle paradigm is from his wave-packet one. Ploog reminded the participants of the classic studies by Klüver (1936, 1965), which shattered the illusions of psychologists who had thought it easy to identify the stimulus equivalent to an animal's response or the detail really perceived of a known signal.

In addition, Premack went on, Sarah seems to have the capacity to represent a class of stable perceptive elements by a particular response.

### Sign Sets

Gardner and Gardner (1969) tell how, at an early stage of Washoe's education, she used "flower" to mean not only "blossom"

---

*The Gardners have pointed out that in the case of Washoe the input was not solely visual. Both Washoe and her human companions used auditory adjuncts to the American Sign Language. These, the Gardners feel, were comparable to the normal visual adjuncts of speech.

but also "tobacco pouch," "mentholated ointment," and other scented things, by a conceptual process like the one we recognize in children who are learning new words. Later, her original concept of "flower" was dissociated into two concepts, one a correctly used new "smell" and the other "flower."

### Familiarity of Sign Set

In commenting on the rapidity with which Sarah now learns a new association of object and sign, Ploog asked whether or not new words must nevertheless be pieces of plastic—that is, members of a set already familiar as being possible words. Premack said it was an interesting point that had not yet been tested.

### Memory

Premack reported that at the time of the Conference, Sarah had a vocabulary of more than 125 "words." Should it ever grow too large for her long-term retentive powers, visual aids could be given. The problem of short-term memory during a task was of course obviated by the use of visible plastic words that can be left up on a board. He said evidence of actual forgetting appears to be weak, although it had not been rigorously tested.

Premack thinks that human conversations serve to keep language elements alive in long-term memory while new things are being added. Because Sarah does not take part in conversations, he had been worried about her possible loss of knowledge through forgetting, so he gives her frequent rehearsals of language elements that are not used too frequently in her new training.

In thinking about Sarah's memory, it is perhaps important to remember that Sarah never has her full vocabulary available to her; rather, she has only a few alternatives at any given time. All of her production (or completion and comprehension) tests are "multiple choice" tests among a few word symbols offered as candidates.

### Invention of New Names

Melnechuk asked whether Sarah had ever spontaneously named an object, and Ploog asked if Sarah had ever disagreed with the name being given an object and tried to substitute a different piece of plastic as its name. Premack answered no to both, and said that he had not yet allowed Sarah to try to introduce a new name into the language, either by use of "name of" or by the older associational method. Bruner suggested that she be allowed to generate new words. Ploog felt that invention of names might be important evidence of having language. He stated that Washoe did not use the experimenters' sign "coldbox" in referring to the refrigerator but signed "open fooddrink" instead. However, Brown thought it possible that Washoe was not inventing a new name but merely signaling a wish.

### Errors as Wishes

Premack said that some of Sarah's "errors" in the use of fruit names turned out to be, in essence, requests for some other fruit that she preferred to the one being used in the test. By obtaining independent preference orderings on the various kinds of fruit and on their names—orderings that were in 80% agreement—Premack reassured himself that Sarah really knew which word went with which fruit.

### Colors and Other Properties

Premack described three possible methods for teaching property names, one of which failed, one of which succeeded, and one of which had not yet been tried. In the first method, Sarah was offered pieces of apple dyed either red or yellow. The dye was tasteless and except for the difference in color the pieces of apple were identical. The training consisted simply of arranging that Sarah write "give yellow" and "give red" on appropriate trials. Choice trials revealed that she had learned nothing on the previous trials. This same procedure was tried on shapes and failed there also. The procedure is one in which the subject is confronted with two objects that are identical except for the properties to be named. In one sense it should be ideal; nevertheless it fails. (See Premack, 1971b, for speculations as to the source of failure.)

The successful method was based upon the use of sets of objects that were completely dissimilar except for the property to be named. Thus in teaching "red" and "yellow" Premack used a set of red objects consisting of a ball, toy car, Lifesaver, etc., and a set of yellow objects consisting of a block, flower, crayon, etc. In the presence of all members of the red set, Sarah was required to write "give red" and in the other case "give yellow." The same method was used also to teach "round" and "square," and later "large" and "small." The method that was not tried, but which Premack thought would work, was to introduce the property as a modifier on an already named object, e.g., "give red apple." This seems a reasonable prediction in view of the fact that when property names were introduced essentially as nouns, e.g., "give red," they did transfer successfully as modifiers, e.g., "take red dish," etc.

*Binary Contrasts*

Lenneberg commented that children master color names relatively late, because they find it easier to learn polar distinctions, such as "bright/dark," than to learn a spectrum. Accordingly, he has tried making false polarities, such as "red/green," and found that children learn the colors sooner. Believing with Bruner that a bipolar distinction facilitates learning, he recommends teaching inverse relationships at the same time.

To Bruner's question whether Sarah learns binary contrasts most easily, Premack answered that he did not know, but that he has taught virtually all words in opposing pairs. Pairs have sometimes involved negations, e.g., "name of" versus "not-name-of," sometimes simple opposites, e.g., "same" versus "different," and sometimes words that were neither but that at least belonged to the same class, e.g., "apple" versus "banana." No word pairs have been taught that are still more weakly related, e.g., "red" versus "large," because the negative instances generated by such cases produce sentences in which predicates take illegitimate arguments (e.g., "red is not the size of apple," suggesting, regrettably, that it may be the size of something else; see Premack, 1971b). In addition to teaching most words in pairs, two positive and two negative instances of the concept are generally used in the teaching; for instance, "red color of apple," "yellow color of banana" as positive instances, and "red not color of banana" and "yellow not color of apple" as negative instances. Are two such instances necessary? Premack has not yet tested the matter but reports a case in which only one positive and one negative instance were used and failed; learning occurred when a second positive and negative instance were added (Premack, 1971b).

*Synonyms*

Some of the synonyms that Sarah knows were taught her inadvertently, when one training assistant would make an arbitrary piece of plastic the word for some object class without knowing that another trainer had already established some other piece of plastic as a name for that class.

## Internal Representation

*Object Versus Name*

Sarah is not at all confused when she is asked to put an object's name, instead of the object, into a container. She can tell an object from its name; e.g., "Put apple (in) dish" versus "Put name of apple (in) dish."

*Feature Analysis*

Sarah was given a series of trials in which she was presented with an apple and a pair of alternatives. On each trial she had to show which alternative she considered to be more like the apple. The alternatives used are shown in Figure 4. Sarah chose red, round, stemmed square, and round. Then the features analysis was repeated, with the object apple replaced by its plastic name. Although this was a blue triangle, she assigned it the same properties she had assigned its referent. Her analysis of the word was not of its physical form but of that which the form represents. She has also been given the same kind of test on other words, e.g., "caramel" and "Mary," in some cases being tested first with the word as the sample and then with the referent, and she has succeeded in these cases, too. The few errors she made occurred late in the session.

*Testing Perceptual Transforms*

The matching technique permits testing the animal's capacity for perceptual transforms, as in matching three-dimensional objects to

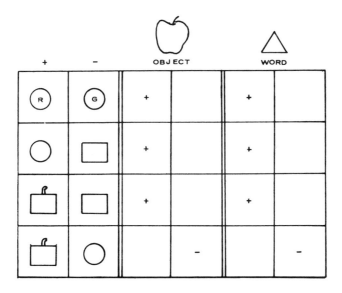

Figure 4. Features analyses of apple and "apple." [Premack, 1971a]

their two-dimensional representations and vice versa, or in matching one orientation to another. At an early stage, Sarah proved unable to match photographs of apples taken from different magazines, but testing was brief, as the experimenter's interest was in other matters. Premack could not explain Sarah's failure but mentioned that the pictured apples were of different sizes and that one had a background while the other did not. Melnechuk suggested that perhaps she would have succeeded if she had first been taught the word "picture of."

DeVore pointed out that Washoe can match photographs of cats from many sources. The Gardners confirm that Washoe has done what appears to be a task equivalent to the one Sarah failed (Gardner and Gardner, 1971a), and they mention that Viki did also (Hayes and Hayes, 1953). They point out that unlike Sarah, Viki and Washoe were exposed to picture books and magazines from early infancy. (See also Davenport and Rogers, 1971.)

*Abstractions*

After Sarah learned four color names, she was taught the supposedly more abstract concept "color of." Other higher-order abstractions, such as "shape of" and "size of," were taught in the same manner.

*Quantifiers*

Since the Conference, Sarah has been taught the quantifiers "all," "some," "one," and "none." At first she was taught these exclusively with reference to the dimension of shape, e.g., "All crackers are round." Later, she was able to transfer the quantifiers also to the dimensions of color and size, as in "Some disks are green" or "One cracker is large."

These quantifiers were taught to Sarah only in the context of declarative sentences, as in "All crackers are round." However, she also proved able to comprehend them in the context of declarative sentences such as, "Sarah take one candy."

*Metalinguistics*

Melnechuk asked if Sarah has been taught words for "word" and "sentence." Premack answered no, even though Sarah has been taught the word "name of." However, he thought that by putting named instances of verbs against named instances of nouns, he could teach her the concepts of verbs and nouns, so that she then would classify newly learned words by one of these parts of speech.

## Syntactic Capacities

With the mention of the metalinguistic concept "name of" and parts of speech, the discussion entered the realm of syntax, concerned with the organizing principles of strings of words. Motivational factors continued to be intermingled with cognitive ones.

### Word Order

Sarah's mastery of arranging words in correct order went through at least two stages. In the first of these, as already mentioned, she put the correct words on the board but in incorrect order, and then arrived at the correct order in one or two rearrangements. She did this despite the fact that the trainer always produced sentences in the correct order. At a later stage, Sarah largely abandoned production through rearrangement, and, like the trainer, produced sentences in their final order. On occasion she still reverts to rearrangement. Premack pointed out that in the written case the grammar could be stated in terms both of the order of production and of the final product. In fact, it is stated only in terms of the final product, a sentence being considered correct no matter how the final order was achieved. In the spoken and gestured case, only the final order is visible, so that the grammar could only be stated in terms of this order. Premack urged as another advantage of the written language, externalization of the operations that go into sentence production.

Brown said that because Washoe apparently does not use word order, it is impossible to tell whether she has the corresponding cognitive linguistic capacity.*

### Possessive and Attributive

Premack has not tried teaching Sarah the possessive, in either the "Sarah's apple" or "apple of Sarah" form. Instead, he uses "Sarah apple." However, such attributive phrases as "apple red" and "banana yellow" were taught early.

Garrett asked whether Sarah had sufficient synthetic lability so that if she were presented with a green apple, she would say "Apple not red." Premack thought that if the alternatives presented were the names of other items of food, she would probably call the apple an apple, but

*The Gardners disagree with this statement; see Appendix I.

that if the alternatives were color names, she might call the apple "green" rather than "apple." He also thought he could train her to accept a green apple as an apple, through the usual method of small steps.

## Interrogative

Brown said that Washoe has the signs that human signers use, sometimes unwittingly, to mark the ends of sentences.

> In the declarative case, the hands of the signer return to the position of repose from which they started when he began to sign. In the interrogative case, the hands remain, for a perceptible period, in the position of the last sign or even move out toward the person being interrogated. (Brown, 1970)

Sarah is able to apply the concept of interrogation broadly, but only in the comprehension mode, not generatively, for Premack has not yet given her an opportunity to ask questions in a training session; he has not yet thought of a simple condition in which to make the test.

## The Verb "Is"

Premack considers the manner in which Sarah was taught "is" (in the sense of subset or class membership) to have been unsatisfactory, even though Sarah finally mastered the usage. He described a procedure which he considered preferable: The names of the fruits should be made temporarily unavailable; she should be given the unknown word "fruit" along with the known words "give," "Sarah," and "Mary," in the presence of apple, banana, orange, etc., leading her to write on all these occasions "Mary give fruit Sarah." This would serve to establish a functional equivalence between the word "fruit" and the already existing names "apple," "banana," "orange," etc. A similar procedure would serve to establish the new word "candy" as a functional equivalent of the already existing "chocolate," "caramel," and "gumdrop." These words would permit introducing "is" in a completely standard way: (1) as the only unknown in a string of known words, and (2) on the basis of two positive and two negative instances. For example, "apple is fruit," "caramel is candy," as two positive instances, and "apple is not candy" and "caramel is not fruit" as two negative instances. Premack believes that no more than the normal number of errors should occur in the learning with this procedure.

The actual procedure used to teach Sarah "is" was "red is color" and "round is shape" as the two positive instances, and "color is not red" and "shape is not round" as the negative. Although Sarah made numerous errors, she finally learned in the important sense of being able to apply the word to nontraining cases, e.g., "triangle is not color," etc. Premack believes she might have learned "is" even on the basis of these rather abstract exemplars with no more than normal errors if they had been legitimate exemplars. But Premack pointed out that he had neglected to teach her a genitive particle or the possessive, and thus was not able to give her the desired string "red is color" but only the rather crude approximation, "red is color of." With one of the several new animals now being worked with, he plans to introduce the genitive particle, and thus to be in a position to teach "is" with abstract but legitimate exemplars. In this way, he can test his conjecture concerning the source of Sarah's difficulty. The other new animal will be taught "is" with the more functional exemplars (fruit and candy) as a partial basis of comparison.

Premack noted that he might have distinguished between the analytic and synthetic uses of "is," even though English does not make this distinction, by introducing different forms for the two cases. Premack defined as "analytic" those cases in which answers to questions do not require examination of a nonlinguistic state of affairs, "synthetic" as cases in which such an examination is required. For example, "? red is color" (is red a color?) is analytic, while "? apple is big" (is apple big?) is synthetic, since apples can vary in size and inspection of the apple in question is required before the question can be answered correctly. (Intermediate cases can be considered but need not be for the present point.)

Sarah was also taught a plural marker (in keeping with the rest of the system that relied on markers and word order rather than inflectional changes) so that she could distinguish between, for example, "apple is fruit" and "apple, banana is pl ('pl' being the plural marker) fruit," as well as "red is color" and "red, green, blue is pl color." The question arose as to the basis of Sarah's use of the plural: Was it simply number of words in a particular position in the sentence, or something more akin to compound subject? Premack reported that at least a few tests have been run on this point, indicating that Sarah uses the plural marker in the case of, for example, "apple, banana is pl fruit," but, appropriately, not in the case of "red apple is fruit." There is at least the suggestion therefore that she may be able to discriminate between multiple words in comparable positions in the sentence that do and do not constitute plural subjects.

### Negative

#### *Declarative*

Sarah, like a child, makes more negative than positive errors. Premack does not altogether know why, but he suspects that the trouble is motivational, not cognitive. Some of Sarah's difficulties with the negative come about because she prefers making assertions to making denials; like a child, she will sometimes change the situation in order to be able to answer a question positively.

If, for example, the trainer places a red card on a yellow one, and then asks "? red is on green" (is red on green?), about 65% of the time she will answer "no" correctly. On the other 35% she will modify the world and then answer "yes"; that is, she will replace the yellow card with a green one, and then answer "yes, red is on green." The basis of this action is unclear. Does she prefer the use of "yes" or have an aversion for the negative particle, which was introduced originally as an injunction against carrying out desired actions? Does she prefer congruence between linguistic and nonlinguistic states of affairs to the point where she is willing to modify the world to get the congruence? Does she misinterpret interrogative sentences as imperative ones? The fact that she does not modify the world on most occasions but answers "no" correctly would tend to rule out the last possibility.

#### *Compound Imperative*

Sarah still has difficulty of a sort with negative instructions. Given a sentence like "If Sarah take apple, then Mary no give Sarah chocolate," she may do neither alternative with the apple but merely point to it. As Hirsh put it, this would seem to indicate that the response set is not seen as closed, but it also seems to show a comprehension of what is not to be done. However, Bigelow thought it might mean a failure to understand whether the prohibition applies to "take" or "apple."

This was the first instance given Sarah of a negative embedded in a clause. The form has since been tested successfully in other situations; e.g., the negative particle has been embedded in the antecedent clause.

Ploog suspects that Sarah would be incapable of conceiving a double negative as a positive, which is the normal thing to do in English, although in Russian a second negative in a clause merely confirms or intensifies the negation. Rather early in training Sarah was taught to produce "different" by negating "same" and vice versa; e.g., "A no same B" and "A no different A."

### Dislike of Descriptions of Disliked Outcomes

Sarah's preference for making assertions rather than denials has been mentioned. Sarah also dislikes writing sentences that describe situations she disfavors. This tendency can be overcome by arranging appropriate contingencies. Thus, to get Sarah to write "Mary give apple Jim," by which she would deny herself the apple, she is given a tidbit she prefers to apple. As Premack has written (1970a), "Altruism can become reliable if it is rewarded."

### Material Conditional

This is the logician's term for the concept of "if A, then B."* Sarah needed a large number of trials to learn this term. It is not known whether this resulted from the difficulty of the concept, or its incompatibility with her previous experience, or from an accidental change in procedure midway through the training. (While Premack was away and progress was discouraging, the trainer switched the test, using the if/then particle sometimes with one fruit, sometimes with another. Instead of confusing Sarah, this switching seemed to put across to her the meaning of the particle.)

Musing on the material conditional, Garrett wondered whether Sarah would have replied with "then" if the trainer had presented her with "if," if Premack had taught it in two parts. Garrett also suspected that the biconditional would prove easier for Sarah to learn than the material conditional had been.

---

*Logicians distinguish the *material* conditional "If A, then B" from the *bi*conditional "B if and only if A." In the first case, B always occurs with A, but could occur without A, whereas in the second case, B occurs only with A.

## Semantic and Syntactic Potentialities

Having depicted the semantics and syntax that Sarah had already learned, Premack helped to complete the profile of her capacities by discussing aspects of syntax that she had not yet been taught.

### Next Steps

Premack reminded the conferees that he was more interested in understanding why some training methods succeeded while others failed, than he was in merely succeeding with an effective procedure. For this purpose, failures are more revealing than successes. Accordingly, he intends to try teaching a number of semantic and syntactic forms that express difficult concepts.

*Tense*

The mapping of the tenses of verbs is difficult, because the stimulus is so subtle. (Ploog commented that verb tenses are one of the two most difficult aspects of language to teach autistic children.) Premack suspects that a grasp of time may be a logical requisite not only for a grasp of tense but also for a grasp of causality.

*Causality*

This is another concept that Premack wants to try to teach. Sarah's vocabulary does not now contain "why?" because she is never taught a question without an answer and she does not yet have a name for the concept of "because." There is more to it than that; he foresees having to try to teach two types of causality—nonpsychological and psychological. A nonpsychological example would be "Apple gone because Sarah ate apple"; a psychological example would be "Mary no give Sarah apple because Sarah bad." The latter type intuitively seems more difficult to teach. And how would he interpret a failure?

*Morality*

Premack would like to introduce a small moral system—a moral microcosm—and then let it grow; that is, to establish a "good/bad" dichotomy of judgments and then have Sarah transfer to other classes in the usual way. Values would have to be explicit and graded: Stealing a banana would be bad, but stealing a pomegranate would be worse, for pomegranates are rarer and more expensive than bananas. Sarah would have to sum over a series of previous acts. She could progress to a fairly complicated set of ethical judgments. Unless a more or less symmetrical set of "points for good behavior" were also established, the slate would have to be wiped clean from time to time, for long-term accumulation of wrongdoing would disbar Sarah from all rewards.

Bigelow predicted that added difficulties would stem from inconsistencies on the part of the trainer. The valuation that "stealing pomegranates is worse than stealing bananas because pomegranates are rarer and more expensive than bananas" can (probably) only be taught as "more punishment (less reward) for stealing pomegranates" so that the animal learns a correspondence between her acts and meted punishment. It would come no closer to ethics than this. And when an animal cannot keep track of where it is with reference to a state of grace, it behaves arbitrarily, not metrically.

### Comparative

Premack would also like to see if Sarah could be taught some generalized ideas of preference, i.e., one thing is better than another, has higher value than another, etc. Bigelow commented that experience using computers in attempts to translate human languages by machine showed that the programming of indefinite comparatives is very difficult. Definite concepts in context, like "longer than" or "warmer than" are usually possible to formalize, but the abstract concept of "better than" is impossible because it has a different interpretation in different instances. He wondered how Premack hoped to do it.*

### Arithmetic

Premack's next goal is to teach Sarah arithmetic, or at least the terms "plus," "minus," and "equals" necessary for doing addition and subtraction. Bigelow said that these concepts might prove very difficult and subtle to teach in the experimental environment in which Sarah is being taught. They involve the idea of summing a total, and of properties exhibited only by a total collection and not direct interpretations of the acts Sarah has shown she could do. This is true of the idea of "equality" and of "the sum is the totality of its parts." On the other hand, counting can be reduced to quite a simple operation, in a specially prepared experimental context, because in such a context

---

*This proposal is discussed further in Premack, 1971a,b.

counting can be reduced to the carrying out of an inductive rule over an ordered set, the rule being to make repeatedly a one-to-one correspondence between the items to be counted and the items in the ordered set. Specifically, if an "ordered set" of "words" were arranged on the game board, Sarah might be taught to take the items in a pail one at a time and pair them with (number) items on the prepared ordered set, and to deliver the last item to the trainer. This would be a form of counting. Bigelow granted that the difficulty would depend on what the true primitives were taken to be, or on what you can reach from where you already are. Premack had in mind the use of the "give" and "take" already in Sarah's vocabulary, with the possible use of new primitive concepts.

## Other Grammar to Try

In response to a Premack question, the conferees suggested two types of grammatical concepts for further testing of Sarah's ability to learn. The first type, given in this subsection, was an extension of the grammar he had already taught; the second type, given in the next three subsections, would represent new paradigms of reality for Sarah.

### Order of Introduction

Brown pointed out that there was a certain element of necessity to the order in which to teach new concepts. For example, names of objects had to be taught before definite and indefinite articles, and names of actions had to precede tenses.

### Pronouns

Brown mentioned Washoe's signs, described as "you" and "I-me," and suggested teaching explicit personal pronouns to Sarah. Ploog said that pronouns were, like verb tenses, the most difficult grammatical thing to teach to autistic children; Premack reported that he had avoided teaching pronouns, because they were so cumbersome to teach. He seeks a set of sentences that will teach as relativistic a concept as the difference between "me" and "you."

The difficulty seems to lie in the subtlety of the stimulus. Therefore one strategy might be to emphasize the stimulus; for example, in teaching the "you/me" dichotomy, one person in the couple could wear a cap. This blatant adjunct stimulus could then be faded out. Another strategy might be to paraphrase; for example, to pair "we" with "Sarah and Mary," "Sarah and Jim," and so forth.

Garrett thinks that Sarah may already have implicit pronouns buried in her use of names. He suggested starting with "it," because, substituting across so many classes, it would demonstrate a still greater capacity for generalization than Sarah had yet shown.

Sarah has since been taught "this" and "that," words that share with pronouns the critical property of being defined relative to the speaker. Premack suggested that some degree of parallel can be seen or, if necessary, deliberately imposed between space markers, time markers, and pronouns; that is, "here," "now," "I"; "there," "past," "you" (proximal "other one"); "over there," "later" or "future," "they" (distal "other one"). Although the parallels were imperfect, he felt that definition-relative-to-speaker, which is common to all three sets of distinctions, is the critical property, and the fact that Sarah had learned to use this property in the case of "this" and "that" augured well for her ability to learn the other distinctions.

### Declarative

Bruner suggested teaching Sarah what logicians call the existence operator, "There is an x, such that ... " as one way to broaden her syntactic capacities.

### Interrogative

Brown said that the "what" in Premack's question set was not as general as the "what" in English, which could substitute for any noun phrase of whatever degree of complexity. He suggested broadening its range.

### Imperative

Brown said that the narrow range of values permitted to the direct object of the verb in the action paradigm ("Mary give apple Sarah!"), while the subject, verb, and indirect object were held essentially constant, made it possible that the phenomenon was one of rote learning, or of solution by ad hoc routines possible only because the range is narrow. He suggested generalizing more; the direct object should not always be an item of food, and the agent, action, and indirect object ought also to take a wider range of values, cutting across classes. For example, the verbs "take" and "see" ought not always to have "Sarah" as their subject, although he foresaw problems if the verb "see" were to have, say, "Mary" as subject.

Bigelow remarked that the verbs in Premack's system seemed more restricted than the nouns, which he saw as being more like free variables. He suggested exploring the mapping of the action space at least as richly as the object space had been explored, both by providing more specific verbs, more general verbs, and more synonyms of verbs.

*Conjunctions*

Garrett suggested trying, besides "if/then," a conjunction that implies a proposition about the two sentences that are being connected as clauses. For example, one such conjunction is "but," which implies a negative claim.

*Context Shift*

Bigelow observed that a typical test situation seems to present Sarah with a sentence that arrives in a set of similar sentences. He asked whether Sarah can cope with massive shifts of context from one sentence to another. Premack answered yes, and successive sentences can differ in type from "what?" to "yes-no" questions without disconcerting her. No systematic investigation has yet been made of this capacity. However, he expects that he would find some impairment of learning if he presented her with a series of sentences that shifted in context and increased in complexity at the same time.

Brown mentioned his own difficulty in studying Japanese as an adult: He could manage sentences of the same construction in a given vocabulary, but could not easily handle simultaneous shifts in sentence structure and vocabulary.

**Other Conceptual Forms**

Here reported is a series of experiments suggested by participants that have to do with the subtle mix of semantic and syntactic capacities required for linguistic situations with less simplicity or immediacy than those that Sarah has shown she can understand.

*Semigrammatical Forms*

Bruner suggested that Sarah might be ready to cope with ill-formed sentences. Premack was concerned that Sarah might learn the wrong thing as right. He said that heretofore he has never deliberately taught her anything that would subsequently be negated, although this principle has inadvertently been violated by a failure of communication between trainers. However, he felt that it might indeed be safe to teach her the wrong thing now, and told an anecdote bearing on Sarah's flexibility in this regard. It seems that the plastic "words" for Mary and Sarah are larger than those for inanimate objects. (This was generally true of all proper nouns in the system, for no particular reason; it was an accident that was generalized.) In order to fit a long sentence on the wall board, Mary once diminished the size of her own sign, which did not disturb Sarah.

### Ambiguity

Garrett suggested the use of ambiguity—for example, adjectives that could relate either to subject or object— to learn if Sarah could provide conditions for resolving the interpretation.

Premack thought he might also try to fashion a compound sentence with an ambiguous clause made unambiguous by another clause.

### Metaphor

Sarah having shown her ability to detect the defining characteristics of at least some sets, and Aristotle having claimed that the essence of creative imagination is the capacity to perceive a similarity in dissimilars,* Bigelow suggested testing Sarah's comprehension of metaphors, perhaps by extending a verb like "carry out" from "water in a pail" to "instructions."

Premack replied that while there are individual differences in cognitive styles, every metaphorical leap might be reducible to a set of propositions, and that he had been thinking about the verb "give" in this way. But even as simple-seeming a verb as "do" or "give" has many frames of reference that seem to be metaphors of one another. Brown said that "do" is indeed a tremendous abstraction and comprises a family of concepts. Lenneberg remarked that every generalization has a metaphorical aspect.

### Semantic Anomalies and Tests of Comprehension

Bigelow asked if Sarah can comprehend semantic anomalies. Premack answered yes, within reason, even if the situation strikes her as odd, as, for example, "chocolate *on* pail" instead of the more usual "chocolate *in* pail."

Garrett suggested trying more striking absurdities, such as a personal name as the direct object of a verb, or a transposition of donor and recipient, to see if Sarah would reject them.

*Aristotle, *Poetics* XXII, paragraph 9: Literally, "... a command of metaphor ... is the mark of genius, for to make good metaphors implies an eye for resemblance."

*Deviation from Actuality*

Hirsh suggested teaching the concept pair "possible/impossible" and then testing such a sentence as "pail in banana."

Lenneberg suggested trying to teach Sarah the difference between a fact and an opinion.

Melnechuk suggested testing Sarah's ability to lie. Premack thought she would be able to tell a lie if it would profit her to do so. DeVore said that chimps, whether or not they can lie, certainly can deceive, as by hiding, and as by enticing victims within spitting distance, in some instances maintaining the deception for half an hour until the victim is within range. Premack mentioned an apparent attempt by Sarah to deceive her trainer by passing over the correct word in a completion test that she spontaneously gave to him.

Having in mind the interest of children in making up and being told stories, Melnechuk suggested testing Sarah's ability to invent a fantasy and to describe it. A failure to do the latter would not necessarily disprove her ability to do the former.

## Other Paradigms

Premack said he found these suggestions helpful, but that he wanted to pose another question concerning paradigms of reality other than the "agent-action-object-recipient" ("Mary give apple Sarah") and the "subject-predicate adjective-object" ("Apple bigger than cherry"). Sarah had accepted these and minor variations of them, and he was now curious to learn if there were paradigms used by human beings that she could not accept.

*Paradigm of State*

Proceeding from simple state descriptions of the sort "red on green," "apple in pail," that Sarah can do, Melnechuk suggested that Premack try to teach descriptions of differences and changes of state. Premack said he found such descriptions as, for example, of the fact that the greenness of leaves varies in time and space, far more difficult to map in language than agent-action-object.

Brown agreed that states go over into language with more difficulty than do deeds, for states reflect variations in a perception. He pointed out that there are many situations that cannot be described in terms of either the action paradigm or of the state paradigm, as, for example, "I hear music" or "I need money."

As we all know, language is only one of a number of human symbol systems, each of which seems especially well suited for the communication of certain aspects of reality. Thus, language is good for communicating discrete deeds, but music seems better for communicating continuously varying and ambivalent feeling states, as graphs are better for communicating simultaneous and continuous interrelationships of many abstract variables. This idea underlies not only grand opera but also some current experiments in multimedia communication.

## Subjective Experience

Melnechuk suggested trying to teach Sarah how to express her subjective states, as, for example, with words for emotions. Premack replied that in a sense he is already working on these, in that Sarah's alarm cries have been recorded. He plans to play them back to her and hopes to solicit her opinion regarding them.

Since the time of the Conference, Sarah has been taught two psychological state words, "prefer" and "want." They were decidedly easier to teach than "think" or "know," although Premack is not certain that the latter introduces qualitatively new problems. "Want" was taught her by using deprivation for X and an essentially ad lib supply of Y to establish the difference between "Sarah want X" and "Sarah no-want Y." One could argue that "want" does not always mean desire through deprivation and that one might want something for other reasons. Premack suggested that in teaching the animal words that map her presumed internal conditions, e.g., states of desire that are and are not specific to the parameter of deprivation, the distinctions can be arranged in any way we like. "We can seek a high correspondence with the vernacular, or we can ignore it and fall into correspondence with it only because the vernacular inevitably reflects many, perhaps most, of the distinctions of which we are capable" (Premack, 1971b).

Premack considered it of interest to ask whether he should teach Sarah psychological state words at all. He noted that it would be possible to strand the animal in an operational format. Lacking psychological state terms, she would have no alternative but to say, for example, "It's been a long time since I had bananas," or "If I were given bananas, I would eat them promptly and fast," or "The probability that I would eat bananas is high."

A last possibility is to introduce psychological state words, not by manipulating conditions that presumably alter the states mapped by the words, but rather by explicit definition. For instance, teach "want" as the explicit equivalent of some or all of the above descriptive statements. "Sarah want banana" would then be the same as "If Sarah had banana she would eat them promptly and fast."

## Profile of Capacities

### Sarah

Lenneberg called Premack's results "spectacular and unprecedented." In a paper presented nine months before this Conference, Lenneberg had proposed an

> ... imaginary primitive language that contains some fundamental features of human language. Any animal that can understand sentences in this language, i.e., execute commands or answer questions by signalling yes or no, may be said to have at least some of the basic capacities for language. (Lenneberg, 1970)

In Lenneberg's opinion, the primitive language that Premack has devised for Sarah is at least as rich as his own imaginary language of 7 object words, 3 action words, 4 attribute words, 7 relational words, and 3 syntactic markers; and so he feels that Sarah must be granted possession of parts of some of the concepts and operations on which human language is based.

However, Lenneberg feels that the general cognitive capacity to form stable perceptual categories is not too significant, for even rats and pigeons can do this to some extent.* As he once put it,

> ... it is clear that there is no formal difference between man's concept-formation and animal's propensity for responding to categories of stimuli. There is, however, a substantive difference. The total possibilities for categorization are clearly not identical across species.

---

*For example, see Siegel and Honig (1970) for a report on the ability of pigeons to learn, through operant conditioning, to discriminate with a high degree of success between projections of slides that differed only in that one picture contained one or more of a set of different human beings. Presumably, the pigeons could be said to have formed the broad concept of "human being."

Moreover, he granted that

> Most primates and probably many species in other mammalian orders have the capacity to relate various categories to one another and thus to respond to *relations* between things rather than to things themselves; an example is to respond to the largest of any collection of things. (Lenneberg, 1967, pp. 331 and 332)

He is therefore quite willing to agree with Premack that, with one exception, the concepts to which Sarah associated plastic "words" had not been inculcated in her by Premack; that, instead, training merely provided her with names for concepts already existent in her. The exception was the concept of "name of," which Premack had pointed out was probably not a concept that Sarah had possessed before being trained.

Premack had said earlier that if the concepts are for the most part not instilled by a course of language training but that the course merely provides names, one could reasonably expect the trained animal to be able to generalize from a few instances to other members of the class, as he claims Sarah is indeed able to do. However, Lenneberg was not convinced that Sarah could generalize a concept as broadly as even a child can; as he wrote in 1967,

> ... it is possible to train a hunting dog to "point," ... But it does not appear to be possible to teach a dog to do the "name-specific stimulus generalization" that every child does automatically. ... There is no convincing evidence that any animal below man has ever learned to relate any given word to the same range of stimuli that is covered by that word in common language-usage. (Lenneberg, 1967, p. 329)

And he is still of that opinion.

Bigelow granted that the chimp learns to respond separately to apples, bananas, etc., but did not think this sufficient evidence for claiming that the chimp perceived these as stable elements. Moreover, he felt that it had yet to be demonstrated whether or not the chimp can correctly use "food" as a stable concept, or even the less general "fruit" as a stable concept, apart from its elements.

Most words, of course, label open domains of concepts, rather than physical things.* In Lenneberg's view, concepts are not static.

*"When names have unique referents, such as Michelangelo, Matterhorn, Waterloo, they may be incorporated into discourse but are not considered parts of the lexicon." (Lenneberg, 1967, p. 332)

> Words are not the labels of concepts completed earlier and
> stored away; they are the labels of a *categorization process or family of*
> *such processes.* Because of the dynamic nature of the underlying
> process, the referents of words can so easily change, meanings can be
> extended, and categories are always open. (Lenneberg, 1967, p. 333)

He did not think Sarah had shown that she or her species cognitively
deals with the environment in just the same way as does man.

Another cognitive difference between chimpanzee and man, in
Lenneberg's view, lay in their preparedness for semantic and syntactic
operations. Thus, Sarah still required the devising of almost foolproof
situations in order to grasp the equivalence of the various forms of
nonquestions. As he stated,

> The essence of language is its productivity; in the realm of perception
> and understanding of sentences, it is the capacity to recognize struc-
> tural similarities between familiar and entirely novel word patterns.
> Thus our criterion for knowing language is not dependent upon
> demonstrations that an individual can talk or that he goes through some
> stereotyped performance upon hearing certain words, but upon evi-
> dence that he can analyze novel utterances through the application of
> structural principles.(Lenneberg, 1967, p. 330)

In brief, in Lenneberg's opinion, Sarah (like Washoe) has not yet shown
evidence of having anywhere near the full human complement of
cognitive and linguistic capacities necessary for language—a difference
that may not be simply quantitative but may in fact hide important and
systematic deviations from the human propensity for language.

### Washoe

Brown agreed that because so many linguistic competences had
not been demonstrated in Sarah's case, he could not yet assert that her
remarkable accomplishments had demonstrated her to be capable of
language. He felt that while Sarah's case was the more novel, Washoe's
was the more important, because of her spontaneity in using signs.
Brown granted that Washoe did not, as Sarah did, observe rules of word
order in her strings,* but thought that her failure to do so might have
had any of several causes:

---

*The Gardners do not agree with this statement; see Appendix I.

1. The Gardners and their assistants may not have always used signs in what corresponds to normal word order.

2. The Gardners may not have corrected Washoe's departures from proper order.

3. The Gardners may find word order more frequently correct than incorrect in the frequency comparisons they now have under way.

4. The situations in which Washoe has communicated are sufficiently structured so that she was understood; no matter what the order, she emitted contentives (content signs), which showed that she felt no communication pressure to use correct word order. In Brown's opinion, the question of Washoe's capacity for correct word order is still open and may emerge if she ever needs to communicate the difference between such equally plausible situations as "car hit truck" and "truck hit car."

### Both Chimpanzees

At this point in the Conference, it was clear that even with the most generous acceptance and interpretation of Sarah's and Washoe's performances, neither of the two chimpanzees could yet be said to have anywhere near "a good grasp of the language" with regard to English as spoken by a native. The absence of tense and the lack of various other aspects of language as employed by human beings would seem to have permitted only a negative answer to the question "Are Apes Already Using Language?" except for the participants' awareness of a group of human beings who, admittedly, are using language, even though their semantic and syntactic limits seem similar in some respects to those of the two chimpanzees—namely, children.

In opening the Conference, Ploog had suggested that any eventual profile of chimpanzee capacities be compared to that of human children at various stages of development, to establish similarities and differences, and the participants turned to this question.

### Comparison of Chimps and Children

In a paper written before the publication of any of Premack's results, Bronowski and Bellugi had attempted

> ... to compare Washoe to children learning spoken language. There are grounds for arguing that the Gardners' method of signing makes this an

appropriate comparison. ... Since the growth rate of chimpanzees is faster than that of children, it seems reasonable to compare her development with that of children of the same age. These and other details have been discussed. (Bronowski and Bellugi, 1970a)

## Pathological Conditions and Speech Development in Children

At first thought, it might be supposed that the children with linguistic capabilities most resembling those of the two chimpanzees would be those suffering some neurological or environmental deficit. As Bronowski and Bellugi (1970a) wrote, "It might be held that ideally Washoe's progress should be compared with that of a deaf child of deaf parents who is learning sign as a native language."

Brown reports that Ursula Bellugi-Klima, at the Salk Institute, now has in progress a study of a deaf child's acquisition of Sign Language. This study should yield important comparative data. (See also Schlesinger, 1971.)

At the Conference, Lenneberg summarized the differences in the development of language between the deaf child and the child with Down's syndrome (previously called "mongolism"). The child with Down's moves along the same developmental sequence as the normal child, but at a decelerating, slower rate, and stops before completing normal development. No special training is known that will accelerate his rate of language learning at any stage; this finding strongly argues the existence of some readiness curve. In contrast, deaf children who are introduced to language at school age have a backlog of readiness of several stages, so that they catch up rapidly to the level of language normal for their age, but in a mixed-up way, not in the usual order. (For details, see Lenneberg, 1967, pp. 304-324.)

Premack referred to a third possibility, analogous to pathology, which he has described as follows:

> Several years ago, in working with psychotic children, R. Metz and I devised some tests of language comprehension to determine whether the severely impaired speech production, which characterized these children, was owed to performance factors or to something deeper. Having earlier confirmed Lenneberg's (1967) surprising claim that the feeble-minded child is grammatical, I was distinctly surprised when some of the psychotic children failed all of the comprehension tests (while at the same time performing adequately on nonlanguage

tests). They had what we ended up calling "word knowledge" but, so far as our tests could determine, little else. (Premack, 1970a)

In Premack's opinion, Sarah's mastery of word order* demonstrated a syntactic grasp that was not consistent with a view of her as "psychotic-like." To Brown, this ability, and her ability to handle predication and the conditional, made her resemble "a precocious and discontinuous" child.

### Early Stages of Human Language Development

Lenneberg in 1967 published a table of "Milestones in Motor and Language Development"; its entries are reproduced in Table 4.

Brown reported on a more recent study utilizing the mean (or average) length of children's utterances (MLU), which he considers to be a better index than chronological age of early grammatical development. In a recent paper (Brown, 1970), he states this argument as follows:

> We have information on an initial period which is bounded by an MLU of 1.0, the threshold of syntax, and an MLU of 2.0. ... We shall call this full interval "Stage I." It seems to correspond fairly exactly with the period for which the Gardners have provided data on Washoe: the age of 12 months through the age of 36 months ... Table [5] sets out some of Washoe's strings or sentences; they are drawn from the Gardners' fifth and sixth summaries which appeared in 1968. ... The classification ... is the Gardners' own. ... How do these multi-sign sequences compare with the first multi-word combinations produced by children learning American English and other languages? ... Table [6] sets out some of structural meaning which, among them, characterize the majority of two-word sentences in Stage I. [These] structural meanings account for about 75% of all multi-word utterances. ... In sum we find that Washoe does not seem to express all of the operations and relations of Table [6], but she does seem to express some of them. Specifically: Recurrence, Agent-Action, Action-Object, Action-Locative (and Agent-Action-Object; Agent-Action-Locative). To these we might add the signs used to name pictures in a picture-book as elliptical Nominative sentences. All of which suggests that Washoe has a simpler version of child syntax

---

*The absence of this mastery, i.e., "the failure so far to develop any form of sentence structure," was cited by Bronowski and Bellugi (1970a) as "the most subtle yet crucial way in which Washoe's performance falls short of a hearing child."

TABLE 4

Developmental Milestones in Motor and Language Development
[Lenneberg, 1967]

| At the completion of: | Motor Development | Vocalization and Language |
| --- | --- | --- |
| 12 weeks | Supports head when in prone position; weight is on elbows; hands mostly open; no grasp reflex | Markedly less crying than at 8 weeks; when talked to and nodded at, smiles, followed by squealing-gurgling sounds usually called *cooing*, which is vowel-like in character and pitch-modulated; sustains cooing for 15–20 seconds |
| 16 weeks | Plays with a rattle placed in his hands (by shaking it and staring at it), head self-supported; tonic neck reflex subsiding | Responds to human sounds more definitely; turns head; eyes seem to search for speaker; occasionally some chuckling sounds |
| 20 weeks | Sits with props | The vowel-like cooing sounds begin to be interspersed with more consonantal-sounds; labial fricatives, spirants and nasals are common; acoustically, all vocalizations are very different from the sounds of the mature language of the environment |
| 6 months | Sitting: bends forward and uses hands for support; can bear weight when put into standing position, but cannot yet stand with holding on; reaching: unilateral; grasp: no thumb apposition yet; releases cube when given another | Cooing changing into babbling resembling one-syllable utterances; neither vowels nor consonants have very fixed recurrences; most common utterances sound somewhat like ma, mu, da, or di |

*Note:* In presenting this table, Lenneberg would emphasize that the comparison of language development with motor development is vital for the argument that language development is *paced* by maturational events. For purposes of comparison, motor-language synchronous development highlights the difference between man and chimpanzee

TABLE 4 (Continued)

| At the completion of: | Motor Development | Vocalization and Language |
|---|---|---|
| 8 months | Stands holding on; grasps with thumb apposition; picks up pellet with thumb and finger tips | Reduplication (or more continuous repetitions) becomes frequent; intonation patterns become distinct; utterances can signal emphasis and emotions |
| 10 months | Creeps efficiently; takes side-steps, holding on; pulls to standing position | Vocalizations are mixed with sound-play such as gurgling or bubble-blowing; appears to wish to imitate sounds, but the imitations are never quite successful; beginning to differentiate between words heard by making differential adjustment |
| 12 months | Walks when held by one hand; walks on feet and hands—knees in air; mouthing of objects almost stopped; seats self on floor | Identical sound sequences are replicated with higher relative frequency of occurrence and words (mamma or dadda) are emerging; definite signs of understanding some words and simple commands (show me your eyes) |
| 18 months | Grasp, prehension and release fully developed; gait stiff, propulsive and precipitated; sits on child's chair with only fair aim; creeps downstairs backward; has difficulty building tower of 3 cubes | Has a definite repertoire of words—more than three, but less than fifty; still much babbling but now of several syllables with intricate intonation pattern; no attempt at communicating information and no frustration for not being understood; words may include items such as thank you or come here, but there is little ability to join any of the lexical items into spontaneous two-item phrases; understanding is progressing rapidly |

TABLE 4 (Continued)

| At the completion of: | Motor Development | Vocalization and Language |
|---|---|---|
| 24 months | Runs, but falls in sudden turns; can quickly alternate between sitting and stance; walks stairs up or down. one foot forward only | Vocabulary of more than 50 items (some children seem to be able to name everything in environment); begins spontaneously to join vocabulary items into two-word phrases; all phrases appear to be own creations; definite increase in communicative behavior and interest in language |
| 30 months | Jumps up into air with both feet; stands on one foot for about two seconds; takes few steps on tip-toe; jumps from chair; good hand and finger coordination; can move digits independently; manipulation of objects much improved; builds tower of six cubes | Fastest increase in vocabulary with many new additions every day; no babbling at all; utterances have communicative intent; frustrated if not understood by adults; utterances consist of at least two words, many have three or even five words; sentences and phrases have characteristic child grammar, that is, they are rarely verbatim repetitions of an adult utterance; intelligibility is not very good yet, though there is great variation among children; seems to understand everything that is said to him |
| 3 years | Tiptoes three yards; runs smoothly with acceleration and deceleration; negotiates sharp and fast curves without difficulty; walks stairs by alternating feet; jumps 12 inches; can operate tricycle | Vocabulary of some 1000 words; about 80% of utterances are intelligible even to strangers; grammatical complexity of utterances is roughly that of colloquial adult language, although mistakes still occur |
| 4 years | Jumps over rope; hops on right foot; catches ball in arms; walks line | Language is well-established; deviations from the adult norm tend to be more in style than in grammar |

TABLE 5

Some of Washoe's Sign Sequences
As Classified by R. A. Gardner and B. T. Gardner.
[Adapted from Brown, 1970]

A.    Two Signs.

    1.    Using "emphasizers" (please, come-gimme, hurry, more):

        Hurry open
        More sweet
        More tickle
        Come-gimme drink

    2.    Using "specifiers":

        Go sweet (to be carried to fruitbushes).
        Listen eat (at sound of supper bell).
        Listen dog (at sound of barking).

    3.    Using names or pronouns:

        You drink.
        You eat.
        Roger come.

B.    Three or More Signs.

    1.    Using "emphasizers":

        Gimme please food.
        Please tickle more.
        Hurry gimme toothbrush.

    2.    Using "specifiers":

        Key open food.
        Open key clean.
        Key open please blanket.

    3.    Using names or pronouns:

        You me go-there in.
        You out go.
        Roger Washoe tickle.

## TABLE 6

Ten Structural Meanings of 75% of the
First (Two-Word) Sentences in Child Speech
[Adapted from Brown, 1970]

I. Operations of reference:

| | |
|---|---|
| Nominations: | That (or It or There) + book, cat, clown, hot, big, etc. |
| Notice: | Hi + Mommy, cat, belt, etc. |
| Recurrence: | More (or 'Nother) + milk, cereal, nut, read, swing, green, etc. |
| Nonexistence: | Allgone (or No-more) + rattle, juice, dog, green, etc. |

II. Relations:

| | | |
|---|---|---|
| Attributive: | Adjective + Noun | (Big train, Red book, etc.) |
| Possessive: | Noun + Noun | (Adam checker, Mommy lunch, etc.) |
| Locative: | Noun + Noun | (Sweater chair, Book table, etc.) |
| Locative: | Noun + Verb | (Walk street, Go store, etc.) |
| Agent-Action: | Noun + Verb | (Adam put, Eve read, etc.) |
| Agent-Object: | Noun + Noun | (Mommy sock, Mommy lunch, etc.) |
| Action-Object: | Verb + Noun | (Put book, Hit ball, etc.) |

at [Stage] I. ... While I am prepared to conclude that Washoe has not demonstrated that she intends the structural meanings of Table 6, I do not conclude that it has been demonstrated that she lacks these meanings. ... The question of Washoe's syntactic capacity is still quite open.

Garrett commented that this first stage of human language also transcends what Sarah has done to date. Premack remarked that he had not attempted to teach Brown's list of structural meanings to Sarah as such. Although the attributive had actually been taught, and rather early, he had not tried teaching the possessive, but it did not look to him as if it would be a difficult thing to teach to Sarah.

Following the Conference, Garrett commented on Premack's work as follows:

Premack has taught Sarah a particular and impressively complicated kind of unchimplike behavior. An assumption of the attempt to relate this behavior to human language is that the formal system that describes the competence underlying Sarah's behavior captures an important aspect of the logic underlying human language behavior. Notice that this is a move to reduce linguistically complex objects, whose relations we want to explicate, to an analysis in terms of a "simpler" system of more "primitive" concepts. It is very likely that such a move cannot succeed. Though there is admittedly controversy over the details of argument among linguists and logicians, there can be little doubt that the logical systems underlying human abilities to make valid inferences about relations among sentences will prove to be nearly as complex as is the natural language itself.* Though it may seem impossibly exigent, it may well prove to be the case that no performance on the part of an infrahuman organism short of one which matches that of, say, a normal four-year-old child, will suffice.

### Are Apes Already Using Language?

The comparison of the chimps to children implied that if what the chimps do is really to be language, then it must develop further. Otherwise, the animals will have shown themselves to be as minimally capable of language as severely retarded human beings (Lenneberg, 1967).

Whether or not Sarah and Washoe will continue to show progress is of course still to be seen.† However, the participants discussed evidence that bears on the probability of such progress. Their deliberations are reported in the next, concluding section.

---

*For a discussion of the relations between logic and syntax, see Fodor, 1970.

†Recently (1971), the Gardners sent Washoe to Norman, Oklahoma, where Drs. William B. Lemmon and Roger Fouts of the University of Oklahoma are training eight other chimpanzees in American Sign Language.

## IV.  ARE  APES  LIKELY  TO  USE  LANGUAGE?

The question of whether Sarah or Washoe (or any other comparable animal) was likely ever to show a qualitatively greater linguistic competence than they had already shown led into a discussion of the motivational, developmental, and biological matrix in which linguistic cognition is embedded.

Bruner raised the general question of how natural, to an ape, language could be, cognitively speaking. He pointed out that although Premack and the Gardners differed strategically, in that one provided controlled instances of supposedly essential paradigms of language while the others provided free use of supposed language, both approaches introduced something into the environment of their animals, as though a latent capacity, once awakened by this artificial intervention, would remain; however, he suspects the opposite will be true, and that the need for intervention manifests a fundamental deficiency in the animals.

### Developmental  Questions

**Readiness to Learn**

Lenneberg commented that it was surprising that Sarah seems ready to learn whatever Premack teaches her, whereas a child is not always so ready to learn everything; human linguistic capacities manifest themselves in a strict order. Premack said Sarah had not always been ready to learn the plastic words, that rather she had gone through stages in her learning of them.

As for Sarah's current readiness to learn things in an order different from the specific human order, Ploog suggested that she might not be bound to standard stages because she was already nearly adolescent.

Premack also suggested that if a child were exposed to his training methods, the child might prove quite able to learn things that he does not grasp when acquiring language in the "natural" way; indeed, if you do not accept this possibility, then you must concede that the chimp can learn the conditional when, for example, the child cannot. He felt, in brief, that the comparison is invalid unless the training procedures are equivalent.

*Critical Period*

This raised the question of whether or not there is a "critical period" for the primary acquisition of language aspects in chimpanzees as there is in children; for

> There is evidence that the primary acquisition of language is predicated upon a certain developmental stage which is quickly outgrown at puberty. ...
>
> Automatic acquisition from mere exposure to a given language seems to disappear after this age, and foreign languages have to be taught and learned through a conscious and labored effort. Foreign accents cannot be overcome easily after puberty. However, a person can learn to communicate in a foreign language at the age of forty. This does not trouble our basic hypothesis on age limitations because we may assume that the cerebral organization for natural language learning as such has taken place during childhood, and since natural languages tend to resemble one another in many fundamental aspects ... the matrix for language skills is present. (Lenneberg, 1967, pp. 142 and 176)

Premack doubted that there was a critical period in the chimpanzee; he thought that the learning of language elements and paradigms was independent of age, although the underlying concepts might mature with age.

Ploog said a chimp begins to mature sexually between 6 and 8 years and then takes 4 to 6 years to become adult; an arrest of maturation, after showing first signs of sexuality, can occur in caged animals (Goodall, 1968; Riopelle and Rogers, 1965). Washoe's training began when she was between 8 and 14 months old, and the observations discussed at this Conference were made during the first 24 months of her training (or in the second and third years of life). Sarah, on the other hand, is over 7 years old and, Premack says, sexually mature; she was about 5 when she first began using plastic words.

Lenneberg (1967) wrote as follows on the developmental differences in ratios of brain to body weight in the two species:

> ... at birth man's brain weight is only 24% of the adult weight, whereas the chimpanzee starts life with a brain that already weighs 60% of its final value. ... By extrapolation, we may assume that the maturational events of the chimpanzee brain during childhood differ from those in man in that at birth his brain is probably much more mature and all parameters are probably more stabilized than in man. This would

indicate that the facility for language learning is not only tied to a state of flux but to a maturational history that is characteristic for man alone.

He added:

> We should not suppose that we might be able to train a chimpanzee to use a natural language, such as English, simply by delaying the animal's physiological development. ... When physiological delay occurs in man (as in mongolism) it also protracts his speech development. (Lenneberg, 1967, p. 174)

The development of Washoe and Sarah in captivity has presumably been somewhat retarded. Yet Premack reports that Sarah began to menstruate at about 6 or 7 years, while the average age for the onset of menstruation in captive chimpanzees is 8.8 years, according to the data reported by Riopelle and Rogers (1965). If not precocious, Sarah may be older than originally estimated.

### Prior Interest in Objects

Bruner asked if language development must be preceded by an interest in objects outside the self. If so, he suspected that tool using might be a type of object involvement especially likely to predispose an ape to a system of hierarchical syntactic rules. Brown thought that other behaviors describable in terms of rules might also be important to look at, as well as those behaviors in which objects were involved.

DeVore cited Schaller's monographs (1963, 1964) on free-ranging mountain gorillas, which reported their apparent lack of interest in objects. They brush aside even interesting things deliberately put in their pathway. But, in contradiction, he also referred to a description of gorilla behavior by Fossey (1970), which disputes this alleged disinterest in objects.

Premack said that a lack of interest in objects does not characterize chimps. Sarah's ability to describe situations ("red on green") in which she played no direct part—although admittedly she was in a social situation and was rewarded for a correct description— seems to argue her possession of a certain objectivity. Bigelow wished it were possible to test this objectivity in the absence of trainer and reward by somehow having her comment as a third-party observer when not required to do so; Premack replied that the proper comparison, in assessing Sarah's "objectivity," was not to a logical positivist but to a four-year-old child.

Ploog said that in squirrel monkeys, interest in objects depends on age. Young squirrel monkeys play with all kinds of objects. Their interest decreases after the age of 3 or 4, unless an object is frightening. He added that, as Japanese work on macaques has shown, this early phase of vivid interest in objects—a general trait in most mammals, especially in monkeys and apes—may be important for innovations such as one troupe's newly acquired habit of washing sweet potatoes, which is an indication of processes of subcultural propagation (Carpenter and Nishimura, 1964; Kawamura, 1963).

*Self-Recognition*

Ploog asked whether or not another requisite for the development of language is the capacity for self-recognition. He drew attention to Gallup's proof (1970) that chimpanzees, but not monkeys, are capable of self-recognition in a mirror.

*Prior Socialization*

Ploog said that the communication system of monkeys and apes breaks down if their critical period of socialization is disturbed. Sociability seems to be a prerequisite for communicative competence (Ploog, 1970). However, as DeVore pointed out, socialization has not been followed or accompanied by language in any species but the human.

Ploog still felt that socialization was an important precursor of communication, not only ontogenetically but phylogenetically. He asked the conferees to consider the long natural history of the evolutionary development of social signals, the ritualization of these signals, and the phyletic process of "shaping" both the signaler's physical structures and the recipient's internal organization so that increasingly complex communication systems evolved. The conspecifics had to produce and to perceive their signals in order to achieve two-way communication. In the case of the mammals, and specifically the monkeys, this meant enormous structural refinements in the signaling devices, such as the facial muscles and the vocal apparatus. But more than that, it required an increasing competence in the recognition of relations in order to communicate appropriately in an ever-changing context of social events. To a partner A it makes a difference whether partner B is distant or near, and whether he is close to C, and whether both B and C are far from D, and whether D is between A and B or behind them, etc., not to speak of the temporal factors involved in these relations. It makes a difference whether A is young or old, a male or a female; and not only what sort of relationship he has had with B, C, and D in the past but also what relationships have existed between the others.

Ploog even felt that the capacity to perceive social relationships may be related to the capacity to use the logical and syntactic relationships of language.

However, Lenneberg felt that language may have had cognitive antecedents that did not have social functions at one time. In this connection, Bruner pointed out the quasi-syntactic organization of manipulative behavior, as, for example, in the case of Köhler's famous ape Sultan, who appeared to have an experience of insight into a successful sequence of clause-like steps by which he assembled a pole long enough to unhook bananas hanging high overhead, from pieces of bamboo each too short to reach the hook (Köhler, 1925). Ploog recalled that Miller had discussed such a notion at the prior Work Session in which the basic idea is that the plans that guide behavior may resemble the phrase structures that organize our grammatical utterances. The notion is not new; Karl Lashley argued for it in 1951, and it was advocated again a decade later (see Miller et al., 1960; Miller, 1964). Miller distinguished a behavioral "lexicon," or repertoire of alternative behavioral units, and a behavioral "grammar," or rules that guide the sequential patterning of the behavioral units in the lexicon of a species. (See Ploog and Melnechuk, 1969.)

## Motivational Questions

### Reward, Encouragement, and Social Contact

Premack believes that the frequency of utterance, though not its content, style, or syntax, depends on the contingency of the "pay-off." To study what Sarah's reward for learning words really is, Premack would have to extinguish at least some of Sarah's knowledge, which he does not want to do. However, the tangible items that have been given Sarah are items of food, usually one of the long list of fruits, fresh and treated, that she likes. He said that the first words taught her were the names of such fruits, which she ate in the process, and that in her four-word imperative sentences the direct object of the verb was almost always an article of food, and only occasionally a toy, etc.

Ploog pointed out that tickling and games served as rewards for Washoe, and he asked whether any of Sarah's rewards were of a social character. Premack answered yes, and outlined some social rewards.

Like a young dependent child, Sarah needs explicit encouragement—such as patting and saying "Good girl!" in an approving manner. To ensure her the highest possible degree of success, Premack never asks Sarah questions for which there are no answers or that she lacks the knowledge or the words to answer, and he makes the incremental steps in her training as small as possible.

Close social contact is necessary between chimp and trainer. Sarah will not work if the trainer leaves the room, or goes to its other end, or even lets his attention wander. To Ploog this fact is extremely important. Perhaps the falling-off in Sarah's performance when she was trained by a "dumb" trainer was not caused by his being unschooled (dumb) but by his being new to her in this test situation.

### Conversation

Ploog said that monkeys and apes, unlike any other animal, including the dog, are almost constantly prone to social interactions, which appear to be intrinsically rewarding. Children, too, use language socially. He asked whether Sarah initiated or maintained conversation with her trainers, as Washoe did. Premack answered no, he had never allowed her to have the plastic words except in programmed training situations. However, she has occasionally reversed her role with the trainer. At such times, she has put on the board a declarative sentence (e.g., "red is on — — — — —") lacking one word, together with a set of words containing the missing word, and has pointed to each possible word in turn, passing by the correct word rapidly, as if to fool the trainer, to whom she appears to be giving a "fill-in-the-blank" test. Premack thinks she does this when she is bored with being trained and tested. Brown commented that Sarah's role-reversal behavior resembled that of a child, but Lenneberg thought it might be not a role-reversal at all, rather a sign of having misunderstood her instructions.

Bruner suggested trying to involve Sarah in a dialogue by giving her the option of a systematic demand for information.

### Spontaneous Use

Ploog asked if Sarah made much spontaneous use of her plastic words, the way a child babbles or Washoe makes signs of her own volition. Premack answered that Sarah once stole the test materials after a training session and went on both to produce many of the questions she had been taught and to answer them. He also predicted that "presleep monologue" activity would be found in chimpanzees as in children.

### Interference with Cognition

Premack pointed out that questions of cognitive capacity are complicated by motivational factors.* A child with a capacity for writing poems might be too shy to do so. Garrett felt it would certainly be desirable to be able to test cognition as distinct from motivation.

## Cognitive Questions

### Awareness of Utility

Bruner wondered whether either chimp had showed some sign of becoming aware of the usefulness of language as a tool, in the way that Köhler's Sultan had become aware of the usefulness of coupled bamboo poles as a tool.

Did Sarah, for example, use language as an aid to memory, writing notes with which to remind herself of things later? Premack said that she did not have free access to the plastic words.

Brown asked generally why, if anthropoid apes were capable of language, had there not been a great change in their life style? Why did they not use a symbol system for denotation, for example?

One way in which Premack thought the use of language might change the life style of an ape would be in de-emotionalizing the animal. The use of sentences might somehow interfere with the acting out of impulses.

Bruner wondered how one could determine if Sarah ever did or ever could reflect on the nature of her rules.

### Lack of Internal Representation

To Worden, the most striking feature of the Premack results is the contrast between the complexity of the syntax and semantics that

---

*See Section II, subsection, "Toward Competence in the Imperative Mode" and Section III, subsections, "Negative: Declarative" and "Dislike of Descriptions of Disliked Outcomes."

Sarah can master and use successfully with her trainer, and, on the other hand, her total lack of any interest in language as a powerful tool. To illustrate, she does not ask what the names of things are, she does not seem eager for more and more lessons, more and more vocabulary, nor does she show any strong inclination to initiate conversation; she does not even show the most egocentric use of language—as a tool for getting what she wants. This suggests that there may be profound differences between what a "word" is for a chimp and what a "word" is for a human; it is easy for the human to discover the word as a tool, while it may be almost, if not entirely, impossible for the chimp truly to discover in a self-reflective way the word as a tool, even though behaviorally she can manipulate the word to accomplish things that look as if they would have to be mediated by processes similar to those in the human. Specifically, the word for apple for the chimp may not have an autonomous independent status in internal representation whereby it can be generalized across all apples no matter what their color, size, or other real qualities, including various representations in photographs, drawings, etc. The human capacity so to generalize the word as a symbol depends upon a clear distinction that the human perceives between the word as an internal mental representation and the entire category of things for which the word stands, as well as the differentiation between the word as such and other internal aspects of the mental contents.

Another way of stating this thought: Sarah can perhaps discover "words" only in the form of external real objects but cannot make the step of discovering words as internal representations; that is, as psychic objects that have a definite, objective quality with reference to the rest of Sarah's experience. Thus, the internal representation of the external plastic word may not be accessible to discovery by the self-reflectiveness of Sarah, just as, for the human, certain nonverbal communications go on outside the awareness of the person and cannot be manipulated in the same self-reflective way in which words can be manipulated. The power of words for humans lies precisely in the fact that they become autonomous internal "objects," mentally manipulated and no longer dependent on physical, tangible, or external things like the written word or the original objects, the perceptions of which form the basis of words.

## "Ping-Pong Problem"

In Brown's opinion, the Conference was now dealing with a problem equal in importance to the "Clever Hans" problem: Whether or not Sarah's use of plastic "words" in ordered "sentences" really demonstrated her possession (not necessarily in consciousness) of human language knowledge, as opposed to an impressive, but only approximate, simulation of it.*

Brown termed this the "pigeon ping-pong problem," referring to Skinner's (1962) ingenious demonstration that pigeons could be trained to play something that looked very much like table tennis—except that they did not keep score or develop strategies for misleading one another.

Brown felt that to the reader of articles about Sarah, the plastic pieces were mesmerizing—that once you granted their equivalence to words, you read more significance into their regular use than perhaps was warranted. He pointed out a number of considerations that seemed to him to add weight on the doubtful side:†

1. Sarah's dependability was at best 75% to 80%, whereas other chimps achieve 90% success levels on other kinds of tests.[1]

2. This modest figure must be interpreted in the light of the fact that her response options were often severely limited.[2]

3. The "features analysis" test described earlier, as Miller has pointed out, meant little or nothing if it was not done at the first trial, even if Sarah was not trained for it.[3] Only four attributes were tested, and this was done just after the test for naming a real apple; the test should be given for every noun in her vocabulary.

4. She failed to identify as apple the apples shown in photographs from magazines.

5. Sarah is not reported to have passed the "name of" test for every object named in her vocabulary.

6. Sarah seems not to have been asked to master, as a man could, substitutions for every "word" in the "sentence," "Sarah insert banana dish apple pail," but only for some of them.

*As Robert Galambos later put it, even if a latterday Pygmalion were to train his primate Galatea to simulate all of English, could such a Professor Higgins justly say of his simian fair lady, "By George, I think she's got it!'"?

†Premack has commented on Brown's remarks:

1. Juvenile chimps that have not been deprived of food are notorious for doing less than 100%; they are frequently outperformed by monkeys, rats, and pigeons.
2. Sarah performed at about the same level whether her alternatives numbered 2 or 10.
3. It *was* done at the first trial and Sarah was indeed *not* trained for it.

As a reductio ad absurdum Brown argued that you could not claim that a rat knew "the concept of the imperative" if, after being trained to "jump to the left when presented with a triangle," he was presented with a triangle and consistently jumped to the left. The range of an animal's responses to a stimulus must be narrowed to the unique.

Ploog added that one might arrive at specific solutions through ad hoc routines. He wondered whether one could indeed preserve the essence of the linguistic process when using a paradigmatic approach.

### A Second Animal

Melnechuk asked whether a second animal would be trained, and if so, whether Premack thought Sarah would talk to it by means of the plastic words, or would perhaps even teach their use to her fellow chimp?

Premack replied that there was indeed a second chimp, but that he does not know whether the two will converse in the artificial language, though he thinks teaching one chimp to teach it to another would be the most interesting task of all.*

Melnechuk thought that if the concept of "name of" had indeed been inculcated, and if the inculcation were not an instance of second-order simulation, then it might be difficult for one chimp to teach another chimp those concepts that Premack had called "meta-linguistic."

Worden thought that the occasional human difficulty with the personal pronouns "me" and "I" may shed light on this problem. If language is a way of mapping concepts that are already present in the ape, then the concept of "I" as a subject, versus "me" as an object, might not be amenable to being mapped in language unless the ape actually has such a self-aware or self-conscious percept.

In children, and in certain patients, conditions occur in which the capacity to distinguish between one's self ("I") as a subject and one's self as an object ("me") is blurred or not yet acquired. Thus a child will refer to himself in the third person. The hypothesis that apes do not have an internal representation of words that is sufficiently autonomous for them to reflect upon the words as they would reflect upon a stick-tool would predict that the ape will never transfer the use of the acquired language to communication with other apes and will

---

*Premack (1970a) calls Sarah's communication system "Chimp English." Melnechuk suggests he consider calling it "Chimpanzeenglish" or "Chimpanglais."

never show eagerness for conversation or recognition of the power that the use of language could give them. In a sense, this is merely asserting that the "word," which is so visible to the human being, is in a psychic sense invisible to the ape, except in terms of those behavioral manipulations of plastic words that lead to successful performance in the Premack training situation.

Lenneberg also felt that the chimps were very unlikely ever to achieve communication by means of language, because it does not come to them as naturally and completely as it does to a normal child, who, as Lenneberg put it, finds language no work to learn and who grows up, like Molière's M. Jourdain, surprised at having spoken prose all his life.

## Biological Questions

If, as Lenneberg believes, the ability to speak a natural language is species-specific, the question naturally arises as to the differences in function, structure, or composition of primate brains that account for the specificity.

Geschwind (1970) is a leading exponent of the view that structures and arrangements unique to man's brain account for his unique linguistic capability. Lenneberg prefers physiological to anatomical uniqueness:

> It is not so much one or the other specific aspects of the brain that must be held responsible for the capacity of language acquisition but the way the many parts of the brain interact. Thus it is mode of function rather than specific structures that must be regarded as the proper neurological correlate of language. ...

> ... man is unique among vertebrates in the functional asymmetry of neurophysiological process within the adult brain. Only man has hemispheric dominance with lateralization of function and marked preference with respect to side in the use of limbs and sensory organs. (Lenneberg, 1967, pp. 170 and 174)

Lenneberg (1970) suggests a series of experiments concerned with the timing of cerebral events correlated with speech, and Melnechuk suggests a corresponding electrophysiological study of neuroelectric correlates of Sarah's and Washoe's use of their languages.

Not indicated as especially promising are experiments to test the language capability of male chimpanzees, or of other anthropoid apes, if, as the conferees seemed to feel, Sarah, and possibly Washoe too, did not really understand her own use of an artificial language. For that reason, however, it did seem important to try Premack's technique on organisms lower on the evolutionary scale.

## A Whorfian Question

Bruner called the question, "Are Apes Likely to Use Language?", "a Whorfian question," referring to B. L. Whorf, the originator of the term "language relativity," although, as Lenneberg explains it,

> ... similar notions had been expressed before by many others (Basilius, 1952). In his studies of American Indian languages, Whorf was impressed with the general difficulties encountered in translating American Indian languages into European languages. To him there seemed to be little or no isomorphism between his native English and languages such as Hopi or Nootka. He posed the question, as many had done before him, of whether the divergences encountered reflected a comparable divergence of thought on the part of the speakers. He left the final answer open, subject to further research, but it appears from the tenor of his articles that he believed that this was so, that differences in language are expressions of differences in thought. It has been pointed out since (Black, 1959; Feuer, 1953; but see also Fishman, 1960) that there is actually little *a priori* basis for such beliefs. ...
>
> Since the use of words is a creative process, the static reference relationships, ... as they are recorded in a dictionary, are of no great consequence for the actual use of words. However, the differences between languages that impressed Whorf so much are entirely restricted to these static aspects and have little effect upon the creative process itself. (Lenneberg, 1967, p. 363)

It would appear that Whorf's idea may fit differences between species better than it fits differences within species.

## Applications in Therapy and Education

Even if, as seemed possible, the case of Sarah was indeed a "pigeon ping-pong problem," Brown thought that Premack's training procedure could perhaps be used to shift the heretofore apparently invariant order in which normal children at home develop linguistic capabilities.

Brown and Bigelow also felt that Premack's method might prove useful for the linguistic training of retarded children and the linguistic retraining of aphasic patients. Since the Conference, Premack has in fact applied his procedures to language-deficient children and, in conjunction with Gazzaniga, to the retraining of global aphasics (Gazzaniga et al., 1971).

## APPENDIX I:
## A Discussion by R. Allen Gardner and Beatrice T. Gardner

### Word Order

The chief disagreement that we Gardners have with any point of fact raised at the Conference concerns Brown's assertions there is not yet evidence that Washoe can use word order as well as a child at a comparable level of development. Brown has made available to us his tabulations of the utterances sampled by himself and his associates for children. From these tabulations it is evident that the rate of error observed in Washoe's combinations is not markedly greater than the average rate of error in Brown's observations of children at Stage I.

### Chimps Versus Children

The critical turning point in the Conference was that at which the conferees turned their attention to a comparison of the actual performance of Washoe with the actual performance of young children. The material available to the conferees concerning Washoe's development covered the first 24 months of the project, which left Washoe at about 34 months of age. At this point in her skeletal development, the last molars of her first set of teeth had only just appeared (thus confirming our initial estimate of her birth date). She was not to lose her first deciduous teeth and begin to acquire her adult teeth until the 50th month of the project when she was nearly 5 years old.

A child who is so immature in skeletal development as Washoe was at 34 months is considered to be very far from reaching intellectual maturity. Thus, we are certain that as long as Washoe is maintained in a suitably stimulating environment, further intellectual gains will be made. The discrepancies already noted between Washoe and very young human children cannot indicate that a chimpanzee will never reach the competence of a human child, but merely suggest that a chimpanzee may never reach the competence of a human adult. Even so, we are probably underestimating chimpanzee development, because Washoe did not enter her intellectually stimulating environment until she was nearly one year old, and during the early phases of her training, her human companions were, themselves, only beginning to become competent in ASL* and in the ways of rearing a young chimpanzee. It is more

*American Sign Language.

than likely that at comparable ages the chimpanzees who follow Washoe will acquire a great deal more linguistic competence than she did.

### Natural Versus Artificial Language

An important difference between Washoe and Sarah that needs to be stressed is that Washoe's is a natural* language, ASL (Stokoe et al., 1965), which is used by a great many human beings and which is the only language exposure received by the preschool deaf children of deaf parents. Therefore, the existence of a population of child controls is guaranteed. Furthermore, available reports (Schlesinger, 1971) already indicate many points of similarity between deaf children and Washoe.

### "Clever Hans" Problem

Washoe's competence has been thoroughly tested under double-blind conditions that preclude any possibility of "Clever Hans" error. The methods used are described in detail in Gardner and Gardner (1971a). The test used most extensively depended upon the fact that Washoe can describe photographs to an observer who cannot himself see the photographs. The use of photography makes it convenient to present material to Washoe that she has never seen before, thus precluding the possibility that she could answer correctly by memorizing routines.

In addition, we have been able to conduct our tests in such a way that double-blind observers can read Washoe's signs through one-way glass. Thus, these additional observers can be strangers to Washoe and "Washoese," yet familiar with ASL. Two deaf observers who have participated so far were able to read Washoe under these conditions with about 70% accuracy the first time they were exposed to "Washoese" and with about 95% accuracy on their second exposure. This is what one would expect if Washoe signed with a mild-to-strong chimp accent.

Much of the data for human children that was cited during this Conference (by Lenneberg in particular) was obtained under conditions that were in no way blind and that permitted the maximum possible

---

*Linguists usually apply "natural" only to the spoken languages, such as French or Japanese, that subgroups of man have developed spontaneously.

amount of "Clever Hans" error. The fears expressed by the conferees regarding the danger of "Clever Hans" error in the evaluation of linguistic competence apply with great force indeed to the extravagant claims made by psycholinguists about the linguistic feats of very young children.

### "Ping-Pong Problem"

The usefulness of Brown's "pigeon ping-pong problem" is much attenuated by his failure to define the game of ping-pong, or the game of language. With respect to ping-pong, he suggests that pigeons cannot play this game, because he does not know whether or not they keep score. Certainly, human beings often play ping-pong without keeping score. The example of the recent visit of an American ping-pong team to mainland China comes to mind. There must be hundreds of millions of people who are certain that ping-pong was played by the teams, but who never asked whether or not scores were kept.

Brown offers us no criterion at all by which we could tell whether or not a very young child or a chimpanzee is using language, and we Gardners strongly suspect that he has none. Consequently, his ping-pong problem can be reduced to one of the questions with which Ploog opened the Conference: Can we address ourselves to the problem of this Conference without having arrived at a widely accepted theory of language? Clearly, the results of the Conference yield an affirmative answer to this question. It is equally clear that knowledge is not advanced by suggesting that chimps are *not* really playing ping-pong if no one can say just what it is that makes us believe that children are playing ping-pong.

## APPENDIX II:
### Further Comments by David Premack

On the whole, I detect a certain amount of scrambling to protect the uniqueness of man. Able investigators indulge in inconsistencies, lapses in operationalism, and reliance on phrases that are sustained more by alliteration than substance. Lenneberg (1970) had proposed a set of tasks for which, if an animal were to demonstrate competence, it would be said to have some of the basic capacities for language. Sarah's performance goes substantially beyond the tasks that Lenneberg proposed, a fact to which he apparently accedes. Yet the conclusion he draws is that "language is species-specific" and therefore that Sarah's performance is not (could not be) evidence of the "basic capacities for language."

Brown invents the phrase "pigeon ping-pong problem." The substantive issue apparently is the venerable one besetting all simulation. An organism and a machine, two machines, or two organisms carry out comparable functions: Do they do so on the basis of the same underlying processes? In order to answer that question, it is necessary to be able to give an account of the processes in both cases. Thus, until we can give an account of how man produces and understands sentences, we can hardly ask whether when a chimp carries out these functions it does so in a comparable way. Indeed, we are not yet fully able to answer that question for the comparison between adult and child. Though important, the issue of process is formidably difficult. If two devices carry out comparable functions, and there is no evidence to the contrary, parsimony recommends assuming equivalent underlying processes. Parsimony is not nearly so gratifying as evidence, but simply to assume or imply differences in process is unwarranted in the absence of supporting evidence. Moreover, as the number of functional similarities pile up, one must either find a functional dissimilarity or begin to entertain the possibility that the underlying processes are, in fact, comparable.

Worden raises two points, the first having to do with the motivation of the chimp to use language, the second with the psychic status of a word. He does not hesitate to conclude that the chimp and child differ radically in that Sarah is bored by her lessons, does not reach out hungrily for new linguistic knowledge, does not attempt to use language as a tool—this in contrast to the child. The comparison and conclusion are unwarranted. Worden should ask, are the motivational opportunities in the two cases equated? Unless they are, the relative apathy of the one organism may simply reflect characteristics of the training situation. Sarah could not "attempt to use language as a tool" because she was not given words on any occasion other than those on which a lesson took place, and then only those words considered to be relevant to the lesson.

Rather than being interested in the question, "Will a chimp behave like a child if it is put in a child's situation?", my interest was in analyzing language into an ordered set of constituents and of devising a training program for each constituent. We are still largely at the stage of going down the checklist, marking constituents off as they are instilled. This scholastic approach is nevertheless not entirely incompatible with the kind of research that is needed in order to answer Worden's question. After the individual constituents have been instilled, the question arises, can they be put together so as to enable the subject to converse, to use language to solve problems, or to get things from the environment? Some of this has been done, albeit on an extremely limited basis. For instance, having taught Sarah the conditional or "if/then" on one occasion, and the quantifiers on another, they were combined, giving rise to forms such as "Sarah take all banana ⊃ Mary give Sarah chocolate," etc.

Nevertheless, the motivational questions will remain difficult to answer in a laboratory. After working three years on this question, we found that tile walls and a 3 x 6 m mesh cage did not offer the material basis for a rich semantic. We are in the process of adapting Sarah to a "typewriter," the keys of which are marked with the plastic words and the output of which occurs on colored TV. Among the advantages are automatic recording. This will make it possible to give Sarah full access to her vocabulary and to record her output (not easily possible with plastic words). But I regret having to keep this typewriter in the laboratory. I would prefer to see the device snugged against a banyan tree, another one like it 5 or 10 miles off, perhaps a third one still farther into the forest. Then, for the first time, one group of chimps could tell a second group what foraging conditions were like in the papaw grove, who was sexually receptive at the moment, and so on. Only in circumstances of this kind would it seem possible to equate the motivational opportunities for chimp and man, upon which a serious answer to Worden's question depends. The alternative is to reduce the motivational opportunities of man to those of the caged chimp. But we already know something about the apathy of caged men; there may not be an overwhelming need to do that half of the experiment.

What about the psychic status of Sarah's words? It is too easy to fault Worden for a lack of operationalism on this point and in doing so to lose sight of the important substance of his point. What I take Worden to be asking in part at least is, "Can Sarah think in her language?" To give a full answer to that question is not possible, but we can grapple informatively with one small part of it. To think successfully with language requires being able to generate the meaning of words in the absence of their external representations. For Sarah to be able to match "apple" to an actual apple, or "Mary" to a picture of Mary, may indicate that she knows the meaning of these words, but it does not prove that when she is given the word "apple" and no apple is present, she can think apple, i.e., mentally represent the meaning of the word to herself. The latter goes beyond the former because it frees the use of language from a dependence upon the external sstates of affairs that language can be used to map. It involves the sort of thing Hockett has called "displacement" and has listed as one of the design features of language.

Fortunately, there is already evidence that Sarah is able to understand the meaning of words in the absence of external referents. The first evidence of this kind is very simple and occurred early in training. Sarah was given a piece of fruit and two words, one of them the name of the fruit, and required to use it to get the fruit. Her performance on these early choice trials ran disappointingly close to chance. The possibility dawned on us that her errors might not be a failure in association so much as a request for a preferred fruit not present on a given trial. A preference order was obtained for the fruits and subsequently for the words; their concordance was better than 80%. The preference order for the words was obtained in the absence of the fruits. This suggests that she could generate the meaning of the fruit names in the absence of the fruits.

More classical displacement-type evidence was obtained at a later stage of the experiment. For example, she was given the instruction "brown color of chocolate," as a means of introducing the word "brown," the other two terms in the string being known to her. Later she was given four disks only one of which was brown, and given the instruction, "take brown." She performed correctly on this and similar tasks (Premack, 1971a). Because the instruction was given her in the absence of chocolate, the word "chocolate" must have sufficed for her to generate or picture the properties of chocolate. Whether she can actually think in this language involves many more issues, not to mention the difficulty of explicating what is involved in thinking. But we already have reason to believe that she is capable of generating the meaning of words in the absence of their external representations.

It does not make too much sense to compare the kinds of sentences Sarah has been taught with those found to occur in the child's early speech, mainly because no attempt was made to simulate the child. In addition, when Sarah began to learn language she was much older (5 or 6) than the child is when it begins to learn. I would expect this to give Sarah an advantage—on the grounds that the acquisition of language is the mapping of semantic relations that have a maturational course themselves—and this expectation was not confounded. The first sentence type we taught Sarah was the dative, e.g., "Mary give apple Sarah," and the dative is by no means the first sentence type the child learns. On the other hand, the child learns the possessive sentence type relatively early, and Sarah did not have this type. But the reason she did not was simply because we did not attempt to teach it to her. There is no indication, either in the kind of training the possessive would entail or in the concept of possession itself, that Sarah would have any difficulty with this form. Forms more likely to pose difficulties are those involving subtle stimuli, such as the pronouns and space markers, forms that are defined relative to the speaker. Yet Sarah has been successfully taught "this" and "that," words that are defined relative to the speaker (suggesting that the pronouns too will be within her reach). Among the more complex sentence types she has been taught are the conditional, e.g., "If Mary take red then Sarah take apple," which is surely not found in the early speech of the child. Although comparison between Sarah and the child is of limited value for the reasons stated, Sarah does not suffer by the comparison. On the contrary, the main thing the comparison suggests is that if the child were taught by procedures analogous to Sarah's, it could be brought to Sarah's level and presumably beyond.

It is not the case that Sarah cannot learn except by *one-to-one substitution*. The latter represents what may be the simplest possible training procedure (Premack, 1971a). When each new word is taught by arranging it so that its introduction at a marked location in a string of known words has the effect of completing the sentence, three primary sources of difficulty are eliminated: (1) Only one new word will be present, so the subject cannot err in choice of words. (2) The blank location in the (potential) sentence is marked (with the interrogative marker) so the subject cannot err as to where in the sentence to put the word. (3) The completing operation always consists of addition, rather than addition plus the possibility of deletion and/or rearrangement.

If one-to-one substitution is the simplest language-teaching procedure, then by systematically imposing complications upon it we can test the limits of the language-acquisition capacities of different species: (1) We can leap from one-to-one to *many-to-one* substitution and examine the subject's ability to negotiate the leap. (2) Incorrect elements can be inserted in incomplete sentences, and completion of the sentence made to depend not only upon addition of new elements but upon *deletion* of existing ones. (3) Correct words can be incorrectly ordered, and sentence production made to depend upon *rearrangement.*

Can Sarah learn only by one-to-one substitution? We have not yet tried her with other methods, so the question cannot be answered. However, what is already of interest is that although she was taught primarily with one-to-one substitution, when tested on many-to-one substitution and the comprehension of deletion, she proved capable of both of these and of other linguistic operations as well (Premack, (1971a).

It is premature to assert what infrahuman and pathological human populations can and cannot learn. All such claims are at the mercy of the methodologies used in the training. In 1917 the chimp was incapable of demonstrating its grasp of the middleness concept by learning to indicate which one of the objects in a row was the central one. By about 1935, the chimp had overcome this inability, provided that the elements in the sequence did not exceed seven. By 1950, it had overcome even this limitation and was indicating middleness on sequences of 20 or more, roughly the capacity of the apparatus. Is the rate of evolution of chimp intelligence greater than we had supposed? The serious point is that we have not yet produced a formal theory of pedagogy even for our own species. Until we can identify simplest training procedures, failures are indeterminate. Who failed, teacher or species?

# BIBLIOGRAPHY

This bibliography contains two types of entries: (1) citations given or work alluded to in the report, and (2) additional references to pertinent literature by conference participants and others. Citations in group (1) may be found in the text on the pages listed in the right-hand column.

Page

Altmann, S.A. (1967): The structure of primate social communication. *In: Social Communication Among Primates,* Altmann, S.A., ed. Chicago: University of Chicago Press, pp. 325-362.    635,639

Basilius, H. (1952): Neo-Humboltian ethnolinguistics. *Word* 8:95-105.    683

Black, M. (1959): Linguistic relativity: the views of Benjamin Lee Whorf. *Phil. Rev.* 68:228-238.    683

Black, M. (1968): *The Labyrinth of Language.* New York: Frederick A. Praeger, Inc., p. 61.    634

Bronowski, J. and Bellugi, U. (1970a): Language, name, and concept. *Science* 168:669-673.    664,665

Bronowski, J. and Bellugi, U. (1970b): Washoe the chimpanzee. *Science* 169:328.

Brown, R. (1970): The first sentences of child and chimpanzee. *Psycholinguistics.* New York: Macmillan Company, pp. 208-231.    612,639, 640,648,665, 669,670

Brown, R. (1971): *A First Language; The Early Stages.* Cambridge, Mass.: Harvard University Press. (In press)

Bruner, J.S. (1957): Neural mechanisms in perception. *Psychol. Rev.* 64:340-358.

Bruner, J.S. (1959): The cognitive consequences of early sensory deprivation. *Psychosom. Med.* 21:89-95.

Bruner, J.S. (1964): The course of cognitive growth. *Am. Psychol.* 19:1-15.

Calder, N. (1970): *The Mind of Man.* New York: Viking Press.    631

Carpenter, C.R. and Nishimura, A. (1969): The Takasakiyama colony of Japanese macaques (*Macaca fuscata*). *In: Proceedings, 2nd International Congress of Primatology, Vol. I, Behavior.* New York: Karger, pp. 16-30.    675

Chomsky, N. (1965): *Aspects of the Theory of Syntax.* Cambridge, Mass.: M.I.T. Press.    636

Page

Chomsky, N. (1967): Appendix A. The formal nature of language. *In: Biological Foundations of Language.* Lenneberg, E.H., ed. New York: John Wiley and Sons, pp. 397-442.     637

Davenport, R.K. and Rogers, C.M. (1971): Perception of photographs by apes. *Behaviour* 39:318-320.     646

DeVore, I., ed. (1965): *Primate Behavior: Field Studies of Monkeys and Apes.* New York: Holt, Rinehart and Winston.

DeVore, I. (1969): Behavioral differences between primates and man (discussion). *Neurosciences Res. Prog. Bull.* 7(5):459. Also *In: Neurosciences Research Symposium Summaries, Vol. 4.* Schmitt, F.O. et al., eds. Cambridge, Mass.: M.I.T. Press, p. 142.

Drake, F.D. (1961): Project Ozma. *Physics Today* 14(4):40-46.

Eimas, P.D., Siqueland, E.R., Jusczyk, P., and Vigorito, J. (1971): Speech perception in infants. *Science* 171:303-306.     617

Feuer, L.S. (1953): Sociological aspects of the relations between language and psychology. *Phil. Sci.* 20:85-100.     683

Fishman, J.A. (1960): A systematization of the Whorfian hypothesis. *Behav. Sci.* 5:323-339.     683

Fodor, J.D. (1970): Formal linguistics and formal logic. *In: New Horizons in Linguistics,* Lyons, J., ed. London: Pelican, pp. 198-214.     671

Fossey, D. (1970): Making friends with mountain gorillas. *Nat. Geog.* 137:48-67.     674

Fossey, D. (1971): More years with mountain gorillas. *Nat. Geog.* 140:574-585.

Freudenthal, H. (1960): *Lincos: Design of a Language for Cosmic Discourse.* Amsterdam: North-Holland Publishing Co.

Gallup, G.G., Jr. (1970): Chimpanzees: self-recognition. *Science* 167:86-87.     675

Gallup, G.G., Jr. (1971): Chimpanzees and self-concept: it's done with mirrors. *Psychology Today* 4(10):58-61.

Gardner, R.A. and Gardner, B.T. (1966-1970): Films of Washoe. Department of Psychology, University of Nevada, Reno, Nevada. (In preparation)     631

Gardner, B.T. and Gardner, R.A. (1968): How a young chimpanzee was toilet trained. *Laboratory Primate Newsletter* (Schrier, A.M., ed.) 7:1-3.     610

Gardner, B.T. and Gardner, R.A. (1971a): Two-way communication with an infant chimpanzee. *In: Behavior in Nonhuman Primates, Vol. 4.* Schrier, A. and Stollnitz, F., eds. New York: Academic Press, pp. 117-184.     610,646, 686

Page

Gardner, R.A. and Gardner, B.T. (1969): Teaching sign language to a chimpanzee. 605,610,
*Science* 165:664-672. 612,613,616,
634,641

Gardner, R.A. and Gardner, B.T. (1971b): Communication with a young chimpan- 610
zee: Washoe's vocabulary. *In: Modèles Animaux du Comportement Humain.*
Chauvin, R., ed. Paris: CNRS. (In press)

Garrett, M. and Fodor, J. (1968): Psychological theories and linguistic constructs.
*In: Verbal Behavior and General Behavior Theory.* Dixon, T.R. and Horton,
D.H., eds. Englewood Cliffs, N.J.: Prentice-Hall, pp. 451-477.

Gazzaniga, M.S., Velletri, A.S., and Premack, D. (1971): Language training in 684
brain-damaged humans. *Fed. Proc.* 30:403.

Geschwind, N. (1970): The organization of language and the brain. *Science* 682
170:940-944.

Goodall, J. van L. (1968): The behaviour of free-living chimpanzees in the Gombe 673
Stream Reserve. *Animal Behaviour Monographs* 1:161-311.

Goodall, J. van L. (1970): *My Friends, the Wild Chimpanzees.* Washington, D.C.:
National Geographic Society.

Greenberg, J. H., ed. (1966): *Universals of Language.* 2nd Ed. Cambridge, Mass.: 639
M.I.T. Press.

Greenberg, J.H. (1969): Language universals: a research frontier. *Science* 639
166:473-478.

Hayes, C. (1951): *The Ape in Our House.* New York: Harper & Brothers. 610

Hayes, C. (1970): A chimpanzee learns to talk. *In: Psychological Studies of Human* 610
*Development.* Kuhlen, R. and Thompson, G., eds. New York: Appleton-
Century-Crofts, pp. 331-339.

Hayes, K.J. and Hayes, C. (1950): *Vocalization and Speech in Chimpanzees.* 610
Audio-Visual Services, The Pennsylvania State University, University Park,
Pennsylvania. (16 mm silent film)

Hayes, K.J. and Hayes, C. (1951): The intellectual development of a home-raised 610,612
chimpanzee. *Proc. Amer. Phil. Soc.* 95:105-109.

Hayes, K.J. and Hayes, C. (1952): Imitation in a home-raised chimpanzee. *J. Comp.* 610,612
*Physiol. Psychol.* 45:450-459.

Hayes, K.J. and Hayes, C. (1953): Picture perception in a home-raised chimpanzee. 610,646
*J. Comp. Physiol. Psychol.* 46:470-474.

Hayes, K.J. and Hayes, C. (1954a): The cultural capacity of chimpanzee. *Human* 610
*Biol.* 26:288-303.

Page

Hayes, K.J. and Hayes, C. (1954b): *The Mechanical Interest and Ability of a*          610
*Home-Raised Chimpanzee*. Audio-Visual Services, The Pennsylvania State University, University Park, Pennsylvania. (16 mm silent film, 4 parts)

Hockett, C.F. (1960): Logical considerations in the study of animal communica-          639
tion. *In: Animal Sounds and Communication*. Lanyon, W.E. and Tavolga, W.N.,
eds. Washington, D.C.: American Institute of Biological Sciences, pp. 392-430.

Hockett, C.F. (1963): The problem of universals in language. *In: Universals of*          639
*Language*, Greenberg, J.H., ed. Cambridge, Mass.: M.I.T. Press, pp. 1-22.

Hockett, C.F. (1968): Reply to review of *Current Trends in Linguistics, Vol. III:*          639
*Theoretical Foundations*. Sebeok, T.A., ed. The Hague: Mouton (1966). *Current*
*Anthropology* 9:171-174.

Hockett, C.F. and Altmann, S.A. (1968): A note on design features. *In: Animal*          639
*Communication: Techniques of Study and Results of Research*. Sebeok, T.A.,
ed. Bloomington, Ind.: Indiana University Press, pp. 61-72.

Horowitz, N.H. (1966): The search for extraterrestrial life. *Science* 151:789-792.          608

Jacobsen, C.F., Jacobsen, M.M., and Yoshioka, J.G. (1932): Development of an          610
infant chimpanzee during her first year. *Comp. Psychol. Monogr.* 9(1):1-94.

Jay, P.C. (1968): Primate field studies and human evolution. *In: Primates: Studies*
*in Variation and Adaptation*. Jay, P.C., ed. New York: Holt, Rinehart and
Winston, pp. 487-503.

Kawamura, S. (1963): The process of sub-culture propagation among Japanese          675
macaques. *In: Primate Social Behavior*. Southwick, C.H., ed. New York: Van
Nostrand, pp. 82-90.

Kellogg, W.N. (1932a): *Ape and Child*. II. Comparative Tests on a Human and a
Chimpanzee Infant of Approximately the Same Age. Audio-Visual Services, The
Pennsylvania State University. University Park, Pennsylvania. (16 mm silent
film)

Kellogg, W.N. (1932b): *Ape and Child*. I. Some Behavior Characteristics of a
Human and a Chimpanzee Infant in the Same Environment. Audio-Visual
Services, The Pennsylvania State University, University Park, Pennsylvania. (16
mm silent film)

Kellogg, W.N. (1968): Communication and language in the home-raised chimpan-          609,610
zee. *Science* 162:423-427.

Kellogg, W.N. and Kellogg, L.A. (1933): *The Ape and the Child: A Study of*          610,612
*Environmental Influence Upon Early Behavior*. New York: McGraw-Hill. (2nd
Ed., 1967. New York: Hafner.)

Klüver, H. (1936): The study of personality and the method of equivalent and          641
non-equivalent stimuli. *Character and Personality* 5:91-112.

Page

Klüver, H. (1965): Neurobiology of normal and abnormal perception. *In: Psycho-*            641
*pathology of Perception.* Hoch, P.M. and Zubin, J., eds. New York: Grune and
Stratton, pp. 1-40.

Koehler, O. (1954a): Vom Erbgut der Sprache. *Homo* 5:97-104.                    639

Koehler, O. (1954b): Vorbedingungen und Vorstufen unserer Sprache bei Tieren.            639
*Zool. Anz. Suppl.* 18:327-341. See also: *Verh. Deut. Zool. Gesell. Tübingen,* pp.
327-341.

Köhler, W. (1925): *The Mentality of Apes.* London: Kegan, Paul. (2nd Ed., 1959.            676
New York: Vintage)

Kučera, H. and Francis, W.N. (1967): *Computational Analysis of Present-Day*            617
*American English.* Providence, R.I.: Brown University Press.

Ladygina-Kohts, N.N. (1935): *Infant Ape and Human Child.* Moscow: Museum
Darwinianum.

Lancaster, J.B. (1968): Primate communications systems and the emergence of            635
human language. *In: Primates: Studies in Adaptation and Variability.* Jay, P.C.,
ed. New York: Holt, Rinehart and Winston, pp. 439-457.

Lapp, R.E. (1961): *Man and Space: The Next Decade.* New York: Harper and Row.

Lashley, K.S. (1951): The problem of serial order in behavior. *In: Cerebral*            676
*Mechanisms in Behavior.* (The Hixon Symposium) Jeffress, L.A., ed. New York:
John Wiley and Sons, pp. 112-136.

Lenneberg, E.H. (1967): *Biological Foundations of Language.* New York: John        607,661,
Wiley and Sons.                                662,664,665,
                                         666,671,673,
                                         674,682,683

Lenneberg, E.H. (1969a): On explaining language. *Science* 164:635-643.                639

Lenneberg, E.H. (1969b): Problems in the systematization of communicative            635
behavior. *In: Approaches to Animal Communication.* Sebeok, T.A. and Ramsay,
A., eds. Paris: Mouton, pp. 131-137.

Lenneberg, E.H. (1970): Brain correlates of language. *In: The Neurosciences:*        635,660,
*Second Study Program.* Schmitt, F.O., editor-in-chief. New York: Rockefeller        682,688
University Press, pp. 361-371.

Lenneberg, E.H. (1971): Of language knowledge, apes, and brains. *J. Psycholinguis-*    638,639
*tic Res.* 1:1-29.

Lieberman, P. (1968): Primate vocalizations and human linguistic ability. *J.*        611,640
*Acoustic Soc. Amer.* 44(6):1574-1584.

Page

Lieberman, P.H., Klatt, D.H., and Wilson, W.H. (1969): Vocal tract limitations on   611,640
the vowel repertoires of rhesus monkey and other nonhuman primates. *Science*
164:1185-1187.

Linton, M.L. (1970): Washoe the chimpanzee. Letter to the editor. *Science*
169:328.

Lyons, J. (1970): *Noam Chomsky*. New York: Viking Press.   637

Marler, P. (1969): Discussion. *Fifth Nobel Conference (Communication)*. (Gustavus   639
Adolphus College, St. Peter, Minnesota.) Roslansky, J.D., ed. Amsterdam:
North-Holland Publishing Co.

Miller, G.A. (1964): Communication and the structure of behavior. *In: Disorders of*   676
*Communication*. (Proceedings of the Association for Research in Nervous and
Mental Diseases, Vol. 42.) Rioch, D.M. and Weinstein, E.A., eds. Baltimore:
Williams & Wilkins, pp. 29-37.

Miller, G.A., Galanter, E., and Pribram, K. (1960): *Plans and the Structure of*   676
*Behavior*. New York: Henry Holt and Co.

Minsky, M., ed. (1968): *Semantic Information Processing*. Cambridge, Mass.: M.I.T.   608
Press.

Moles, A.A. (1969): The concept of language from the point of view of animal
communication. *In: Approaches to Animal Communication*. Sebeok, T.A. and
Ramsay, A., eds. Paris: Mouton, pp. 138-145.

Morris, W., ed. (1969): *The American Heritage Dictionary of the English Language*.
Boston: American Heritage and Houghton Mifflin.

Ohmann, R. (1969): Grammar and meaning. *In: The American Heritage Dictionary*
*of the English Language,* Morris, W., ed. Boston: American Heritage and
Houghton Mifflin, pp. xxxi-xxxiv.

Oliver, B.M. (1962): Radio search for distant races. *Internat. Sci. Technol.*
10:55-60.

Ploog, D. (1970): Social communication among animals. *In: The Neurosciences:*   675
*Second Study Program,* Schmitt, F.O., editor-in-chief. New York: Rockefeller
University Press, pp. 349-361.

Ploog, D. and Melnechuk, T. (1969): Primate communication. *Neurosciences Res.*   605,606,
*Prog. Bull.* 7(5):419-510. Also *In: Neurosciences Research Symposium Sum-*   608,635,
*maries, Vol. 4.* Schmitt, F.O. et al., eds. Cambridge, Mass.: M.I.T. Press, pp.   636
103-190.

Premack, D. (1970a): A functional analysis of language. *J. Exp. Anal. Behav.*   610,616,
14:107-125.   629,631,633,
  651,665,681

# INDEX

# CUMULATIVE NAME INDEX*

Page numbers immediately following a name indicate discussion or mention of that individual's work. "Cit." indicates a reference to a specific publication.

*Each chapter in this book is followed by its own index of subjects and names. As a further aid to the reader, the names in each chapter index have been collated in this cumulative index.

# NRP CENTER STAFF

*Chairman*
Francis O. Schmitt

*Executive Director*
Frederic G. Worden

*Communications Director*
Theodore Melnechuk

*Managing Editor & Librarian*
George Adelman

*Business Manager*
L. Everett Johnson

*Administrative Officer*
Katheryn Cusick

*Assistant to the Chairman*
Harriet E. Schwenk

*Staff Scientists*
Glenn G. Dudley
Eduardo Eidelberg
Eric H. Lenneberg
Frederick E. Samson, Jr.

*Secretaries*
Nellie Rae Burman
Heather H. Campbell
Evelyne A. Commoss
Beverly Gonsalves
John Potthast
Jane I. Wilson

*Staff Writer-Editors*
Joanne Belk
Dorothy W. Bishop
Yvonne M. Homsy
Ava B. Nash

*Library Assistant*
Nancy Robie

*Audio-Visual Technician*
Wardwell F. Holman

*Former Staff/Resident Scientists*

Michael A. Arbib
Curtis C. Bell
Julian H. Bigelow
Kao Liang Chow
Charles M. Fair
Richard Hirsh
Masao Ito
Raymond T. Kado
Arnold L. Leiman
Ljubodrag Mihailović
Lewis C. Mokrasch
Frank Morrell

Jacques Mouret
Tomio Nishihara
Robert G. Ojemann
Joe M. Parks
Detlev Ploog
Gardner C. Quarton
Kurt Rosenheck
Frederick E. Samson, Jr.
Victor E. Shashoua
John R. Smythies
Howard H. Wang
Wolfgang Wechsler